THE ROUTLEDGE HANDBOOK OF METHODOLOGIES IN HUMAN GEOGRAPHY

The Routledge Handbook of Methodologies in Human Geography is the defining reference for academics and postgraduate students seeking an advanced understanding of the debates, methodological developments and methods transforming research in human geography.

Divided into three sections, Part I reviews how the methods of contemporary human geography reflect the changing intellectual history of human geography and events both within human geography and society in general. In Part II, authors critically appraise key methodological and theoretical challenges and opportunities that are shaping contemporary research in various parts of human geography. Contemporary directions within the discipline are elaborated on by established and emerging researchers who are leading ontological debates and the adoption of innovative methods in geographic research. In Part III, authors explore crosscutting methodological challenges and prompt questions about the values and goals underpinning geographical research work, such as: Who are we engaging in our research? Who is our research 'for'? What are our relationships with communities?

Contributors emphasize examples from their research and the research of others to reflect the fluid, emotional and pragmatic realities of research. This handbook captures key methodological developments and disciplinary influences emerging from the various sub-disciplines of human geography.

Sarah A. Lovell is a Senior Lecturer in the Faculty of Health at the University of Canterbury, New Zealand.

Stephanie E. Coen is an Associate Professor in the School of Geography at the University of Nottingham, UK.

Mark W. Rosenberg is a Professor of Geography in the Department of Geography and Planning and cross-appointed as a Professor in the Department of Public Health Sciences at Queen's University in Kingston, Ontario, Canada.

THE ROUTLEDGE HANDBOOK OF METHODOLOGIES IN HUMAN GEOGRAPHY

Edited by Sarah A. Lovell, Stephanie E. Coen and Mark W. Rosenberg

LONDON AND NEW YORK

Cover image: 'City', an original mixed media painting by Carla Lam

First published 2023
by Routledge
4 Park Square, Milton Park, Abingdon, Oxon OX14 4RN

and by Routledge
605 Third Avenue, New York, NY 10158

Routledge is an imprint of the Taylor & Francis Group, an informa business

British Library Cataloguing-in-Publication Data
A catalogue record for this book is available from the British Library

Library of Congress Cataloging-in-Publication Data
Names: Rosenberg, Mark W., editor. | Lovell, Sarah, editor. | Coen, Stephanie, editor.
Title: The Routledge handbook of methodologies in human geography /
Sarah Lovell, Stephanie E. Coen and Mark W. Rosenberg
Other titles: Handbook of methodologies in human geography
Description: Abingdon, Oxon ; New York, NY : Routledge, 2022. |
Series: Routledge international handbooks |
Includes bibliographical references and index.
Identifiers: LCCN 2022009361 (print) | LCCN 2022009362 (ebook) |
ISBN 9780367482527 (hardback) | ISBN 9781032313795 (paperback) |
ISBN 9781003038849 (ebook)
Subjects: LCSH: Human geography–Research. |
Human geography–Methodology. | Human geography–History.
Classification: LCC GF26 .R67 2022 (print) |
LCC GF26 (ebook) | DDC 304.2072–dc23/eng20220716
LC record available at https://lccn.loc.gov/2022009361
LC ebook record available at https://lccn.loc.gov/2022009362

ISBN: 9780367482527 (hbk)
ISBN: 9781032313795 (pbk)
ISBN: 9781003038849 (ebk)

DOI: 10.4324/9781003038849

Typeset in Bembo
by Newgen Publishing UK

COVER ARTIST

CONTENTS

List of Illustrations   *xi*

List of Contributors   *xiii*

Introduction   1
Sarah A. Lovell, Stephanie E. Coen and Mark W. Rosenberg

PART I

Origins, Reflections and Debates   **7**
Mark W. Rosenberg

1   The Great Debate in Mid-Twentieth-Century American Geography:
     Fred K. Schaefer vs. Richard Hartshorne   9
     Trevor Barnes and Michiel van Meeteren

2   The Archive and the Field: Methodological Procedures and Research
     Outcomes in the Work of Carl O. Sauer (1889–1975)   24
     W. George Lovell

3   The Quantitative Revolution   39
     Mark W. Rosenberg

4   Towards Interdisciplinary: The Relationship between GIS/GIScience/
     Cartography and Human Geography   47
     Alberto Giordano

5   Reflections on Human Geography's Methodological 'Turns'   61
     Robin Kearns

Contents

6 For an Intersectional Sensibility: Feminisms in Geography 70
Karen Falconer Al-Hindi and LaToya E. Eaves

7 Making Space for Indigenous Intelligence, Sovereignty and Relevance
in Geographic Research 83
Chantelle Richmond, Brad Coombes and Renee Pualani Louis

8 Geohumanities: An Evolving Methodology 94
Sarah de Leeuw

PART II
Methodologies of Human Geography's Sub-Disciplines **105**
Sarah A. Lovell

9 Affective Landscapes: Capturing Emotions in Place 109
Ronan Foley

10 Geography's Sexual Orientations: Queering the Where, the What,
and the How 123
John Paul Catungal and Micah Hilt

11 Political Geographies: Assemblage Theory as Methodology 134
Jason Dittmer, Pooya Ghoddousi and Sam Page

12 Indigenous Geographies: Researching and De-colonising
Environmental Narratives 144
Meg Parsons and Lara Taylor

13 Storytelling in Anti-Colonial Geographies: Caribbean Methodologies
with World-Making Possibilities 161
Shannon Clarke and Beverley Mullings

14 Historical Geographies: Geographical Antagonism and Archives 173
David Beckingham and Jake Hodder

15 Black Geographies: Methodological Reflections 183
Renato Emerson dos Santos and Priscilla Ferreira

16 Digital Geographies and Everyday Life: Space, Materiality, Agency 196
Casey R. Lynch and Bahareh Farrokhi

17 GIScience: Addressing Aggregation and Uncertainty 207
Hyeongmo Koo and Yongwan Chun

18 Health Geography and Big Data Adventures: Methodological
   Innovations, Opportunities and Challenges   227
   Malcolm Campbell and Lukas Marek

19 Geographies of Disability: On the Potential of Mixed Methods   244
   Sandy Wong and Diana Beljaars

20 Methodologies for Animal Geographies: Approaches Within and
   Beyond the Human   257
   Guillem Rubio-Ramon and Krithika Srinivasan

21 Urban Geographies: Comparative and Relational Urbanisms   270
   Kevin Ward

22 Economic Geographies: Navigating Research and Activism   279
   Kelly Dombroski and Gerda Roelvink

23 Geographies of Education: Data, Scale/Mobilities, and Pedagogies   295
   Yi'En Cheng and Menusha De Silva

24 Children's Geographies: Playing with Participatory Methods   306
   Nicole Yantzi and Janet Loebach

25 Anarchist Research Within and Without the Academy: Everyday
   Geographies and the Methods of Emancipation   322
   Richard J. White and Simon Springer

PART III
Cross-Cutting Issues in Human Geography Methodologies   **337**
Stephanie E. Coen

26 Politics, Institutions and Place: Researching Sensitive Subjects in
   Urban Contexts   339
   Peter Hopkins and Robin Finlay

27 Navigating Ruralities in Human Geography Research: Reflections from
   Fieldwork in Complex Rural Settings   348
   Moses Kansanga, Elijah Bisung and Isaac Luginaah

28 Participatory Geographies: From Community-Engaged to
   Community-Led Research   358
   Heather Castleden and Paul Sylvestre

Contents

29 The Methodological Implications of Integrating Lived Experience in
   Geographic Research on Inequalities   371
   Claire Thompson

30 What Role for More-Than-Representational, More-Than-Human
   Inquiry?   381
   Richard Gorman and Gavin Andrews

31 Dear Feminist Collective: How Does One Take Up Slow Scholarship
   (in the Midst of Crises)?   395
   Jenna M. Loyd, Stepha Velednitsky, Ileana I. Diaz, Sameera Ibrahim,
   Carla Giddings, Kela Caldwell, Roberta Hawkins, Alison Mountz and
   Anne Bonds

32 Refining Research Methodologies to Make a Difference in Policy   407
   Carolyn DeLoyde and Warren Mabee

Index   418

ILLUSTRATIONS

Figures

1.1 Richard Hartshorne defends the regionalist position at the 1960
Lund seminar in urban Geography, 16 August 1960   11

1.2 Kurt Schaefer's 1933 ID he brought with him as a refugee   14

2.1 Man in Nature   30

2.2 Illustrations by Antonio Sotomayor   32

2.3 Sotomayor's "animated or living map of North America in
Indian days"   33

4.1 The dispersed stage of the Budapest ghetto   51

4.2 The International ghetto and the Pest ghetto   52

4.3 Measures of concentration: Location of ghetto houses   53

4.4 Walking distances to market halls   54

12.1 The Four R's   145

12.2 Decolonising the environmental research sector through empowering
Indigenous knowledge, people, and resources   149

12.3 One framework for bridging the divide between different worldviews
from Aotearoa New Zealand   150

12.4 Continuum of Indigenous involvement in research   150

17.1 Visualization of uncertainty using (a) side-by-side mapping and
(b) bivariate mapping   211

17.2 Three configurations for uncertainty mapping with bivariate mapping   213

17.3 An illustration of 'confidence level' between two normal distributions   215

17.4 An illustration of the separability index mapping using the median
household income in Texas from the 5-year (2009–2014) ACS data.
(a) a choropleth map and (b) a plot of the classification result with the
attributes and their 95 percent confidence intervals (horizontal bars)   216

17.5 Optimal map classification with uncertainty for the median household income in Texas from the 5-year (2009–2014) ACS data. (a) an illustration of a network structure for map classification, (b) an optimal map classification result with the class separability index, and (c) an optimal map classification result with Bhattacharyya distance 217

17.6 A simulation result of Moran's *I* values using attributes from the 5-year (2009–2014) ACS data. (a) median household income for Texas counties and (b) median household income of Hispanic households for census tracts in Dallas county 218

17.7 Moran scatterplot of the median household income in Texas from the 5-year (2009–2014) ACS data 221

18.1 Fixed (static) and mobile (dynamic) exposure spaces: Authors analysis and data 229

18.2 Static (fixed) and Dynamic (mobile) exposures 231

18.3 Static (fixed) and Dynamic (mobile) exposures with temporal constraints 233

18.4 Apple mobility data, Auckland, New Zealand during COVID-19 restrictions, 2020 236

22.1 The economy iceberg 284

22.2 Floating coconut economy 285

24.1 The Eureka moment 307

24.2 Meaningful roles for young people 311

24.3 Purple's photograph 314

24.4 Environmental audit map produced by student co-researchers 316

24.5 Environmental audit map produced by student co-researchers 316

24.6 Activity map of play yard activity produced by student co-researchers 317

32.1 Modifying the scientific method to facilitate better policy uptake 409

Table

22.1 The diverse economy of Tūaropaki Geothermal Enterprise 286

CONTRIBUTORS

Editors

Sarah A. Lovell is a senior lecturer in the Faculty of Health at the University of Canterbury, New Zealand. She is a health geographer and her PhD at Queen's University in Canada examined the use of community based participatory research. In the years since, her funded research projects have addressed access and delivery of care in the community, with a particular focus on sexual and reproductive health. Dr Lovell has published widely in journals such as *Social Science & Medicine* and *Critical Public Health*. She has over ten years of experience supervising postgraduate students and teaching qualitative research methods.

Stephanie E. Coen is a health geographer using participatory and arts-based approaches to interrogate how micro-scale environments of everyday life are implicated in health inequities. Her research focuses largely on gendered inequities in physical activity and socio-environmental influences on young people's health, along with a parallel focus on creative methods as a substantive research area in relation to questions of rigour and knowledge translation. She is Associate Professor in the School of Geography at the University of Nottingham (UK).

Mark W. Rosenberg is Professor of Geography in the Department of Geography and Planning and cross-appointed as a Professor in the Department of Public Health Sciences at Queen's University in Kingston, Ontario, Canada. He is the Tier 1 Canada Research Chair in Aging, Health and Development. His research is on aging, health, health care, and health and the environment. While he began his career as a quantitative researcher, he became an early proponent of mixed methods research in human geography. Publications from his research can be found in the leading journals of geography, gerontology, social science and medicine.

Contributors

Karen Falconer Al-Hindi is Professor of Geography at the University of Nebraska at Omaha, where she directs the women's and gender studies program. Her research and teaching explore feminism, identities, social structures, disadvantage and materiality in different contexts. She uses collective biography as an explicitly feminist methodology for investigating power and survival. She is keenly interested in intersectionality as theory and methodology and its role in centering Black feminism in feminist geography. Her recent research argues that intersectionality is crucial for understanding and addressing the effects of COVID19; other recent work examines feminism and agency in the academy, in communities, and in families.

Gavin Andrews is a health geographer based at McMaster University in Canada. His empirical interests include aging, holistic medicine, health care work, phobias, fitness cultures, health histories of places, and popular music. Much of his work is theoretical and positional considering the state-of-the-art and future of health geography. In recent years he has become interested in the potential of posthumanist and non-representational theory for conveying the emergence and performance of health and wellbeing. His current projects examine a range of concepts used in health contexts and studies, and explore their further explanatory potential, including 'world', 'entropy' and 'onflow'.

Trevor Barnes is Professor of Geography at the University of British Columbia where he has taught since 1983. His research interests are in economic geography and the history of geographical thought. His most recent book is jointly edited with Eric Sheppard, *Spatial Histories of Radical Geography* (2019). He has been keenly interested, occasionally obsessed, by the Hartshorne-Schaefer debate since he was an undergraduate. He is glad at last to get it off his chest.

David Beckingham is an Associate Professor of Cultural and Historical Geography at the University of Nottingham. His research investigates the historical geographies of governance and regulation, primarily connected to alcohol consumption and control in Great Britain in the nineteenth and early twentieth centuries. David's publications have examined social, legal, and medical responses to alcohol, analysing the temperance movement, policing and alcohol licensing, and dedicated reformatory regimes.

Diana Beljaars is a Research Fellow at the Geography Department of Swansea University, UK. She approaches her interest in the geographies of culture, disability, and health through postphenomenological and posthumanist theories in human geography in combination with the medical humanities and the Tourette syndrome-related biomedical and clinical sciences. She currently works full-time on the EU-funded project COVINFORM to analyse the effectiveness of Covid responses, in particular with reference to vulnerable groups in Wales. She co-edited *Civic Spaces and Desire* (Routledge), and her monograph *Compulsive Body Spaces* (Routledge) is due to be published by late 2021.

Elijah Bisung is an Assistant Professor in the School of Kinesiology and Health Studies at Queen's University, Kingston ON, Canada. He is a health geographer whose primary area of research focuses on social and environmental production of health and wellbeing.

Anne Bonds is an Associate Professor in the Department of Geography at the University of Wisconsin-Milwaukee. She is a feminist urban economic geographer whose work focuses on race and racialization, feminist political economy, critical poverty studies, and carceral and abolition geographies. She is an editor of Urban Geography and chair of the Urban Geography Speciality Group of the American Association of Geographers (AAG).

Kela Caldwell is a M.S. student in the Department of Geography at the University of Wisconsin-Madison. Her research explores Black geographies and encounters of disaster and crisis.

Malcom Campbell is an Associate Professor in human geography at the University of Canterbury in New Zealand. His expertise is in health and medical geography (spatial epidemiology) as well as regional science/regional analytics. Dr. Campbell is working on a series of projects which attempt to examine and understand social and spatial inequalities in different contexts. He also has an interest in developing and applying novel methods to geographical problems.

Heather Castleden is a Professor and Impact Chair in Transformative Governance in the School of Public Administration at the University of Victoria. Her research program is community-based and participatory, primarily in partnership with Indigenous peoples, and focuses on relational ethics and the politics of knowledge production in environment and health justice.

John Paul Catungal is an Assistant Professor in Gender, Race, Sexuality and Social Justice in the Institute for Gender, Race, Sexuality and Social Justice at the University of British Columbia, where he is currently Interim Director of the Asian Canadian and Asian Migration Studies program. A queer, first-generation Filipinx Canadian settler living in unceded Coast Salish territories, JP is an interdisciplinary scholar trained in the nexus of critical human geographies and intersectional feminist and queer of colour theorizing. His research, teaching and public facing work generally concern the community organizing and cultural production practices of migrant, racialized and LGBTQ+ communities, with particular interests in the fields of sexual health, education and social services. JP was co-editor of the landmark 2012 volume *Filipinos in Canada: Disturbing Invisibility* (University of Toronto Press), as well as special issues of *ACME: International Journal of Critical Geographies* and *TOPIA: Canadian Journal of Cultural Studies* on sexuality, race and nation in Canada. He was a past faculty fellow of UBC's Public Humanities Hub and the Green College Leading Scholars Program. He has been a member of the editorial collective of *ACME: International Journal of Critical Geographies* since 2017.

Yi'En Cheng is Research Fellow in Asia Research Institute at the National University of Singapore. His research interests lie in the intersection across education, youth, and mobilities in Asian cities. He is currently researching on how international student mobilities in East and Southeast Asia are being reconfigured through shifting cultural and geo-politics of the Belt and Road Initiative and the COVID-19 pandemic. More information at chengyien.wordpress.com.

Yongwan Chun is an Associate Professor of Geospatial Information Sciences at the University of Texas at Dallas. His research interests are in Geographic Information Science (GIS) and spatial statistics methodologically and lie in urban geographical issues as substantive research area. Specific research topics include quantitative methods (especially, spatial statistical approaches) for geographical research, geographic data uncertainty, population migration, public health, and urban crime. His research has been supported by funding agencies including the US National Science Foundation and the US National Institutes of Health.

Shannon Clarke is a PhD candidate in human geography at Queen's University, in Canada. She is researching urban governance, migration, and anti-colonial geographies, with a focus on Caribbean thought.

Brad Coombes is a lecturer and researcher at the School of Environment, University of Auckland, Aotearoa/New Zealand. Kati Mamoe and Ngati Kahungunu are his iwi (tribes). A geographer and lawyer by training, he researches at the interface between indigenous livelihoods, political ecology and environmental justice, and regularly contributes to the Waitangi Tribunal's settlement process for Maori land and resource claims.

Sarah de Leeuw is a cultural-historical geographer, anti-colonial feminist activist, award-winning poet and literary essayist, and Canada Research Chair (Humanities and Health Inequities) with the Northern Medical Program, a distributed site of the University of British Columbia's Faculty of Medicine. In 2017 she was appointed to the Royal Society of Canada's College of New Scholars, Artists, and Scientists.

Carolyn DeLoyde is a Postdoctoral Fellow at Queen's University. Carolyn is a Member of the Canadian Institute of Planners and a Registered Professional Planner. Her research is in geography, urban planning, and cities with a focus at the intersection of the natural environment and the urban form.

Menusha De Silva is a lecturer at the Department of Geography, National University of Singapore. Her research examines the intersections of transnational migration and ageing, within the context of Sri Lankan migrants' later-life mobility and negotiations of transnational citizenship, and eldercare relations within transnational families. She teaches modules in social and cultural geography, and her recent work on pedagogy focuses on online teaching and learning, and collaborative approaches.

Ileana I. Diaz is an Afro-Caribbean-Latinx feminist transdisciplinary researcher, thinker, and learner currently living on the traditional territory of the Anishinaabeg, Haudenosaunee, and Neutral peoples. She is a PhD candidate in the Department of Geography and Environmental Management at the University of Waterloo. Broadly her academic work explores the politics of race, gender, imperialism, and environmental issues. She is a SSHRC Bombardier Doctoral Scholar, American Geographical Society Council Graduate Fellow, and the recipient of five teaching awards.

Jason Dittmer is a Professor of Political Geography and Head of Geography at University College London. He writes about diplomacy, heritage, and militarism using assemblage theory. His most recent books are *Diplomatic Material: Affect, assemblage and foreign policy* (2017) and *Popular Culture, Geopolitics, and Identity*, Second Edition (with Daniel Bos, 2019).

Kelly Dombroski is an Associate Professor of human geography at Te Kunenga ki Pūrehuroa | Massey University, in Aotearoa New Zealand. She co-chairs the research cluster for Community and Urban Resilience, and is a member of the Community Economies Institute and the Community Economies Research Network. She recently co-edited *The Handbook of Diverse Economies* (2020) with J.K. Gibson-Graham, and undertakes and supervises research in feminist economic geography and community development throughout the Asia-Pacific region.

Renato Emerson dos Santos is a Professor of Human Geography in the Urban and Regional Planning Research Institute (IPPUR/UFRJ) at Federal University of Rio de Janeiro (Brazil). His research is focused on spatialities of social movements, cartographical activisms and Brazilian Black Movement's struggles. He has published books in Brazil like *Diversity, Space and ethnic/race relations: blacks in Brazilian Geography* (2007), *Social Movements and Geography: the spatialities of action* (2011) and *Racism and Urban Questions* (2012). He was president of the Brazilian Geographers Association (2012-2014).

LaToya E. Eaves is a native of Shelby, North Carolina. She earned her PhD in the Department of Global and Sociocultural Studies at Florida International University in Miami. Her small-town, Southern upbringing informs her research, which combines insights from Black geographies, Black feminisms, queer studies, and southern studies in order to engage spatial processes of home, community, and belonging. Her work has been supported by the National Science Foundation. LaToya is currently an assistant professor in the Department Geography at the University of Tennessee, Knoxville. She is a founder and past chair of the Black Geographies Specialty Group with the American Association of Geographers (AAG).

Bahareh Farrokhi is a PhD student in the Department of Geography at the University of Nevada, Reno. Prior to joining UNR she received her BA and MA in geography and urban planning from the University of Tehran. Her research interests include urban planning, sustainability, social justice, GIS, and health geography.

Priscilla Ferreira, Ph.D., is an Assistant Professor of Geography and Latinx and Caribbean Studies at Rutgers University, New Brunswick, New Jersey. She was a Postdoctoral Fellow in Black and Latinx Studies in the Department of African, African American and Diaspora Studies and the Dept. of Mexican American and Latina/o Studies at the University of Texas in Austin (2019–2021). Dr. Ferreira is an engaged geographer and sociologist whose scholarship is informed by longstanding grassroots organizing, and popular education experiences. Her research focuses on Black community economies, Black urban geographies, Afro-Latinx geographies, and decolonial research and pedagogical praxis. She has published on cognitive injustice, Black community economies, and feminist solidarity economics.

Robin Finlay is a Postdoctoral Research Associate in the School of Geography, Politics and Sociology at Newcastle University, UK.

Ronan Foley is an Associate Professor in Health Geography and GIS at Maynooth University, Ireland, with specialist expertise in therapeutic landscapes and geospatial planning within health and social care environments. He has worked on a range of research and consultancy projects allied to health, social and economic data analysis in both the UK and Ireland. He is on the Editorial Board of *Health & Place*, edited *Irish Geography* between 2015 and 2022 and was an Erskine Fellow at the University of Canterbury (NZ) in 2015. His current research focuses broadly on relationships between water, health and place, including two authored/co-edited books, *Healing Waters* (2010) and *Blue Space, Health and Wellbeing: Hydrophilia Unbounded* (2019) as well as journal articles on auxiliary hospitals, holy wells, spas, social and cultural histories of swimming and 'blue space'.

Carla Giddings is an occupational therapist and PhD Candidate in the Department of Geography, Environment and Geomatics at the University of Guelph on the Treaty lands and

Territory of the Mississaugas of the Credit First Nation. Her research explores the experiences of care and belonging through the Private Sponsorship of Refugees (PSR) program. She currently works as a mental health clinician at the Canadian Mental Health Association and is a co-chair of the Occupational Justice for Newcomers Network.

Alberto Giordano is a Professor in the Department of Geography at Texas State University, a former President of UCGIS, the University Consortium for Geographic Information Science, and a Fulbright Specialist in the Department of History at the University of Vilnius in Lithuania. Alberto is a founding member of the Holocaust Geographies Collaborative, a network of researchers and scholars interested in bringing geographical approaches, methods, and perspectives to the study of the Holocaust and other genocides. He is also involved in projects related to spatial applications to forensic anthropology, most recently on migrant deaths at the U.S-Mexico border.

Pooya Ghoddousi is ESRC Post-doctoral Fellow in Human Geography at Queen Mary University of London. He was a Teaching Fellow in Global Migration at University College London where he also did his PhD on transnational assemblages of identity, belonging and collective action among Iranians in London. His research, activism and writing are inspired by the concepts of assemblages and nomadism by Deleuze, Guattari and Ibn Khaldun.

Richard Gorman is a more-than-human health geographer currently based at Brighton and Sussex Medical School in the UK. Rich's work is concerned with how, and for whom, matters of care come to be enacted, how different knowledges interact, and how different interests are spoken for. Previously, this has involved examining nature and animal-assisted therapies, practices of laboratory animal research, and cultures of patient involvement. Rich's engagement with posthumanist theory and interest in the potentials of a 'more-than-representational' geography have led to an enthusiasm for finding creative and sensitive methodological approaches for encountering liveliness and vitality within multispecies worlds.

Roberta Hawkins is an Associate Professor in the Department of Geography, Environment and Geomatics at the University of Guelph. Her research uses principles from feminist geography to examine ethical consumption campaigns and their discursive and material connections to the environment, social justice, and international development. She also theorizes and advocates for slow scholarship and the possibilities of a feminist academia.

Micah Hilt is a PhD candidate in Geography at the University of British Columbia. He is an urban geographer working at the intersection of governance, mobile forms of networked policy and expertise, and queer geographies across the global North and South. His work contributes to evolving understandings of urbanization across urban studies, STS, queer theory, and urban political ecology. His current work examines the contested development of urban space in Vancouver's West End as well as the development of resilience policy. Micah's interest in critical urban theory and geography is inspired by and grounded in his work as the lead seismic policy planner for the City of Vancouver. Prior to this, and to UBC, Micah worked as an urban planner and deputy resilience officer for the City and County of San Francisco and has received masters degrees in both urban planning and photography in the Bay Area. His research has focused on urban resilience policy implementation in Durban, South Africa, Huangshi, China, and Los Angeles, California. His photographic practice is similarly driven by his interests in global urbanisms and is focused on capturing urban typological forms and the urban experience.

Jake Hodder is Assistant Professor in the School of Geography at the University of Nottingham. His research examines the entangled histories of two key geographical concepts: internationalism and race. His previous work has explored the role of black internationalism in the postwar civil rights, anti-colonial and peace movements and more recent research has investigated the relationship between black activism and the emergence of global governance in the twentieth century, particularly in relation to the League of Nations.

Peter Hopkins is a Professor of Social Geography in the School of Geography, Politics and Sociology at Newcastle University, UK, and Distinguished International Professor at Universiti Kebangsaan Malaysia.

Sameera Ibrahim is a M.S. student in the Department of Geography at the University of Wisconsin-Madison.

Moses Kansanga is an Assistant Professor of Geography and International Affairs at the George Washington University. He is a critical geographer whose research focuses on questions at the intersection of sustainable food systems and natural resource management from a political ecology perspective. Dr. Kansanga has worked extensively with smallholder farmers in different countries in sub-Saharan Africa.

Robin Kearns is a Professor of Geography in the School of Environment, University of Auckland. His PhD at McMaster in Canada was supported by a Commonwealth Scholarship and examined the urban city experience of impoverished psychiatric patients. In the decades since he has continued to explore links between health and place. He is an editor of the journal of the same name. Papers on the nature of place in medical/health geography in *Professional Geographer* (1993) and *Progress in Human Geography* (2002) are among the most cited in the subdiscipline. As a methodologist he has a particular interest in experiential and observational approaches.

Hyeongmo Koo is an Assistant Professor in the Department of Geoinformatics at the University of Seoul, Korea. His research interests lie in Geographic Information Science (GIS) and spatial data analysis. Specific research topics are visualizing and modeling spatial data uncertainty, developing spatial data analysis methods, exploring urban issues including public health, housing, crime, and population migration, and analyzing hydrological and environmental models incorporating data and model uncertainties.

Janet Loebach is the Evalyn Edwards Milman Assistant Professor for Child Development in the Department of Design + Environmental Analysis at Cornell University. Dr. Loebach also serves on the Board of Directors of the International Play Association (Canada), the Editorial Board of the journals Children, Youth & Environments, Cities for Health and PsyEcology, and as the Co-Chair of the Children, Youth & Environments Network of the Environmental Design Research Association (EDRA). Dr. Loebach is also the lead editor on the 2020 *Routledge Handbook of Designing Public Places for Young People: Processes, Practices and Policies for Youth Inclusion*.

Renee Pualani Louis is a Kanaka 'Ōiwi associate researcher at the University of California at Davis Native American Studies and the University of Kansas Institute for Policy & Social Research. She theorizes as a Kanaka Hawai'i scholar of Indigenous cartographies, Indigenous geographies and Indigenous research methodologies.

W. George Lovell is a Professor of Geography, emeritus, at Queen's University in Kingston, Ontario and visiting professor in Latin American history at the Universidad Pablo de Olavide in Seville, Spain. Central America, Guatemala in particular, has been the regional focus of much his research. In 1995, the Conference of Latin American Geography honoured him with its Carl O. Sauer Distinguished Scholarship Award and, in 2018, with its Preston E. James Eminent Career Award.

Jenna M. Loyd is an Associate Professor in the Department of Geography at the University of Wisconsin-Madison. She is a feminist geographer whose work focuses on health politics, carceral and abolition geographies, and the politics of asylum, refugee resettlement, and deterrence in U.S. migration policy. She is the co-editor (with Matthew Mitchelson and Andrew Burridge) of *Beyond Walls and Cages: Prisons, Borders, and Global Crisis*; author of *Health Rights Are Civil Rights: Peace and Justice Activism in Los Angeles, 1963-1978;* and co-author (with Alison Mountz) of Boats, Borders, and Bases: Race, the Cold War, and the Rise of Migration Detention the United States.

Isaac Luginaah is a Professor of Geography and Environment at the University of Western Ontario, Fellow of the African Academy of Sciences and Member of the College of the Royal Society of Canada. For over 20 years, Dr. Luginaah has led several projects in smallholder farming contexts across Africa.

Casey R. Lynch is an Assistant Professor in the Department of Geography at the University of Nevada, Reno. His research examines the politics of urban social and technological change, with a focus on emerging digital technologies and competing visions of the futures they enable. His current work considers the development and deployment of socially-interactive robots in the spaces of everyday life, as well as the use of blockchain technology in remaking urban economies and government.

Warren Mabee is a Professor and Canada Research Chair at Queen's University, where he serves as Director of the School of Policy Studies and Associate Dean, Arts & Science. His research is in renewable energy and its role in our transition to a sustainable, low impact, inclusive economy.

Lukas Marek is a Postdoctoral Fellow at the University of Canterbury in New Zealand. Currently, he is working on the Sensing City project where he is exploring the possibilities of smart city and real-time monitoring to improve people's health.

Michiel van Meeteren. Following Peter Gould's advice that a geographer should always resist specialization, Michiel van Meeteren's research agenda covers urban, economic, financial geography and the discipline's post-1930 history. After studying human geography in Amsterdam (BA, MA) and Ghent (PhD), and passing through Brussels, he currently is a lecturer in Human Geography at Loughborough University (UK). He is particularly interested in how the different traditions of human geographical praxis can be integrated in a contemporary disciplinary and transdisciplinary critical social science.

Alison Mountz is a Professor of Geography and Canada Research Chair in Global Migration at Wilfrid Laurier University. Her research explores how people cross borders, access asylum, survive detention, resist war and incarceration, and create safe havens. Mountz's books include

Seeking Asylum: Human Smuggling and Bureaucracy at the Border (Minnesota); *Boats, Borders, and Bases: Race, the Cold War, and the Rise of Migration Detention in the United States* (with Jenna Loyd); and *The Death of Asylum: Hidden Geographies of the Enforcement Archipelago* (Minnesota). Mountz directs Laurier's International Migration Research Centre and edits the journal *Politics & Space*.

Beverley Mullings is a Professor in the Department of Geography and Planning at the University of Toronto, whose work is located within the field of feminist political economy and engages questions of labour, social transformation, neoliberalism, and the politics of gender, race and class in the Caribbean and its diaspora. She is interested in the ways that evolving racial capitalist regimes are recasting and transforming work, divisions of labour, patterns of urban governance and ultimately, responses to social and economic injustice. Her research has appeared in a number of journals including the *Annals of the Association of American Geographers*, *Gender, Place and Culture*, the *Journal of Economic Geography*, *Antipode*, *Review of International Political Economy*, *Small Axe* and *Geoforum*. Beverley is currently engaged in three major research projects: one examines the financialization of Caribbean remittance economies; the second explores the possibilities that diasporic dialogue holds for reviving Caribbean Radical Traditions; the third project traces the impact of the Black middle-class on social transformation in post-Plantation Economies.

Sam Page is a political geographer and independent research living in Helsinki. His PhD research at University College London studied the UK Labour Party's 2015 General Election campaign and the Deleuzo-Guattarian concepts of assemblage and affect has contributed a novel approach to political parties and electoral geography. He has also written on the topics of Donald Trump's presidency of the US, and on the recent resurgence of breweries in London.

Meg Parsons is of New Zealand Māori (Ngāpuhi), Lebanese, and Pākehā/European heritage, Parsons' is a historical geographer whose research adopts transdisciplinary and decolonizing approaches to examine how Indigenous communities understand and respond to intersecting processes of social and environmental changes. The majority of her research and teaching focuses on bringing a decolonial lens to theories, policies, and practices surrounding climate change adaptation, environmental governance and management, and sustainable transformations. She is a Senior Lecturer at the University of Auckland and a contributing author to the Sixth Assessment of the Intergovernmental Panel on Climate Change.

Chantelle Richmond (Anishinaabe Biigtigong) is an Associate Professor in the Department of Geography & Environment at Western University in London, Ontario, where she holds the Canada Research Chair on Indigenous health and environment. Her research is based on a community-centered model of research that explores the intersection of Indigenous people's health and knowledge systems within the context of local and global environmental change.

Gerda Roelvink is a member of the Community Economies Institute and the Community Economies Research Network. She is the author of *Building Dignified Worlds: Geographies of Collective Action* (2016) and co-editor of *Making Other Worlds Possible: Performing Diverse Economies* (2015) with Kevin St. Martin and J.K. Gibson-Graham. Her research on collective action and hybrid collectives brings theories of affect and embodiment into economic geography. She also holds an adjunct position in the School of Social Sciences at Western Sydney University.

Guillem Rubio-Ramon is a PhD researcher in Geography at the University of Edinburgh. His work combines perspectives from animal studies, political ecology, the environmental humanities and geographies of national identity. His research is grounded in Catalonia and Scotland where he investigates a diversity of animal nationalisms through cases ranging from animal agriculture to biodiversity conservation and environmental conflicts. He has collaborated with the Centre for Contemporary Culture of Barcelona and earned his MPhil in Development, Environment, and Cultural Change at the Centre for the Development and the Environment in Oslo.

Simon Springer is a Professor of Human Geography, Head of Discipline for Geography and Environmental Studies, and Director of the Centre for Urban and Regional Studies at the University of Newcastle, Australia. His research agenda explores the social, political, and economic exclusions and violence of capitalism. Simon's books include *A Primer on Anarchist Geography* (2021), *Fuck Neoliberalism* (2021), *The Anarchist Roots of Geography* (2016), *The Discourse of Neoliberalism* (2016), *Violent Neoliberalism* (2015), and *Cambodia's Neoliberal Order* (2010). His edited books include *Energies Beyond The State* (2021), *Inhabiting The Earth* (2021), *Undoing Human Supremacy* (2021), *Vegan Geographies* (2021), *The Handbook of Neoliberalism* (2016), *The Handbook of Contemporary Cambodia* (2016), *The Practice of Freedom* (2016), *The Radicalization of Pedagogy* (2016), and *Theories of Resistance* (2016). He also serves as Editor for the Transforming Capitalism book series published by Rowman & Littlefield.

Krithika Srinivasan's research and teaching interests lie at the intersection of political ecology, post-development politics, animal studies, and nature geographies. Her work draws on research in South Asia to rethink globally established concepts and practices about nature-society relations. Through empirical projects on street dogs and public health, biodiversity conservation, animal agriculture, and non-elite environmentalisms, her scholarship focuses on decolonizing and reconfiguring approaches to multispecies justice.

Paul Sylvestre is a SSHRC postdoctoral fellow at the Institute of Criminology and Criminal Justice at Carleton University. His current research examines how innovations in settler colonial policing help manage settler state's legitimacy crises in relation to the material and discursive demands of Indigenous social movements.

Lara Taylor is of New Zealand Māori (Ngāti Tahu, Te Arawa, Ngāti Kahungunu, Kai Tahu), Pākehā and Dutch descent. Taylor is an Indigenous Māori researcher supporting and enabling Māori, and equitable partnerships between Māori and others, in various spaces of environmental planning, policy, and practice. Taylor focuses on holistic governance and management at multiple scales, across real and imaginary boundaries. She contributes towards hopeful and transformative philosophical, attitudinal and behavioural changes in the way humans interact with and care for our environment.

Claire Thompson is a Senior Research Fellow at the University of Hertfordshire and works on the NIHR Applied Research Collaboration (East of England). She completed a PhD in Human Geography at Queen Mary University of London. Claire is a qualitative health researcher working largely within the disciplines of geography, sociology, and public health. Her research interests lie in the lived experiences of urban health inequalities and particularly around the topics of the food and alcohol environments, food poverty, and urban regeneration.

In terms of methods, Claire is interested in Discourse Analysis, Qualitative Longitudinal Research (QLR) and visual methods.

Stepha Velednitsky is a PhD student in the Department of Geography at the University of Wisconsin-Madison. She conducts participatory research with a workers' rights organization supporting migrant caregivers in Israel/Palestine, particularly those from the former Soviet Union. In her academic work, Stepha explores how economies of social reproduction shape differential experiences of embodiment and dis/ability among migrant caregivers. Her research interests include regimes of labor migration, Israel/Palestine, social reproduction, settler colonialism, disability, embodiments of trauma, and care.

Kevin Ward is a Professor of Human Geography and Director of the Manchester Urban Institute at the University of Manchester. He has published numerous books and journal articles over the years. Currently he is editing a book with others, entitled *Infrastructuring Urban Futures*, as well as researching and writing on US municipal finance under COVID 19. Kevin is the current editor in chief at *Urban Geography*.

Richard J. White is a Reader in Human Geography at Sheffield Hallam University, UK. Greatly influenced by anarchist praxis, his work explores a range of ethical, economic, and activist landscapes rooted in questions of social and spatial justice. Richard has published his research widely, including contributions to *A Historical Scholarly Collection of Writings on the Earth Liberation Front* (2019); *Education for Total Liberation* (2019); *Animal Oppression and Capitalism* (2017); *Critical Animal Geographies* (2014) and *Defining Critical Animal Studies* (2014). He has co-edited *Vegan Geographies* (forthcoming); *The Radicalization of Pedagogy* (2016); *Theories of Resistance* (2016); *The Practice of Freedom* (2016) and *Anarchism and Animal Liberation* (2015).

Sandy Wong is an Assistant Professor in the Department of Geography at Florida State University, U.S. As an interdisciplinary health geographer, she specializes in disability, accessibility, and environmental influences on wellbeing; and applies geospatial, statistical, and qualitative techniques to the study of health inequities. She has published in journals such as *Health & Place, Journal of Transport Geography, Social Science & Medicine*, and *The Professional Geographer*.

Nicole Yantzi is an Associate Professor in the Department of Geography, Geomatics and Environment at the University of Toronto Mississauga. Her research examines the accessibility and inclusiveness of children's environments, and the positive and negative influences on children's environmental health. This research features valuable insights from children and youth about their experiences, perspectives, and uses of their daily environments. Her work has been published in several international journals including *Children's Geographies* and *Children, Youth and Environments*. She is first author of *Childhood Gardens for Health Learning and Play*, a forthcoming book in Routledge's series on Health and the Built Environment.

INTRODUCTION

Sarah A. Lovell, Stephanie E. Coen and Mark W. Rosenberg

As the chapters in this volume were submitted, we observed our authors elaborate on the philosophical commitments that shape their conceptualisations and practices of research. Ontological reflection leads us to question the nature of being and its implications for our research assumptions (Blaser, 2014; Joronen and Hakli, 2016). It guides our ethical commitments to our participants, our conceptualisations of what data is and can be, and our "understanding of what is possible in the world" (Dombroski and Roelvink, Chapter 22). The ontological emphases of our authors depict a disciplinary research agenda concerned with the blurring of lines between the subject and object of research, the interconnectedness between the researcher and the researched, and the governance of research subjects/objects.

Research methodology emerges at the intersections between the ontological commitments of the researcher and the subject/object of research. Bryman (2008, p.160), explains that one's chosen methodology guides the "practices and assumptions" that underpin one's research. Conventional understandings of research methodology equate the term with the lens through which we undertake research, particularly the selection and deployment of research methods. The chapters in this volume speak to the growing presence of participatory methodologies (Chapters 23, 24, 28) and ethnography (Chapters 16, 23) but also forms of post-qualitative inquiry (Chapters 8, 30) (see McPhie, 2019). Critics of methodology argue it is a Western Science construct that seeks to add validity to research by prescribing the selection and deployment of research methods (McPhie, 2019). Increasingly, we are seeing creative methods infuse human geography research driven by the advent of new technologies, innovations in participatory methods, and the blending of new theories/methodologies (e.g., Ream, 2021)—trends which are apparent throughout this collection.

This volume speaks to the ethical shifts that have been made in human geography research. In *Doing Cultural Geography,* Shurmer-Smith (2002, p. 96) contrasted their immersive approach to ethnographic research with "dashing into 'the field' with a dictaphone and a schedule of interviews…" In this provocative critique, Shurmer-Smith entwined their relationships in the "field," for lack of a better term, as supporting both a deepened understanding of the subject and ethical commitment to their participants. Layered in this critique is an understanding of interviews as capturing participants' representations of the world, a position shared by a growing number of non-representational theorists in human geography (Chapters 11 and 30).

DOI: 10.4324/9781003038849-1

In 2009, Tuck similarly concerned with the subjectification of research participants, called for an ontological shift in our understanding and approach toward research with Indigenous peoples. They wrote:

> I want to recognize that, particularly in Native communities, there was a need for research that exposed the uninhabitable, inhumane conditions in which people lived and continue to live... I have boundless respect for the elders who paved the way for respectful, mutually beneficial research in Indigenous communities. I appreciate that, in many ways, there was a time and place for damage-centered research. However, in talking with some of these elders, they agree that a time for a shift has come, that damage-centered narratives are no longer sufficient. We are in a new historical moment—so much so that even Margaret Mead probably would not do research like Margaret Mead these days.
>
> *Tuck, 2009*

Tuck's (2009) recognition of the need for research that aligns with the values of Indigenous peoples is taken up in Chapters 7 and 12. These chapters by Richmond et al., and Parsons et al., examine methods through which Indigenous research is currently undertaken and the complex navigation of cultural values, traditions and hierarchies that are involved with being an Indigenous researcher. These positions align with political debates that human geography has been tackling for two decades, specifically, "how to make sense of and empower marginalized human subjectivities that are conceived from the start as multiple, fractured, and fluid and that defy efforts to impose order on them" (Knopp, 2004, p. 122). As our authors demonstrate, the solutions are complex and varied aligning with the subjects, values, and commitments of the researcher and the researched (see, for example, Chapters 7, 15, 22, 28).

Throughout this volume, the consideration of decision making in research as a political act that serves to reinforce or resist existing power structures emerges as a predominant theme. Research can be a tool for social and political change. Rationalist models of decision-making emphasise the influence of generaliseable research that provides evidence in support of a policy solution, usually through research based on large datasets (e.g., censuses or population health surveys) (see Chapter 32). Yet, the ontological commitments spoken to in this volume increasingly consider the practice of research as a political act that, in itself, should be a commitment to redressing silences and omissions (see Chapters 7, 12, 15). Thus, we witness researchers embracing experimentation, moving away from prescriptive methodologies and carving out space for activism and ontological reflection. Increasingly, human geographers are making note of an ontological 'turn' in the discipline characterised initially as a post-humanist acknowledgement that the human and non-human constitute an assemblage with the potential to affect change (e.g., Gorman and Andrews, Chapter 30). Less certain is the future policy influence of human geography as we move further from the quantitative orientations that are more widely accepted in the public consciousness and transferable to government policy.

What This Volume Is/Is Not

This volume engages with the key methodological debates and developments that are transforming research in human geography. In its entirety, the edited collection reminds the reader that contemporary developments within human geography are shaped by the ontological, ethical, and political debates and practices of the past. Like Shurmer-Smith (2002), we acknowledge that the way a researcher approaches their work "becomes a part of what is being studied." As a result, we did not set out to create a book that would be a 'how to' guide to best practice in research methods. Instead, we acknowledge the influence of the ethical dispositions of the researcher, the pragmatics of conducting research in challenging settings, and the empirical implications of contemporary thinking in geography.

In the well-recognised words attributed to Anais Nin: "We don't see things as they are, we see them as *we* are." As editors, each of us brings a lens to research that is shaped by the dominant paradigms of the day and our observations of the changing discipline of human geography. When I (Mark) commenced my undergraduate degree at the University of Toronto, the Quantitative Revolution was at its peak. I was part of a third generation of quantitative geographers having been trained by such notable urban and economic geographers as Les Curry, Larry Bourne, James Simmons and Allen Scott. By time I left the London School of Economics, the first wave of critics such as James Blaut, Derek Gregory, David Harvey, Doreen Massey, Richard Peet, Milton Santos and Andrew Sayer were already calling for the end of quantitative research as objective and apolitical and proposing new research agendas based on the inequalities and inequities of capitalism. Their critiques planted the seeds for a post-modern adoption of qualitative methodologies in human geography that were to follow.

When I (Sarah) embarked on my research journey in human geography, qualitative researchers had grappled with debates over structure and agency and were in the midst of a poststructuralist turn, that recognised societal structures as a human product (Kobayashi, 2010). Kobayashi (2010) argues that "the most important result of poststructuralism, with its emphasis on human social outcomes, has been recognition of the importance of studying the ways that human beings, including geographers, create difference through social, cultural, political, and economic practices." As I undertook research into women's access to health care, feminist post-structuralist methodologies offered an understanding of women as the central subject of research; the 'truth' of their experiences was a response to the organisation of health services, the provision of funding and their own personal histories. Some years later while undertaking a PhD amidst the rapid growth in participatory research approaches, methodologies centred on documenting women's experiences no longer seemed sufficient for untangling power structures (re)produced by traditional research approaches.

Human geography has arguably been engaging with different ways of knowing through creative practices for decades (e.g., Rose, 2001 on visual methodologies), but when I (Stephanie) began my PhD the so-called 'creative turn' in human geography was beginning to take an even firmer hold. This movement toward the deepening of more creative geographies was consolidated with the launch of the journal *GeoHumanities* by the American Association of Geographers in 2015. Geohumanities, now a burgeoning sub-field, reflects work at the intersections of geographical scholarship/practices with arts and humanities scholarship/ practices (see Chapter 8 by de Leeuw). I was drawn to the potential of creative approaches to offer new windows into research questions, as methods that comprised both *process* (the act of creating a 'thing') and *product* (the 'thing') (Guillemin, 2004; Coen, 2016). Further, the potential for such approaches to be disruptive (of traditional research power imbalances) and transformative (in terms of including previously excluded voices and experiences) resonated with the critical aims of my work on gendered geographies of physical activity (de Leeuw & Hawkins, 2017; Velasco et al., 2020). This prompted my foray into creative methodologies, starting with drawing during my Ph.D., to poetry during my postdoc, and mostly recently theatre and film as interventions for social change. As Harriet Hawkins (2019: 979) put it, "We can and should understand the possibilities of the creative (re)turn as an inspiration not just for how it opens up possibilities to do research differently, but also for the resources it offers us to remake worlds, our own academic worlds included."

Organisation

The edited collection is divided into three parts. The first reviews the major methodological and paradigmatic shifts in contemporary human geography and positions these developments relative to the changing intellectual history of human geography and events both within human geography and society in general (Part I). Among the series of chapters that make up Part I, Chapter 4 examines how feminist influences provided the ontological space that facilitated the widespread adoption of qualitative methods in human geography. Chapter 7 similarly engages with the epistemological and ontological bases of knowledge, bringing together Indigenous scholars from across the world to examine Indigenous ways of knowing and decolonising methodologies in geography. Part I also critically documents the confluence of actors, social movements, and theoretical developments influencing research within the discipline.

In Part II, authors critically appraise key methodological and theoretical challenges/opportunities shaping contemporary research in various sections of human geography. Contemporary directions within the discipline are elaborated on by established and emerging researchers who are leading ontological debates and the adoption of innovative methods in geographic research. The identification of chapters in this section pays attention to the sub-disciplines that are influencing research across human geography. This provides postgraduate students and more senior academics alike with reference points for summative methods discussions within some of the core sub-disciplinary areas. In Parts II and III, authors emphasise examples from their research and the research of others to reflect the fluid, emotional, and pragmatic realities of research. In this way, our goal with these sections is to capture key developments and disciplinary influences emerging from the various sub-disciplines of human geography.

In Part III, authors take on debates that human geographers implicitly and explicitly engage with in their thinking around methodology and the practice of research. The chapters contend with familiar (and perennial) questions and issues in human geography methodologies from ethics, to representation, to participation, and impact. Yet, in their treatment of these cross-cutting issues, the authors move our discussions forward by offering new ways of contending with and conceptualising these issues in practice. As a whole, this section addresses key challenges researchers face when designing studies in human geography with authors drawing on their own empirical experiences and their methodological influences.

The chapters in this volume capture key developments and contemporary methodological debates in human geography, which are moving forward disciplinary thinking. Collectively, this volume holds a mirror up to the axiological assumptions, research methods, and the roles of researcher and participant in research. In doing so, we contribute to the current state of methodological maturity within human geography and hopefully, open the door to new methodological developments in the future.

References

Blaser M. (2014). Ontology and indigeneity: On the political ontology of heterogeneous assemblages. *Cultural Geographies*, 21(1), 49–58. doi:10.1177/1474474012462534

Bryman, A. (2008). *Social research methods* (Third ed.). New York. Oxford University Press.

Coen, S.E. (2016). What can participant-gendered drawing add to health geography's qualitative palette? In: Fenton, N., Baxter, J. (Eds.), *Practising Qualitative Methods in Health Geographies*, pp. 131–152. Oxon & New York: Routledge.

de Leeuw, S., and Hawkins, H. (2017). Critical geographies and geography's creative re/turn: Poetics and practices for new disciplinary spaces. *Gender, Place & Culture*, 24(3), 303–324.

Guillemin, M. (2004). Understanding illness: Using drawings as a research method. *Qualitative Health Research*, 14(2), 272–289.

Hawkins, H. (2019). Geography's creative (re)turn: Toward a critical framework. *Progress in Human Geography*, 43(6), 963–984.

Knopp, L. (2004). Ontologies of place, placelessness, and movement: Queer quests for identity and their impacts on contemporary geographic thought. *Gender, Place & Culture*, 11(1), 121–134.

Kobayashi, A. (2010). People, place, and region: 100 years of human geography in the annals. *Annals of the Association of American Geographers*, 100(5), 1095–1106. https://doi.org/10.1080/00045608.2010.523346

McPhie, J. (2019). *Mental health and wellbeing in the Anthropocene: A posthuman inquiry*. Palgrave Macmillan. https://go.exlibris.link/BjcTZsmd

Ream, R. (2021). Rambling with(in) the Wairarapa: Revisioning arcadia through affective creativity. Area (London 1969), 53(2), 211–218. https://doi.org/10.1111/area.12446

Rose, G. (2001). Visual methodologies: An introduction to the interpretation of visual materials. (2nd ed.). London; Thousand Oaks: Sage.

Shurmer-Smith, P. (2002). *Doing cultural geography*. London: Sage. https://go.exlibris.link/2yR

Tuck, E. (2009). Suspending damage: A letter to communities. *Harvard Educational Review*, 79(3), 409–427. https://doi.org/10.17763/haer.79.3.n0016675661t3n15

Velasco, G., Faria, C., and Walenta, J. (2020). Imagining environmental justice "across the street": Zine-making as creative feminist geographic method. *GeoHumanities*, 6(2), 347–370.

PART I

Origins, Reflections and Debates

Mark W. Rosenberg

The concept of a *method* is easily understood as a systematic practice of collecting data defined in the broadest sense and carrying out an analysis of the data collected. Doing a regression analysis on geocoded census data or carrying out focus groups and analysing the content for themes are examples of quantitative and qualitative methods respectively. What is meant by *methodology* is more abstract. The *Oxford Dictionary of Human Geography* defines methodology as:

> How a study is conceived and operationalized. While methods are the specific techniques used for data generation and analysis, methodology is the wider approach and organization through which those methods are employed. It provides the justification for the types of questions asked, how they are asked, and how they are analysed and interpreted, as well as practical issues such as how data is sourced and the sampling framework used. The methodology adopted by a researcher is informed by their wider ontological, epistemological, and ideological beliefs as these define what is the most appropriate and valid way to make sense of the world. In other words, methodology is not simply about the practicalities of undertaking a study, but is a deeply philosophical endeavour.

While not every author in Part I or indeed in the other two parts of this collection necessarily draws the distinctions between method and methodology in such a clearcut fashion, what is indeed important in reading the chapters in Part I and beyond is how the authors reflect on key geographers, groups, and the debates that took place as the palette of methods used in human geography evolved and expanded. As a section, the overall goal of Part I is to provide readers with a sense that the current methods being used and the methodologies with which they are associated are not isolated constructs and processes but have their origins in the intellectual development of human geography over time and within the context of broader technological and societal changes that took place especially post World War Two.

It is important to acknowledge that prior to World War Two, there were important methodological, epistemological and ontological changes in geography. Accounts of them can be found in various books that have been written on the history and philosophic underpinnings of human geography (e.g., Cresswell, 2012). Part I does not try to cover all of the methodological

DOI: 10.4324/9781003038849-2

debates that have taken place in human geography. We purposefully chose to start with the debate between Hartshorne and Schaeffer (Chapter 1) because of the profound changes that resulted methodologically in human geography as an increasing number of human geographers rejected regional approaches and adopted quantitative approaches in their studies. In Chapter 2, Lovell reflects on the importance of Carl Sauer's ideas emphasizing the importance of landscape in shaping human geography, while Rosenberg (Chapter 3) argues how the Quantitative Revolution was really a methodology based on models and systems thinking. Paradoxically, Rosenberg also argues that the theoretical and methodological limitations of the Quantitative Revolution spurred the theoretical critiques and the decline of statistical and mathematical modelling in favour of qualitative methodologies on the one hand and geographic information systems on the other hand. Giordano (Chapter 4) traces similar paths and connections among historical research, cartography and GiScience in a provocative chapter that draws on his research on the Holocaust. Chapters 5, 6, 7 and 8 can be read as alternative ways human geographers have rejected quantitative research in favour of alternative methodologies, epistemologies, and ontologies. In a chapter driven by personal reflection, Kearns (Chapter 5) takes the reader on a journey of "turns" that led him to embrace qualitative methods and ultimately focus on "embodied mobility and engaged visual methods." Central to Chapter 6 by Al-Hindi and Eaves is the role that intersectional feminist thinking increasingly plays in human geography. In a unique mode of presentation, the three authors of Chapter 7 each provide a sense of Indigenous ways of knowing from their own Indigenous backgrounds in Canada (Richmond), Hawaii (Pualani) and Aotearoa (Coombes) and their sense of the challenges of seeing human geographies from their Indigenous ways of knowing. In Chapter 8, DeLeeuw shows us how the methods of the arts and humanities offer another avenue for exploring human experience and places from a geographic perspective.

As such, Part I interrogates the epistemological and ontological groundings of human geography, to understand how geographers make sense of the world and how these perspectives influence contemporary research in the discipline. For example, while some aspects of qualitative research have their starting point in the rejection of quantitative research by feminist geographers, other qualitative methods (e.g., historical/archival research) date back to the beginnings of modern human geography. Part I is really a choice of appetizers that hopefully encourages the reader to delve more deeply into the methodologies and methodological challenges that are the choices for main course in Part II and the desserts in Part III.

Bibliography

Cresswell, T. (2012) *Geographic Thought: A Critical Introduction*. London: Wiley-Blackwell.

Rogers, A., Castree, N. and Kitchin, R., eds. (2013) *Oxford Dictionary of Human Geography*. Oxford University Press, (online version).

1

THE GREAT DEBATE IN MID-TWENTIETH-CENTURY AMERICAN GEOGRAPHY

Fred K. Schaefer vs. Richard Hartshorne

Trevor Barnes and Michiel van Meeteren

Introduction

Since geography's establishment as a university-based discipline in Western Europe during the mid-to-late nineteenth century its methodology was primarily descriptive. Geography relied on the gathering and recording of raw empirical detail, displayed as factually laden essay-style prose, assorted typologies, tables of numbers and perhaps most importantly maps. There was very little of what we would now call theory, explanation or analysis. The heyday of geography's descriptive methodology was during the 1920s and 1930s. In an era when the Internet was a distant future, compilation and curation of relevant facts on a subject was a valid academic endeavor and arduous work. And in many places in the world geographers commanded respect for their ability to synthesize knowledge about distant places. Nevertheless, there grew an increasing sense of an intellectual gap between descriptive geography and the methods of other disciplines. This was certainly the case for the physical sciences that were rigorously theoretical, quantitative, and often experimental, but also social sciences like economics, political science, sociology, and psychology. In the social sciences, change had been especially afoot from the 1930s, and once the Second World War ended, those disciplines became increasingly like the physical sciences, deploying formal theory, pursuing generalization, carrying out measurement using quantitative data and strictly defined statistical techniques for testing and verification (Mirowski, 2001; Abbot and Sparrow, 2007). Carl Schorske (1997, 295) called this move by those social sciences, "the new rigorism", a "passage ... from range to rigor, from loose engagement with a multifaceted reality historically perceived to the creation of sharp analytical tools that could promise certainty where description and speculative explication had prevailed before".

But not so much in geography, despite its claim that it was the discipline *par excellence* to bridge natural and social worlds. The discipline seemed rather lightweight and naïve, continuing to peddle "mere" description, compiling dull gazetteer lists of facts. Neil Smith (1989) likened the state of at least American geography until the early 1950s to a museum-like existence; in effect, it was preserved under glass, unchanging, stuck in a past era. The methodological debate

DOI: 10.4324/9781003038849-3

that we review in this chapter set in the early post-Second World War period changed all that. It was an iconoclastic moment when the museum display cases were smashed open, and the present came storming in. The change that the debate wrought was fundamental. For those participating, it did not involve just tweaking the existing methodological position around its edges, but invoked vast wholesale transformation. Louis Althusser (1972, p. 85) used the (geographical) metaphor of discovering a new continent to denote a fundamental alteration in knowledge acquisition. The debate we review here was about discovering a new continent by a young generation of geographers, opening territory they thought was never explored before, producing possibilities never previously imagined.

Lasting between 1953 and 1959, the debate was between Fred K. Schaefer (1904–1953) at the University of Iowa and Richard Hartshorne (1899–1991) at the University of Wisconsin 180 miles (290 kms) up the road. For these two protagonists, the dispute was about the very soul of geography. It turned on fundamental questions around the nature of geographical practice, the correct method, and the definition of disciplinary progress. Bill Bunge (1928–2013) (1968, p. 12), an acolyte of Schaefer, but early on a student of Hartshorne, thought the controversy was *the* historic battle of the discipline, likening the clash to that "between Michelson and Newton, or Hegel and Feuerbach".

For those like Bunge who framed Schaefer versus Hartshorne as geography's "great debate", the central issue was about the appropriate geographical method. Should it stay with its historically descriptive method, known as the idiographic approach, that focused on the assembly and careful arrangement of unique geographical facts, Hartshorne's view? Or, instead, should it be like the physical sciences and some social sciences and seek laws expressed mathematically and as rigorously defined abstractions, a nomothetic approach, and Schaefer's view?[1]

In reviewing Anglo-American geography's mid-twentieth century *great methodological debate*, we divide the chapter into three sections. First, we discuss the state of Anglo-American geography before the debate broke out in the early 1950s as well as introducing its *dramatis personae*, Schaefer and Hartshorne. Second, we provide an account of the dispute, starting with Schaefer (1953), followed by Hartshorne's (1954, 1955, 1958, 1959) responses. Finally, we discuss the aftermath and disciplinary consequences.

Context and *Dramatis Personae*

A key methodological dualism in geography is between "systematic" and "regional" geography first articulated by Varenius in 1650 (Hartshorne, 1939, p. 41). If geography is about understanding variation on the earth's surface, should the geographer prioritize general mechanisms causing geographic difference – systematic geography – or begin by detailed documenting of areal variation – regional geography?

Systematic geography was originally defined by analogue to "systemic science" as the search for "general laws" (Hartshorne, 1959, p. 109). As the discipline specialized, systematic geography was subsequently understood as thematic specialization such as economic, transportation, or political geography (Ackerman 1945), although the older causal connotation remained resonant (Schaefer, 1953). Both Hartshorne (1939, p. 456–459) and Schaefer (1953) acknowledged that both systematic and regional perspectives mattered for geographical practice. They differed, though, in their assessment about which approach should be the ultimate disciplinary goal and justify geography as a "science".

During the second quarter of the twentieth-century, American geography was dominated by regional geography. The earlier search for causal geographical laws had been discredited around the 1930s, associated with a crude environmental determinism linked to racist and

Western supremacy theories (Martin 2015). The prime methodological goal for Hartshorne's generation became the preservation of geography as a science while purging a malignant environmental determinism. To do so they changed the object of study of geography from the causal relationship between the environment and humans to the synthetic study of the region. But what kind of scientific object was the region, and methodologically how should it be studied? These were key questions for Hartshorne and his peers (Martin 2015).

Their problem was that the synthetic "science of regions" did not always produce appealing studies. John Leighly (1937, p. 127) lamented that regional geography was often "a vision of the whole surface of the earth plastered with topographic descriptions – like the luggage of a round-the-world tourist with hotel stickers … . [It] terrified even the most tolerant reader of regional descriptive literature". Instead, Leighly (1937, p. 131) proposed that regional geography be an "art" in which the "skill and intellectual integrity" of the geographer-artist created a synthesis of regional features.

Hartshorne disagreed with Leighly that geography's methodological problem could be resolved by an appeal to art. Instead, he argued resolution would best come from studying recent developments in German geography. In fact, American geography had been largely erected on German foundations, but direct intellectual exchange between geographers from the two countries dried up after the First World War (Martin 2015). Hartshorne, who spoke German and met many German geographers during a year of fieldwork in Germany in 1931 (Martin 2015, p. 890), thought Germanic foundations remained useful in addressing geography's key methodological problem. That this usefulness was not already recognized, he believed, was a result of bad translations of German works and inept interpretations. After voicing his dissatisfaction to Derwent Whittlesey, the editor of the *Annals of the Association of American Geographers,* Hartshorne was invited, "to write a "statement … it can be brief" (Hartshorne 1979, p. 63).

Hartshorne got down to work. He believed once he finished, there would be no more misunderstandings. But his statement got quickly out of hand. In April 1938, it was 61 manuscript pages (Hartshorne 1979, 70). By July 1938, on the eve of Hartshorne's academic leave to Vienna where he planned to study boundary issues in the mid-Danube region, it had grown to 194 pages (Hartshorne 1979, 71). The Nazi *Anschluss* that occurred in March 1938 put paid to Hartshorne's plans for fieldwork in that country. Instead, he hit the library stacks at

Figure 1.1 Richard Hartshorne defends the regionalist position at the 1960 Lund seminar in urban Geography, 16 August 1960 (Photos: Chauncy D. Harris)

the University of Vienna, giving him access to many German-language sources unavailable in the US (Martin 1995, p. 906). When the manuscript was finally completed in April, 1939, in Meilen, Switzerland, where Hartshorne had gone for safety fearing a German war with Poland (and for good reason), it weighed in at over 600 pages (Hartshorne 1979, p. 73). This hefty manuscript became *The Nature of Geography: A Critical Survey of Current Thought in the Light of the Past*, whose publication required two full issues of the *Annals of the American Association of Geographers* (Hartshorne 1939). The volume meticulously explicated, rigorously justified, and genealogically fixed the discipline like no other English language volume before it.

Nature conceived geography as a science, although Hartshorne recognized that it was a different kind of science compared to the systematic sciences that he variously also called the "exact sciences," "natural sciences," or "physical sciences" (Hartshorne 1939, p. 367). Hartshorne leaned on the German definition of science, *wissenschaft*. It implied a broad systematic pursuit of knowledge, learning and scholarship, but not the imposition of a singular methodology or set of ends. The Anglophone conception was narrower, conceiving science as concerned with generalization and the ultimate generalization, a law. For Hartshorne, under the broader German *wissenschaft* definition, geography was a science but it did not necessarily pursue generalizations and laws. As he elaborated in *Nature*, the science of geography was defined by the objective understanding of complex geographical phenomena made possible by precise description, meticulous citation practice, and expertise in relevant methods including fieldwork, cartography and statistics.

Given that definition, Hartshorne (1939, p. 468) believed that the science of geography was best realized by regional geography; that is, by the description of regional or areal difference. Systematic geography could help in identifying factors that contributed to areal differentiation, but as a pursuit it was secondary to regional geography.

Regions therefore became the building blocks of the discipline, with all geographical information organized through them. Note, though, that for Hartshorne the regional division of the world was not a "naively given fact" (Hartshorne 1939, p. 275). Rather, the researcher imposed regional demarcation. As Hartshorne (1939, p. 275) put it:

> Regional entities [...] are ... in the full sense mental constructions; they are entities only in our thoughts, even though we find them to be constructions that provide some sort of intelligent basis for organizing our knowledge of reality.

Specifically, Hartshorne defined a region as a complex combination of hard facts and causal relations objectively described (Entrikin, 1981). He called those combinations an "element complex". Importantly, their combinatorial character made each region unique, not found anywhere else in the world (Sack, 1974, p. 441). To use Hartshorne's illustration, while several regions might share one common element, say, producing grain – for example, the Po plain, the Middle Danube plain, and the American Corn Belt – this did not then make those regions identical. This is because none of them shared in exactly the same combination all the other objective geographical elements found in those specific grain producing regional complexes (Hartshorne, 1939, p. 392). Accordingly, each region "occurs but once on the earth" (Hartshorne, 1939, p. 393). Therefore, "regional geography ... is ... concerned with the description and interpretation of unique cases...." (Hartshorne, 1939, p. 449). While the region is the core idea in Hartshorne's approach, he nonetheless recognizes that systematic geography might still be useful in determining the components and their relationship within them (Hartshorne, 1939, p. 457).

A critically important methodological corollary stemmed from Hartshorne's claim of regional uniqueness. As discussed, unlike the German *wissenschaft*, the Anglophone tradition equated science with the formulation of general laws. Such laws, however, cannot be expressed for combinatorial entities like Hartshorne's region. For a scientific law to be stated, the entities on which the law bears must be homogenous. Because regions are not a uniform class of phenomena – each is unique – then no law can ever be derived.

Because for Hartshorne regions are never the same, the making of law-like statements that define Anglophone systematic science cannot apply; it has no purchase. As a result, as Hartshorne (1939, p. 446) noted, "we arrive, therefore, at a conclusion similar to that which Kroeber has stated for history: 'the uniqueness of all historical phenomena.... No laws or near laws are discovered.' The same conclusion applies to the particular combination of phenomena at a particular place." Thus, for regional geography, unlike systematic geography, we cannot explain, or predict, or knowingly intervene but only describe: "Regional geography, we conclude, is literally what its title expresses: ... a descriptive science" (Hartshorne 1939, p. 449).

Although Hartshorne (1939, p. 430) surmised that a generic classification of regions might one day progress "to the statement of general principles [i.e., laws]", he also warned that "any principles we attempt to develop can have no more validity than the 'objects' [i.e. regions] we have constructed as their foundation".

The publication of Hartshorne's book represented a crossing of the Rubicon for English-language geography. Never before had there been a book published like it with the ambition both to define the discipline historically and to lay out a meticulously justified methodology. Hartshorne's star necessarily rose. In 1940 he moved from the University of Minnesota to become full Professor at the University of Wisconsin, Madison. The next year he was appointed Chief of the Geography Division at the Research and Analysis Branch of America's military intelligence organization, the Office of Strategic Services (OSS) (Barnes and Farish 2006). In 1949 he became President of the Association of American Geographers, the professional organization of the discipline within the US. Consequently, by 1953 at the start of the Schaefer-Hartshorne debate, he was likely America's most well-known geographer. That said, Hartshorne's methodological position enunciated in *Nature* was not uniformly accepted or praised. Some of the younger geographers, including those he hired at OSS, such as Edward Ackerman (1911–1973), thought Hartshorne's regional geography made geographers a jack of all trades and master of none (Ackerman, 1945).

Hartshorne's fame was in stark contrast to the man who debated him, Fred K. Schaefer. Very few had heard of him. Hartshorne (2004a, p. 277) later said he was "an essentially unknown professor of geography". Schaefer was trained as a metal worker in Berlin immediately after the end of World War I. In 1925 he went back to school as a mature student and in 1931 he graduated from the University of Berlin with a Diploma primarily in economics. His politics were left-wing. When the Nazis came to power in 1933 that was enough for him to be blacklisted, then arrested and sent briefly to a concentration camp. The writing was on the wall. Later that same year he obtained a permit to travel to Switzerland for a skiing trip. He never returned. A political refugee in England, 1933–1938, he worked primarily for various left-wing organizations collecting data and carrying out statistical analyses. In 1938 he left for America, moving to the Scattergood Hostel for European refugees run by a Quaker society, the American Friends Service Committee, in Iowa City, Iowa. After a short stint as a house painter, in 1939 he began to teach part time at the University of Iowa, Iowa City, and because of his training in economics he lectured at the College of Business. Harold McCarty (1901–1987), a Professor at the College of Business, and the future founding chair of the Geography

Department, invited him to teach courses on Eastern Europe. In 1943 Schaefer was hired full time in Business, and appointed Assistant Professor. Fluent in English, Russian and German, interested in geopolitics, and having taken courses in political and economic geography in Berlin, in 1946 Schaefer was asked by McCarty to become a member of the brand-new Iowa Department of Geography.

While Schaefer may have been a novice geographer, it did not deter him from being David, trying to slay the disciplinary methodological Goliath, Richard Hartshorne. Schaefer's sling-shot was a version of logical positivism developed by members of the Vienna Circle during the 1920s (see Sigmund, 2017, for a situated introduction). Logical positivism claimed that there were only two kinds of meaningful knowledge. The first were empirical truths verified by methods of the systematic sciences. The second were truths that were correct because of the very meaning of the terms in which they were expressed such as logic or mathematics. Logical positivists believed that unless claims to knowledge met either of these two criteria, knowledge was unreliable and spurious. Further, for logical positivists, the highest form of empirical know-ledge was a universal law. It took the logical form: if cause A, then for all time and all places the same effect, B. For example, Newton's law of gravity says *if* a pair of planetary masses *then* for eternity and everywhere the gravitational force between them is proportional to their respective sizes divided by the square of the distance that separates them.

It was logical positivism that Schaefer critically brought to bear on Hartshorne's *Nature*. In doing so, he got inside help from a colleague in the University of Iowa's Philosophy and Psychology Department, Gustav Bergmann (1905–1987) (Heald, 1992). Bergmann was a

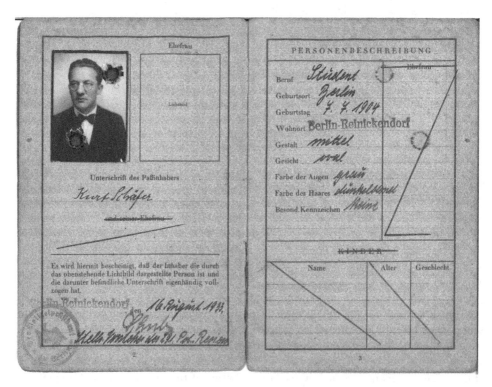

Figure 1.2   Kurt Schaefer's 1933 ID he brought with him as a refugee (Collection: American Geographical Society, AGS-NY AC1 - Box 338 - Folder 7)

member of the Vienna Circle, originally a mathematician, later a philosopher. Smoothing the relationship between the two men was also that they were a similar age (Schaefer was a year older), were native German speakers, were political refugees from the Nazis – Schaefer because of his left-wing political views, Bergmann because he was Jewish – were employed by the University of Iowa in the same year, 1939, and perhaps most importantly, were unreservedly champions of systematic science.

Going to this last point, for Schaefer the fundamental problem with Hartshorne's approach to geographical knowledge, as understood by logical positivism, was its unscientific character. And if it was not scientific, necessarily it was unreliable and spurious. For Schaefer, Hartshorne had denied the possibility of scientific geographical knowledge when he denied geographical laws. In doing so, he consigned the subject to a study of the unique and the exceptional, an idiographic discipline. In contrast, Schaefer aimed to make the subject a systematic science, capable of formulating its own laws, a nomothetic discipline. For Schaefer, the scientific task of geographers was to discover laws that explained and predicted the spatial distribution of phenomena. They would complement the laws in economics, sociology and other social sciences (Schaefer, 1953, p. 248). From Schaefer's perspective, Hartshorne's (1939, p. 551) claim that geography was scientific because of its unique "point of view, a method of study" was rejected. If regional geography could not generate laws, it was simply not a science. Claiming to be a special discipline was an appeal to "exceptionalism". For this reason, regional geography could not be the scientific core of a discipline. It was at best only a data vault to inform systematic geographical inquiry (Schaefer, 1953).

The Debate

Strictly speaking the Schaefer-Hartshorne debate was not a debate. By the time Schaefer's (1953) "Exceptionalism in geography: a methodological examination" was published in the *Annals of the Association of American Geographers* in September, 1953, its author had been dead for three months. Consequently, Schaefer was in no position to hold up his end in any argument. He suffered a fatal heart attack while attending a matinee at an Iowa City movie theatre in early June of that year. While Schaefer may not have gotten to defend himself against the series of ferocious rebuttals that Hartshorne subsequently published in the *Annals* in 1954, 1955, 1958 and in another monograph in 1959, the consensus is that he carried the day anyway, at least for a while. Clyde Kohn (1972, p. i), likely one of the original referees of Schaefer's paper at the *Annals,* reflected in 1972:[2]

> Schaefer's article must be cited as one of the more important contributions to geographic methodology in the history of our discipline. The "revolution" which followed in the late 1950s, and the continued search since then for laws containing spatial variables, demonstrate the vitality and challenge of the ideas set forth by Schaefer in 1953.

Schaefer initially met Hartshorne the first year he was a geography professor at Iowa. They were on a panel together at a regional economics conference in Chicago talking about the Soviet Union. Hartshorne remembered only pleasantries over a cup of coffee, with Schaefer saying that he was interested in his "work in methodology which he would like sometime to discuss ... at length."[3] That opportunity came four years later when Hartshorne was invited by McCarty to the Iowa Department of Geography to participate in among other things Schaefer's graduate seminar on methodology where he was asked to speak about *The Nature of Geography*. Schaefer acted as a facilitator, with Hartshorne kept on his toes by sharp but always polite questions from the well-informed graduate students.[4] Schaefer wrote to Hartshorne after the event congratulating him on his "fortitude" and "splendid response to the questioning."[5] Hartshorne later reflected, "we all seemed to enjoy ourselves."[6]

Hartshorne's next encounter with Schaefer was less enjoyable. In October, 1953, he picked up from his mail box at Science Hall at the University of Wisconsin, Madison, a copy of the latest edition of the *Annals* to read Schaefer's paper that was just published. Schaefer submitted his paper on December 4th, 1952. According to Bunge (1968, p. 20), before Schaefer sent off the paper, he had walked into Harold McCarty's office, and with his hands "trembling" put the paper on his desk and said, "This is my existence in geography". Although this anecdote may be apocryphal, it expresses at least for some the elemental power of Schaefer's paper and its significance for the discipline.

It was only once Hartshorne began reading the paper that he first learned of Schaefer's death. It did not soften his reaction, however. He was incandescent with rage, but also nonplussed. The paper violated all the rules on methodological debate he had so meticulously set out (Hartshorne, 1948). There were so many things that he did not understand including the word "exceptionalism" in Schaefer's very title. He tried to look it up in the *Encyclopaedia for the Social Sciences* but "there were four or five different meanings, so that wasn't very helpful" (Hartshorne, 2004b [1986], p. 294). Schaefer, though, was clear in his paper about its definition. It was the view that because of their subject matter some disciplines were unable to formulate scientific laws about the material they studied. Of course, this was Hartshorne's view about regional geography.

In the standard model of explanation and prediction in systematic sciences, the existence and identification of scientific laws are crucial. Once a law is established it is used either to explain or predict. Explanation and prediction are flip sides of one another, both the consequence of having established a scientific law. But because Hartshorne denied for regional geography the possibility of scientific laws, there could be neither explanation nor prediction. It was against such exceptionalism that Schaefer's paper railed.

Specifically, Schaefer's critique of Hartshorne followed two strategies. The first was to counter Hartshorne's interpretation of the German historical writings that justified the definition of the region as unique. If Schaefer could show that those German authors on whom Hartshorne relied did not define the region as unique, he could argue that there was no reason to abide by exceptionalism. Regional geography could then join with other systematic sciences and seek laws, as well as explain and predict and not merely describe.

The second was to make explicit the kind of regional geography that was possible once the discipline's "exceptionalism [was] disposed of" (Schaefer, 1953, p. 242). It would be a regional geography unified with, not separate from, the systematic sciences, law-seeking and concerned with explanation and prediction. Here Schaefer (1953, p. 239) imagined the kinds of laws that geographers might discover:

> Spatial relations among two or more selected classes of phenomena must be studied all over the earth's surface in order to obtain a generalization or law. Assume, for instance, that two phenomena are found to occur frequently at the same place. A hypothesis may then be formed to the effect that whenever members of the one class are found in a place, members of the other class will be found there also, under conditions specified by the hypothesis.

The prose is stilted, but Schaefer is setting out here the structure of a geographical law conforming to the classic configuration within logical positivism, if A, then B. What makes a law geographical is that the instances making up the homogenous classes A and B are indexed by location. In Schaefer's example, every case of A occurs at the same geographical location as every case of B. That is, there is something about the geographical location that ensures, if an instance of A, then an instance of B. It is the indexing the instances of A and B by location that makes the law a geographical law.

Schaefer also says that the type of laws geographers would most likely draw up are "morpho-logical", that is, "containing no reference to time and change" (Schaefer, 1953, p. 243). They would take the form, if spatial pattern A, then spatial pattern B. There is no temporal process in this formulation. It involves only a spatial relation. Schaefer thought geographers might also refer to "process laws" (i.e., those involving time), but morphological laws highlighted the discipline's special interest and expertise.

Hartshorne was having none of it, however. A week after Hartshorne read Schaefer's paper he wrote to the editor of the *Annals*, Henry M. Kendall, cataloguing Schaefer's "major errors and mis-statements."[7] A week later he was even madder. Publishing Schaefer's paper, Hartshorne charged, had "create[d] a mess, for me, and for American geography... ." The paper was "a palpable fraud, consisting of falsehoods, distortions and obvious omissions."[8] Hartshorne never got over it. In 1955 Hartshorne was still writing to the *Annals'* editor, by then, Walter Kollmorgen: "In whatever sense it is possible for a learned journal to commit a crime ... The *Annals* has committed a crime unparalleled in its history" (quoted in Martin, 1989, p. 76). Even at age 89, two years before he died, Hartshorne continued to fume, in this case writing cor-rective letters to both Derek Gregory and Fred Lukermann for giving credence in their writings to Schaefer's paper when it was so obviously demonstrably wrong.[9]

Hartshorne wrote several formal replies to Schaefer. The 1954 reply was a two-page letter outlining his main criticisms and anticipating the full-blown critique to follow the next year in a second paper. The letter was a blistering attack on Schaefer's scholarship, or more precisely, the absence of scholarship. It pulled no punches. The *Annals* editorial assistant, sister of the Editor, Walter Kollmorgen, was forced to "excise the color words" from the letter before it was deemed acceptable for publication.[10]

The 1955 reply was 40 printed pages long, with over 100 footnotes. On many pages the footnotes occupied more of the page space than the main text. In the reply, Hartshorne concerned himself exclusively with defending himself from Schaefer's critique of his historical scholarship, while at the same meting out corrosive criticisms of Schaefer's own scholarship. Hartshorne was extraordinarily well prepared to undertake both tasks. He spent that 1938–1939 sabbatical year at the University of Vienna Library reading all there was to read in German on geographical methodology. Further, he often read those German texts with the University of Vienna's Professor of Geography, Johan Sölch, literally by his side. With Sölch, Hartshorne made exact translations of key texts, or at least as exact as was possible given it was German academic prose (Hartshorne, 1979).

Drawing on his extensive knowledge of German geography accumulated in the preparation of writing *Nature*, Hartshorne made mincemeat of Schaefer's argument that he had misun-derstood German geographical scholarship. In contrast, Hartshorne with relish as the great corrector systematically, line-by-line, word-by-word, demonstrated Schaefer's own shoddy textual interpretations that were filled with misunderstandings, misquotations, and mistaken citations; in short, it was Schaefer's scholarship that was egregious, sloppy and slapdash. None of Hartshorne's textual corrections have been ever challenged, even by some of Schaefer's most fanatical supporters, and they could be crazily fanatical.

But it did not matter. Schaefer's shocking historical scholarship was beside the point. American geographers who took up Schaefer's cause – Bill Bunge, but other "space cadets" who will be discussed in the next section – were not concerned one whit with a set of long-dead German geographers. Indeed, in the immediate post- World War II environment, associating yourself with German thought was a liability and especially in geography that had a lot of dirty laundry (Michel, 2016). Moreover, the imperative to professionalization was moving the discipline away from regional geography, viewed as amateurish (Ackerman, 1945).

As Bunge wrote to Hartshorne in 1959: "I do not care about the historical scholarship. I consider it irrelevant" (quoted in Martin, 1989, p. 79). The critically important point was that Schaefer's paper opened up geography to the methods of systematic science, allowing explanation, prediction, and the search for laws as well as corollary practices such as measurement, quantification and theorization. Bunge was saying, in contrast to Hartshorne, that there was nothing about the kind of material that geographers studied that prevented them from drawing on the methods of systematic sciences.

All this criticism was not entirely fair. Hartshorne tried to adapt during the 1950s and clarified and restated himself in other publications, including an abridged and revamped version of the argument developed in Nature, *Perspective on the Nature of Geography* (Hartshorne, 1959). He courted and encouraged the mathematically oriented geographers in Schaefer's old department to deepen their methods in a way that was compatible with Hartshorne's methodology.[11] Moreover, at the end of his 1959 reformulation Hartshorne (1959, p. 182) admitted that "as in any science, [geography] seeks to secure that approach to certainty and universality of knowledge that is made possible by the construction of generic concepts and laws of interrelations among factors".

Nonetheless, he could not help holding on to his earlier position:

the manifold variety of different and incommensurable factors involved in many features of our object of study, the complex world of the earth surface permits interpretation only of a part of our findings by that desired method.

Hartshorne 1959, p. 182

But Hartshorne's 1959 qualification to his argument no longer mattered. The modern era of "sputnik" had begun, filled with young people experimenting with new technology and possibilities (Van Meeteren, 2019). During this period epitomized by divided generations, there was no one more "old guard" imaginable than Richard Hartshorne promoting long-dead German scholars. Young scholars preferred learning FORTRAN than German as a second language. In that world, math skills mattered more than access to a Viennese library. As Davies (1966, p. 127) put it: "acceptance of geography as a science can only be made if it places itself within the current methodological conception of a science, and uses its techniques, instead of harking back to the science of other eras". Almost at the end of his life, Hartshorne seemed to finally get it. In a 1986 interview he realized that he "hadn't met his [Schaefer's] logical thesis" (Hartshorne, 2004b [1986], p. 278). But that was the thesis that mattered, that propelled the aftermath.

The Aftermath

The force of Schaefer's thesis convinced at least a number of younger American geographers during the 1950s to abandon the methodological injunctions of *Nature* and to do geography differently; to do it as systematic science. All of those geographers at least until the late 1960s were male. In part, this reflected the historically discursive character of geography established as a "manly science" as well as more general structural barriers that obstructed and discriminated against women from entering Mary McCarthy's "groves of academe" (Monk, 2004). In the

United States it was further propelled during the early post-war years by passage of the GI Bill that gave large numbers of especially young men who had been enlisted in the military cheap loans to attend university including graduate school (Abbot and Sparrow, 2007, p. 292–293). American universities, including Departments of Geography, were flooded with young male students, while the limited number of talented women that graduated often left academic research to pursue teaching, government and administrative careers (Monk, 2004).[12]

One of the young male geographers who served as a navigator for USAF bombers in the Pacific Theatre, and central to what came after the publication of Schaefer's paper, was William Garrison. He had completed a PhD in geography on the GI Bill at Northwestern University during the late 1940s drawing on Hartshorne's regionalist method. As he put it in a memoir, his dissertation research had involved just "a lot of walking around, … classification and description" (Garrison, 2002, p. 103). He was so dissatisfied with the result that he subsequently stole his dissertation from Northwestern's library to prevent anyone ever reading it. His first appointment in 1950 was at the Geography Department at the University of Washington, Seattle. There he began to undertake a different kind of geography, rigorous, systematic, logical, quantitative and scientific. Schaefer's paper profoundly resounded with him. In 1955, he wrote exuberantly about Schaefer's paper to a colleague at the Department, Edward Ullman, another of Schaefer's positive reviewers (Martin, 1989), who had similar intellectual leanings, but who also had a longstanding relationship with Hartshorne from the OSS:

> I was and still am excited by Schaefer. Now you may present me with formal proofs (1) that all German geographers are deaf, dumb, and unable to write and (2) that Schaefer was cruel to little children, and I would still be excited by Schaefer. Excited simply because Schaefer seemed to know in some crude way of the world of science of which geography is a part.
>
> cited in *Martin 1989, p. 77*

For Garrison, Schaefer was the brand-new exciting future of the discipline, Hartshorne its moribund past. Something had gone seriously awry in the intellectual core of geography he believed. As evidence, Harvard had closed down its geography department in 1948, its university president, the chemist, James Conant, declaring that geography was "not a university discipline" (quoted in Smith, 1987, p. 159). In his paper, Schaefer used the terms "isolationist," "complacent" (p. 226), "apologetic" (p. 227) and "somewhat lacking" (p. 227) to describe the discipline. He thought it was not practised as a serious subject, confined by Hartshorne to "mere description" (p. 227). But it did not need to be that way. That was what so excited Garrison about Schaefer's paper. It showed that geography could link to contemporary philosophical discussions about science that extended from logical positivism. And even more importantly, it showed that as a scientific discipline of spatial relations, spatial science, geography could join methodologically with other social sciences that had made similar moves to connect themselves with science such as economics, psychology, sociology, and political science (Schorske, 1997). All of this was possible Garrison believed. "There is nothing so powerful as an idea whose time has come, and the time had come for Schaefer by the decade of the '50s", he said.[12]

Schaefer of course could not take up the mantle because of his premature demise. Garrison could, though, and did at the University of Washington. In *annus mirablus* 1955, Garrison along to a lesser extent with Edward Ullman and the Department's cartographer, John Sherman, began to assemble a group of young male graduate students, the "space cadets", who were to pioneer the new scientific geography that Schaefer anticipated. They included Brian Berry, Ronald Boyce, Richard Morrill, and John Nystuen, who joined an earlier student of Garrison's,

Duane Marble. Within two years Art Getis, Waldo Tobler, and Bill Bunge, a refugee from the University of Wisconsin, Madison, where Hartshorne had flunked him in his doctoral comprehensive exams, also joined that group.[13]

Collectively they committed to geography as science, to Schaefer's vision of the discipline. Richard Morrill (1984, p. 59) remembers: "we were introduced to [Schaefer's article] quite soon – I think in that first year '55-'56 by Garrison." But Morrill also says, "Hartshorne was what we were against". To that larger end of doing scientific geography, they took courses in mathematics and statistics. In 1955, Garrison gave the first ever statistics course (Geog 426: Quantitative Methods in Geography–a "baptism of fire"; Morrill, 1984, p. 60) in an American geography department. They also taught themselves how to operate a computer. This was no mean feat. The first computer ever on the University of Washington campus arrived in 1955, an IBM 650, and was located in the attic of the Chemistry Building. It came with no formal programming language, and no hard disk for memory storage. Using the computer was a bootstrap operation, learning by doing, and usually for the space cadets that happened in the very early hours of the morning when no one else on campus wanted to use it. And finally, and maybe most importantly, they tried to practice an exact, formal scientific theory, with the ultimate goal of achieving Schaefer's aspiration of geographical (morphological) laws.

In 1959 it seemingly all came together in a volume Garrison and his students wrote, a "revolutionary book" according to Morrill (1984, p. 61), one of its co-authors, *Studies of Highway Development and Geographic Change* (Garrison et al., 1959). It was a remarkable publication, crammed with calculations, data matrices, statistical techniques, costs curves and demand schedules, and conventional maps overlaid with numbers, arrows, starburst lines, and balancing equations. The real revolution, however, was the changed conception of the region. It was no longer conceived as a Hartshornian element complex, but as a scientific theoretical object after Schaefer, capable of explanation, prediction and law-like statements.

Conclusion

As the Washington students (and indeed Garrison himself) left Seattle for new jobs, they took with them the new conception of geography as a science that later came to permeate and irrevocably alter American geographical thought. Schaefer won the debate, not by proving Hartshorne was wrong in his historical sources or interpretations, but by profoundly changing disciplinary thought and practices (Davies, 1966). Geographers increasingly came to live in Schaefer's world, not Hartshorne's. The Hartshorne-Schafer debate, as interpreted by Schaefer's followers, travelled, influencing other national geographical traditions, stimulating similar transformations elsewhere in the world. Eventually, it contributed to re-writing geography's history in Germany itself (Harvey and Wardenga 1998). The victory tour lasted until Schaefer's world met its own limits, when the search for spatial laws came up empty-handed (Barnes, 2004). One might think that would have vindicated Hartshorne, but it did not. Although there have been recurring calls to renew regional geography (Gregory, 1978; Johnston et al., 1990), there have been few calls to renew Hartshorne. He remained trapped in historiography, viewed as a past figurehead of a dreary regional description. Ironically, Hartshorne is now remembered less as a champion of the region than as an opponent of Schaefer's geographical scientific revolution. His masterpiece is seldom interpreted beyond Schaefer's scathing verdict, and with knowledge of the discipline's German roots squarely beyond the rear-view mirror. Such can be one's fate in methodological debates within geography.

Notes

1 The distinction between idiographic and nomothetic was first made in the late nineteenth century by two German philosophers, Wilhelm Windelband and Heinrich Rickert (Staiti 2013). They divided disciplines into two kinds: the idiographic was concerned with the unique (history was their exemplar); and the nomothetic concerned with making generalisations, the ultimate of which was a scientific law (their examples were chemistry and physics).

2 There is no definitive proof that he submitted a review. Unlike the other two reviewers, Edward Ullman and Stephen Jones, Kohn's review if it existed, has not survived (Martin 1989).

3 "Summary," August, 1969, p. 1, Richard Hartshorne Papers (**RHP**), Box 192, Schaefer and the origins of exceptionalism, file E, Library of the American Geographical Society, University of Wisconsin, Milwaukee.

4 The essay of one of Schaefer's students, Miss Martha Corry, made such an impression on Hartshorne that her paper was taken back to Madison to be discussed and commented by Hartshorne's own graduate students, including David Lowenthal, who was studying at the history department. Correspondence from 1950 shows that much of her critique echoes the concerns that Schaefer would publish three years later. Subsequent communication after the debate broke suggests that Hartshorne even contemplated that Schaefer had plagiarized his student's essay. See **RHP**, Box 6, Folder 8, Schaefer's seminar at Iowa, 1950-1954.

5 Fred K. Schaefer to Richard Hartshorne, May 17, 1950, **RHP**, Box 192, Schaefer and the origins of exceptionalism, file B.

6 "Summary," August, 1969, p. 1, **RHP**, Box 192, Schaefer and the origins of exceptionalism, file E

7 Richard Hartshorne to Henry M. Kendall, 29 October, 1953, **RHP**, Box 192, Responses – Origins of exceptionalism, file G.

8 Richard Hartshorne to Henry M. Kendall, 6th November, 1953, **RHP**, Box 192, Responses – Origins of exceptionalism, file G.

9 See Richard Hartshorne to Fred Lukerman, July 26, 1989, **RHP**, Box 194, Correspondence – Fred Lukerman, File S; and Richard Hartshorne to Derek Gregory, October 19, 1989, **RHP**, Box 195, Correspondence/ D-E, File D.

10 Richard Hartshorne to Harold McCarty, September 4, 1969, **RHP**, Box 6, folder 7, personal views, 1952-1969.

11 Richard Hartshorne to Harold McCarty, April 30, 1956, **RHP,** Box 5, folder 8, Correspondence Miscellaneous, A-Z, 1930-1992.

12 William L. Garrison, no title, no date, **RHP**, Box 192, Responses – origins of exceptionalism, File J.

13 None of the space cadets were women. Women were actively discouraged from doing quantitative work with its connotation of masculine rationality. In 1967, when the geographer Susan Hanson entered graduate school at Northwestern University, one of the holy sites of quantitative geography, she was discouraged by the Chair of the Department, Ed Espenshade. As Hanson (2002) put it, Espenshade could not understand 'why you would be in graduate school if you were female and already had a child. He just couldn't understand it. … When he spoke, he only asked how the family were, but never asked about scholarly work. Realistically as a woman in grad school at that time, one did not expect anything different! We knew very well that we were entering male turf.'

References

Abbot, A. and Sparrow, J.T. (2007) Hot war, cold war: The structures of sociological action, 1940–1955. In *Sociology in America: A History*, ed. C. Calhoun, 281–313. Chicago: University of Chicago Press.

Ackerman, E.A. (1945) Geographic training, wartime research, and immediate professional objectives. *Annals of the Association of American geographers*, 35(4), 121–143.

Althusser, L. (1972) *Politics and History: Montesquieu, Rousseau, Hegel, Marx*. London: New Left Books.

Barnes, T.J. (2004) Placing ideas: *Genius loci*, heterotopia and geography's quantitative revolution. *Progress in Human Geography*, 28(5), 565–595.

Barnes, T.J., and Farish, M. (2006) Between regions: Science, militarism, and American geography from World War to Cold War. *Annals of the Association of American Geographers*, 96(4), 807–826.

Bunge, W. (1968) Fred Schaefer and the science of geography. *Harvard Papers in Theoretical Geography, Special Papers Series*, Paper A.

Davies, W.K.D (1966) Theory, science and geography. *Tijdschrift voor Economische en Sociale Geografie*, *57*(4), 125–130.

Entrikin, J.N. (1981) Philosophical issues in the scientific study of regions. In *Geography and the Urban Environment: Progress in Research and Applications, Volume IV*, eds. D. T. Herbert and R. J. Johnston, 1–27. London: John Wiley & Sons.

Garrison, W.L. (2002) Lessons from the design of life. In *Geographical Voices: Fourteen Autobiographical Essays*, eds. P. Gould and F. R. Pitts, 99–123. Syracuse: Syracuse University Press.

Garrison, W.L., Berry, B.J.L., Marble, D.F., Nystuen, J.D. and Morrill, R.L. (1959) *Studies of Highway Development and Geographic Change*. Seattle, WA: University of Washington Press.

Gregory, D. (1978) *Ideology, Science and Human Geography*. London: Hutchinson.

Hanson, S. (2002) Interview with Trevor Barnes. Worcester, MA, May.

Hartshorne, R. (1939) *The Nature of Geography: A Critical Survey of Current Thought in the Light of the Past*. Lancaster, PA: Association of American Geographers.

Hartshorne, R. (1948) On the mores of methodological discussion in American geography. *Annals of the Association of American Geographers, 38*(2), 113–125.

Hartshorne, R. (1954) Comment on "Exceptionalism in geography." *Annals of the Association of American Geographers*, 44(1): 108–109.

Hartshorne, R. (1955) "Exceptionalism in geography" re-examined. *Annals of the Association of American Geographers*, 45(3): 205–244.

Hartshorne, R. (1959) *Perspective on the Nature of Geography*. Chicago: Rand McNally.

Hartshorne, R. (1979) Notes towards a bibliography of *The Nature of Geography*. *Annals of the Association of American Geographers*, 69(1): 63–76.

Hartshorne, R. (2004a [1986]) Interview with Maynard Western Dow, Minneapolis, MN, May 5, 1986. *Geographers on Film Transcriptions*, 271–284. Available at: https://textarchive.ru/c-2977382-pall.html (last accessed 8th November, 2020).

Hartshorne, R. (2004b [1986]) Interview with Maynard Western Dow, Minneapolis, MN, May 6, 1986. *Geographers on Film Transcriptions*, 285–297. Available at: https://textarchive.ru/c-2977382-pall.html (last accessed 8th November, 2020).

Harvey, F. and Wardenga, U. (1998) The Hettner-Hartshorne connection: Reconsidering the process of reception and transformation of a geographic concept. *Finisterra, 33*, 131–140.

Heald, W. (1992) From positivism to realism: The philosophy of Gustav Bergmann. Books at Iowa 56: 25–46. https://ir.uiowa.edu/bai/vol56/iss1/4/ (last accessed October 1st, 2020).

Johnston, R.J., Hauer, J., and Hoekveld, G.A. (eds) (1990) *Regional Geography: Current Developments and Future Prospects*. London: Routledge.

Kohn, C. F. (1972) Preface. *The Fred K. Schaefer Memorial Lecture Number 1*. Department of Geography, the University of Iowa, Iowa City, Iowa.

Leighly, J. (1937) Some comments on contemporary geographic method. *Annals of the Association of American Geographers, 27*(3): 125–141.

Martin, G.J. (1989) *The Nature of Geography* and the Schaefer-Hartshorne debate. In *Reflections on Richard Hartshorne's The Nature of Geography*, eds., J.N Entrekin, and S.D. Brunn, 69–90. Washington DC: Association of American Geographers.

Martin, G.J. (2015) *American Geography and Geographers: Toward Geographical Science*. Oxford: Oxford University Press.

Michel, B. (2016) "With almost clean or at most slightly dirty hands". On the self-denazification of German geography after 1945 and its rebranding as a science of peace. *Political Geography, 55*: 135–143.

Mirowski, P. (2001) *Machine Dreams: Economics becomes a Cyborg Science*. Cambridge: Cambridge University Press.

Monk, J. (2004) Women, gender, and the histories of American geography. *Annals of the Association of American Geographers, 94*(1): 1–22.

Morrill, R.L. (1984) Recollections of the "quantitative revolution's" early years: The University of Washington, 1955-65. In *Recollections of a Revolution: Geography as Spatial Science*, eds., M. Billinge, D. Gregory, and R. Martin, 57–72. London: MacMillan.

Sack, R.D. (1974) Chorology and spatial analysis. *Annals of the Association of American Geographers* 64(3): 439–52.

Schaefer, F.K. (1953) Exceptionalism in geography: A methodological introduction. *Annals of the Association of American Geographers* 43(3): 226–49.

Schorske, C.E. (1997) The new rigorism in the human sciences, 1940-60. *Daedalus* 126(1): 289–309.

Sigmund, K. (2017) *Exact Thinking in Demented Times: The Vienna Circle and the Epic Quest for the Foundations of Science*. New York: Basic Books.

Smith, N. (1987) 'Academic war over the field of geography': The elimination of geography at Harvard, 1947-1951. *Annals of the Association of American Geographers*, 77(2): 155–172.

Smith, N. (1989) Geography as a museum: private history and conservative idealism in *The Nature of Geography*. In *Reflections on Richard Hartshorne's The Nature of Geography. Occasional Publications of the Association of American Geographers*, eds. N. Entrikin, and S. D. Brunn, 91–120. Washington DC, Association of American Geographers.

Staiti, A. (2013) Heinrich Rickert. In *The Stanford Encyclopedia of Philosophy,* ed. E. N. Zalta. http://plato.stanford.edu/archives/win2013/entries/heinrich-rickert/ (last accessed October 1st, 2020).

Van Meeteren, M. (2019) Statistics do sweat: Situated messiness and spatial science. *Transactions of the Institute of British Geographers, 44*(3), 454–457.

2

THE ARCHIVE AND THE FIELD

Methodological Procedures and Research Outcomes in the Work of Carl O. Sauer (1889–1975)

W. George Lovell

A half-century after his death, Carl O. Sauer's inspirational presence continues to hover over the discipline of geography, his legacy unmatched by any other twentieth-century geographer of standing. He started out early – an "Outline for Field Work in Geography" (co-authored with Wellington D. Jones) was published in 1915 – and finished late, *Seventeenth-Century North America*, the last of twenty book titles, appearing posthumously in 1980. While three musings on methodology discussed below are noteworthy – the first, "The Morphology of Landscape" (1925), not only advanced the construct of a "cultural landscape" but spelled out a modus operandi – Sauer was avowedly concerned more with geographical practice than pontificating about procedure. For him, creativity of endeavour counted above all. His lifetime accomplishment was bountiful, the fruit of an inquisitive bent and investigative drive that combined archival foraging – assiduous, patient, persistent – with dogged fieldwork. Sauer's vast output is here scrutinized with methodological matters foremost in mind, his strategies of how to conduct scientific inquiry related to specific outcomes, including the production of a little-known text for elementary schoolchildren besides scholarly monographs cited still in the literature to which they pertain.

Beginnings, Schooling, and Academic Career

Born on Christmas Eve, 1889, of German immigrant stock in Warrenton, Missouri, at the time a farming community some 50 miles west of St. Louis, Sauer's European roots did much to shape him, as did being raised a Christian by deeply religious parents. His father, William, was one of the first teachers (of music and French) to hold a position at Warrenton's Central Wesleyan College. William's belief that a good education began with the best of elementary schooling saw him take his family – wife Rosetta Johanna Vosholl, first-born son Albert, and nine-year-old/second-born Carl – to Calw (near Stuttgart) in Germany, where the young Sauer was a pupil for two years at the local gymnasium. Its strict regimen, he later acknowledged, was not entirely to his liking, but its emphasis on learning how to learn served him well. Upon returning to Warrenton in 1901, Sauer's intellect was of such a calibre that, aged 16, he enrolled at Central Wesleyan College, in 1906 its youngest-ever student. His studies continued first in

DOI: 10.4324/9781003038849-4

geology (two semesters only) at Northwestern University in Evanston, Illinois, and then the University of Chicago, where his graduate career in geography began in September 1909. He defended his doctoral dissertation, "The Geography of the Ozark Highland of Missouri," on December 15, 1915, describing the work as "the outgrowth of a long acquaintance with the area and of a deep affection for it" [1]. A seven-year spell at the University of Michigan in Ann Arbor, which made him a full professor at age 33, ended when Sauer accepted an offer from the University of California at Berkeley to found and chair its Department of Geography, which he did for over thirty years. It was at Berkeley that Sauer's star assumed a meteoric trajectory [2].

The Morphology of Landscape

Upon arrival in California in 1923, Sauer was keen to engage firsthand the America that lay south of the Río Grande, but that would have to wait. He had unfinished business, of sorts, to settle with the American Midwest that had once been home but no longer was, his penchant to move on as much intellectual as it was residential, if not more so. Old ways of thinking about geography, Sauer soon would urge, must be replaced.

Sauer had begun graduate studies in geology before moving to geography, its human subjects capturing his imagination more than its physical components, though these would always be taken into account. Human geography then was very much dominated by the tenets of environmental determinism, whose emblematic champion, Ellen Churchill Semple (1863–1932), was Sauer's most esteemed teacher at the University of Chicago [3]. The highest of regard in which he held her, however, was tempered by Sauer's conviction that Semple's reading of Friedrich Ratzel's *Anthropogeographie* (1882) was misconstrued, that her advocation of the geographic method and the understanding it encapsulated, was overly deterministic [4]. His "Morphology of Landscape" (1925) was a definitive rejoinder, a treatise that emphasized environmental possibilism as opposed to environmental determinism, the agency of humankind not the primacy of nature. Laying out an alternate agenda for human geography, Sauer ([1925] 1963, p. 343) writes:

> The cultural landscape is fashioned from a natural landscape by a culture group. Culture is the agent, the natural area is the medium, the cultural landscape the result. Under the influence of a given culture, itself changing through time, the landscape undergoes development, passing through phases and probably reaching ultimately the end of its cycle of development. With the introduction of a different – that is, an alien – culture, a rejuvenation of the cultural landscape sets in, or a new landscape is superimposed on remnants of an older one. The natural landscape is of course of fundamental importance, for it supplies the materials out of which the cultural landscape is formed. The shaping force, however, lies in the culture itself. Within the wide limits of the physical equipment of area lie many possible choices … . This is the meaning of adaptation, through which … we get the feeling of harmony between the human habitation and the landscape into which it so fittingly blends. But these, too, are derived from the [human] mind, not imposed by nature, and hence are cultural expressions.

Henceforth, the study of cultural landscapes assumed a pivotal role in human geography, if not its primary focus. Sauer never conceived of "The Morphology of Landscape," however, as a programmatic directive, that following its lead amounted to some kind of dictate. However, many whose formulations he sought to challenge, and thereby offer alternatives to, saw things differently, interpreting Sauer's views – as Michael Williams (2014, p. 59) puts it – as dogmatic and prescriptive, "what geography should not and should be," the very opposite of his intent. "Several of the Middle Western geographers," Williams contends, "apparently spent more time [mis]reading 'The Morphology of Landscape' than Sauer had in writing it."

Possibly the words that peeved his critics most had to do with the supposed objectivity of science and the scientific method. Alluding to four notable German geographers, Sauer ([1925] 1963, p. 344) asserts:

A good deal of the meaning of area lies beyond scientific regimentation. The best geography has never disregarded the esthetic qualities of landscape, to which we know no approach other than the subjective. Humboldt's "physiognomy," Banse's "soul," Volz's "rhythm," Gradmann's "harmony" of landscape, all lie beyond science. These writers seem to have discovered a symphonic quality in the contemplation of the areal scene, proceeding from a full novitiate in scientific studies and yet apart therefrom.

The most conservative of practitioners, at this stage in the argument, must have found their ideas about the pursuit of geography already challenged, if not refuted. They would have been even less prepared for, and surely not disposed to, what came next. Sauer ([1925] 1963, p. 344–345) declares:

To some, whatever is mystical is an abomination. Yet it is significant that there are others, and among them some of the best, who believe, that having observed widely and charted diligently, there remains a quality of understanding at a higher plane that may not be reduced to formal process.

No prophet was Sauer, nor did he wish to be so considered. His sentiments, however, anticipated the postmodern "cultural turn" in the humanities and social sciences (Jameson 1998) by half a century, and the "cultural turn" in geography (Jackson 1997) by a quarter-century more.

Foreword to Historical Geography

Having promoted landscape as "the unit concept of geography" – its German designation, *Landschaft*, evokes the process of shaping key to Sauer's schema ([1925] 1963, p. 321) better than the word does in English – it was time for ideas to be put into practice. His portrayals of the Ozarks and the Upper Illinois Valley (Sauer 1916) had pioneered the cause, but it was Sauer's forays into Mexico, which began in 1926, that afforded him a setting where he could gather data and put theory to the test. Field sojourns in Mexico, almost an annual affair until 1950, were conducted in the guise of historical geography, considered by Sauer ([1925] 1963, p. 344) as "the series of changes [that] cultural landscapes have undergone," thereby involving "the construction of past cultural landscapes." The temporal factor was crucial. "We cannot form the idea of landscape," Sauer ([1925] 1963, pp. 344) stressed, "except in terms of its time relations as well as of its space relations." Aspects of time – evolution, change, continuity, sequence, and succession – are fundamental criteria in landscape analysis, and constitute the bedrock of historical geography.

His second methodological musing, "Foreword to Historical Geography," delivered as a presidential address to the Association of American Geographers, Sauer (1941, p. 1) prefaced as an "apologia" and posed as "a confession of the faith that has stood behind one's work." In many ways it is an "afterword" to historical geography because, while he chose *not* to "present data and conclusions from my own work in Mexico," Sauer (1941, p. 9) draws nonetheless on lessons learned there to take stock and offer counsel on how to further geographical inquiry. "Retrospect and prospect are different ends of the same sequence," he states.

> Today is therefore but a point on a line, the development of which may be reconstructed from its beginning and the projection of which may be undertaken into the future. Knowledge of human processes is attainable only if the current situation is comprehended as a moving point, one moment in an action that has beginning and end.

There are echoes in these two sentences not only of Newtonian principles of physics but also the poetics of Sauer's fellow Missourian, a contemporary of his, T. S. Eliot [5].

For Sauer, a "three-point underpinning for geography" that would hold students in good stead as they conducted research entailed (1) awareness of the history of geography; (2) being prepared to engage, if and when appropriate, aspects of physical geography; and (3) familiarity with geography's "sister discipline of anthropology," deemed (Sauer 1941, pp. 5–6) "methodologically the most advanced of the social sciences." Two noted anthropologists who became kindred spirits at Berkeley were Alfred L. Kroeber and Robert H. Lowie, affinity and bonding between the three men palpable, their students the beneficiaries thereof. Close friendships were forged, Williams (2014, pp. 79–83) informs us, as they traversed common ground. "The forms of material culture with which the anthropologist deals," Sauer (1941, p. 6–8) maintained, "are identical with those of human geography" – which he equated "as culture-historical geography."

Two points of communion, not three, underpinned historical geography: the archive and the field. "The first step in reconstruction of past stages of a culture area," Sauer (1941, p. 13–14) insisted, "is mastery of its written documents." Gaining such command, he conceded, "takes much time and search." But the rewards could be as immense as the documentary base itself, particularly for the region that had lured him south to research and write about:

> There is an embarrassment of riches in the old Spanish records for New Spain [Mexico], from parish records up to summary reports that were sent to the king in Spain. There are diaries and accounts of early explorations, the *visitas* made by inspecting officials who reported in detail on condition of the country, letters of missionaries, the so-called geographic relations [*relaciones geográficas*] ordered for all Spanish America at several times in the sixteenth and eighteenth centuries, records of payments of taxes and tributes, data on mines, salines, and roads. Perhaps no other part of the New World has as elaborate a documentation on settlements, production, and the economic life of every part as do the Spanish colonies.

Bountiful though written sources may be, the field too has enduring allure, just as enticing and rewarding – especially when, in terms of methodological application, what a document has to say can be read in situ to illuminating effect. Sauer (1941, p. 14) reflects:

> Let no one consider that historical geography can be content with what is found in archive and library. It calls, in addition, for exacting field work. One of the first steps is the ability to read the documents in the field. Take into the field, for instance, an account of an area written long ago and compare the places and their activities with the present, seeing where the habitations were and the lines of communication ran,

where the forests and the fields stood, gradually getting a picture of the former cultural landscape concealed behind the present one. Thus one becomes aware of nature and [the] direction of changes that have taken place. There comes a time in such study when the picture begins to fit together, and one comes to that high moment when the past is clear, and the contrasts to the present are understood. This, I submit, is genetic human geography.

The geography in Sauer's genes produced books by the score and articles, book chapters, commentaries, and reviews numbering in the hundreds. He also supervised thirty-seven doctoral dissertations, almost half of them on topics pertaining to Latin America. His legacy lives on in a genealogy now six generations in the making, its Latin Americanist line of descent a notable characteristic [6].

The Education of a Geographer

Another presidential address to the Association of American Geographers, on this occasion an honorary one delivered a year before he became professor emeritus in 1957, sees Sauer (1956, p. 297) look back and muse, gainfully and incisively. Good practice is again to the fore, despite him believing the trail he blazed "mainly unmarked by any arrows of methodology." He reiterates: "The geographic bent rests on seeing and thinking about what is in the landscape. There is, I am confident, such a thing as a 'morphologic eye,' a spontaneous and critical attention to form and pattern," a "predilection" that lies "at the very heart of our being." Sauer (1956, p. 289–90) elaborates:

> We work at the recognition and understanding of elements of form and of their relation in function. Our forms and their arrangements are grossly macroscopic and infinitely numerous so that we have always to learn about selecting what things are relevant and eliminating the insignificant. Relevance raises the question of why the form is present and how it is related to other forms. Description is rarely adequate and even less often rewarding unless it is tied to explanation. It seems necessary therefore to admit to the geographic bent the fourth dimension of time, interest in how what is being studied came to be.

Historical awareness is again emphasized, as is the importance of fieldwork, of which Sauer was an intrepid, life-long practitioner; Robert C. West (1979, pp. 3–8) documents him as having logged up over 30 excursions of variable duration (one week to six months) between 1926 and 1968. While his trips to the field were far-flung, most of them took him to Mexico – "the historical geographer," Sauer (1941, p. 10) upheld, "must be a regional specialist" – and the American Southwest, often in the company of students accompanied to, and advised in, the location of their site of study. Of this persuasion, Sauer (1956, pp. 295–98) emphasizes:

> Underlying what I am trying to say is the conviction that geography is first of all knowledge gained by observation, that one orders by reflection and reinspection the things one has been looking at, and that from what one has experienced by intimate sight comes comparison and synthesis. In other words, the principal training of the geographer should come, wherever possible, by doing field work.

Somewhat poignantly, he then adds: "Field time is your most precious time; how precious you will know only when its days are past."

In his "Education of a Geographer," Sauer (1956, pp. 296–297) returns, too, to the necessity of being acquainted with the history of geographic thought, "to the circumambient intellectual climates within which geography has lived at different times and places," raising the need to be conversant with languages other than one's own native tongue:

> We, as little as any group, can be content with current literature, or with what is available to us in English. Complacency as to our own language means the exclusion of a great, probably the greater, part of what has been well learned and well thought about. Can anyone choose to remain ignorant because it requires exertion to find out what has been done in other times or written in another language? Scholars [do] not limit [themselves] to what is most convenient, least of all to such arbitrary reduction of knowledge. A monolingual Ph.D. is a contradiction of terms.

Sauer grew up well versed in German as well as English – he corresponded with his parents in the former language, a cache of "over 160 letters, many in old German script," culled for inclusion in the Willams biography of him [7]. Of other languages acquired later, Spanish was his most proficient, skill at translation called upon especially when grappling to capture English renditions for *The Early Spanish Main* (1966) and the *Ibero-Americana* monographs (1932a; 1932b; 1934; 1935; 1948).

A First Book in Geography

Sauer was much drawn to investigating and writing about beginnings, be they concerned with the early development of humankind (1944; 1957; 1962), agricultural origins and dispersals (1952), or European arrival in the Americas from the late fifteenth through seventeenth centuries (1966; 1971; 1980). These research interests of his are acknowledged and recognized. Less well-known is Sauer's involvement in, and orchestration of coordinated effort for, the creation of an elementary school text that would serve grade-four pupils in California and across the United States as "A First Book in Geography" (Figure 2.1). The impetus behind *Man in Nature: America Before the Days of the White Men* (1939) involved making decisions that were bold and innovative, showcasing Sauer in his operational prime, at his inimitable best [8].

Methodologically, the book is a distillation of all that Sauer held dear, and chose to propound: the need to familiarize oneself with, and determine the strengths and weaknesses of, available extant sources; making a critical commentary of these materials, what today we would consider a literature review, identifying lacuna or shortfalls that investigation can fill or make less problematical; the imperative of maps, to locate phenomena in place and show their distribution across space; how a textual narrative, worded simply but elegantly, can be enlivened by having a complementary visual dimension – illustrations of every possible kind. And on every page, resolute querying, an insistence on questioning at all times what propels and sustains the story being told – why is that so, was it always thus, are there links or rifts between then and now, how did such things come to pass, and if not for the good, how might they be changed for the better?

A forgotten classic that engages its subject matter with instructional zest, conceptual clarity, and organizational prowess, *Man in Nature* takes its title from *Man and Nature* (1864) by the

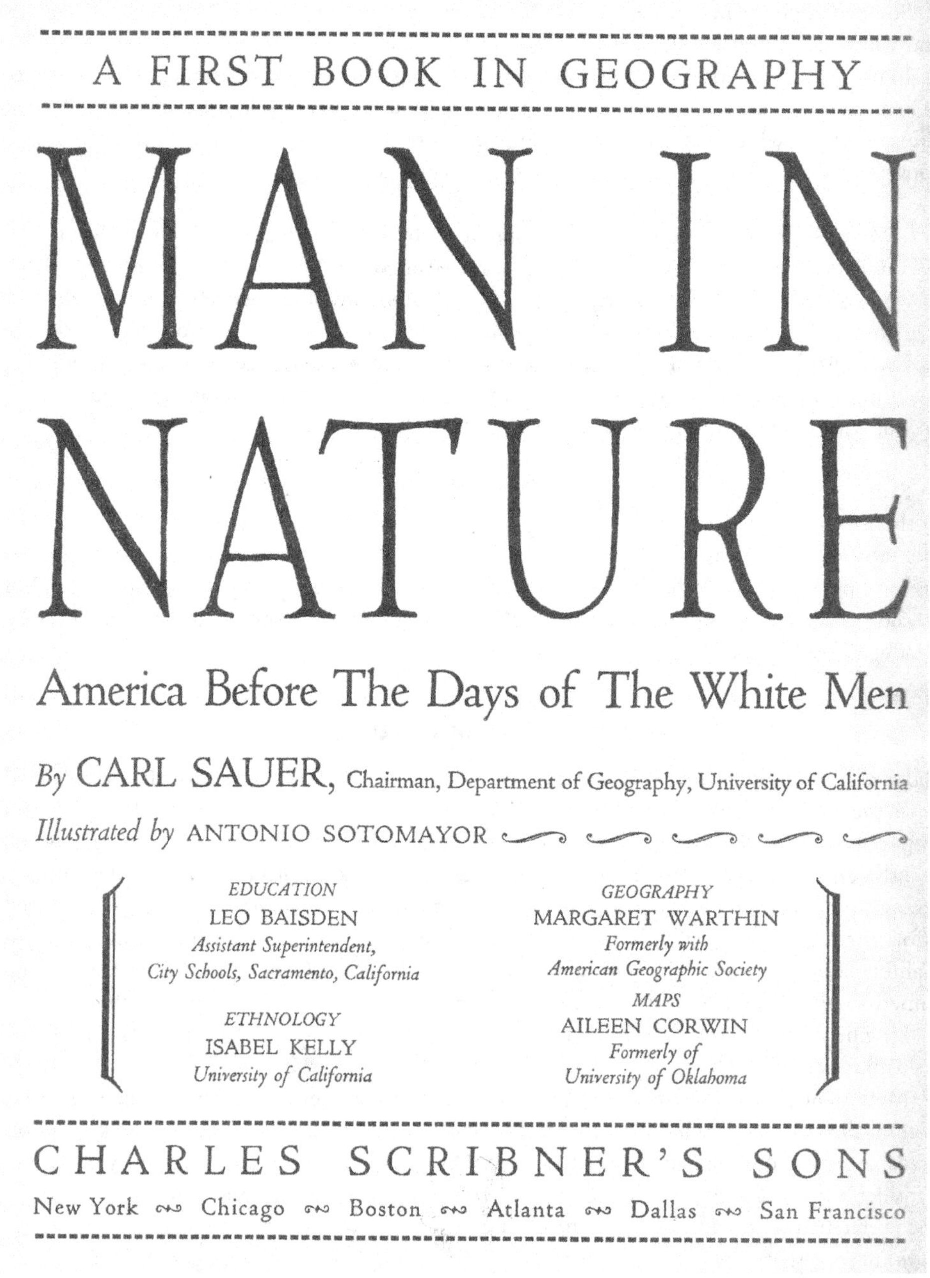

A FIRST BOOK IN GEOGRAPHY

MAN IN NATURE

America Before The Days of The White Men

By CARL SAUER, Chairman, Department of Geography, University of California

Illustrated by ANTONIO SOTOMAYOR

EDUCATION	*GEOGRAPHY*
LEO BAISDEN	MARGARET WARTHIN
Assistant Superintendent,	*Formerly with*
City Schools, Sacramento, California	*American Geographic Society*
	MAPS
ETHNOLOGY	AILEEN CORWIN
ISABEL KELLY	*Formerly of*
University of California	*University of Oklahoma*

CHARLES SCRIBNER'S SONS

New York ᔗ Chicago ᔗ Boston ᔗ Atlanta ᔗ Dallas ᔗ San Francisco

---

Figure 2.1 Title page of *Man in Nature* (Sauer 1939, p. i).

visionary George Perkins Marsh (1801–1882), whose embryonic environmentalism and call for land stewardship Sauer greatly admired and sought to emulate. The book's thrust is hardly what elementary school teachers – certainly in the 1930s but most likely even today – would have expected an introductory text in geography to be. Sauer (1939, p. 8) pulls no punches and refuses to spare his young audience the moral indignation he feels when telling them about the Native American past:

> Our people in settling America have changed nature a great deal. We have let soil wash away on hills that we have farmed. In many places we have made rivers muddy that once were clear. We have cut down forests and plowed up grasslands. We have killed off many animals, like the buffalo. We have built towns and roads and done many things to change the country. Some of these changes are good. Some of them may be bad for us. Before the white men came all the land belonged to the Indians. This book is about Indian days. The Red Man was much more part of nature than we are. By learning how and where the Indian lived, we shall learn what kind of country the white man found. We shall then know better what he has done with it.

Of the fate of Maya peoples in Mexico and Guatemala Sauer (1939: 230) is especially forthright, but takes pains to note cultural survival in the wake of Spanish conquest:

> Like all other Indians, these people were conquered by greedy and war-like white men. Very many of them died; the rest were made to work for the white man. Most of their rulers and priests and teachers were killed. They were forbidden to go to their temples. They were made to learn the ways of the white man. They had no time to carve or write or make the things of which they had been so proud. Soon they lost most of the skills and knowledge that had enabled them to reach such a high place in civilization.
>
> Today this part of the world is still largely inhabited by Maya Indians. They still speak their old languages and know of their old ways. Some of them are again thinking of the days when they were a great people and are hoping again that they can build a fine civilization in the old land of the Maya.

The Vicissitudes of Collaboration

Though *Man in Nature* began as a collaborative project, and for the most part unfolded as such, Sauer in the end assumed primary authorship. At the outset, the idea was for him to furnish detailed chapter outlines that Margaret Warthin of the American Geographical Society would then flesh out. In doing so, Warthin was to be guided in terms of ethnographic content by the Berkeley anthropologist Isabel Kelly, whom Sauer lobbied the publisher, Charles Scribner's Sons, to hire with that goal in mind. In terms of pedagogic fit for nine-year-olds, Warthin was advised initially by Ernest Horn, a professor of education at the University of Iowa, and after he chose not to continue by Leo Baisden, Assistant Superintendent of City Schools in California's state capital, Sacramento. A lover of maps, Sauer advocated that these had to be "built into the process of learning"[9]. To this end, Aileen Corwin, a fine arts graduate of the University of Oklahoma, devoted herself, an appendix of nine double-page maps embellishing the finished product. The artistry of Bolivian Antonio Sotomayor enhances the book considerably; over 200 pen-and-ink renderings of his (Figure 2.2) unify the text visually and thematically – reconstructions of historic events and encounters, cameos of human pursuits and activities, drawings of plants and animals, sketches of topographic features. "It is the only part of the work with which I am quite fully content," Sauer confessed, "and I am not sure that in the end he isn't the most valuable contributor to the book"[10]. Sotomayor's "animated or living map of North America in Indian days" (Figure 2.3) opens up for his youthful clientele, as Sauer (1939, p. 10) words it, "different Indian ways in [the] different regions you will be reading about."

The project got off to a promising enough start, especially the strategic collaboration between Sauer and Warthin and between Warthin and Kelly. Warthin moved from New York to work with Sauer and Kelly on the West Coast. "Miss Warthin and Miss Kelly are hammering out a

Figure 2.2 Illustrations from "A Story Chart" by Antonio Sotomayor (Sauer 1939, xii).

remarkable job," Sauer wrote to Scribner's on November 14, 1934. "I've never seen a better team and I don't believe you could duplicate this particular one in the country" [11]. A month or so earlier, on October 4, 1934, Sauer had written to Warthin in a positive, complimentary manner, saying, "You are doing an excellent piece of work on this thing" [12].

A year later, however, serious doubts had set in. On October 11, 1935, Sauer voiced his disappointment at one of Warthin's drafts: "That prairies unit strikes me as pretty weak. I don't

Figure 2.3 Sotomayor's "animated or living map of North America in Indian days" (Sauer 1939, p. 10).

think that it is up to any of the others. You will see that I have scratched it up sadly and have
written a couple of suggestive pages" [13]. A year after that, disappointment with Warthin had
turned to consternation. On October 6, 1936, he wrote to the publisher:

> I don't understand the situation. I spent a good deal of time on this last year, writing
> out script in detail. I am just afraid it is another combination that failed. I have done
> enough work on that thing to have done it myself, and it would have been a lot easier
> to do so. I'll make another attempt to see whether we can't get some writing into it.
> It has just been a continuous drag not to get the whole thing droning along in mono-
> tone. I suppose it's a difficult spot, this business of having to talk to children interest-
> ingly, pleasantly, and yet respecting the dignity of the child, but it must be done. I'll
> give the material one more going over [14].

He buckled down with customary resolve, but alludes in a letter written on April 8, 1938 to
"the wind-up of the Scribner's text book for the fourth grade" having become "a terrific chore,
because finally I decided that I would have to rewrite the entire thing after Margaret Warthin
had done it once. She worked hard at the job, but did not have the flair for presenting the subject to children" [15]. Warthin was eventually relieved of her duties as Sauer dedicated himself to
finishing the task, though he saw fit to include Warthin's name on the title page upon completion, along with those of his four more stalwart collaborators (Figure 2.1).

Following publication, Howard P. Miller of Scribner's wrote to Sauer with unbridled enthusiasm. "I am greatly pleased with the book, and Baisden is about as proud of it as if it were
his own." Sotomayor too, noted Miller, "is greatly pleased with the illustrations and the job in
general" [16]. On December 4, 1939, Sauer wrote to C.F. Board of Scribner's after *Man in Nature*
began to be reviewed and the first accolades started to roll in. His words serve as both synopsis
and valediction, revealing an intervention of methodological import from an unlikely but cred-
ible source:

> We have gone on a new trail in this book. I wrote [it] as simply as I could. You are
> probably aware of the manner in which the original plan of associate authorship failed.
> When I was beginning to feel desperate about the matter, Howard Miller said to me
> that I should try writing it myself. I did so. He began by taking the early sections of
> it and letting his nine-year-old be the first editorial judge thereof. When it had this
> young editor's approval, I went ahead. Baisden then tried the material out in fourth-
> grade classes at Sacramento, and his teachers reported that it worked very nicely.
> That's about all there is to say [17].

Acclaim and Resonance

There is, in fact, much more that could be said, but two disclosures must suffice by way of conclusion. The first is how the Education Division of the U.S. Office of Indian Affairs responded
to the appearance of *Man in Nature*, which it did in its "fortnightly field letter," *Indian Education*,
on February 1, 1940:

> This is the sort of textbook we have been wanting. It is a picture of America before
> the white man came, presented by a famous geographer, assisted by an educator, an
> anthropologist, a geographer [artist], and a map maker. It shows in simple words the
> two continents of the Americas, with their stretches of northern woodland, of tropical

forest, of plains, mountains, deserts. What grew in these different regions? How could man support himself there? The book gives the answer, with pictures of the animals, plants, and trees in each district. Then it describes the Indians who lived there and how they used their environment for food, for houses, and for clothing. This is geography seen as a background for human life. We predict that the book may be ultimately used in every school in the United States.

A second revelation relates to what Sauer was also working on as the five years invested in *Man in Nature* drew to a close. Shortly before its appearance, Sauer published a short article in the *Journal of Farm Economics*, based on a presidential address given to the Social Science Research Council. He called it "Theme of Plant and Animal Destruction in Economic History" (1938). In the piece, Sauer gives plaintive voice to the concerns and preoccupations at the heart of *Man in Nature,* indicating that the time expended in producing an elementary-school text was anything but a diversion from his rigorous scholarly agenda. If Sauer ([1938] 1963, p. 146–47] had wished *Man in Nature* to be reduced to essence and epigraph, he could not have conjured up more fitting words than these:

> We know of scarcely any record of destructive exploitation in all the span of human existence until we enter the period of modern history, when transatlantic expansion of European commerce, peoples, and governments takes place. Then begins what may well be the tragic rather than the great age of man. We have glorified this period in terms of a romantic view of colonization and of the frontier. There is a dark obverse to the picture, which we have regarded scarcely at all.

For grown-up geographers and children alike, the former at any stage of their careers in search of exemplary methodology and sound procedure, that "dark obverse" is illuminated still by perusal of Carl Sauer's "first book in geography."

Coda

From the standpoint of geography in the twenty-first century, Sauer's impact on the discipline endures and resonates. His geographical practices, and singular achievements, have proven to be pertinent and long-lasting, appealing to diverse constituencies, an oeuvre that transcends fad and fashion, the product of astute industry yielding abiding results and affording perennial reward. In terms of the "great debate" that shook up geography mid-way through the past century – see Barnes and van Meeteren, Chapter 1 – Sauer held views distinctly at odds with those of Richard Hartshorne, his idiographic preference among them, though he cannot be said to have favoured the nomothetic postulations of Edward Schaefer without reservation [18]. While research findings for one part of the world may be applied beneficially to another, Sauer (1956, pp. 297–298) held firm to the notion that "good regional geography is finely representational art," of which he was both an authoritative exponent and gifted exemplar. "And creative art," he adds emphatically, "is not circumscribed by pattern or method: method is means." Description as a means may not lend itself easily to theoretical framing, but best it be done analytically and in explanatory fashion.

Sauer's legions of admirers include not only scholars and academics but also poets and writers, among them Charles Olsen (1910–1970), Ed Dorn (1929–1999), and Barry Lopez (1945–2020), whose appreciation of him James J. Parsons (1996) brought to disciplinary attention. Sauer's critiques of colonialism, his depictions of Indigenous demise as a consequence of European invasion, on the early *Spanish Main* (1966) and afterwards far beyond, nurture post-colonial discourse and continue to contest the canon. He championed what John Leighly (1963, p. 7) described as "a humane use of the earth," as now have done generations of concerned environmentalists. And when Bob Callahan's Turtle Island Foundation reprinted *Man in Nature* (1939) in 1975 and 1980, the reappearance of "A First Book in Geography" saw it adopted by native school boards across Canada and the United States, including First Nations in British Columbia and Nova Scotia and the Cherokee in Oklahoma, the Crow in Montana, and the Zuni in New Mexico. A bridge between cultures, of understanding the past to make better sense of the present, Sauer remains.

Notes

1 The dissertation defended in 1915 was published five years later (Sauer 1920). His acknowledgment of "long acquaintance with the area" and expression of "deep affection for it" (Sauer 1920, viii-xi) arise from doctoral investigations being considered "a study in home geography." He writes: "Later residence outside of Missouri has supplied a more objective viewpoint without destroying the old familiarity. To consider the region as an outsider has been impossible and will always be."

2 Williams (2014) is by far the best biographical account of Sauer we have, the author's manuscript at the time of his death (2009) polished to completion by the labors of David Lowenthal and William M. Denevan. Sauer himself penned an insightful self-reflection of his work, "Chart of My Course," deployed by editor Bob Callahan as a Foreword to *Seventeenth-Century North America* (Sauer 1980, 9–12). The rumination also appears in the selection of readings and commentaries edited by Denevan and Mathewson (2009, 324–329). Collections of essays edited by Kenzer (1987) and Mathewson and Kenzer (2003) also afford insight into Sauer's remarkably productive life.

3 Williams (2014, 20–23) records Sauer as considering Semple's lectures "more gripping than any that I have had. She expresses herself in a masterly style. Many public speakers would envy her." He admired Semple's "poetic sensibilities as well as learning, which is a comparative rarity." She "stirred the imagination and feeling with her fervor and eloquence," traits to which he himself would aspire.

4 Semple's *Influences of Geographic Environment* (1911) is a work of mammoth complexity, and of "distinctively encyclopedic character," as noted in the *American Historical Review* 17, 2: 355 (1912). "Man is no longer the conqueror of natural environment," the reviewer reads Semple as asserting, "nor is he considered the passive creature of physiographic influences. The political, social, and industrial evolution of a community is shown to be a resultant of forces acting upon man in every conceivable proportion and degree."

5 As eminent a twentieth-century poet as Sauer was a geographer, Thomas Stearns Eliot (1888–1965) revelled in the meditation of time and place, nowhere more sublimely so than in his *Four Quartets* (1943).

6 Mathewson et al., (2020) chart the progeny throughout Latin America. Urquijo Torres, Segundo, and Bocco (2020) focus for the most part on Mexico. A proud great grandson, the author is a third-generation offspring, his doctoral supervisor (John F. Bergmann, 1928–1983) having been supervised by Henry J. Bruman (1913–2005), who was Sauer's eleventh successful PhD graduate, filing his dissertation ("Aboriginal Drink Areas in New Spain") in 1940. A book (*Alcohol in Ancient Mexico*) appeared sixty years later (Bruman 2000).

7 Williams (2014, xvii). Perusal of these letters, and other materials in the Sauer Papers held by the Bancroft Library at the University of California in Berkeley, helped Williams "distinguish between the man and the legend, until he sometimes became a very alive, vivid person."

8 See Lovell (2003) for fuller discussion.

9 Sauer to Horn, April 21, 1936 (Sauer Papers, Bancroft Library).

10 Sauer to Scribner's, October 28, 1936 (Sauer Papers, Bancroft Library).

11 Sauer to Scribner's, November 14, 1934 (Sauer Papers, Bancroft Library).

12 Sauer to Scribner's, October 4, 1934 (Sauer Papers, Bancroft Library).

13 Sauer to Scribner's, October 11, 1935 (Sauer Papers, Bancroft Library).

14 Sauer to Scribner's, October 6, 1936 (Sauer Papers, Bancroft Library).

15 Sauer to Gladys Wrigley of the American Geographical Society, April 8, 1938 (Sauer Papers, Bancroft Library).

16 Miller to Sauer, June 12, 1939 (Sauer Papers, Bancroft Library).

17 Sauer to C. F. Board, December 4, 1939 (Sauer Papers, Bancroft Library).

18 Hartshorne (1939, 183) writes: "While the interpretation of individual features in the geography of a region will often require the student to reach back into the geography of past periods, it is not necessary that the geography of a region be studied in terms of historical development." Sauer (1941, 2) was quick to assert his contrary position. "Geography, in any of its branches," he insists, "must be a genetic science – that is, must account for origins and processes." In the same critique, "Foreword to Historical Geography," his presidential address delivered to the Association of American Geographers in Baton Rouge in 1940, Sauer (1941, 2) takes Hartshorne to task for not grasping fully the ideas of the German geographer Alfred Hettner, whom he held in high regard.

References

Bruman, H.J. (1940) "Aboriginal Drink Areas in New Spain." Ph.D. dissertation. Berkeley: Department of Geography, University of California.

Bruman, H.J. (2000) *Alcohol in Ancient Mexico*. Salt Lake City: University of Utah Press.

Denevan, W.M., and Mathewson, K. eds. (2009) *Carl Sauer on Culture and Landscape: Readings and Commentaries*. Baton Rouge: Louisiana State University Press.

Eliot, T.S. (1943) *Four Quartets*. London: Faber and Faber.

Hartshorne, R. (1939) *The Nature of Geography*. Lancaster, PA: Association of American Geographers.

Jackson, P. (1997) "Geography and the Cultural Turn." *Scottish Geographical Magazine* 113, 3: 186–88.

Jameson, F. (1998) *The Cultural Turn: Selected Writings on the Postmodern, 1983–1998*. London and New York: Verso.

Jones, W.D. and Sauer, C.O. (1915) "Outline for Field Work in Geography." *Bulletin of the American Geographical Society* 47, 7: 520–25.

Kenzer, M.S. ed. (1987) *Carl O. Sauer: A Tribute*. Corvallis: Oregon State University Press.

Leighly, J. ed. (1963) *Land and Life: A Selection from the Writings of Carl Ortwin Sauer*. Berkeley, Los Angeles, and London: University of California Press.

Libby, O. G. (1912). *Influences of Geographic Environment, on the Basis of Ratzel's System of Anthropo-Geography*. By Ellen Churchill Semple (New York: Henry Holt and Company; London: Constable and Company. 1911. pp. xvii, 683.), *The American Historical Review*, 17, 2: 355–357. https://doi.org/10.1086/ahr/17.2.355

Lovell, W. G. (2003) "A First Book in Geography: Carl Sauer and the Creation of *Man in Nature*." In Kent Mathewson and Martin Kenzer, eds., *Culture, Land, and Legacy: Perspectives on Carl O. Sauer and the Berkeley School of Geography*, 323–38. Baton Rouge: Louisiana State University Press.

Marsh, G. P., (1864). *Man and Nature*. London: S. Low, Son and Marston.

Mathewson, K.M. and Kenzer, M.S., eds. (2003) *Culture, Land, and Legacy: Perspectives on Carl O. Sauer and the Berkeley School of Geography*. Baton Rouge: Louisiana State University Press.

Mathewson, K.M., Allen, A.L., Grismore, A., Lagos, M., Simms, J.R., and Spencer, B. (2020) "The Sauer Tree in Time and Place." *Journal of Latin American Geography* 19, 1: 84–97.

Parsons, J.J. (1996) "Mr. Sauer and the Writers." *The Geographical Review*, 86, 1: 22–41.

Ratzel, F. (1882) *Anthropogeographie, oder Grundzüge der Anwendung der Erdkunde auf die Geschichte*. Stuttgart: J. Engelhorn.

Sauer, C.O. (1915) "The Geography of the Ozark Highland of Missouri." PhD dissertation. Chicago: Department of Geography, University of Chicago.

Sauer, C.O. (1916) *Geography of the Upper Illinois Valley and History of Development*. Illinois State Geological Survey, Bulletin 27.

Sauer, C.O. (1920) *The Geography of the Ozark Highland of Missouri*. Geographic Society of Chicago, Bulletin 7.

Sauer, C.O. [1925] (1963) "The Morphology of Landscape." *University of California Publications in Geography* 2, 2:19-54, in John Leighly, ed., *Land and Life: A Selection from the Writings of Carl Ortwin Sauer*, 315–50. Berkeley, Los Angeles, and London: University of California Press.

Sauer, C.O. and Brand, D. (1932a) *Aztatlán: Prehistoric Mexican Frontier on the Pacific Coast*. Ibero-Americana 1. Berkeley: University of California Press.

Sauer, C.O. (1932b) *The Road to Cíbola*. Ibero-Americana 3. Berkeley: University of California Press.

Sauer, C.O. (1934) *The Distribution of Aboriginal Tribes and Languages in Northwestern Mexico*. Ibero-Americana 5. Berkeley: University of California Press.

Sauer, C.O. (1935) *Aboriginal Population of Northwestern Mexico*. Ibero-Americana 10. Berkeley: University of California Press.

Sauer, C.O. (1938) "Theme of Plant and Animal Destruction in Economic History." *Journal of Farm Economics*, 20, 765–775. https://doi.org/10.2307/1231378

Sauer, C.O., Sotomayor, A., Baisden, L., Corwin, A., Kelly, I., and Warthin, M. (1939) *Man in Nature: America Before the Days of the White Men. A First Book in Geography*. New York: Charles Scribner's Sons.

Sauer, C. O. (1941) "Foreword to Historical Geography." *Annals of the Association of American Geographers* 31, 1: 1–24.

Sauer, C.O. (1944) "A Geographic Sketch of Early Man in America." *The Geographical Review*, 34, 4: 529–73.

Sauer, C.O. (1948) *Colima of New Spain in the Sixteenth Century*. Ibero-Americana 29. Berkeley and Los Angeles: University of California Press.

Sauer, C.O. (1952) *Agricultural Origins and Dispersals*. Bowman Memorial Lectures, Series 2. New York: American Geographical Society.

Sauer, C.O. (1956) "The Education of a Geographer." *Annals of the Association of American Geographers* 46, 3: 287–99.

Sauer, C.O. (1957) "The End of the Ice Age and Its Witnesses." *The Geographical Review* 47, 1: 29–43.

Sauer, C.O. (1962) "Seashore: Primitive Home of Man?" *Proceedings of the American Philosophical Association*, 106: 41–47.

Sauer, C.O. (1966) *The Early Spanish Main*. Berkeley and Los Angeles: University of California Press.

Sauer, C.O. (1971) *Sixteenth-Century North America: The Land and the People as Seen by the Europeans*. Berkeley and Los Angeles: University of California Press.

Sauer, C.O. (1980) *Seventeenth-Century North America*. Berkeley: Turtle Island Foundation.

Semple, E.C. (1911) *Influences of Geographic Environment, on the Basis of Ratzel's System of Anthropo-Geography*. New York: Henry Holt and Company.

Torres, U., Pedro S., Segundo, P.C., and Bocco. G. (2020) "Geografía latinoamericanista en México: Balance histórico a partir dee la Escuela de Berkeley. *Journal of Latin American Geography* 19, 1: 98–114.

West, R.C. (1979) *Carl Sauer's Fieldwork in Latin America*. Ann Arbor: University Microfilms.

Williams, M., Lowenthal, D., and Denevan, W.M. (2014) *To Pass On a Good Earth: The Life and Work of Carl O. Sauer*. Charlottesville: University of Virginia Press.

3

THE QUANTITATIVE REVOLUTION

Mark W. Rosenberg

This chapter picks up where Barnes and van Meeteren left off (Chapter 1). The original disagreement between Hartshorne and Schaefer as described and analyzed in detail by Barnes and van Meeteren opened the door for the introduction of a monumental methodological shift away from a descriptive regional approach to human geography to one that became increasingly based on applied statistics and mathematics. This chapter does not, however, try to retrace the philosophic debates that took place throughout the 1960s and 1970s that continued among Hartshorne and his adherents on one side and the "space cadets" referred to by Barnes and van Meeteren who were the leading proponents of adopting quantitative methodologies. For those who argued in favour of the Quantitative Revolution, it was mainly about realigning human and physical geography and in some respects unifying them through the adoption of statistical and mathematical models, and taking advantage of new methods in data collection and technology mainly associated with the rapid changes taking place in computing. Everyone who graduated with a degree in human geography can no doubt recall those debates with dread and indeed those debates might still take place in some required courses with weighty titles such as the "Philosophy of Geography" in undergraduate and graduate programs.

Instead, the argument that will be made in this chapter is that the Quantitative Revolution is really a misnomer for what might better have been called a "model and systems" revolution. The remainder of the chapter is divided temporally into four parts. In the first part, Setting the Stage, the argument as to why models and systems thinking came to dominate the Quantitative Revolution is explained. What follows in the next part is a recounting of some of the best known and arguably most important examples of how modelling and systems thinking developed in human geography. The third part, The Beginning of the End, examines how the limitations of models and systems thinking evoked reactions that were both philosophic and methodological rejections of what the Quantitative Revolution had become. The other subtheme that is highlighted is how changes in technology (e.g., the shift from mainframe computing to personal computing), also weakened the importance of statistical and mathematical modelling in favour of geographical information systems (GIS). The conclusion of the chapter is that the Quantitative Revolution was the starting point for a series of methodological shifts in human geography that were tied to philosophic and technological shifts that are likely to continue to shape human geography in new ways in the future.

DOI: 10.4324/9781003038849-5

Setting the Stage

To this day, the use of quantitative methodologies in human geography is often simplistically associated with logical positivism (see Sheppard 2014). Logical positivism as articulated by the philosophers of the "Berlin" and "Vienna Schools" in its essence was a desire to reduce the material world to a series of laws written as parsimonious mathematical expressions. While no doubt influential in the thinking of early quantitative geographers (see Barnes and van Meeteren, Chapter 1), logical positivism was arguably far less influential in human geography than Kuhn's (1962) *The Structure of Scientific Revolutions* with its emphasis on paradigms, and other philosophers of science such as Skilling's (1964) definition of models, and Apostel's (1960) views on systems.

Just how influential Kuhn, Skilling, Apostel and indeed other philosophies of science were in thinking about paradigms, models and systems in geography was epitomized by Chorley and Haggett's (1967) classic collection, *Models in Geography*. At the end of their opening chapter entitled, "Models, paradigms and the new geography," Haggett and Chorley (1967, pp. 38–39), wrote:

> We have looked at the traditional paradigmatic model of geography and suggested that it is largely classificatory and that it is under severe stress. We have tentatively suggested an alternative model-based approach.

In truth, their 816-page collection in 18 chapters by many of the leading geographers of the time was neither "tentative" nor a "suggestion" but an extended argument that all parts of academic geography (not just human geography) should adopt a model-based approach to research and teaching. Part 1 included two chapters where authors laid out the philosophic arguments for "a model-based paradigm" of geography and the various types of models. Part 2 was dedicated to models in physical geography. Part 3 was entitled "Models of Socio-economic Systems." Its chapters were entitled: "Demographic Models and Geography"; "Sociological Models in Geography"; "Models of Economic Development;" "Models of Urban Geography and Settlement Location"; "Models of Industrial Location"; and "Models of Agricultural Activity." The fourth part of the book was three chapters that proposed various ways that human and physical phenomena and behaviour might be integrated into systems through regionalization, ecosystems, spatial evolution or networks. In the last part of the book, the authors covered "Maps as Models," "Hardware Models in Geography" and "Models in Geographical Teaching."

Beyond the theoretical debates that were ensuing in human geography at the time, a Quantitative Revolution might never have taken place without the associated changes in the development of applied mathematics, statistics, and computing. Post-World War Two, the rapid development and improvements in applied mathematics, statistical theory, and mainframe computing, resulted in the creation of statistical software packages such as the first releases of versions of Statistical Package for the Social Sciences (SPSS) in 1968 and Statistical Analysis System (SAS) in 1971. With these advances in software and hardware, human geographers could for the first time take advantage of the largest data sets available (e.g., national censuses) to search for patterns, and build and test mathematical and statistical models of structures and behavior. The advances in applied mathematics, statistics and computing software and hardware also led to the introduction of a new lexicon in human geography. Human geographers were now not only speaking about models and systems, but points, arcs, lines, spaces, hexagons, centroids, intercepts, coefficients, Eigen values, location-allocation, etc. Quantitative research came to dominate the sessions at the major meetings of human geographers and the major national peer-reviewed journals. During this period, specialist international peer-reviewed journals such as *Environment and Planning* (1969), *Geographical Analysis* (1969), *Regional Studies* (1967) and *Urban Geography* were launched to highlight the latest quantitative research.

The other contextual note in assessing the impacts of the Quantitative Revolution on human geography was that there were far fewer divisions in human geography at the beginning of the 1960s than there are today. In many geography departments, the main cleavages were between urban and economic geographers on the one hand and cultural and historical geographers on the other hand. The former group being the main proponents and early adopters of quantitative methodologies while the latter continued focused on regional methodologies. This is not to say that many of the sub-disciplines of human geography did not exist (e.g., economic geography) but they were generally seen as part of one or the other main groupings. It is also important to acknowledge that not all urban and economic geographers embraced the Quantitative Revolution and some human geographers who counted themselves as cultural or historical geographers did indeed embrace the Quantitative Revolution.

Early on in the Quantitative Revolution, human geographers went off in three directions: one direction was a focus on geometry and spatial order; the second focus was on statistical relationships and a third focus was on mathematical models. These methodologies are discussed below.

The Start of the Revolution

Nothing epitomized the start of the revolution so much as the search for spatial order and an understanding of the hierarchy of urban places as proposed in central place theory by Walter Christaller (1933, 1966). Human geographers from Europe and English-speaking countries sought out examples that to the "eye of the beholder" demonstrated hexagonal patterns that could be "mapped" onto landscapes all around the world (King, 1985). For those human geographers who wanted more precise ways of demonstrating the attributes of central place theory (e.g., hexagonal market areas and hierarchical systems), they drew their inspiration from August Lösch (1938, 1954) who provided a justification as to why hexagonal markets would result based on supply, demand, distance, and rational economic behaviour (i.e., economic theory).

To explain the geographic patterns of agriculture, human geographers drew their inspiration from an even earlier German scholar, J.H. von Thünen (1826) who proposed a model based on a theoretical isolated state, different unit value of crops and unit costs of transport from a central location; the theoretical outcome being a series of zones around a central location. This model would be given expression in agricultural geography by Chisholm (1962) and urban land rent models by Alonso (1964).

A third set of models had their origins in the urban ecology school at the University of Chicago. Beginning with the concentric zone model proposed by Ernest Burgess (Park and Burgess, 1925), Homer Hoyt's (1939) sector model and the multiple nuclei model of Harris and Ullman (1945). Methodologically, researchers using these models drew mainly on description and what researchers observed. These ideal models were derived from observation and became the motivation for other researchers to search out data to fit these ideal models. Not many years later, Brian Berry and his followers would unleash a torrent of research on "cities as systems within systems of cities" through the 1960s and 1970s from the Center for Urban Studies at the University of Chicago (see Berry, 1972a).

In his evaluation of many of the models referred to above, Garner (1967, 304–305) observed that they were intended to explain "urban geography and settlement location." As such they shared six common characteristics according to Garner: they reflected how human activity adjusted in an orderly way to distance, location decisions reflect the minimization of distance, some locations are more accessible than others, human activities group together to take advantage of scale economies, the importance of hierarchy in how human activities are organized, and human occupation is fundamental to the other five characteristics in understanding "a system of human organization in space." What the research on the models of the early years of the Quantitative Revolution also shared was a strong belief that order and structure existed based on rational and predictable behaviour. It was these beliefs that would ultimately be challenged as many human geographers rejected the reductionist and capitalist view of the world that they implied and the inability that such beliefs could explain the variability of human agency.

While methodologically the focus on the geometry of space drew heavily on traditional cartographic practices and simple descriptions derived from basic data sets (e.g., census data), a second direction in the quest to model structure and behaviour was founded on the application of increasingly sophisticated statistical and mathematical analysis (see Berry and Marble, 1968). Implicit among the early statistical modelers was a belief that explanation was partial at best. Explanation could best expressed in probabilistic terms or as Curry (1967, 1972, p. 341) wrote, "[T]he random approach ... allows a problem to be approached with the explicit admission of considerable ignorance."

At its core, those who became the proponents of a statistical approach early on were mainly fixated on linear relationships, usually between a small number of basic phenomena (e.g., population size and rank in the urban system, rent and distance from the city centre, or interactions between places as a function of population size and distance). For example, finding relationships between population size and city rankings in various countries and at different geographic scales became the "rank-size" rule (Dacey, 1979); and interaction between places regardless of scale and distance became the gravity or spatial interaction model formulated in its simplest form as an analogue from Newtonian physics.

As urban and economic geographers became more sophisticated in their understanding of mathematics and statistics, there was a shift away from relatively simple linear relationships that focused on spatial structure towards more complex multivariate models and a more sophisticated understanding of probability. Hägerstrand (1968) proposed a diffusion model of how ideas and technology spread in time and space using Monte Carlo simulations as his *modus operandi*. Out of his basic ideas on diffusion would develop a broader set of ideas of time-space geography (see Carlstein, 1982; Thrift and Pred, 1981). The simple probability models of spatial interaction were replaced by the complex ideas of maximum entropy models (see Wilson, 1970). With the creation of input-output tables and the use of Markov models, regional economies were analyzed as part of what Isard (1960) and his colleagues labelled "regional science." In transportation geography, network measures and graph theory were introduced to understand the level of connection among places (see Tinkler, 1979). Location-allocation modelling developed to answer questions about where the optimal sites are for locating everything from factories to hospitals (Scott, 1970).

From the above description of the types of models that were associated with the early stages of the Quantitative Revolution, it should come as no surprise that there was an acknowledgement that geographic research was heavy in its emphasis on spatial structures and flows and the assumption that human behaviour if considered at all was rational and constant. Without abandoning any of its methodological underpinnings as a response to the predominant focus on spatial structure, some human geographers shifted their focus to behavioural modelling drawing a distinction between spatial behaviour (i.e., rules that explain how people make decisions) and behaviour in space (i.e., behavioural patterns within an area that depend on the structure of the area (Davies, 1972; Rushton, 1968). Olsson and Gale (1971) proposed a general model of spatial behaviour while others focused on models of specific types of spatial behaviour. For example, Brown and Moore (1971) proposed a model of how people search for housing within a city while Wolpert's (1972) model sought to explain migration between cities. Golledge and Brown (1972) conceptualized the market decision process between producers and consumers as spatial learning behaviour.

By the early 1970s, quantitative research dominated human geography practice. Three goals, however, remained elusive among human geographers who were the proponents of the Quantitative Revolution: how to provide a holistic view of the region; finding methods that took account of the simultaneity of how human behaviour structures space and how space structures human behaviour (what would later often be referred to as the structure-agency debate in human geography); and a theoretical justification for a human geography based on quantitative methodology. Isard's *Methods of Regional Analysis*, already in its ninth printing in 1973 was the ultimate quantitative response in how to understand the structure of regions in response to those who clung to a traditional regional approach in human geography. It did not, however, resolve the other two goals. In response to the second issue, the major proponents of the Quantitative Revolution sought to adapt general theories from the natural sciences to overcome the division between structure and agency. For example, Berry (1972b) believed structure and behaviour could be modelled using general field theory. At the end of *Explanation in Geography*, Harvey's (1969) penultimate chapter is an argument for the use of systems concepts. To address the third issue, books by such human geography luminaries as Bunge (1966), and Harvey (1969), and the authors in the edited collections of Chorley and Haggett (1967) and Davies (1972) sought to produce theoretical justifications for the Quantitative Revolution.

Ironically, Burton (1963) had declared the Quantitative Revolution over because of the widespread adoption of courses in quantitative methods by the mid-1960s. The beginning of the end of the Quantitative Revolution, however, was to come in 1973 with the publication of *Social Justice and the City* (Harvey, 1973), Harvey's theoretical and methodological repudiation of the Quantitative Revolution, which came only four years after *Explanation in Geography*.

The Beginning of the End?

In many respects, Burton's (1963) reading of the state of methodology in human geography by the mid-1960s is probably a more accurate assessment than most in his suggestion that revolutions are generally short-lived. Either they succeed and this leads to a new *status quo*, or they fail, and the previous regime re-establishes its dominance. For Burton, quantitative research was the new *status quo* but he was also aware that at the time he was writing, much of the research he cited were atheoretical methods without theory. He saw this as the challenge and was optimistic that theory would follow.

The publication of *Social Justice and the City*, however, marked the beginning of a long and sustained critique of the use of quantitative methods in human geography. In many respects, the critique was not a methodological critique *per se*, but a rejection of theory. As others have noted, the critique of the Quantitative Revolution as a rejection of logical positivism was in many respects a simplistic argument at best given how little of the Quantitative Revolution was about creation of laws that *explained* human structure and agency (see Sheppard, 2014). For Harvey and other human geographers who joined him in rejecting the version of human geography that was embodied in the Quantitative Revolution, the critique was mainly a rejection of capitalism and how it structured urban and rural life, and regional, national and global economies. Another group of geographers argued against the Quantitative Revolution because of its failure to take into account individual human experience from a humanistic perspective (see Ley and Samuels, 1978). Many feminist geographers rejected quantitative methodologies as "masculinist" and argued for the need to incorporate a palette of qualitative methodologies that captured the everyday challenges faced not only by women, but increasingly through lenses of gender and intersectionality (see Nast, 1994 and other feminist geographers who contributed to a special section of *The Professional Geographer* in 1994). To this day, the rejection of quantitative methodologies in favour qualitative methodologies is reflected in many of the chapters of Part Two of this book. Alternatively, there are human geographers today creating feminist approaches to quantitative methods and qualitative/critical geographical information systems methodologies.

What is also reflected in the chapters of Part Two is how the changes in technology and data collection have changed the nature of quantitative methodologies. By the second half of the 1980s, the shift from mainframe to personal computing was well on its way. Computer-assisted telephone interviewing was replacing in-person survey data collection methods at government statistical agencies. In 1981, the Environmental Systems Research Institute now simply known as ESRI launched the first commercial geographic information system. The Internet was launched in 1983. Search engines like YAHOO and Google soon followed and early in the 2000s, social media giants like Facebook debuted. Some human geographers took advantage of the increasing amounts of data available, the ever-expanding computing power of personal computers, and the growing user-friendly development of statistical software for personal computers to continue to develop increasingly sophisticated statistical and mathematical models of spatial behaviour and structure (e.g., multi-level modelling). For them the Quantitative Revolution never ended, but their numbers have been in decline as more and more switched to using geographic information systems.

Others like Michael F. Goodchild and Gerard Rushton began their careers as well-known quantitative geographers but are now best known for their contributions to the development of geographical information systems (see Phipps, 2016). They have trained new generations of human geographers who have embraced geographical information systems as their mode of research. Also indicative of this methodological shift has been the decline in the number of geography departments where quantitative methods are offered at all or as required courses, while the number of geographic information systems and qualitative methods courses offered continue to grow. There are even some who have argued that the future of geography is now tied to the development of critical geographic information systems (Wilson and Graham, 2013) just as it was once argued that the future of geography would be tied to the Quantitative Revolution.

Concluding Thoughts

The impact of the Quantitative Revolution on human geography and indeed, on the whole of geography, cannot be over-estimated. The Quantitative Revolution did not lead to the creation of a set of irrevocable laws that explain how people and objects are organized and behave in space and time as its early proponents had hoped and argued it would. We also need to be cognizant of the re-assessments of the underlying values of the researchers who proposed some of the early defining models of the Quantitative Revolution that are now happening (see Barnes and Minca, 2013, Kobayashi, 2014). The Quantitative Revolution did, however, accomplish a shift in methodology from a discipline where research was mainly based on observation and description to a discipline based on modelling human agency and structure through the "rigor" of applied mathematics and statistics. This methodological shift might never have occurred if Schaefer and the early proponents of the Quantitative Revolution had not been prepared to take on the geography establishment at the time (see Chapter 1). It might also never have occurred without the technological changes that took place in how data were gathered and stored (i.e., mainframe computing) and the development of software to analyze large data sets.

Today, there is arguably no methodology that dominates the way quantitative methodologies briefly did in the last quarter of the twentieth century. Geographical information systems as a methodology and the most direct descendent of the Quantitative Revolution, has become the most popular approach to the analysis of data in human geography today. Ironically, geographical information systems incorporate many of the quantitative methods first developed during the Quantitative Revolution (e.g., optimal location methods, spatial statistics, etc.) but in a more sophisticated fashion.

In a sense, the Quantitative Revolution with all its limitations has also provided the basis for a panoply of qualitative methodological counterrevolutions by humanist, Marxist, feminist, activist, LGBTQ+, racialized and Indigenous geographers. While they may not recognize it, they too follow in the footsteps of the Quantitative Revolution with their insistence on rigour and reliability (Baxter and Eyles, 2004) and by taking advantage of new "smart" technologies (see Warf, 2017). What will be the next methodological revolution in human geography is an open question? It will, however, have its origins in a path that can be traced back to the Quantitative Revolution.

References

Alonso, W. (1964) *Location and Land Use: Toward a General Theory of Land Rent*. Cambridge, MA: Harvard University Press.

Apostel, L. (1960) Towards the formal study of models in the non-formal sciences. *Synthese* 12, pp. 125–161. https://doi.org/10.1007/BF00485092

Barnes, T.J. and Minca, C. (2013) Nazi spatial theory: The dark geographies of Carl Schmitt and Walter Christaller. *Annals of the Association of American Geographers*, 103, pp. 669–687.

Baxter, J. and Eyles, J. (2004) Evaluating qualitative research in social geography: Establishing 'rigour' in interview analysis. *Transactions of the Institute of British Geographers*, 1997, 22, pp. 505–525.

Berry, B.J.L. (1960) The impact of expanding metropolitan communities upon the central place hierarchy. *Annals of the Association of American Geographers*, 50, pp. 112–116.

Berry, B.J.L. (1972a) Cities as systems within systems of cities. In W.D.K. Davies, ed. *The Conceptual Revolution in Geography*. London: University of London Press, 312–330. Originally published in *Papers and Proceedings of the Regional Science Association*, 13, 1964.

Berry, B.J.L. (1972b) Interdependence of spatial structure and spatial behavior: A general field theory formulation. In W.D.K. Davies, ed. *The Conceptual Revolution in Geography*. London: University of London Press, 394–416. Originally published in *Papers and Proceedings of the Regional Science Association*, 21, 1968.

Berry, B.J.L. and Marble, D.F., eds. (1968) *Spatial Analysis: A Reader in Statistical Geography*. Englewood Cliffs, NJ: Prentice-Hall.

Brown, L.A. and Moore, E.G. (1971) The intra-urban migration process: A perspective. In L.S. Bourne, ed. *The Internal Structure of the City*. New York: Oxford University Press, 200–209. Originally published in *Geografiska Annaler*, Series B. 52B, 1970.

Bunge, W. (1966) *Theoretical Geography*. Lund: Royal University of Lund, Department of Geography and Gleerup.

Burton, I (1963) The quantitative revolution and Theoretical Geography. *The Canadian Geographer*, 7:151–162.

Carlstein, T. (1982). *Time Resources, Society and Ecology: On the Capacity for Human Interaction in Space*. London: Allen & Unwin.

Chisholm, M. (1962) *Rural Settlement and Land Use: An Essay in Location*. London: Hutchinson University Li.

Chorley, R.J. and Haggett, P., eds. (1967) *Models in Geography*. London: Methuen.

Christaller, W. (1933). *Die zentralen Orte in Süddeutschland*. Jena: Gustav Fischer.

Christaller, W. (1966) *Central Places in Southern Germany.* C.W. Baskin, trans. Englewood Cliffs, NJ: Prentice-Hall.

Curry, L. (1972) Chance in landscape. In W.D.K. Davies, ed. *The Conceptual Revolution in Geography.* London: University of London Press, 341–353. Originally published in *Northern Geographical Essays: in Honour of G. Daysh,* Department of Geography, Newcastle: University of Newcastle upon Tyne, 1967.

Dacey, M.F. (1979) A growth process for Zipf's and Yule's city-size laws. *Environment and Planning A,* 11, 361–372.

Davies, W.K.D. (1972) Geography and behavior. In W.D.K. Davies, ed. *The Conceptual Revolution in Geography.* London: University of London Press, pp. 331–340.

Hägerstrand, T. (1968) *Innovation Diffusion as a Spatial Process.* Translation and postscript by A. Pred, Chicago: University of Chicago Press.

Harris CD, Ullman EL. (1945) The Nature of Cities. *The ANNALS of the American Academy of Political and Social Science* 242(1), pp. 7–17. doi:10.1177/000271624524200103

Harvey, D. (1969) *Explanation in Geography.* London: Edward Arnold.

Harvey, D. (1973) *Social Justice and the City.* Baltimore: Johns Hopkins University Press.

Isard, W. (1960), THE SCOPE AND NATURE OF REGIONAL SCIENCE. Papers in Regional Science, 6, 9–34. https://doi.org/10.1111/j.1435-5597.1960.tb01698.x

Golledge, R.G. and Brown, L.A. Search, learning and the market decision process. In W.D.K. Davies, ed. *The Conceptual Revolution in Geography.* London: University of London Press, 381–393. Originally published in *Geografiska Annaler,* Series B. 49B, 1967.

Hoyt, H. (1939) *The Structure and Growth of Residential Neighborhoods in American Cities.* Washington: Federal Housing Administration.

King, L.J. (1985) *Central Place Theory.* Reprint. G.I. Thrall, ed. Morgantown, West Virginia: WVU Research Repository, 2020.

Kuhn, T.S. (1962) *The Structure of Scientific Revolutions.* Chicago: University of Chicago Press.

Kobayashi, A. (2014) The dialectic of race and the discipline of geography. Annals of the *Association of American Geographers,* 104, pp. 1101–1115.

Ley, D. and Samuels, M.S., eds. (1978) *Humanistic Geography: Problems and Prospects.* London: Croom Helm.

Lösch, A. (1938) The nature of economic regions. *Southern Economic Journal,* 5, pp. 71–78.

Lösch, A. (1954) *The Economics of Location.* H.H. Voglom, trans. New Haven, Yale University Press.

Nast, H.J. (1994) Women in the field: Critical feminist methodologies and theoretical perspectives. *The Professional Geographer,* 46, pp. 54–66.

Olsson, G. and Gale, S. (1972) Spatial theory and human behavior. In W.D.K. Davies, ed. The *Conceptual Revolution in Geography.* London: University of London Press, 354–368. Originally published in *Paper and Proceedings of the Regional Science Association,* 15, 1967.

Park, R. E., Burgess, E. W., McKenzie, R. D., & Wirth, L. (1925). *The city.* Chicago, Ill, The University of Chicago Press.

Phipps, A.G. ed. (2016) *Research Advances in Behavioral, Economic and Health Geography Inspired by Gerard Rushton.* Windsor, Canada: Sociology, Anthropology and Criminology Books. https://scholar.uwindsor.ca/sac-books/1

Rushton, G. (1968) Analyses of spatial behavior by revealed space preferences. *Annals of the Association of American Geographers,* 59, pp. 391–400.

Scott, A.J. (1970) Location-allocation systems: A review. *Geographical Analysis,* 2, pp. 95–119.

Sheppard, E. (2014) We have never been positivist. *Urban Geography,* 35, pp. 636–644.

Skilling, H. (1964) An operational view. *American Scientist,* 52, pp. 388A–396A.

Thrift, N. and Pred, A. (1981) Time-geography: A new beginning. *Progress in Human Geography,* 5, pp. 277–286.

Thünen, J.H. von (1826) *Der Isolierte Staat in Beziehung auf Landwirtschaft un Natonalökonomie.*

Tinkler, K.J. (1979) Graph theory. *Progress in Human Geography,* 3, pp. 85–116.

Warf, B., ed. (2017) *Handbook on Geographies of Technology.* Cheltenham: Edward Elgar.[

Wilson, A.G. *Entropy in Urban and Regional Modelling.* London: Routledge.

Wilson, M.W. and Graham, M. (2013) Guest editorial: Situating neogeography. *Environment and Planning A.* 45, pp. 3–9.

Wolpert, J. (1972) Behavioral aspects of the decision to migrate. In W.D.K. Davies, ed. *The Conceptual Revolution in Geography.* London: University of London Press, 341–353. Originally published in *Geografiska Annaler* 49B, 1967.

4

TOWARDS INTERDISCIPLINARITY

The Relationship between GIS/GIScience/ Cartography and Human Geography[1]

Alberto Giordano

Introduction

In this chapter, I will concentrate on aspects of the intersection between human geography—a way of understanding and making sense of the world—and geographic information systems, geographical information science, and cartography—methods, tools and modes of representation that inform how geographers measure and describe the world. I will do that from a personal perspective based on my background as a geographer, cartographer, and geographic information scientist (GIScientist) that has worked and is interested in questions related primarily to history and historical geography, and in particular in the geographies of the Holocaust. Note that the intersection between GIS and geographic information science (GIScience) on one side, and human geography on the other, is at the same time quite broad and quite narrow, for while the methods and tools of GIScience and GIS can be deployed across the various subdisciplines that compose human geography, at their core these techniques do one thing: measure the value of quantitative data. This not only leaves out, at least for the most part, qualitative data, but also underscores the fact that the meaning of these measurements is often outside the subject matter of GIS and GIScience and needs to be interpreted according to models and theories which reside elsewhere: GIS and GIScience unearth relationships between data, but do not explain them. The case of cartography is somewhat different, as maps are both a way of measuring and representing the world and also offer a window into the times, ideologies, social norms, and worldview of the cultures and societies that produced them. It has been argued that the same is true for GIS and GIScience, as I will explain in the next section. In this chapter, I will discuss topics—for example, historical geography—that are covered elsewhere in this book. When I do so, my purpose is to illustrate the intersection between that topic with GIS, GIScience, and cartography, and in doing so I will simplify and generalize concepts and ideas that are more fully developed elsewhere.

DOI: 10.4324/9781003038849-6

Basic Concepts and Definitions

As the *Dictionary of Human Geography* (2013a) puts it, human geography "concentrates on the spatial organization and processes shaping the lives and activities of people" and is composed of a variety of subfields, including economic geography, historical geography, political geography, social geography, and others. Furthermore, "what distinguishes human geography from other related disciplines…" (economics, history, political sciences, and sociology, to continue with the example above) "are the application of a set of core geographical concepts to the phenomena under investigation, including space, place, scale, landscape and nature." Three of these five concepts—space, place, and scale—are especially relevant to this chapter's objectives. Much has been written by geographers about what exactly space, place and scale are; here, I will highlight only some of these ideas, again in the context of the objectives of this chapter. To start with, space can be thought of as abstract or relative. To return to the Dictionary (2013b), space is the "geometric container in which life takes place and matter exists… for physicists, space is…" where "objects move and interact…," and these interactions and movements "can be understood through Euclidean geometry and mapped." This is a definition a GIScientist would recognize and agree with. In this abstract view of space—one of quantitative analysis and methods—one kilometer is one kilometer, regardless of what is located at the start and end points of the measurement, regardless of what one finds along the way, and regardless of what this all means and what causes and explains these differences.

Of course, this is not the only definition of space: to continue with the Dictionary (2013b), starting in the 1970s radical geographers started to argue that space "was not a given, neutral, and passive geometry, essentialist and teleological in nature; rather it was relational, contingent, and active, something that is produced or constructed by people through social relations and practices." This relativistic view is quite different from the abstract definition of space that is the object of study for GIScientists. (Interestingly, scholars in the humanities also tend to understand space in a relativistic sense, which can make collaborating with GIScientists a complicated affair.) As is the case for space, there are also many definitions of place. While a full treatment of the concept is beyond the scope of this chapter (see Adams, 2017 for a review; see also Giordano and Cole, 2019), one way of looking at place is as a dynamic entity, a product of social processes, individually experienced—and thus similar to the relativistic view of space—and constituted by the triad of location, locale, and sense of place (Agnew and Duncan, 1989). This definition incorporates the idea of space and the physicality of place (location); the social, cultural, political and economic dimensions of place (locale); and the behavioral and emotional component of place (sense of place, as theorized by Yi-Fu Tuan, 1977). I believe this idea of place as a process, as a dynamic entity, together with the triad of location, locale, and sense of place, is especially interesting for the study of the intersection of human geography with GIScience, GIS, and cartography. I will return to this point later.

A third key concept is scale. As it applies to cartography, scale is simply the ratio between the map and the real world, so that at the scale of 1:25,000, a distance of one kilometer in the real world equals a distance of four centimeters on the map. In this sense, scale is a quantitative measurement. But there is another view of scale—a qualitative one—that is relevant to our discussion. As is the case for space and place, there are many theorizations of scale (for a review, see Sheppard and McMaster, 2004; Clifford et al., 2009; Herod, 2011); one important aspect of this debate concerns the ontological status of scale. Some regard scale as the material product of political, social, and economic processes, and thus as something real; others understand scale as a conceptual device—a tree, a network, a hierarchy—that allows the researcher to study and make sense of the characteristics of complex events and topics in a systematic way (Cox, 1998;

Jones, 1998; Brenner, 2001; Marston, Jones, and Woodward, 2005; Jessop et al., 2008, Herod, 2011). In this sense, which I believe is the most useful given my objectives here, scale is both a metaphor and a method, a way of framing the world to render its structure and meaning intelligible and measurable. To give one example, for the past several years I have been part of an interdisciplinary group of researchers known as the Holocaust Geographies Collaborative (holocaustgeographies.org) working to understand the geographies of the Holocaust. From the beginning of our collaboration, we chose scale as an organizing principle for the research (Knowles, Cole, and Giordano, 2014). This makes sense given the totality and finality of this event for the Nazis, who built ghettos at the urban scale, moved their victims across Europe and built a system of camps at the continental scale, organized their murderous activities in different places depending on the social and cultural context they encountered and based on practical considerations, and so on. Most importantly—and most intimately—the body was the ultimate scale of the Holocaust, as the Nazis systematically proceeded to exterminate all those who were deemed undesirable, with no possibility of redemption or escape from one's condition (it is in this sense that the Holocaust was unique). The Nazi regime was also explicitly guided by an ideology with roots in geographic thought and notions, including *Lebensraum*, which distinguished Aryan versus non-Aryan space, and the planning and designing of "Germanified" cities, Jewish ghettos, and concentration camps, which all involved the physical destruction of what was there before—witness the disappearance of the *shtetl* in eastern Europe—and the construction of a new built environment.

The discussion above assumes that the reader is familiar with GIS, GIScience, and cartography, but it is perhaps useful to explore how these three terms relate to each other. Let us start with GIS and cartography. Developed beginning in the 1970s, the technology of geographic information systems, or GIS, has found applications well beyond its original intent as a tool to aid natural resource management, and has become a worldwide multibillion dollars industry. Note that I have used the term "technology," as this GIS is a computerized tool for storing, managing, analyzing, and visualizing geographic data. By geographic data I mean any known fact or piece of information about the world whose location can be plotted on a coordinate system—i.e., that can be georeferenced. The act of georeferencing allows a certain location to be compared with other locations. For example, which of two cities is larger, how far are they from each other, and so on. One might argue that a paper map also allows this type of measurement. While this is true, the biggest advantage of GIS over a traditional paper map is that in the former, geographic data and its representation are kept separate, while in the latter the two are inextricable from each other. This is possible exactly because GIS is a computer technology: in a GIS, data are stored, accessed, explored, and edited via a database management system that ultimately relies on tables composed of records and variables, while the data are visualized as maps following the rules and conventions of a cartographic language. In this way, a GIS user manipulates either the map or the table, never the two at the same time, at least not directly. (Think of your experience using a GIS software if you are not convinced.) One of the consequences of this separation of labor is that editing the data—for example, changing the value of one of the variables for a certain record, or changing a symbol on the map—does not require the entire map to be redrawn. In other words, in a GIS it is possible to edit a single piece of geographic data without modifying any other part of the dataset. Note also that this is "merely" a technological advancement, a very useful one indeed but one that leaves unchanged and untouched the process and logic by which we abstract the world for representing and making sense of it. In a GIS, data are still organized by layers ("rivers," "roads," "cities," and so on), as they are on a map, and according to a model whose rationale resides outside of the GIS. For example, both in a GIS and on a map a river is rendered visually as a squiggly line with a name and a color (usually blue, even though very few rivers are actually blue): of all the myriad characteristics of a river, we extract those we believe are relevant, and only those, and we use a visual language to display them that is based on cartographic rules and conventions.

As defined above, GIS is a container and a tool for the application and implementation of the methodologies of GIScience, which, in a widely influential essay, Michael Goodchild (1992) defines simply and succinctly as "the science behind the systems," the spatial analytical quantitative tools and methods implemented as functionalities in the GIS. This definition has been worked on, modified, and argued about in the last thirty years, but it is now substantially understood and agreed upon, at least within geography, for outside of it, and especially the field I am most familiar with—history—"GIS" often refers to both the technology and the science. To continue with the Dictionary of Human Geography (2013c), geographic information science is "an interdisciplinary field... the science underpinning the development of GIS and how GIS analyse the world." This is a key point, as GIScience—just like GIS—both owes its principles and methods to multiple disciplinary traditions and can be applied to the exploration of the physical as well the human world. GIScience is a hybrid field, with roots in computer and information science, cartography and geography, mathematics and geometry, the cognitive sciences, and statistics. As is the case for GIS, GIScience has found applications in a multitude of fields—basically, any field in which location matters—including human geography, although this relationship has been conflictual at times (for more extensive discussion, see Giordano 2019, Miranker and Giordano 2020, Giordano and Cole 2018), especially in the early days, the 1980s and the 1990s (Schuurman, 2000; O'Sullivan, 2006; Thatcher et al., 2016). Under the general umbrella of "critical GIS," Pickles (1995), Curry (1998), and Obermeyer (1995)—among others—questioned the ideological foundations of the field and the worldview promoted by the technology, including its military applications. Self-reflection on the ontology and epistemology of GIS was alive and lively debated in the GIScience community as well, as shown by a series of Technical Papers published by the National Center for Geographic Information and Analysis (NCGIA) in the same two decades, including a range of foundational contributions on the language of spatial relations, the relationship between cognitive science and GIS, the ontology of GIS, and public participation GIS.[2] While the debate has subsided to a certain extent in recent years, this conversation has been beneficial to both human geography and GIScience, and has led to a flourishing of studies on alternatives to the then dominant paradigm of GIS—including feminist GIS (Kwan, 2002) and qualitative GIS (Cope and Elwood 2009)—and more generally has resulted in an increased awareness by GIScientists of both the history of the field and of the assumptions and implications behind the methods, tools, and the model of reality implied by GIS and GIScience. The increased use of computational technologies and big data methods in the digital and spatial humanities (Bodenhamer, Corrigan, and Harris, 2010; Travis, 2015; Dear et al., 2011; Giordano, Shaw, & Sinton, 2020) has also contributed to bringing new voices and new approaches into GIScience.

GIS, GIScience, and Cartography: From Space to Place

In their edited volume on the spatial humanities, Bodenhamer, Corrigan, and Harris (2010:x), note that "the humanities pose far greater epistemological and ontological issues [to GIS]" compared to the social sciences and the physical sciences. Perhaps the major obstacle arises from the emphasis of the humanities on qualitative data and methods, as opposed to the quantitative data and methods of GIScience. Put in different terms, while the former is preoccupied with questions related to place, the latter's object of study is space, intended is in *abstract* form. In the rest of this chapter, I aim to illustrate challenges and opportunities, issues and pitfalls in the application of GIS, GIScience, and cartography to human geography in general and historical geography—the subfield of human geography I am more familiar with—in particular. I will suggest mixed methods analysis and an anchoring in theories of space *and* place as possible ways to take advantage of the contributions of both quantitative and qualitative methodologies. Again, the context of my considerations is the geography of the Holocaust. More specifically, the example I will refer to is the ghettoization of the Jews of Budapest, the capital of Hungary. Ghettos were a defining moment and instrument in the Nazi's plan to annihilate the Jews of Europe, but they were not enacted everywhere, and they took a different form in different places. Even so, ghettos were a feature of the urban geography of hundreds of towns and cities in occupied central and eastern Europe. In Hungary, Jews were placed into urban ghettos following the German occupation of the country in the spring of 1944. In the country's capital, Budapest, a highly dispersed form of ghetto was adopted in June 1944, when the city's close to 200,000 Jews were ordered to move into just under 2,000 separate apartment buildings, each one marked with a large yellow star, dispersed across every district of the city. Issued on June 16, the first list of apartment buildings was amended and finalized on June 22 (Figure 4.1). This was not the end of the story, though, as a radically different model of two ghettos—one a looser group of houses (the International ghetto), the other a closed area around a few streets (the Pest ghetto)—was adopted in late November 1944 (Figure 4.2), shortly before the Red Army reached the city at the end of December 1944 and liberated it in February 1945.

The Historical GIS of the Budapest Ghetto (HGIS) maps the changing shapes of the ghetto, including its apartment buildings and public places (market halls, hospitals, restaurants, etc.),

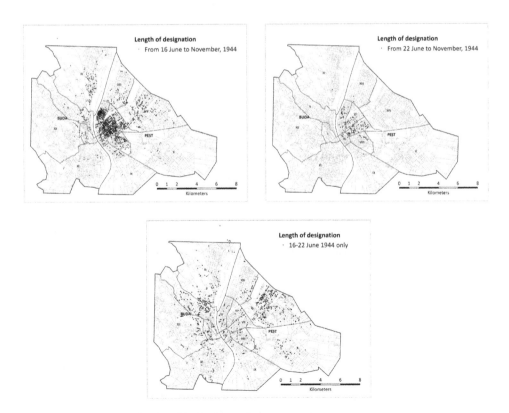

Figure 4.1   The dispersed stage of the Budapest ghetto.

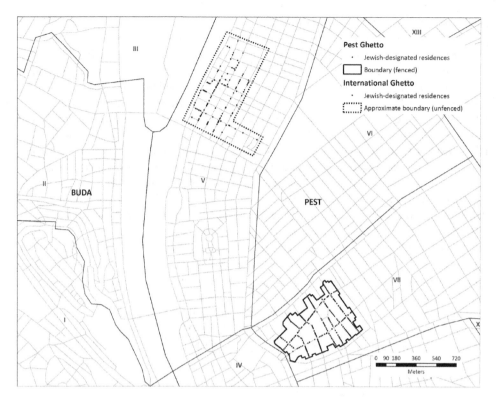

Figure 4.2   The International ghetto and the Pest ghetto.

as well as the International and Pest ghettos (Cole and Giordano, 2014). Individual Jewish-designated residences are symbolized by equally sized dots plotted on a base map of 1944 Budapest along the historic road network of the city. Note that this representation, which simply plots the presence of a ghetto residence, masks the variety of Budapest's housing stock: at the time, the city was—and still in part is—characterized by the presence of large multi-story apartment buildings in the center of the city through to smaller single-family homes in the suburbs. To try and bring this variety into the analysis, Cole and Giordano (2014) worked with census data from 1941 to weigh the size of buildings by district and thus show the spatial variability of the phenomenon. Rather than simply eyeballing these weighted maps, GIS enabled the adoption of a range of spatial analytical tools to analyze the situation on the ground, including the *mean center* to identify the geographic center of the distribution of ghetto houses, the *directional distribution* to show their overall directional pattern, and the *standard distance* to measure their degree of concentration. Running these measures on a mapping of the ghetto weighted by building size on June 16 and June 22 (bottom part of Figure 4.3), showed only a moderate shift in the degree of concentration of the ghetto houses—and by extension, of the Jews who were forced to move there—which appeared instead more pronounced on the unweighted data (top part of Figure 4.3).

What was most striking was the degree of continuity between these two dates, despite the seeming radical cutting of the number of ghetto houses from over 2,600 to just over 1,900. In reality, the reductions were largely confined to smaller dwellings in the outlying suburbs rather than the larger apartment buildings in the central districts of the city. To confirm this

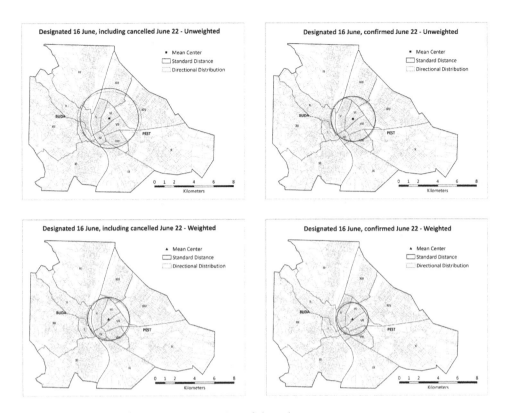

Figure 4.3 Measures of concentration: Location of ghetto houses.

initial observation—of spatial stability rather than change—Cole and Giordano (2014) shifted the scale of analysis from analytical tools that return one summary measure for the entire study area—the city of Budapest—to a representation at a much larger scale via kernel density analysis. Kernel density analysis assigns continuous values to cells of various resolution (15 to 60 meters in our case), with larger values identifying hot spots—areas of high geographic concentration of the variable studied (here, weighted population)—and smaller values identifying cold spots. Running this more nuanced method returned very similar results to the previous analysis, thus confirming our initial findings.

I discussed this example in some detail as it is illustrative in at least three ways of the relationship between GIS, GIScience, cartography, and human geography. My first point concerns the importance of scale: my coauthor Tim Cole and I set out to answer questions related to the changing shapes of the Budapest ghetto seeking confirmation of our findings across a variety of scales, from the entire city to Buda and Pest (on the opposite sides of the Danube), to the individual district—fourteen at the time—and down to the individual cell. The fact that continuity in the shape of ghettoization was a feature of the analysis at multiple scales reassured us in the reliability of our conclusions. The second point concerns the spatial analytical tools used, and by extension the centrality of GIScience to the research: the mean center, the directional distribution, the standard distance, and kernel density analysis all pointed at continuity, a conclusion we could not reach with unweighted data or via simple visual exploration of the digital or analog maps. The third—and perhaps most important—point concerns the nature of the research process itself. Ours is an interdisciplinary endeavor, which benefits from the collaboration of a geographer and GIScientist with an historian. The collaboration extends to the refinement and rethinking of the research question itself, as illustrated by the decision to systematically study the Holocaust in Budapest at multiple scales, thus adopting a fundamental geographic concept as a guiding principle of our research. Another example comes from measuring the proportion of the streets in the city that were 'ghetto streets' during the second half of 1944, an exercise that led us to reframing our understanding of ghettoization. Given the nature of this highly dispersed ghetto, Jews could and did leave their residences for a limited number of hours each day in order, for example, to shop for foodstuff. Working with the layer of the street network, an average estimated walking speed of 1.3 meters per second, and network analysis techniques, we were able to calculate 30- and 60-minute walking distances from the ghetto residences to a number of sites we knew to be critical in 1944 (Figure 4.4), including market halls accessible to Jews. A striking feature of this analysis was that it made clear that within the ghetto individuals and families living in different places had greater or lesser access to market halls depending on where they lived, suggesting that the dispersed nature of the ghettoization did not simply divide Jews from non-Jews—as is generally imagined by the concept of segregation—but also divided Jews from other Jews, as well as their access to critical resources, and in different ways depending on the resource considered (market hall, hospital, etc.). This ultimately led us to imagine 'invisible walls' within this dispersed ghetto, which in turn became a new focus in our work.

Note that the map in Figure 4.4 is not a representation of historical realities, as are the maps in Figures 4.1 and 4.2, but experimental models created in the context of an interdisciplinary

Figure 4.4 Walking distances to market halls.

research process. By explicitly referring to human behavior, this example brings us to an important point. Reflecting back on the HGIS of Budapest we became increasingly aware that the tool of GIS and the methods of GIScience are especially useful to tell the story of the Holocaust from the perspective of the perpetrator—in the case of Budapest, by mapping address by address the shifting spatial structure of the Budapest ghetto as designed and recorded in the perpetrators' sources. This does not mean that GIS and GIScience are intrinsically, inherently, or by design, tools of—and for—oppression. On the contrary, GIS and GIScience can uncover critical aspects of the geographies of the Holocaust by showing how widespread and systematic the spatio-temporal patterns of victimization were, at a variety of scales. But this is not the whole story. As we shifted from working primarily with perpetrator sources—e.g., Jewish-designated residences in Budapest—to victim sources—primarily, testimonies of Holocaust survivors—that reveal "spatial strategies" of survival and experiences of place (Cole, 2016), it became clear that the spatial analytical and quantitative methods of GIScience were lacking (see Cole and Giordano, 2021). In the final section of the chapter, I will discuss recent developments in GIScience and in cartography to include the integration of place and space through mixed methods analysis and an interdisciplinary research model, concluding with an overview of what seem to me promising avenues of research.

Conclusions: Towards a GIS of Space AND Place

The term "platial GIS" (Goodchild, 2015), or GIS of place, is relatively new to GIScience, but discussions concerning the relationship between GIS and place are not new in the field and date back to the already mentioned debates on "critical GIS" of the 1980s and 1990s. More recently, the discourse about space, place and GIScience and the use of qualitative methods—or, rather, mixed methods—have resurfaced in the context of the spatial humanities and the geohumanities, human geography, historical GIS, and qualitative GIS (Cope and Elwood, 2009; Bodenhamer, Corrigan, and Harris, 2010; Drucker, 2012; Dear et al., 2011; Giordano and Cole, 2018). Recent symposia on the topic (Westerholt et al., 2018; Giordano, Shaw and Sinton, 2020) offer different perspectives and approaches to the theory and practice of a GIS of place.

A GIS of place—or, as it is implicit in this definition, a "GIS of place AND space"—is, first of all, a collaborative enterprise, as many possible implementations of GISs of places may be envisioned, rather than a monolithic and general-purpose system (Giordano and Cole, 2018, 2019). This is why anchoring the GIS of place to a specific view of place is essential. Yi-Fu Tuan, in his classic study of space and place, observed that "home" is not necessarily anchored in place (Tuan, 1977): it can reside in a person, or a personal relationship, such as the deeply comforting bond that can exist between parent and child, or any two people who trust and rely upon one another. In other words, place and social relations are inextricably intertwined and one cannot be studied without the other. For example, in order to explore place and social networks as key strategies of survival as described by Holocaust survivors, one needs to enter, metaphorically at least, the buildings and the camps and listen to the voices of those who were there. This requires new sources (testimonies and memoirs), new digital methods (appropriate for studying narratives and testimonies), and a shift from spatial quantitative methods to spatial qualitative methods—or rather the integration of quantitative and qualitative methods into a mixed methods analysis. A GIS of place also needs to be by definition topological: it is the existence of a relation between people and places that take the center stage in a GIS of place. The object of study in a GIS of place is therefore the relational and the topological. This is the deepest meaning of statements such as "the social is spatial" and "social relationships are explicitly spatialized, i.e., they are inextricably and explicitly described as happening in space." As a way of spatializing social networks—or, if one prefers, adding the social dimension to spatial networks—several authors have proposed methods to integrate social networks analysis and GIS: see, for example, Andris (2016), especially the idea of *anthrospace* and ways to measure it; and, in the context of the transmission of disease, Emch et al., (2012) and Bian et al., (2012); as it applies to the Holocaust, see Mariot and Zalc (2017) for France, Chronakis (2018) for Greece, and Giordano and Cole (2011) on the social and spatial networks of the Budapest Ghetto. In the context of the Holocaust, a research agenda on a GIS of place answers questions such as: how are the places of the Holocaust formed? What kind of relationships shape and define these places? What happens in these places? How are these places connected? Crucially, how does this all relate to human survival and resilience? Building a GIS of place will require technical (e.g., new visualization methods) as well as conceptual and methodological (model of reality, data model, analytical functionalities) solutions: for example, the ability to study simultaneously social networks and spatial networks and how the two interact together and at scale—the scales of Budapest, for example. The real question then becomes how to integrate the powerful spatial-analytical capabilities of GIS, which lead to the unearthing of patterns as shown in the preceding section, and at the same time are able to include the experiences of the victims, operationally defined as spatial strategies of survival that are the effect and results of the deployment of social relations in place. The point where the policies of the perpetrator meet and interact with the victim is at the resolution of the individual. It is here, I think, that the abstract view of space—the perpetrator's policies—meets the place-making activities of victims as well as of perpetrators (e.g., a camp guard and a prisoner).

How does one go about building a GIS of place? While no comprehensive approach has been proposed, I want to review three methods, quite different from each other, that I see as part of a future toolbox for a GIS of place (for a more extended discussion, see Giordano and Cole 2018, Miranker and Giordano 2020). The first—deep maps—is a framework and a conceptual model, rooted in cartographic theory. The second—qualitative spatial reasoning, or QSR—is a language for formalizing and analyzing qualitative spatial relationships, with origins in computer science and GIScience. The third—corpus linguistics (CL) and natural language processing—is a big data technique that originates in linguistics and computer science and has been used in digital humanities projects.

To start with, the idea of a "deep map" has been proposed by Bodenhamer et al., (2015) to bridge the gap between the humanities and GIS and GIScience. As Bodenhamer et al., put it (2015, p. 3), a

> deep map is a finely detailed, multimedia depiction of a place and the people, animals, and objects that exist within it… Deep maps are not confined to the tangible or material, but include the discursive and ideological dimensions of place… A deep map is simultaneously a platform, a process, and a product.

Examples of applications are offered in Bodenhamer et al., (2015). As are Kwan's "geo-narratives" (2008) as well as ESRI Story Maps and other examples. Deep maps are steps in the direction of the integration of GIS and GIScience in the humanities that are, in my opinion and at least potentially, especially useful to historians and literary scholars. However, deep maps need to be further conceptualized and more strongly and clearly differentiated from more traditional multimedia interactive maps. From within computer science and artificial intelligence, the goal of qualitative spatial reasoning (QSR) is to design representations that can answer qualitative queries without much numerical information and to derive and manipulate qualitative spatial representations that abstract important fundamental aspects of the underlying data (Bailey Kellog and Zhao, 2004; Cohn and Hazarika, 2001 and 2007). QSR can effectively trace relationships among places that are vaguely or ambiguously defined as opposed to place names that are identified by a set of geographic coordinates. For example, statements such as "a river running through a town," or "a park on the border of a neighborhood" (Stell, 2019, p. 4) are immediately understood by humans, but not by a GIS, as in most cases these places are vaguely defined and cannot be assigned a clear and unambiguous—"my neighborhood" is a qualitative statement—set of *quantitative* geographic coordinates, or even a geometry. QSR has the capacity of representing *qualitative* spatial relationship of the kind above, and it does so with a series of formal logical statements. In an example from Holocaust Studies, Cole and Hahmann (2019) use QSR on survivors' testimonies to show how the trajectories of a Holocaust survivor's movements could be organized and visualized regardless of specificity in time or space.

Finally, corpus linguistics (CL) and natural language processing (NLP) refer to a set of computational methods for analyzing large bodies of text across corpora. They are used most notably in literary studies but—in a trajectory similar to that of GIS and GIScience—they have found application in other subfields of the humanities and the social sciences. As applied to the study of the Holocaust, CL and NLP can help the researcher explore the narratives and testimonies of survivors at scale (i.e., not a single testimony but dozens or hundreds, thus contributing to a "distant reading" (Moretti, 2013) of the archive). For example, Knowles et al., (2021) have used CL/NLP to unearth the spatial experiences of forced labor and to investigate gendered differences of Holocaust survivors.

Overall, QSR and CL/NLP provide a framework for searching for key terms or themes throughout multiple textual sources, narratives, and testimonies, highlighting and visualizing the presence of spatial relationships that are not necessarily mappable in Cartesian coordinate space and in a traditional GIS setting (Miranker and Giordano, 2020; Knowles et al., 2014, 2021; Pavlovskaya, 2018). In a sense, CL/NLP and corpus linguistics are analogous in the context of a GIS of place of what kernel density analysis, the mean center, network analysis, and other established spatial analytical techniques are to traditional GIS. Whether these techniques will evolve into fundamental components of a still-to-be-defined GIS of place—one that in my opinion has a lot to contribute to human geography—remain to be seen.

To close the chapter, I return to cartography. The combination of place and space, the integration of quantitative and qualitative methods of analysis, a historical and critical perspective on the deontology and ethics of the profession—to some extent, all of these themes are an echo of earlier developments and earlier debates in the history of cartography. In other words, before there was a "critical GIS," there was a "critical cartography." Central to this conversation, and especially relevant to human geographers, is David Harley's discourse on the relationship between maps, knowledge, and power (1988a, 1988b, 1989). In this view, maps are social constructs that reflect the worldviews, the values, and the power structures of the society that produced them. Harley's work, while not exempt from criticism (Andrews, 2001; see also Rose-Redwood 2015), has stood the test of time, especially—in my opinion—because his arguments are rooted in a critique of the map itself, intended as a unique artifact, a material product, an object of art, an expression of the principles and rules of cartographic design. Harley's influence as the original "critical cartographer" cannot be underestimated, and has opened the gates to a much broader, comprehensive, and to a degree inclusive approach to studying the history of cartography and examining the role maps play in society—including an attention to indigenous cartographies (Pearce 2014, Rose-Redwood et al., 2020), efforts to decolonizing the maps (Ackerman 2017), the discourse around maps and persuasion (or, propaganda maps) (Pickles, 1991, Muehlenhaus, 2011, Kent, 2016), and others.

Notes

1 The topics discussed in this chapter are based in part on research published elsewhere, including the following: a) Giordano, A. 2019. GIS&T and the Digital Humanities. In *The Geographic Information Science & Technology Body of Knowledge* (John P. Wilson, ed.). DOI: 10.22224/gistbok/2019.2.8.; b) Cole, T. and Giordano, A., 2021. Geographies of Rescue: Spatial Patterns and Social Geographies of Jewish Rescue in Budapest, 1944. *Journal of Historical Geography*, 71:63–72; c) Cole, T. and Giordano, A., forthcoming. Digital Mapping. In *Doing Spatial History* (Riccardo Bavaj, Konrad Lawson, Bernhard Struck, eds.). London, Routledge, Chapter 15; d) Giordano, A. and Cole, T., 2019. Places of the Holocaust: Towards a Model of a GIS of Place. *Transactions in GIS, Special Issue: Modelling and Analysing Platial Representations*, 24(4):842–857; e) Giordano A. and Cole, T. 2018. The Limits of GIS: Towards a GIS of Place. *Transactions in GIS*, 22(3):1–13; f) Miranker, M. and Giordano, A. 2020. Text Mining and Semantic Triples: Spatial Analyses of Text in Applied Humanitarian Forensic Research. *Digital Geography and Society*. https://doi.org/10.1016/j.dig geo.2020.100005.
2 A full list is available at www.ncgia.ucsb.edu/pubs/pubslist.html.

References

Ackerman, J.R., ed. (2017) *Decolonizing the Map: Cartography form Colony to Nation*. Chicago, IL: University of Chicago Press.

Adams, P.C. (2017) Place. In *International Encyclopedia of Geography: People, the Earth, Environment and Technology* (Richardson, D., Castree, N., Goodchild, M.F., Kobayashi, A., Liu, W., and Marston, R.A., eds.). Washington, DC: Wiley-Blackwell, 5072–5084.

Agnew, J.A. and Duncan, J.S. (1989) *The Power of Place: Bringing Together Geographical and Sociological Imaginations*. London: Routledge.

Andrews, J.H. (2001) Introduction: Meaning, Knowledge, and Power in the Map Philosophy of J. B. Harley. In *J. B. Harley. The New Nature of Maps: Essays in the History of Cartography* (Laxton, P., ed.). Baltimore: Johns Hopkins Press, 1–32.

Andris, C. (2016) Integrating Social Network Data into GISystems. *International Journal of Geographic Information Science*, 30(10):2009–2031.

Bailey-Kellog, C. and Zhao, F. (2004) Qualitative Spatial Reasoning: Extracting and Reasoning with Spatial Aggregates. *AI Magazine*, 24(4):47–60.

Bian, L., Huang, Y., Mao, L., Lim, E., Lee, G., Yang, Y., Cohen, M., and Wilson, D. (2012) Modeling Individual Vulnerability to Communicable Diseases: A Framework and Design. *Annals of the Association of American Geographers*, 102(5):1016–1025.

Bodenhamer, D.J., Corrigan, J., and Harris, T.M. (2015) *Deep Maps and Spatial Narratives*. Bloomington, IN: Indiana University Press.

Bodenhamer, D.J., Corrigan, J., and Harris, T.M. (2010) *The Spatial Humanities: GIS and the Future of Humanities Scholarship*. Bloomington, IN: Indiana University Press.

Brenner, N. (2001) The Limits to Scale? Methodological Reflections on Scalar Structuration. *Progress in Human Geography*, 25(4):591–614.

Chronakis, P.P. (2018) From the Lone Survivor to the Networked Self. Social Networks Meet the Digital Holocaust Archive. *Quest. Issues in Contemporary Jewish History*, 13:52–84.

Clifford, N.J., Holloway, S.L., Rice, S.P, Valentine, G., eds. (2009) *Key Concepts in Geography*, Los Angeles: Sage.

Cohn, A.G. and Hazarika, S.M. (2001) Qualitative Spatial Representation and Reasoning: An Overview. *Fundamenta Informaticae*, 46:1–29.

Cohn, A.G. and Renz, J. (2007) Qualitative Spatial Representation and Reasoning. In *Handbook of Knowledge Representation* (van Harmelen, F., Lifschitz, V., and Porter, B., eds.). Amsterdam: Elsevier, 551–596.

Cole, T. (2016) *Holocaust Landscapes*. London: Bloomsbury.

Cole, T. and Giordano, A., forthcoming. Digital Mapping. In *Doing Spatial History* (Bavaj, R., Lawson, K., and Struck, B., eds.). London: Routledge, Chapter 15.

Cole, T. and Giordano, A., (2021) Geographies of Rescue: Spatial Patterns and Social Geographies of Jewish Rescue in Budapest, 1944. *Journal of Historical Geography*, 71:63–72.

Cole, T. and Giordano, A. (2014) Bringing the Ghetto to the Jew: The Shifting Geography of the Budapest Ghetto. In *Geographies of the Holocaust* (Knowles, A., Cole, T., and Giordano, A., eds.). Bloomington, IN: Indiana University Press, 121–157.

Cole, T. and Hahmann, T. (2019) Geographies of the Holocaust: Experiments in GIS, QSR, and Graph Representations. *International Journal of Humanities and Arts Computing*, 13(1–2):39–52.

Cope, M.S. and Elwood, S. (2009) *Qualitative GIS: A Mixed Methods Approach*. Thousand Oaks, CA: Sage.

Cox, K. (1998) Spaces of Dependence, Spaces of Engagement and the Politics of Scale; or, Looking for Local Politics. *Political Geography*, 17(1):1–23.

Curry, M.R. (1998) *Digital Places: Living with Geographic Information Technologies*. London: Routledge.

Dear, M., Ketchum, J., Luria, S., and Richardson, D. (2011) *GeoHumanities: Art, History, Text at the Edge of Place*. London: Routledge.

Drucker, J. (2012) Humanistic Theory and Digital Scholarship. In *Debates in the Digital Humanities* (Gold, M.K., ed.). Minneapolis, MN: University of Minnesota Press, 85–95.

Emch, M., Root, E.D., Giebultowicz, S., Ali, M., Perez-Heydrich, C., and Yunus, M. (2012) Integration of Spatial and Social Network Analysis in Disease Transmission Studies. *Annals of the Association of American Geographers*, 105(5):1004–1015.

Giordano, A., Shaw, S-L., and Sinton, D. (2020) The Geospatial Humanities: Transdisciplinary Opportunities. *International Journal of the Humanities and Arts Computing (IJHAC)*, 14–2. Proceedings of the 2019 UCGIS Symposium. Edinburgh: Edinburgh University Press.

Giordano, A. (2019) GIS&T and the Digital Humanities. In *The Geographic Information Science & Technology Body of Knowledge* (Wilson, J.P., ed.). DOI: 10.22224/gistbok/2019.2.8.

Giordano, A. and Cole, T. (2019) Places of the Holocaust: Towards a Model of a GIS of Place. *Transactions in GIS, Special Issue: Modelling and Analysing Platial Representations*, 24(4):842–857.

Giordano A. and Cole, T. (2018) The Limits of GIS: Towards a GIS of Place. *Transactions in GIS*, 22(3):1–13.

Giordano, A. and Cole, T. (2011) On Place and Space: Calculating Social and Spatial Networks in the Budapest Ghetto. *Transactions in GIS*, 15(s1):143–170.

Goodchild, M.F. (1992) Geographical Information Science. *International Journal of Geographical Information Systems*, 6(1):31–45.

Goodchild, M. (2015). Space, Place and Health. *Annals of GIS*, 21(2): 97–100.

Harley, J.B. (1989) Deconstructing the Map. *Cartographica*, 26(2):1–20.

Harley, J.B. (1988a) Maps, Knowledge, and Power. In *The Iconography of Landscape*, (Cosgrove, D., Daniels, S., eds.). Cambridge: Cambridge University Press, 289–290.

Harley, J.B. (1988b) Silences and Secrecy: The Hidden Agenda of Cartography in Early Modern Europe. *Imago Mundi* 40:57–76.

Herod, A. (2011) *Scale*. London: Routledge.

Jessop, B., Brenner, N. and Jones, M. (2008) Theorizing Sociospatial Relations. *Environment and Planning D*, 26(3):389–401.

Jones, K. (1998) Scale as Epistemology. *Political Geography*, 17(1):25–28.

Kent, A. (2016) Political Cartography: From Bertin to Brexit. *The Cartographic Journal* 53(3):199–201.

Knowles, A., Cole, T., and Giordano, A., eds. (2014) *Geographies of the Holocaust*. Bloomington, IN: Indiana University Press.

Knowles, A., Jaskot, P., Cole, T., and Giordano, A. (2021) Mind the Gap: Reading Across the Holocaust Testimonial Archive. In *Lessons and Legacies XIV: The Holocaust in the 21st Century: Relevance and Challenges in the Digital Age* (Cole, T. and Gigliotti, S., eds.). Evanston, IL: Northwestern University Press, 216–241.

Kwan, M-P. (2002) Feminist Visualization: Re-envisioning GIS as a Method in Feminist Geographic Research. *Annals of the Association of American Geographers*, 92(4):645–661.

Kwan, M-P., and Ding, G. (2008) Geo-Narrative: Extending Geographic Information Systems for Narrative Analysis in Qualitative and Mixed-method Research. *The Professional Geographer*, 60(4):443–465.

Mariot, N. and Zalc, C. (2017) Reconstructing Trajectories of Persecution: Reflections on a Prosopography of Holocaust Victims. In *Microhistories of the Holocaust* (Zalc, C. and Bruttmann, T., eds.). New York: Berghahn, 85–112.

Marston, S., Jones, J.P., and Woodward, K. (2005) Human Geography without Scale. *Transactions of the Institute of British Geographers*, 30:416–432.

Miranker, M. and Giordano, A. (2020) Text Mining and Semantic Triples: Spatial Analyses of Text in Applied Humanitarian Forensic Research. *Digital Geography and Society*. https://doi.org/10.1016/j.diggeo.2020.100005.

Moretti, F. (2013) *Distant Reading*. London: Verso.

Muehlenhaus, I. (2011) Genealogy that Counts: Using Content Analysis to Explore the Evolution of Persuasive Cartography. *Cartographica* 46(1):28–40.

Obermeyer, N.J. (1995) The Hidden GIS Technocracy. *Cartography and Geographic Information Systems*, 22(1):78–83.

O'Sullivan, D. (2006) Geographical Information Science: Critical GIS. *Progress in Human Geography*, 30(6):783–791.

Pavlovskaya, M. (2017) Qualitative GIS. In *The International Encyclopedia of Geography* (Richardson, D., Castree, N., Goodchild, M.F., Kobayashi, A., Liu, W., and Marston, R.A., eds.). Washington, DC: AAG and Wiley Blackwell, 5432–5442.

Pearce, M.W. (2014) The Last Piece is You. *The Cartographic Journal* 51(2):107–122.

Pickles, J. (1995) *Ground Truth: The Social Implications of Geographic Information Systems*. New York, NY: The Guilford Press.

Pickles, J. (1991) Texts, Hermeneutics and Propaganda Maps. In *Writing Worlds: Discourse, Text and Metaphor in the Representation of Landscape* (Barnes, T.J. Duncan, J.T., eds.). London: Routledge, 193–230.

Rogers, A., Castree, N., and Kitchin, R. (2013a) Human Geography. *A Dictionary of Human Geography*. Oxford: Oxford University Press.

Rogers, A., Castree, N., and Kitchin, R. (2013b) Space. *A Dictionary of Human Geography*. Oxford: Oxford University Press.

Rogers, A., Castree, N., and Kitchin, R. (2013c) Geographic Information Science (GISc). *A Dictionary of Human Geography*. Oxford: Oxford University Press.

Rose-Redwood, R., Barnd, N.B., Hetoevéhotohke'e Lucchesi, A., Dias, S., and Patrick, W., eds. (2020) Decolonizing the Map. *Special Issue of Cartographica*, 55(3):151–62.

Rose-Redwood, R., ed. (2015) Deconstructing the Map: 25 Years On. *Special Issue of Cartographica*, 50(1):50–53.

Schuurman, N. (2000) Trouble in the Heartland: GIS and its Critics in the 1990s. *Progress in Human Geography*, 24(4):569–590.

Sheppard, E. and McMaster, R.B., eds. (2004) *Scale & Geographic Inquiry*. Malden, MA: Blackwell.

Stell, J.G. (2019) Qualitative Spatial Representation for the Humanities. *International Journal of Humanities and Arts Computing*, 13(1–2):2–27.

Thatcher, J., Bergmann, L., Ricker, B., Rose-Redwood, R., and O'Sullivan, D. (2016) Revisiting Critical GIS. *Environment and Planning A*, 48(5):815–824.

Travis, C. (2020) Historical and Imagined GIS Borderlandscapes of the American West: Larry McMurtry's *Lonesome Dove* Tetralogy and LA Noirscapes. In *The Geospatial Humanities: Transdisciplinary Opportunities* (Giordano, A., Shaw, S-L., and Sinton, D. 2020, eds.). *International Journal of the Humanities and Arts Computing (IJHAC)*, 14(2):134–153.

Travis, C. (2015) *Abstract Machines: Humanities GIS*. Redlands, CA: ESRI Press.

Tuan, Y-F. (1977) *Space and Place: The Perspective of Experience*. Minneapolis, MN: University of Minnesota Press.

Westerholt, R., Mocnik, F-B., and Zipf, A., eds. (2018) On the Way to Platial Analysis: Can Geosocial Media Provide the Necessary Impetus? *Proceedings of the 1st Workshop on Platial Analysis (PLATIAL'18)*. doi.org/10.5281/zenodo.1475269.

5

REFLECTIONS ON HUMAN GEOGRAPHY'S METHODOLOGICAL 'TURNS'

Robin Kearns

To everything (turn, turn, turn)
There is a season (turn, turn, turn)
Pete Seeger, 1959

Introduction

This chapter offers a critical reflection on the methodological 'turns' experienced within human geography over recent decades and their implications for research practice. It is written from a self-consciously personal perspective. This is for two reasons. First, I draw upon first-hand experience as a window through which to make critical observations, drawing on almost four decades since beginning doctoral research. Second, I offer personal recollections because to do otherwise and adopt a conventionally 'detached' scientific stance would be to risk perpetuating an unhelpful trope in geography: the vantage point of the disembodied and dispassionate observer.

I work with the metaphor of 'turns' in the chapter. What do we mean by 'turn'? Seasons turn: usually gently and imperceptibly, but sometimes with unsettling haste. All the while, this change – in seasons or the weather – is out of our control. We also 'turn' in our daily travels, and perhaps this is the more apt metaphorical reference in this methodological context. Turning a corner, whether walking or driving, involves complicity and a conscious assent to move in a new direction than that experienced when seasons change. In these situations, there can be novelty (within limits, we can choose the trajectory of our turns) but more usually we are following a trend of others doing similarly. Turning takes us in a direction. It gets us somewhere. We see things differently along the way for having turned. So too with research methods.

There was a time when the notion of diverse methodological turns would have been anathema to the sense of progress in human geography. A linear trajectory underwritten by logics of one discovery informing the next was deeply embedded and coloured by modernist and positivist assumptions of progress (Barnett, 2002). The stimulus-response formula that underlies the questionnaire and interview format of behavioural geography is a classic example of this linear thinking. To this extent, the social scientist's practice echoed closely that of the

DOI: 10.4324/9781003038849-7

physical scientist and, along the way, rendered a whole suite of nuances associated with human experience and interaction invisible. The broadly sketched qualitative turn has constituted something of what Thomas Kuhn (1962) described as a 'paradigm shift': a foundational change in the underlying concepts and practices of an academic discipline. This change has heralded a radical openness to developing new and co-produced knowledge. These approaches are less concerned with explanation and more with understanding (Hay, 2016).

The appearance and acceptance of diverse research methods lying within this broadly sketched 'qualitative turn' in human geography has been rapid and wide-ranging. For instance, reflecting on disciplinary changes in the time between publication of their original text in 1996 and a second edition nine years later, Martin and Flowerdew (2005) remark on the need for a significantly expanded volume. Their addition of chapters on, for instance, focus groups and participatory methods, signals the rapid uptake of new approaches to method. In this chapter, I review two such newer 'turns' within the wider embrace of qualitative approaches – the bodily and visual turns. First, however, I turn to four significant moments in my own methodological journey.

Four Anecdotes That Presaged Personal 'Turns'

Method implies a systematic sequence of stages or, as one definition states, a procedure for accomplishing or approaching something. A recipe is a type of method. How many of us follow recipes? I suspect the world of cooks falls into two camps: those who meticulously follow the steps and ingredients outlined in cookbooks and those, like me, who do not. Such people know what works for them and, when asked, retrospectively try and outline how a particular meal came to be. But the moments in the kitchen when one's chosen ingredients not quite gelling are – metaphorically – times when one's practice turns and adapts. In the following sections, I offer four recollections of encounters that hastened turns in my research journey. Later in the chapter, I then place them within a wider context of methodological unfolding.

A Turn in Perspective

While a doctoral student at McMaster University in Hamilton, Ontario, Canada, the opportunity to travel to the Annual Meetings of the Association of American Geographers was always a highlight. Much of the appeal was both experiencing a new US city and hearing disciplinary luminaries speak in person. However, in retrospect, the majority of papers and keynote addresses were forgettable; cookie-cutter presentations rolling out the currently popular keywords and theories. In four years of these pilgrimages as a PhD student, one presentation is etched in my recollection for its impact, however. The technology *de jour* was the overhead projector. A fellow graduate student, whose name and affiliation I cannot recall, sought to strongly signal that vantage point matters in how we see and understand the world. Mid-way through her talk, she took hold of the overhead projector and, flipping the device on its end, projected the remainder of the images onto the ceiling. In that one masterstroke of creative disruption, the presenter demonstrated that what we see is profoundly shaped by where we stand (or, as the case may be, sit). By flipping the projector, she had confirmed for me the critical importance of positionality.

1. **David's story**

   During the data collection period for my doctoral research on the downstream effects of deinstitutionalisation for mental health care users, I interviewed a sample of community-based psychiatric patients in Hamilton. I was interested in a range of their experiences living in the so-called 'boarding house ghetto' after discharge from the soon-to-close hospital. Given the methodological times and the recommendation of my supervisor, my question schedule included rating scales among more open-ended semi-structured 'Tell me about...' type interview prompts.

After a successful round of interviews, I re-contacted all participants six months later with the purpose of exploring what circumstances had occurred to improve or hinder their lives in the inner city. When it came to re-interviewing 'David', he distractedly doodled on a piece of paper, avoiding my eyes. As we moved through the questions, he quietly provided insightful responses. Then I came to "Tell me about your housing situation. Have you moved house since we last spoke?". "Moved", he said. "Can you please tell me why?", I asked. "Because you talked with me earlier in the year", he said. "That made me think I could find a better place". I had no code for this response. I was expecting David to tell me about his world, not find that I had become part of it.

2. The researcher is more than a recording instrument

Upon my return to New Zealand in 1988, I had the opportunity to spend time in the rural and somewhat remote Hokianga district of the northernmost region of the country. Here, a primary care-based health system serves the needs of patients at ten community clinics. Deciding to try and discern their social significance as gathering places, I visited each clinic and attempted to blend into the gathering of attendees who were mainly waiting to see the doctor. My improvised strategy was to try and inconspicuously observe events under the guise of completing the crossword puzzle in the daily newspaper. I had only vaguely heard of participant observation at the time, but the approach seemed to make sense. However, in this predominantly Māori area, I was clearly *Pakeha* ('white') and a visitor. On one occasion all others present in the waiting room were both locals and Māori. They clearly knew each other well.

It was immediately evident that going to the doctor involved more than medical interactions. The occasion appeared to provide an opportunity for locals to exchange news, tell stories and reflect not only on their own well-being but also on that of their families and friends. I observed patients arriving well before their appointment time (sometimes in the company of others who had no intention of consulting the doctor or nurse) and lingering afterwards 'having a yarn'. My embodiment as a researcher was central to the construction of this knowledge. Within the waiting area, I had to position myself in such a way as to neither be obtrusive (and thus inhibit conversation by gazing at others) nor be overly welcoming of conversational engagement (hence my 'hiding' behind a newspaper). This delicate balance was bound to break down and eventually did.

On one occasion, two locals offered to help with the crossword puzzle I was half-heartedly completing and, on another, a *kuia* (older Māori woman) entered, kissed, and welcomed all present to 'her' clinic. She walked around the circle of seats in the tiny room and I could feel her getting closer. I had no option but to put down the newspaper, stand up and embrace this elder. My embodied presence within the observed arena of social interaction clearly rendered binary constructs of researcher/researched and subject/object thoroughly permeable. In this *kuia*'s clinic, my conceptions of being an 'autonomous self' were (re)embodied within their rightful web of social relations (adapted from Kearns 1991; Kearns 1997).

3. Encountering Beth

As part of a postdoctoral project on return to Auckland, I interviewed community-based patients who had formerly been resident of the city's large, old and recently-closed psychiatric hospital. In accordance with the study's ethical agreement, psychiatric nurses established the willingness of potential participants to be interviewed and then passed onto me their contact details. When it came to contacting Beth, she asked me to meet her at her workplace. Her job, she said, was at 'Val's Place' on 'K Road'. From the street name, I immediately surmised it was part of the sex industry for which that street was, at the time, well-known.

I felt uneasy and was tempted to courteously decline this interview. But in fidelity to the research design and Beth's willingness to talk, I summoned courage go to her work-place. This was aided by asking my partner to come along. Neither of us had entered such a place before. On arrival, we were met with brusque curiosity by the respectable-looking woman at reception and told we would find Beth in the parlour. In the low light, she came towards us hesitatingly and led us up into an equally dimly lit room. It was furnished with a bed and a bath, along with whips and skulls with incense sticks protruding from their eye sockets. I followed procedure and asked her to sign a consent form. She reluc-tantly did so but asked why I had brought someone else. She said that having someone else with me made it look like we were inspectors. I sat, anxiously, perched on the edge of the bed, asking what suddenly seemed to be very banal questions. All the while, I could feel the beads of sweat rolling down my back. "Most of the girls here are from the hos-pital", she said.

I descended the stairs with my partner, having omitted some of the questions on the interview schedule. I wondered to what extent I had made things more difficult for her by bringing someone along whose company was to make me a degree more comfortable.

Two 'Turns' in Light of Field Experience

In this second part of this chapter, I focus on two contemporary turns and how they shape quali-tative methodologies in human geography. Underlying this section is an acknowledgement that the foundational methods in human geography have involved (either or both) questionnaires and interviews. At the heart of both these methods of data collection is a dynamic expressed in psychology as SR: the researcher provides a *stimulus* and reasonably expects a *response*. This formula was at work in my first-year psychology course when we fed grains of wheat to rats in return for compliant behaviour; so too, it mirrors the very linear chain reaction experienced when playing pool: a carefully aimed knock with a well-chalked cue leads one ball to hit another, ending up successfully in the pocket. In part, what makes skilled pool playing possible is the absolutely level felt surface and the totally round and smoothly surfaced balls. To expect compliant and easily categorisible outcomes of asking people questions was always going to be at best illusory: people neither live on a level playing field of life's circumstances, nor are they of homogenised predictable characteristics.

As the unrealistic assumptions of behavioural geography were increasingly acknowledged, the 'semi-structured interview' became the method of choice for its flexibility and capacity to be as much conversational as interrogative. Yet, despite its ongoing orthodoxy among those participating in the 'qualitative turn', there has been growing dissatisfaction with the potential of interviews that feature 'tell me about...' prompts to satisfactorily give voice to all that is *really* happening in worlds people are experiencing.

The two turns I discuss here are the embrace of embodiment and the visual in method-ology. Both are at least indirectly an outgrowth of the influence of humanism and the emphasis on human agency in geography. Other roughly contemporaneous philosophical vantage points include behaviourism (epitomised by stimulus-response thinking) and versions of pol-itical economy, inspired by Marxism, that are reliant on mechanistic assumptions of conduct. Both these vantage points, stressing retrospectively the individual and societal structures, were critiqued in the 1980s by proponents of humanistic thinking (Buttimer, 1999; Cresswell, 2013). Humanism was less a coherent philosophy and more a collection of ideas centered on acknowledging the self-evident humanity of individuals (Ley and Samuels, 1978). A recognition of agency is key: people can be creative (or destructive) as well as (potentially, at least) reflective and morally aware (Flowerdew and Martin, 2005). The consequence of acknowledging the full scope of being human – inclusive of intentions, emotions, and dispositions – is an inevitable moving on from any method that reduces or limits the humanity of the research encounter and the data it yields. Within an uncompromising political-economic analysis, people are conceived as being the – sometimes unaware – bearers of structures. In contrast, within the humanistic viewpoint people are the creators of their own structures; not necessarily with complete freedom, but within the constraints imposed within society and economy (the constrained choice proposed by Giddens, see Dyck and Kearns, 2006). Hence, someone walking to work might choose a route perceived to be the safest or most aesthetically pleasing and replete with memories rather than the most logical and direct route. In doing so, they are expressing their feelings and inclinations, and these can be accessed in a research encounter through walking with them or asking them to draw aspects of their walk.

The complexity, diversity and nuance in peoples' embodied, sensory and emotional experience of the world makes, in retrospect, the uni-dimensionality of the assumptions of behavioural geography blatantly naive. What allowed the necessary separations of lived reality and research assumptions was the adoption of what were almost cardboard cut out 'models ' of humanity, the *homo economicus* (economic man), for instance, who rationally makes purchases and travels according to carefully weighed up profit-incentivised considerations. Yet, recently, there has been an increasing interest in a self-conscious inclusion of the body in the field encounter, being with the participant and seeing the encounter as a moving experience.

1. The embodied engagement turn

My earlier anecdote of a thwarted attempt to be inconspicuous in a rural waiting room setting was fundamentally reflected in my naiveté in thinking that my own bodily difference and comportment could be somehow removed (rather than fully incorporated into) the field setting. In the years since that encounter, the body has been firmly brought into geography through studies of experiences such as fatness (Longhurst, 2005). One way the body has been more widely acknowledged and actively included in research encounters is through mobile methods and, specifically, 'go along' interviews. These approaches can be regarded as part of a turn towards a creative intersection between mobilities thinking (Cresswell, 2010; Hannam et al., 2006) and acknowledgement of methodological embodiment (Vannini and Vannini, 2018). Mobile methods are research interactions that involve accompanying a participant on a small journey or more generally talking while being in motion (Buscner et al., 2011; Hein et al., 2008) This approach offers heightened attention to the sometimes ephemeral and relational encounters involving both people and the non-human world which together co-constitute geographies of everyday life in, and through, place (Riley et al., 2021).

As a subset of mobile methods, 'go-along' interviews are a fieldwork approach in which participants walk (or, more recently swim, or cycle) the researcher through their place experiences (e.g., Paren, 2016). After pioneering work developing and extending walking interviews as a mobile method (Carpiano, 2009), a much broader range of mobile methods has more recently been embraced. In one such innovation, the quest to understand wheelchair basketball (an adaptive version of the sport played by participants experiencing a range of abilities and disabilities) was addressed by the researcher accepting an invitation to be a guest on a team and playing her first game as part of fieldwork while engaging with other players on the court (Bates et al., 2019).

The distinctive quality of 'go-alongs' is that, according to Foley et al., (2020, 43), they offer a window into the nuanced and diverse 'embodied, emotional, and physical transformations that unfold as people transition through the networked spaces and places that constitute everyday life'. This expression of the embodied 'turn' in methods can be aligned with the wider mobilities 'turn' within the social sciences at large. As an early proponent, Carpiano (2009) indicated that observation and questioning while engaged in embodied movement allows an examination of practices within mutually acceptable paces and places.

These mutual engagement approaches counter the doomed-to-fail efforts of inconspicuousness epitomised by my anecdote at the rural clinic example earlier. It is the immediacy of these approaches that potentially appeals to both reseachers and particpants. Just as our bodies rarely have much of a delayed reaction to sensory and emotion-triggering inputs, so too this expression of the 'embodied turn' allows for a witnessing of an encounter in the moment of experience rather than a seeking of recollections (Anderson and Jones 2009). The advantages of such embodied immediacy in mobile methods include the dissolving of some of the barriers between researcher and participants (as, for example, joining players on a wheelchair basketball court), the collection of multi-sensory insights from being with the fully-embodied and engaged participant (Bell, Leyshon, and Phoenix 2019), and removing some of the pressure to engage in conversation. The melding of concerns for embodiment and emotion leads Vannini and Vannini (2017) to say that the challenge is to move beyond any vestige of undertaking interviews 'on the move' and rather seek a deeper immersion in the dynamics of place through trying to feel, sense and encounter.

One route to more deeply engaging with place within the domain of mobile methods is to be attentive to the matching of researchers and participants. In a study of mobilities and experience of the urban landscape, Carroll et al., (2015) asked child participants to accompany local high-school students who trained as interviewers on neighbourhood walking interviews. These children became key informants and co-producers of knowledge. They reported on their neighbourhoods, discussed what they liked and disliked, spoke of safety concerns, and offered suggestions for more 'child-friendly' neighbourhoods. In light of my earlier recollection of the presenter upturning the overhead projector, this approach of accompanying participants and, in the case of Carroll et al., (2015), recruiting age-appropriate research assistants, enables the seeing of an alternative vantage point. Walking with, rather than asking questions of, participants constitutes a methodological movement forward; a turn to more intimately engage with people's everyday experiences.

2. The visual turn

A key element of my recollection of that encounter with David, the psychiatric patient in inner-city Hamilton, was of him doodling on the consent form I'd given him. I cannot recall what he drew, but I wished I had paid more attention to it. That drawing may have had some bearing on my interpretation of what he was saying in his hesitant way. In this section, I examine a second 'turn' within the broad field of qualitative methodologies: the visual turn. Along with co-authors, I recently posed a question: "We talk metaphorically about 'drawing conclusions' and 'sketching ideas' but how often do we consider the benefits of asking the people we encounter as we conduct research projects to use images as well as words?" (Kearns et al., 2021, 113).

Visual methodologies concern seeing and acknowledging the potential importance of what can be seen. A long-held saying is 'seeing is believing', and yet there is more to the visual than simply seeing. This can be observing what is already apparent in the landscape, or playing an active role in crafting a non-verbal depiction of something. Arts-based methods engaging with the visual involve a 'handing over' of creative expression to participants encountered in a research project. A common approach would be to elicit a visual depiction of some element of a place or landscape and then ask participants in follow-up interviews to comment on the drawings or sketches they produce during an initial encounter. In a follow-up encounter, participants have frequently been noted as offering more detailed accounts of the memories and experiences they originally shared (Kearns et al., 2021). Art-based research can deepen participants' involvement in a project, allowing the information gathered to be checked and trigger new memories while inviting new insights.

To embrace the visual turn is to depart from expected routines and interactions in fieldwork. This is because various versions of the question-and-answer format of the interview are well-rehearsed in everyday life (e.g. they are evident in situations ranging from market research telephone calls to being questioned by a doctor). To ask someone to draw a picture requires more explanation and is invariably a less common request. Yet drawing as method is increasingly entering the geographical repetoire of methods (Coen, 2016). Drawing has two-fold benefits: it can produce transferable representations and also be a way of 'opening up' encounters and conversations between researchers and their study participants. Engaging in the act of drawing, watching others draw, or looking at finished illustrations, all force us to be watchful for both minute details and the 'larger picture' in which place and self are embedded (Brice, 2018; Eggleton et al., 2017). Through the act of drawing, it is possible to elicit information that might otherwise be avoided, ignored, assumed or devalued in the context of other research strategies, such as photo-based methods. For instance, a study participant might feel self-conscious to verbally express a deep emotion like fear. Fear is a feeling that is not easily photographed. Yet, through choice of colours (e.g., dark shades) it can be represented and then discussed (see Eggleton et al., 2017). This turn to embrace the visual is, in a sense, an admission that while geography might literally be a writing of our worlds, words are a necessary but not a sufficient way of accounting for the geographies of everyday experience.

Of course, every people-facing method has ethical dimensions. In the case of visual methodologies, there is an ever-present risk that a participant may feel uncomfortable being asked to draw and letting others see what they have drawn. In the absence of words, we can sometimes sketch feelings from deep places in our consciousness that potentially speak to past traumas or difficult experiences. One solution reported by Eggleton et al., (2017) is for the researcher themselves to draw with the participant and then discuss each other's drawings. Again this is an example of a turn away from the implied us/them binary that I so acutely felt as I walked into the 'massage parlour' to interview Beth and as I began my interview with David about his daily life as a community-based psychiatric patient. One imperative for this turn to visual methods is ethical. Inviting participants to offer insights into their lives and experiences by creative visual means can assist in allowing the people we encounter to feel safe, included and valued. This is because, as the anecdote concerning Beth indicated, people can find interview schedules intimidating, given the stimulus-response dynamic discussed earlier that underlies this form of engagement.

Conclusion

Research encounters are structured not only by the settings in which they are conducted, but also by the vantage point and worldview of the researchers. This chapter has reflected on the idea of methodological turns, sketching a series of encounters that precipitated my own turns in perspective, and examining two turns that have occurred in the course of qualitative human geography: those towards the mobile body and the visual. I have argued that the move away from stimulus/response-driven methods represents a turn towards more fully human encounters. The more close-up encounters we have with participants in field settings, the more nuanced our methods need to be. The turns – both personal and within wider communities of geographical practice – I have considered in this chapter are, in effect, responses to a recognition that the social world we enter as researchers is complex. Nevertheless, the rewards for stepping out of our comfort zones and engaging with people away from the familiar dynamics of questionnaires or interviews can yield rich dividends.

In contemporary geography, methodological novelty is in part driven by a sense of progress but also fashion (Barnett, 2002). New ways of approaching fieldwork can 'catch on' and become popular, fuelled by researchers' desire to reflect this novelty in their publications and be at the 'cutting edge' of a trend. A further impetus is technological. Cellphones, for instance, have facilitated the near-universal opportunity to take photographs and record conversations while simultaneously noting locations.

The turn away from the 'tunnel vision' of necessarily focusing on only specific behaviour or perceptions in human-geographical research has, of course, come with costs. The losses are the power to explain and even predict, through the accumulation of tidy enumerated data sets that can be subjected to statistical analysis. Researchers also require a more extended period of time and attention with participants than would be the case as 'respondents' in the brief commitment of a questionnaire or interview-based study. Ultimately, as Flowerdew and Martin (2005) point out, most research methods are, however, constrained by time limits, expenses involved, and institutional issues such as compliance to ethics and health and safety requirements.

The anecdotes I offered earlier in the chapter resonate with Becker's (1965) comment that despite plans, research is ultimately designed in the course of its execution. In other words, our personal moments of epiphany in fieldwork can be productive in suggesting or endorsing more widespread turns in practice. Key examples of embodied mobility and engaged visual methods have illustrated examples that have walked the methodological talk, and shown that the old proverb is at least in part the case: that a picture can paint, if not a thousand, then at least many words.

In conclusion, there is a need to find a balance between creativity and conformity. If methods become overly subjective, then do the findings of one's research become too challenging to help make a difference in the world? In whatever turns we embrace as geographers, there is an ongoing need to ask ourselves whether we are achieving not only insight and ethically-sound engagement but also relevance in the world of which we are a part.

References

Afonso, A.I. and Ramos, M.J. (2004) 'New Graphics for Old Stories', In Alfonso, A., Kurti, L. and Pink, S. (Eds) *Working Images: Visual Research and Representation in Ethnography*, Routledge, pp. 72–89.

Anderson, J., and K. Jones (2009) The Difference that Place Makes to Methodology: Uncovering the 'Lived Space' of Young People's Spatial Practices. *Children's Geographies*, 7 (3), 291–303.

Barnett, C. (2002) The Cultural Turn: Fashion or Progress in Human Geography? *Antipode*, 30(4), 379–394.

Bates, L., Kearns, R., Witten, K. and Carrroll, P. 'A Level Playing field': Young People's Experiences of Wheelchair Basketball as an Enabling Place. *Health & Place*, 60.

Becker, Howard S. (1965) Review of *Sociologists at Work: Essays on the Craft of Social Research* by Philip E. Hammond. *American Sociological Review* 30(4), 602–603.

Bell, S. L., C. Leyshon, and C. Phoenix. (2019) Negotiating Nature's Weather Worlds in the Context of Life with SightImpairment. *Transactions of the Institute of British Geographers*, 44(2), 270–283. doi:10.1111/tran.12285

Béneker, T., Sanders, R., Tani, S. and Taylor, L. (2010) Picturing the City: Young People's Representations of Urban Environments, *Children's Geographies*, 8(2), 123–140.

Brice, S. (2018) Situating Skill: Contemporary Observational Drawing as a Spatial Method in Geographical Research. *Cultural Geographies*, 25(1), 135–158.

Buscher, M., Urry, J. and Witchger, K (Eds.). (2011) *Mobile methods*. Routledge: London & New York.

Buttimer, A. (1999) Humanism and Relevance in Geography. *Scottish Geographical Journal*, 115(2), 103–116.

Carpiano, R. (2009) Come Take a Walk with Me: The 'Go-along' Interview as a Novel Method for Studying the Implications of Place for Health and Wellbeing. *Health and Place*, 15, 263–272.

Carroll, P., Witten K., Kearns R. and Donovan P. (2015) Kids in the City: Children's Use and Experiences of Urban Neighbourhoods in Auckland, New Zealand, *Journal of Urban Design*, 20(4), 417–436, DOI: 10.1080/13574809.2015.1044504

Coen, S.E. (2016) What Can Participant-Generated Drawing Add to Health Geography's Qualitative Palette? In N. Fenton & J. Baxter (Eds.), *Practicing Qualitative Methods in Health Geographies* (pp. 131–152). Abingdon, Oxon & New York, NY: Routledge.

Cresswell, T. (2010) 'Towards a Politics of Mobility', Environment and Planning D, 28(1), 17–31.

Cresswell, T. (2013) *Geographic Thought: A Critical Introduction*. Chichester: Wiley-Blackwell.

Dyck, I. and Kearns, R.A. (2006) Structuration Theory: Agency, Structure and Everyday Life. In: *Approaches to Human Geography* (edited by G. Valentine and S. Aitken). Sage, Thousand Oaks, 86–97.

Eggleton, K., Kearns, R., and Neuwelt, P. (2017) Being Patient, Being Vulnerable: Exploring Experience of General Practice Waiting Rooms Through Elicited Drawings. *Social and Cultural Geography*, 18(7), 971–933.

Flowerdew, R. and Martin, D. (eds.) (2005) *Methods in Human Geography: A Guide for Students Doing a Research Project* (2nd Ed.). Pearson/Prentice Hall, London.

Foley, R., Bell, S.L., Gittens, H., Grove, H., Kaley, A., McLauchlan, A., Osborne, T.C., Power, A., Roberts, E., and Thomas, M. (2020) "'Disciplined Research in Undisciplined Settings': Critical Explorations of In-Situ and Mobile Methodologies in Geographies of Health and Wellbeing." *Area*, 52 (3), 514–522.

Hannam, K., Sheller and Urry, J. (2006) Mobilities, Immobilities and Moorings, *Mobilities* 11, 1–22.

Hay, I. (2016). *Qualitative Research Methods in Human Geography*. 4th Edition, Oxford University Press.

Hein, J., Evans, J., and Jones, P. (2008) Mobile Methodologies: Theory, Technology and Practice. *Geography Compass*, 2(5), 1266–1285.

Kearns, R.A. (2016) Placing Observation in the Research Toolkit. In *Qualitative Research Methods in Human Geography* (edited by I. Hay), 4th Edition. Oxford University Press, Don Mills, Canada, Melbourne, 313–333.

Kearns, R.A., Eggleton, K., Van der Plas, A. and Coleman, T. (2021) Drawing and Graffiti-based Approaches. In *Creative Methods for Human Geographers*. Eds Von Benzon, N. and Holton, M. Sage, London, pp. 113–125.

Kuhn, Thomas (1962), *The Structure of Scientific Revolutions* (1st ed.). Chicago, IL: The University of Chicago Press. ISBN 9780226458113

Kusenbach, M. (2003) "Street Phenomenology: The Go-Along as Ethnographic Research Tool." *Ethnography*, 4, 455–489.

Ley, D., & Samuels, M. (Eds.). (1978). *Humanistic Geography (RLE Social & Cultural Geography): Problems and Prospects* (1st ed.). Routledge. https://doi.org/10.4324/9781315819655

Longhurst R. (2005) Fat bodies: developing geographical research agendas. *Progress in Human Geography* 29(3), 247–259. doi:10.1191/0309132505ph545oa

Parent, L. (2016) The Wheeling Interview: Mobile Methods and Disability. *Mobilities*, 11: 521–532.

Riley, M., Turner, J., Hayes, S. and Peters, K. (2021) Mobile Interviews by Land, Air and Sea. In *Creative Methods for Human Geographers*. Eds Von Benzon, N. and Holton, M. Sage, London, pp. 141–152.

Rose, G. (2007) *Visual Methodologies: An Introduction to the Interpretation of Visual Materials*. London: Sage.

Vannini P, Vannini A. (2018) These Boardwalks Were Made for Bushwalking: Disentangling Grounds, Surfaces, and Walking Experiences. *Space and Culture*, 21(1), 33–45. doi:10.1177/1206331217749127

Young, L. and Barrett, H. (2001) Adapting Visual Methods: Action Research with Kampala Street Children. *Area*, 33(2), 141–152.

6

FOR AN INTERSECTIONAL SENSIBILITY

Feminisms in Geography

Karen Falconer Al-Hindi[1] and LaToya E. Eaves

Introduction

Intersectionality is perhaps the pre-eminent exemplar of subjugated knowledge that has arisen from lived experience and found its way into the academy. Intersectionality's earliest formulation was not as research methodology, as such; rather, it began as Black women made sense of their lives under oppressive conditions of enslavement, anti-Black racism, and sexism. Intersectionality inherits more than a century of Black women theorizing, activist organizing, and political intervention, from Anna Julia Cooper's assertion to confront both the "race problem" and the "woman question" (1892) to the concept of "identity politics" from the Combahee River Collective (1977, as cited in Taylor, 2017). From these origins, intersectionality-like thinking today informs much womanist, feminist and queer scholarly work, including much feminist geography. Intersectionality is a power-full idea that has been described variously as a theory, as a methodology, a method, and as a concept. We argue that intersectionality can inform any and all research methodologies, laying bare the ways in which researchers and the research enterprise are always embedded in their historical, social, and geographic contexts. In this chapter we emphasize intersectionality as methodology, always in conversation with epistemology (ways of knowing) and ontology (ways of being). It is as true of feminist geography, as it is of any research endeavor, that research methodologies are intimately connected with epistemology and ontology.

For the chapter, we define intersectionality as a conceptual framework through which complex and cumulative relations of power are analyzed to reveal and redress structures of oppression embedded in/across place and scale. Importantly, the definition does not erase or divorce the term from the Black feminist and womanist trajectory through which intersectionality appears (see Text Box 6.1).

DOI: 10.4324/9781003038849-8

Text Box 6.1: Intersectionality

For the chapter, intersectionality operates as a conceptual framework for understanding compounding relations of power across place and scale. We take seriously Patricia Hill Collins' (2015) wariness of imposing an "imperial definition" of intersectionality so as not to depoliticize its origins, to resist its use to a reductive variable, or to contain its promise for geographic methods and thought.

The origins of intersectionality are found within Black women advocating in defense of their own lives over the past two centuries. Their participation in abolition and anti-slavery campaigns simultaneously challenged ideas of proper womanhood as well as the presence and impact of racism in their lives. As described in the chapter, Sojourner Truth's "Ain't I a Woman" (see Davis, 1983) incorporates a conceptual framework from which intersectionality emerges over a century later. Likewise, Anna Julie Cooper's (1892) *A Voice from the South* draws attention to what she terms "the race problem" and "the woman question" in the eradication of oppression. These articulations continued through the suffrage movement and became explicit once more in the wake of twentieth-century liberation movements, which yielded organizations such as the Black Women's Liberation Committee – later changed to the Third World Women's Alliance – and the Combahee River Collective in response to the racism-sexism of movement life. It is then that the terms "double jeopardy" (Beal, 1970/2008 and "identity politics" (Combahee River Collective, 1977/2017)) are introduced to conceptualize how Black women's knowledges and perspectives can provide frameworks for analyzing and dismantling systems of oppression. Just over a decade later, Crenshaw (1991) writes about how discriminatory practices were legitimized even within sites meant not to be so (i.e. access to shelters and services for domestic violence survivors) and names such thinking as intersectionality.

Non-Black feminist scholars entered the conversation around intersectional sense-making and praxis relatively recently, and feminist geographers even more so. Since the inception of feminisms in academic work, scholars have been concerned with *how we know what we know* and what research methods and techniques will accurately represent the world of experience in scholarly knowledge in order to make progressive change in the world (Harding, 1987; Burt and Code, 1995; Jaggar, 2008). Every research method and technique has an origin story, and feminist geographers tend to be careful as they engage with research methods both to use the methods appropriately and to respect the communities that they research. Such care is all the more important with intersectionality, as it folds these vital considerations together, and comes out of a specific subordinated community at the same time that it makes specific demands of the researcher (Hopkins, 2019). Further, as Black feminism is inherently intersectional, we *center* Black/intersectional feminism (positioning white feminism on the periphery) to argue that only the former can anchor a truly feminist geography (Falconer Al-Hindi, 2020). In this chapter, after Crenshaw (1991), Healey et al., (2011), Mollett and Faria (2018) and Collins and Birge (2020), we argue for an 'intersectional sensibility' in human geography methodologies.

Text Box 6.2: Glossary

a) Black feminism: Black feminism centers the lives and concerns of Black women and gender expansive people, including alliance with Black men.

b) Feminism: Feminism is a term with a neither "stable nor fixed" meaning (Moses 2012). Although bell hooks' definition – "a movement to end sexism, sexist exploitation, and oppression" (hooks, 2000) – is widely accepted, feminism is often associated with upper- and middle-class white women and their interests (hooks 1981).

c) Identity politics: A term introduced by the Combahee River Collective (1977): "This focusing upon our own oppression is embodied in the concept of identity politics. We believe that the most profound and potentially most radical politics come directly out of our own identity, as opposed to working to end somebody else's oppression" (p. 33 as cited in Taylor 2017).

d) Race: Race is both a construct and concept: "which signifies and symbolizes social conflicts and interests by referring to different types of human bodies" (Omi and Winant 1994, p. 55). As such, what is considered racial difference varies over space and time.

e) Third World women: Those who, through birthplace and/or culture, have significant attachments to the Global South. They may live in the U.S.A. or countries of the Global South.

The chapter takes a chronological and historiographic approach to intersectionality, methodologies, and feminist geography, weaving them together with considerations of Black women's thinking, lives and activism; geography's embeddedness in Euro-centric empire-building; and white women's involvement in all three (see key terms in Text Box 6.2). The chapter begins with a section each on three key periods in intersectionality's development alongside research methodologies: the early years (1830–1920); twentieth-century revolutions including the women's liberation movement, the LGBTQ rights movement, and the Third World women's movement (1960–1985); and intersectionality (1991–present). Next comes the argument for an intersectional sensibility in feminist geography's methodologies, accompanied by exemplary studies employing a variety of methodologies. The conclusion emphasizes the value of an intersectional sensibility for feminist geography.

Intersectionality-like Thinking: The Early Years

Intersectionality, like the discipline of geography, has material roots that are intertwined with the legacies of European colonization and global capitalism, including the era of the transatlantic slave trade. But while geography developed by assisting these enterprises, intersectionality emerged from Black feminist and womanist thinking and activism in opposition to them. Intersectionality's evolution over more than a century has prioritized Black women's standpoints, which have always engaged the presence and impact of structures of oppression (Collins, 2009), including how Black women's liberation has been wrapped up with and often accorded second place to that of white women and Black men. Black women's pursuit of the right to vote, and their enfranchisement having been pushed aside by Black men and white women in pursuit of their own voting rights, is a case in point (Davis, 1983). Writings and work by Black feminists and womanists also provided insight into the socio-spatial organization of place, centering the overlapping sites that sustain spatial norms and uphold structural oppression. Ida B. Wells-Barnett's work on lynching spaces illustrates this well (Wells-Barnett 1895/2005).

The earliest formations of Black feminisms can be traced to the abolition movement in the 1830s (Yee, 1992). Black women engaged in anti-slavery campaigns that challenged the politics of "respectable" womanhood of the time as well as the racism within the abolitionist movement itself (Davis, 1983; Yee, 1992). One of the most prominent figures of Black feminist thought and activism is abolitionist and evangelist Sojourner Truth (1797–1883). Born into slavery in New York state, Truth's activism around dismantling slavery and racial-sexual domination was anchored by her quest for her own freedom and her children's. Truth delivered a speech at the 1851 Ohio Women's Rights Convention, later to be known as "Ain't I a Woman." In it, grounded in her personal, embodied experiences:

> I have ploughed and planted, and gathered into barns, and no man could head me! And ain't I a woman? I could work as much and eat as much as a man - when I could get it - and bear the lash as well! And ain't I a woman? I have borne thirteen children, and seen most all sold off to slavery, and when I cried out with my mother's grief, none but Jesus heard me! And ain't I a woman?
>
> *Davis, 1983 p. 61*

With the refrain "Ain't I a woman?", Truth disrupted male-supremacist ideas of women as weak and, therefore, lacking fitness for equal voting rights. At the same time, she commanded white women's attention to her lived experience, which was very different from, yet intimately tied to, theirs by the conditions of slavery and anti-Black racism in the country (Davis, 1983).

Truth is among several prominent Black women of the nineteenth century whose revolutionary politics engaged what scholars now call intersectionality. Such 'intersectionality-like' (Hancock, 2016, in Hopkins, 2019) thinker-activists include Maria W. Stewart, Mary Church Terrell, Ida B. Wells, Harriet Tubman, and Anna Julia Cooper, who all advocated for dismantling the interlocking structures of racism-sexism through wide-ranging campaigns including the abolition of slavery, anti-lynching activism, women's suffrage, access to education, and labor rights. Black women's participation exposed the overlapping structures of racial-sexual domination within each area of human and civil rights and revealed the details of disenfranchisement.

Academic geography's origins in colonization included the labor of white women, or 'lady explorers,' as they were often known (Domosh, 1991). Their contributions to knowledge, in common with those of Black activists, were the product of direct experience:

> As a lady travelling alone, and the first European lady who had been seen in several districts through which my route lay, my experiences differed more or less widely from those of preceding travellers; and I am able to offer a fuller account of the aborigines of Yezo, *obtained by actual acquaintance with them*, than has hitherto been given.
>
> *Bird, 1985 [sic], pp. 1– 2, in Domosh, 1991, p. 99; emphasis added*

The questions with which they approached their travels produced information consistent with their being, in Kimberlé Crenshaw's floor/ceiling metaphor, 'immediately beneath the floor' of their upper-class, white, male peers. Feminist geography is ultimately a product of settler-colonial societies' academic enterprises, and of the ontologies and epistemologies of those so engaged. As a result, "the multiple projects of oppression and exploitation that shaped the academy over earlier centuries are still very much with us today" (Oswin 2020, p. 14). Women's exclusion from geographical societies and from prestigious scholarly positions persisted well into the twentieth century and had the effect of channeling their ambitions into teaching until the social movements of the 1960s and 1970s.

Twentieth-century Revolutions (1960–1985)

While feminist, womanist, women's, gay and lesbian and queer social movements in the U.S. have complex histories and extensive spatial roots, they have in common that each coalesced to some degree during the 1960s and 1970s (Eaves, 2019). Among the reasons for increased solidarity were individuals' experiences of discrimination in the anti-Vietnam War, Civil Rights, and other social movements. Multi-racial, multi-ethnic coalitions of (straight) women, lesbians, and gay men learned – often through consciousness-raising – that they had these experiences in common (Boxer, 1982). Such experiences often arose from and could only be made sense of through bodily experience. Women's bodies in all their diversity, including those marked by racialization or sexual difference, drew reactions that generated knowledge for the person inhabiting that body. The Combahee River Collective, for example, wrote "Black women have always embodied … an adversary stance to white male rule" (1977, in Taylor 2017, p. 29). Embodied as they are, Black women inherently challenge white supremacist patriarchy. A related route to knowing is sensory knowledge, gained through bodily experiences; for instance, feelings of pleasure, pain or even disfigurement. Physical beatings inflicted by San Francisco police officers on transwomen galvanized transgender activism (Currah, 2008). One outcome of the activist movements fueled by an awareness of transphobia's depths was the carving of inroads into the North American academy (regarding women's studies, see Moses 2012; regarding transgender studies, see Khan and Franklin, 2020). The social movements that brought people from marginalized groups onto historically white and male university campuses in the 1960s, touched geography as well. As part of the social justice activism of the 1960s, women along with many progressive men began arguing that more attention be paid to women's lives in geographic space (Zelinsky, 1973).

Crucially to the argument we make in this chapter, despite women's studies' roots in diverse feminist, Black feminist, Third World women's and women's liberation movements, most of these programs and departments of scholarship and activism acclimated and accommodated to the requirements and strictures of the white supremacist institution that was and is U.S. higher education (Nash, 2019). Many foregrounded the concerns of second wave (white) feminism – equal pay, sexual harassment, and so-called reproductive choice – and crowded out knowledge arising from the lives of people who lived with multiple oppressions. The language of 'choice' as part of the reproductive rights movement illustrates how a failure to think intersectionally and to act inclusively impoverished feminist analyses and many women's studies programs. A narrow focus on abortion access ignored needs for freedom from coerced sterilization as well as access to birthing and parenting options for women of color, incarcerated, and impoverished womb-bearing folx (Ross, 2017).

As a result, women's studies programs and departments, journals, and the National Women's Studies Association (the major professional and scholarly association for women's studies) soon drew trenchant criticism from many directions, including from Black and Third World women articulating intersectionality-like positions. Audre Lorde, for example, expressed her intersectional critique as she argued that "there is no hierarchy of oppressions" (1983), and Alice Walker labelled her stance and that of many U.S. American Black women toward (white) feminism's single-axis focus "womanism," which is feminist at the same time that it stands in solidarity with Black men. Further, the appearance of work on pedagogy and research related to Black women and intersectionality-like positions, including the germinal anthology *All the Women are White, All the Blacks are Men, But Some of Us are Brave: Black Women's Studies* (Hull, Bell-Scott, and Smith, 1982), provided further critique to the development of women's studies. Geography too, and feminist geography in particular, was decried for perpetuating the myth of "universal womanhood" that glossed over the lived experiences of racialized and class-diverse women (Sanders, 1990, p. 229).

Black feminist and womanist activism engaged in a politics of interconnected struggle that extended beyond U.S. states into its territories as well as into solidarity with other spaces of the colonized world. For example, the Black Women's Liberation Committee, of the Student Nonviolent Coordinating Committee (SNCC), was renamed the Third World Women's Alliance (TWWA) to recognize the collaboration and solidarity with Puerto Rican women, then morphing into a multiracial collective, in the fight against racism, sexism, class-based oppression, and imperialism (Third World Women's Alliance 1969). TWWA and other collectives were influenced by writings, including Frances M. Beal's pamphlet "Double Jeopardy: To be Black and Female." Beal's argument centers the need to understand Black women's positionality and oppression in order to imagine new worlds. She closes the pamphlet by saying, "Black women must take an active part in bringing about the kind of world where our children, our loved ones, and each citizen can grow up and live as decent human beings, free from the pressures of racism and capitalist exploitation" (Beal 1970/2008).

Along with the publication of Beal's pamphlet, a group of Black feminists formed the Combahee River Collective and produced "A Black Feminist Statement" in 1977, which has become "a touchstone for black feminist engagement with intersectionality's histories" (Nash, 2019, p. 7). In it, they argue for liberation through dismantling interlocking systems of oppression:

> The most general statement of our politics at the present time would be that we are actively committed to struggling against racial, sexual, heterosexual, and class oppression, and see as our particular task the development of integrated analysis and practice based upon the fact that the major systems of oppression are interlocking. The synthesis of these oppressions creates the conditions of our lives.
>
> *Combahee River Collective 1977 in Taylor 2017, p. 28*

U.S. feminism's third wave took these and other criticisms of the second wave on board to generate a gender-, sexuality-, race/ethnicity/nationality, and disability-inclusive, sex-positive movement. Kimberlé Crenshaw's work emerged at precisely the right time and intersectionality "emerged as one of the most important insights of third-wave feminist scholarship and activism" (Grzanka, 2019, p. 1).

Intersectionality (1991–present)

The 1980s ushered in neoliberal policies that circumvented many of the advances of the 1960s and 1970s, focused on the idea of equality via access – to the labor force (particularly for white women), to education (following desegregation), and reproductive rights (including the *Roe v. Wade* decision in the U.S., providing the right to abortion). Moreover, it marked an official shift from the civil rights era into a time where 'colorblindness' became the crux of political policy. Keeanga-Yahmatta Taylor (2016, p. 52) argues, "Instead, 'colorblindness' aided politicians in rolling back the welfare state, allowing Congress and the courts to argue that the absence of racism in the law meant that African Americans could not claim racial harm." The measured disinvestment in the social safety measures and political guarantees of the 1960s and 1970s liberation movements is striking. Within this context Black women's analysis of their lives is named: intersectionality. Intersectionality emerged as a theoretical tool from critical race and juridical scholar Kimberlé Crenshaw. In the work of naming intersectionality, Crenshaw

(1991) examined the complexities of violence against women, specifically as it relates to women of color and the issues of identity politics in relationship to legal protections.

A key element of intersectional thinking is to create an initial understanding of socio-structural difference in critical analyses. Like the Black womanist and feminist thinking before her, Crenshaw positions intersectionality as a way to reveal how multi-axis lived experiences are shaped by relationships of power and identity. Crenshaw centered Black women and explained how multidimensional analyses of Black women's identities can elucidate structural impacts that had not been revealed by mainstream feminist interventions, in her case in the differing lived experiences of women who are victims of domestic violence (Crenshaw, 1991). Another key element of intersectionality is to elucidate how overlapping and mutually exclusive marginalities impact the ways individuals and/or groups experience the sociocultural world. For Crenshaw, Black women and other women of color who are victims of domestic violence have different needs as a consequence of the history of oppressions based on race and gender. Finally, Crenshaw notes that "delineating difference need not be the power of domination; it can instead be the source of social empowerment and reconstruction" (Crenshaw, 1991, p. 1242). Therefore, a final crucial element of intersectionality involves advancing the practices of feminist and anti-racist work, "deployed to highlight unexpected coalitions" (Carbado et al., 2013, p. 311) in the pursuit of justice and liberation.

In spite of the enthusiastic taking up of intersectionality in academia, many women's studies programs remain, like the institutions of higher education of which they are a part, bastions of white supremacy (Nash, 2019; Bilge, 2020). As a result, it is no surprise that many white scholars and students are eager to, and see nothing wrong with, 'find(ing) themselves' in inter-sectional frameworks as well as applying it to others' single-axis oppression. Such a bias towards white feminism can help to explain tensions around intersectionality today, with some scholars arguing that they can use the concept to explain the experiences of people who experience single-axis oppression.

Intersectional thinking was born and thriving outside the academy long before it was taken up within it. We want to acknowledge and celebrate the deep thinking and creativity of multiply oppressed women who, in the deeply difficult circumstances of their lives, birthed such analyses rather than overly concern ourselves with "where," "when," or "who." We insist on honoring intersectionality's roots and these courageous women (Grzanka, 2019, p. 135), alongside those Black women scholars within the academy who have nurtured its development as well as advocated for its transformative potential. Intersectionality is borne of, and funda-mentally concerned with, violence and injustice – of all kinds, at all scales, over short and long durations. The stakes could not be higher for, as Ross argues, "intersectionality is the process; human rights are the goal" (Ross, 2017, p. 293). Intersectionality's foremothers sought to end violence and injustice and improve their lives and those of their communities. And as we have discussed, theory and methodology are so intimately entwined that distinguishing between them is much less the point than orienting oneself and one's research within this perspective (Grzanka, 2019; McCall, 2005).

Drawing on intersectionality to craft a methodological approach requires thoughtfulness and care. It must emphasize the study of power relations with progressive social change as a goal. From the preceding discussion, we argue for an intersectional sensibility in methodolo-gies for feminist geographers that: articulates a standpoint from which relationships of power that undergird oppression are revealed; foregrounds lives lived in multiple axes of oppression (disrupts single-axis thinking); is concerned with ending violence and injustice; and promotes the unlearning of prevailing geographical imaginaries (May, 2015). We expand on each of these tenets and detail our notion of an 'intersectional sensibility' in methodology in the next section.

For an Intersectional Sensibility in Feminist Geography's Methodologies

Today's feminist geography is a product of geographical scholarship as well as feminist thinking from within and from outside geography. Racialized and ethnic standpoints in geography have long been present but have often been under-appreciated or even ignored (Gilmore, 2002; Pulido, 2002; Kobayashi, 2006, 2014; Mahtani, 2014; Oswin, 2020). Valentine's (2007) call to feminists in geography to embrace intersectionality ushered in the current era, in which intersectionality feels omnipresent and at the same time is sometimes mischaracterized, misapplied or even misused.

In just over a decade following Valentine's (2007) challenge to feminist geographers, the concept has become nearly ubiquitous. At the same time, the embrace of intersectionality is controversial, with some scholars arguing for a post-racial interpretation at odds with intersectionality's origins and intentions (Mollett and Faria, 2018). A key concern of geographic inquiry deals with the notion of scale. Intersectionality offers geographic theorizations that interrupt hegemonic structures and deconstruct scales of power. The experiences, particularly related to structural inequalities, are made visible by the provocative value of this framework. Intersectionality also "enables feminists to focus on the social and material aspects of women's lives without abandoning modernist categories of race, class, and gender" (Bryant and Pini, 2011, p. 12). It is noteworthy that intersectionality is a necessary interjection into geographic methods, due to the legacy of enacting oppression and reifying structural inequalities through geographic practice and scientific research methods more broadly. Linda Tuhiwai Smith (2012, p. 1) writes, "It appalls us that the West can desire, extract and claim ownership of our ways of knowing, our imagery, the things we create and produce, and then simultaneously reject the people who created and developed those ideas." Geographers should be concerned with revisiting, reconsidering, and recasting their relationship with power, understanding the kind of power that research imposes and yields.

Feminist geographers have taken up intersectional frameworks in methodological and theoretical provocations, unveiling spatial processes as fundamentally intersectional dynamics. That is to say, feminist geographies,

> ...build on or adapt intersectionality to attend to a variety of context-specific inquiries, including, for example, analyzing the multiple ways that race and gender interact with class in the labor market; interrogating the ways that states constitute regulatory regimes of identity, reproduction, and family formation; developing doctrinal alternatives to bend antidiscrimination law to accommodate claims of compound discrimination; and revealing the processes by which grassroots organizations shape advocacy strategies into concrete agendas that transcend traditional single- axis horizons.
>
> *Cho, Crenshaw and McCall 2013, p. 785*

This approach is consistent with our first tenet for an intersectional sensibility, in that it articulates a standpoint that reveals relationships of power that undergird oppression (see Text Box 6.3). Cahuas (2019) takes up this approach in her research on the multifarious relationships between women and non-binary Latinx community workers and NGOs in Toronto, Canada. She centers Chicana and Latina feminist thought as the foundation for understanding the role of power and social inequities in everyday life for Latinx community workers. Her research focused on Latinx community workers' experiences of economic, mental, and physical wellbeing in relationship to NGOs, even those NGOs whose missions were to reduce harm from systems of oppression. Using intersectionality as a framework, Cahuas' approach led her to shift away from traditional semi-structured interviews, an integral technique in geography, into an approach known as testimonio methodology. The work extends Cho, McCall and Crenshaw's (2013) argument: intersectionality requires that we are innovative with methods, and methods must be shaped to the task.

Text Box 6.3: Tenets of an Intersectional Sensibility in Feminist Geography's Methodologies

- Articulates a standpoint that reveals relationships of power that undergird oppression
- Disrupts single-axis thinking
- Is concerned with ending violence and injustice
- Contributes to unlearning prevailing geographical imaginaries

---

Just as with feminism, one cannot just claim they are doing intersectionality without foregrounding the complexity of the lives at stake in research applications and analyses. Doing so can "sometimes buttress entrenched rationalities by reifying single-axis, gender-universal thinking, subjecting intersectionality to the very forms of epistemic domination it seeks to undo" (May, 2015, p. 95). As such, deepening geographic methodology to use intersectionality requires understanding "the confluence of oppressions, not merely enumerat[ing] diverse identities" (Ross, 2017, p. 288). As we have discussed, intersectionality is rooted in Black feminisms, which are rooted in Black women's struggle and radical politics.

To this end, the second tenet for an intersectional sensibility in geography methodologies is to disrupt single-axis thinking. Ellen Kohl (2019) uses her work in environmental justice governance to do this, as she centers the experiences of Black women working at the Environmental Protection Agency (EPA) in the U.S. Kohl uses "an intersectional lens to argue that although EPA career employees put on their "EPA hats," they never take off the other "hats" that constitute their identities" (p. 645). She asserts that their other "hats" inherently impact the employees' approaches to environmental justice work in the EPA. Kohl draws attention to how the EPA defines environmental justice, troubling the failure to conceptualize its aims, such as "fair treatment," and what it means to be successful in the defined work, with attention to the multidimensionality of lived experiences. Feminist and critical geographers can continue a line of practice similar to Kohl, meaning that in order to operationalize the phenomena they wish to explore through an intersectional lens, geographers should do so with attention to dimensions of oppression and vulnerability.

While traditional methodological approaches in human geography are more often qualitative, a consciousness about the impact of data collection and analyses related to questions of power-difference in all geographic approaches is imperative. **The third tenet of our intersectional sensibility in feminist geography methodologies is that the work be concerned with ending violence and injustice.** In a study of sexually- and racially-diverse men's groups that are focused on ending intimate partner violence, Peretz' (2019) intersectional analysis highlights the spatial context that makes the existence of these groups as well as their work possible. The elements of the spatial context (demographic, geographic, cultural and institutional) that Peretz identified illustrate that geographers who bring an intersectional sensibility to studies of community peace efforts can address questions of power and difference. As concern is raised about geography's pivot away from humanistic elements and towards a specific kind of data-driven spatial science, it becomes more important than ever for geographers to take up intersectional analytics. As Hamilton (2020, p. 301) suggests regarding changes to the U.S. National Science Foundation's shift to the Human-Environment and Geographical Sciences Program (HEGS), formally the Geography and Spatial Sciences Program (GSS),

> So it appears that while the NSF may still consider quantitative human geographical proposals, qualitative, humanistic work is framed as neither 'scientific' nor worthy of advancing the 'health, prosperity, and welfare' of the nation. This strategy of defining radical or humanistic research as 'unscientific' due to its normative commitments has a long history in the sciences and social sciences.

The role of geography and the future of human-oriented geographic approaches will require methodological reconfiguration in alignment with intersectional sensibilities. The integration of qualitative and quantitative data, along with the use of critical GIS, such as Participatory Action Mapping (Boll-Bosse and Hankins 2018), can collectively augment the ways in which geography approaches vulnerability and oppression in the lives most impacted by the complexities of place, inequality, and justice.

Working through intersectionality requires the foregrounding of unequal dynamics that support structural oppression in peoples' lives. The process for doing so is complex, and researchers should understand how subjectivity is interwoven personally and politically in the research process. In doing so, intersectionality can reveal the "fatal couplings of power and difference", as described by Ruth Wilson Gilmore (2002), which underscores the need to examine power differentials, which are ultimately signaled (or as Gilmore says, overdetermined) through naming race, gender, class, and sexuality separately in traditional geographic methods, and even often within feminist geographic traditions. Thinking within the silos of single categories can obscure both difference within categories and similarities between categories. Intersectional thinking moves geographers closer to the complexities of real peoples' engagements with power and oppression. Yet, much geography – even feminist geography research – relies on categories such as 'woman' that hide significant and structurally important differences. Considering difference *first*, alongside questions around researcher investment in the status quo, is vital to advancing intersectional methodologies in feminist geographies.

A core question in applying intersectionality in methodological approaches is: How do our own positionalities, reflexivities, or actions construct the realities or interpretations that are observed by us and narrated to us? Put differently, what makes an analysis intersectional is its adoption of an intersectional way of thinking about the problem of sameness and difference and its relation to power. This framing – conceiving of categories not as distinct but as always permeated by other categories, fluid and changing, always in the process of creating and being created by dynamics of power – emphasizes what intersectionality *does* rather than what intersectionality *is* (Cho, Crenshaw & McCall, 2013, p. 795).

Therefore, **our fourth tenet for an intersectional sensibility in geographical methodologies mandates research that contributes to unlearning prevailing geographical imaginaries**. The knowledge that can shift the understanding of space and place include literary and artistic work in addition to traditional research methods in geography. Literary and artistic work has a long history of informing geographic knowledge and praxis, including through the writings of 'lady explorers' referenced earlier in the chapter. Literary and artistic work can enliven analyses of geographic concepts, such as scale and landscape, as well as power structures embedded in place. We turn to Katherine McKittrick (2000, p. 126), who takes up literary work as an anti-racist feminist geographies project in ways that reveal geographical complexities and the "simultaneously complex and constant" shaping of race, class, and gender.

In analyzing Toni Morrison's novel, *The Bluest Eye*, McKittrick uses notions of scale – home, community, nation – to illustrate material and lived realities as narrated in Morrison's characters and geographic settings. McKittrick reminds us that discrimination, prejudice, and anti-Black racism are "part of the social and built environment" (p. 129) and "that spaces – emotional, bodily, and environmental – produce 'many spaces in the same space'" (p. 139). Using methodological frameworks of anti-racist feminist geographies, as amplified by McKittrick, geographers should look to literary and artistic work as sites for wrestling with and deconstructing power. Further, McKittrick's argument concerning Morrison's work points us toward a fresh reading of Geography's archive. For example, Ellen Churchill Semple's elaboration of Environmental Determinism succeeded largely because of her eloquent and persuasive language (Gelfand 1954) and because her views were consistent with the discrimination, prejudice, and racism widely held by her readers. Unlearning prevailing geographical imaginaries is central to realizing the promise of an intersectional sensibility.

All four tenets of our intersectional sensibility for feminist geography methodologies concern power relations. The first articulates a standpoint relative to knowledge and oppression. The second disrupts the single axis thinking characteristic of analyses that fail to account for differences among women. The third seeks to end oppression and violence, whether immediate or structural (slow). The fourth insists on the unlearning of prevalent geographical imaginaries. Taking up the four tenets requires a shift in worldview for those whose perspective is anything other than that of a standpoint that centers multi-axis oppression.

Conclusion

In this chapter we have shown that intersectionality – as a methodology, set of intellectual commitments, and collection of practices – is firmly grounded in the lives, experiences, and thinking of women of color under the U.S. empire. Contemporaneously with intersectionality's early development, geography as a scholarly discipline was born in European colonization and nurtured in the service of states and capitalist economies. Much more recently, feminism in the academy – and especially feminist geography – has established a tentative hold within the discipline as primarily a white enterprise (Oswin, 2020). It is not a surprise, then, that feminist geography and intersectionality, coming as they do out of two different worldviews that are the products of different activist and scholarly traditions, do not easily find common cause. Yet, in this chapter we have identified shared commitments as the basis for feminist geography's methodologies to embrace a state of mutuality and co-belonging with intersectionality.

We have argued for an 'intersectional sensibility' that is keenly conscious of its origins, foregrounds the complexity of lives lived under structural oppression, applies intersectionality deliberately to research conceptualization, design, and methods, and is thought-full and care-full throughout. As we have argued elsewhere, intersectionality has the potential to transform not only feminist geography but the discipline of geography as a whole along the lines of anti-white supremacy and for the liberation of all people (Eaves and Falconer Al-Hindi, 2020).

Note

1 The authors are equally responsible for this chapter.

References

Beal, F. M. (1970/2008) Double jeopardy: To be Black and female. *Meridians*: 8(8), pp. 166–176.

Bilge, S. (2020) The fungibility of intersectionality: An Afropessimist reading. *Ethnic and Racial Studies*, 43(13), 2298–2326. DOI: 10.1080/01419870.2020.1740289

Bird, I. (1987/1880) *Unbeaten Tracks in Japan*. Beacon Press.

Boll-Bosse, A. J., and Hankins, K. B. (2018) 'These Maps Talk for Us:' Participatory Action Mapping as Civic Engagement Practice. *The Professional geographer*. 70 (2), pp. 319–326.

Boxer, M. J. (1982) For and About Women: The Theory and Practice of Women's Studies in the United States. *Signs: Journal of Women in Culture and Society* 7 (3), pp. 661–695.

Bryant, L. and Pini, B. (2011) *Gender and Rurality*. New York: Routledge.

Burt, S. and Code, L., eds. (1995) *Changing Methods: Feminists Transforming Practice*. Peterborough, Canada: Broadview Press.

Cahuas, M. (2019) Burned, Broke and Brilliant: Latinx Community Workers' Experiences Across the Greater Toronto Area's Non-Profit Sector. *Antipode* 51(1), pp. 66–86.

Carbado, D.W., Crenshaw, K., Mays, V.M., and Tomlinson, B. (2013) Intersectionality: Mapping the Movements of a Theory. *Du Bois Review* 10(2), pp. 303–312.

Cho, S., Crenshaw, K., and McCall, L. (2013) Toward a Field of Intersectionality Studies: Theory, Applications, and Praxis. *Signs: Journal of Women in Culture and Society* 38 (4), pp. 785–810.

Collins, P. H. (2015) Intersectionality's Definitional Dilemmas. *Annual Review of Sociology* 41 (1), pp. 1–20.

Collins, P. H. (2009) *Black feminist thought: Knowledge, consciousness, and the politics of empowerment*. New York: Routledge.

Collins, P. H. and Birge, S. (2020) *Intersectionality*, 2/E. Polity Press.

Cooper, A. J. (1892) *A voice from the South*. Xenia, Ohio: The Aldine Printing House.

Crenshaw, K. (1991) Mapping the Margins: Intersectionality, Identity Politics, and Violence against Women of Color. *Stanford Law Review* 43(6), pp. 1241–1299.

Currah, P. (2008) Stepping Back, Looking Outward: Situating Transgender Activism and Transgender Studies – Kris Hayashi, Matt Richardson, and Susan Stryker Frame the Movement. *Sexuality Research and Social Policy* 5 (1), pp. 93–105.

Davis, A. Y. (1983) *Women, race, & class*. 1st ed. New York: Random House.

Domosh, M., (1991) Toward a Feminist Historiography of Geography. *Transactions of the Institute of British Geographers* 16(1), pp. 95–104.

Eaves, L (2019) The Imperative of Struggle: Feminist and Gender Geographies in the United States. *Gender, place and culture: a journal of feminist geography*. 26 (7–9), pp. 1314–1321.

Eaves, L. and Falconer Al-Hindi, K. (2020) Intersectional Geographies and COVID-19. *Dialogues in Human Geography* 10(2), pp. 132–136. DOI: 10.1177/2043820620935247

Falconer Al-Hindi, K. (2020) Intersectionality. In: *The International Encyclopedia of Geography*, ed. D Richardson, N Castree, M F Goodchild, A Kobayashi, W Liu, and RA Marston. John Wiley. DOI: 10.1002/9781118786352.wbieg0102.pub2

Gelfand, L. (1954) Ellen Churchill Semple: Her Geographical Approach to American History. *Journal of Geography* 53, pp. 30–41.

Gilmore, R. W. (2002) Fatal Couplings of Power and Difference: Notes on Racism and Geography, *The Professional Geographer*, 54, pp. 15–24, DOI: 10.1111/0033-0124.00310

Grzanka, P. R. (2019) *Intersectionality: Foundations and frontiers*, 2/E. New York and London: Routledge.

Hamilton, A. (2020) The White Unseen: On White Supremacy and Dangerous Entanglements in Geography. *Dialogues in Human Geography* 10(3), pp. 299–303.

Hancock, A.M. (2016) *Intersectionality: An Intellectual History*. New York: Oxford University Press.

Harding, S. (1987) *Feminism and Methodology*. Bloomington: Indiana University Press.

Healey, G., Bradley, H., and Forson, C. (2011) Intersection Sensibilities in Analyzing Inequality Regimes in Public Sector Organizations. *Gender, Work and Organization* 18(5), pp. 467–487.

hooks, B. (1981) *Ain't I a Woman: Black Women and Feminism*. Boston, South End Press, 205pp.

Hopkins, P. (2019) Social Geography I: Intersectionality *Progress in Human Geography* 43(5), pp. 937–947. DOI: 10.1177/0309132517743677

Hull, A. G., Bell-Scott, P. and Smith, B. (1982) *All the Women are White, All the Black Are Men, But Some of Us Are Brave: Black women's studies*. New York City: Feminist Press.

Jagger, A. M. (2008) *Just Methods: An Interdisciplinary Feminist Reader*. Boulder: Paradigm Publishers.

Khan, C. and Franklin, K. (2020) Trans Studies. Chapter 5 in: *Companion to Women's and Gender Studies*, ed. N. Naples. John Wiley and Sons Ltd.

Khan, C. and Franklin, K. (2006) Why Women of Colour in Geography? *Gender, Place and Culture: A Journal of Feminist Geography* 13 (1), pp. 33–38.

Kobayashi, A. (2014). PRESIDENTIAL ADDRESS: The Dialectic of Race and the Discipline of Geography. *Annals of the Association of American Geographers*, 104(6), pp. 1101–1115.

Kohl, E. (2019) 'When I Take Off My EPA Hat': Using Intersectional Theories to Examine Environmental Justice Governance. *The Professional geographer*. 71(4), pp. 645–653.

Lorde, A. (1983) There is No Hierarchy of Oppressions. In: *Introduction to Women's, Gender and Sexuality Studies*, eds. L. A. Saraswati, B. L. Shaw, and H. Rellihan. Oxford University Press, p. 76.

Mahtani, M. (2014) Toxic Geographies: Absences in Critical Race Thought and Practice in Social and Cultural Geography. *Social & Cultural Geography*. 15(4), pp. 359–367.

May, V. M. (2015) *Pursuing intersectionality, unsettling dominant imaginaries.* New York, NY: Routledge.

McCall, L. (2005) The Complexity of Intersectionality. *Signs: Journal of Women in Culture and Society* 30(3), pp. 1771–1800.

McKittrick, K. (2000) 'Black and' Cause I'm Black I'm Blue: Transverse racial geographies in Toni Morrison's The Bluest Eye. *Gender, Place and Culture: a Journal of Feminist Geography* 7(2), 125–142.

Mollett, S. and Faria, C. (2018) The Spatialities of Intersectional Thinking: Fashioning Feminist Geographic Futures. *Gender, Place and Culture: A Journal of Feminist Geography* 25(4), pp. 565 – 577.

Moses, C. G. (2012) 'What's in a Name?' On Writing the History of Feminism. *Feminist Studies* 38 (3), pp. 757–779.

Nash, J. C. (2019) *Black feminism reimagined: After intersectionality.* Durham: Duke University Press.

Omi, M., and Winant, H. (1994). *Racial formation in the United States: From the 1960s to the 1990s* (2nd ed.). New York: Routledge.

Oswin, N. (2020) An Other Geography. *Dialogues in Human Geography* 10(1), pp. 9–18.

Peretz, T. (2019) Why Atlanta? A Case Study of How Place Produces Intersectional Social Movement Groups. *Gender, Place and Culture: A Journal of Feminist Geography* 27(10), 1438–1459.

Pulido, L. (2002) Reflections on a White Discipline, *The Professional Geographer* 54:1, pp. 42–49, DOI: 10.1111/0033-0124.00313

Ross, L. J. (2017) Reproductive Justice as Intersectional Feminist Activism. *Souls* 19 (3), pp. 286–314.

Sanders, R. (1990) Integrating Race and Ethnicity into Geographic Gender Studies. *The Professional Geographer*, 42 (2), pp. 228–231.

Taylor, K.-Y. (2017) *How we get free: Black feminism and the Combahee River Collective.* Chicago, IL: Haymarket Books.

Taylor, K.-Y. (2016) *From #BlackLivesMatter to Black liberation.* Chicago, IL: Haymarket Books.

Third World Women's Alliance. (2005) "Third World Women's Alliance: Equal to What?" *The Movements of the New Left, 1950–1975.* New York: Palgrave Macmillan, pp. 131–133.

Tuhiwai Smith, L. (2012) *Decolonizing methodologies: Research and indigenous peoples.* 2nd ed. London: Zed Books.

Valentine, G. (2007) Theorizing and Researching Intersectionality: A Challenge for Feminist Geography. *The Professional Geographer* 59 (1), pp. 10–21.

Wells-Barnett, I. B. (1895/2005) The Red Record: Tabulated Statistics and Alleged Causes of Lynching in the United States. Project Gutenberg eBook #14977 www.gutenberg.net

Yee, S. J. (1992). *Black women abolitionists: A study in activism, 1828–1860.* Knoxville: University of Tennessee Press.

Zelinsky, W. (1973) Women in Geography: A Brief Factual Account. *The Professional Geographer* 25(2), pp. 151–165.

7

MAKING SPACE FOR INDIGENOUS INTELLIGENCE, SOVEREIGNTY AND RELEVANCE IN GEOGRAPHIC RESEARCH

Chantelle Richmond, Brad Coombes and Renee Pualani Louis

Introduction

The purpose of this chapter is to share our perspectives as Indigenous geographers about the significance of Indigenous peoples and knowledges for the wider discipline of geography, and especially for geographic research involving Indigenous communities. If geographic research is to be useful for Indigenous people and matters of concern, a fundamental shift in *how we think about and how we do research* is necessary. Specifically, we advocate that geographic research with Indigenous peoples must demonstrate relevance and be useful for participating communities, it must acknowledge Indigenous intelligence and ways of knowing, and it must support Indigenous leadership and sovereignty. Important and hopeful changes have been achieved within geography, and especially within the sub-discipline of Indigenous geography (Coombes et al., 2014). This includes a steady increase in Indigenous people entering the field, and a growth of critical and decolonial geographies (Richmond and Nightingale, 2021). Buffered partly through the support of our allied colleagues, we are witnessing a broadscale uptake of Indigenous methodologies, whereby research is driven by Indigenous ways of knowing, doing and valuing the world (de Leeuw and Hunt, 2018). Nonetheless, it is also our experience that these exciting developments remain largely at the margins of geographic research. For those students or early career researchers who are supported and engaged with Indigenous professors and mentors, such developments are rich and inclusive; for those without such support, geography can be an isolating and lonely place.

We have too often witnessed the brazen uptake of Indigenous communities and topics as *interesting* case studies, at times for no better reason than that of a strategic funding call, or because the problems experienced by a particular Indigenous community make for a *rich* comparison with non-Indigenous realities. Our Indigenous students are repeatedly put in the awkward and unfair position of having to correct or challenge the boundaries of their *expert*

DOI: 10.4324/9781003038849-9

professors, whose knowledge, experience and narratives fail to acknowledge Indigenous ways of understanding the world. To know that we ourselves experienced those same challenges nearly twenty years ago is deeply frustrating, and points to a systemic, unchanging problem. But should we be surprised? Indeed, the very roots of our discipline are grounded in imperialism, and the belief that the new world and its people and resources were (and are still to be) *discovered* and *helped* is one that persists strongly in the field of geography (Howitt and Jackson, 1998). Such framings support and re-purpose deficit-based framings of Indigenous people (Aldred et al., 2020).

As Indigenous geographers, we come to this work because of our desire to disrupt processes of colonization, particularly those we witness in academia and through the research enterprise. We have witnessed firsthand highly inequitable outcomes of colonization and dispossession. Despite the miles that separate Canada, Hawai'i, and Aotearoa (New Zealand), our everyday and academic realities are shaped in remarkably consistent ways. We also come to this work because of the deep love and responsibilities we have for our families and our communities. We engage in research to support Indigenous healing, empowerment, and self-determination. We want to see, and lead, positive change in our communities. In our roles as professors, mentors, and researchers, we seek to leverage our scholarship to unsettle and transform colonial behaviors and research practices, and to make academia a place that is welcoming and nurturing for many generations of Indigenous peoples (Richmond, 2020).

As contributors to this wider volume, we were invited to write about Indigenous methodologies in geographic research. As we discussed amongst ourselves how such a collective piece might unfold, we found our discussions converging around our own relational experiences with research. For us, the most important methodological discussion points were not about the techniques and practical know-how, but about the *principles we engage in* (and are continually learning from!) to carry out the work we do as Indigenous geographers. Those principles are both relational and community-centered in nature, and they include the ways we come to and carry out our work as geographers, along with the imperatives that guide us in our research. This chapter takes the form of a set of stories. We offer them to share some of our frustrations with the *status quo*, but also to enlighten and challenge geographers to think in more careful and more critical ways about how they contribute – despite their *well-meaning-ness* – to the marginalization of Indigenous people and matters in geographic research. In the first story, Chantelle Richmond addresses the importance of elevating Indigenous voices and perspectives in Indigenous research, and ensuring that Indigenous research is useful, relevant, and applicable for Indigenous communities. In the second story, Renee Pualani Louis highlights the value of elevating Indigenous intelligence, suggesting that when we participate in knowledge exploration using Indigenous methods, everyone benefits. In the third story, Brad Coombes explores an Indigenous demand for leadership within research that implicates Indigenous communities, and he suggests that when such leadership is realized a more creative, spontaneous and inter-subjective approach to research methods will follow. We draw upon our own research journeys to share moments of tension, awakening and learning, and as important experiences of our own *becoming*. We share our stories to support an expanded perspective on concepts central to Indigenous research methodologies including power, privilege, and self-reflection.

'But *How* Will This Research *Be Useful* for the Community?' by Chantelle Richmond

Ten years ago, I was invited to an expert panel on Indigenous food security. The panel consisted of an esteemed international membership of equity-minded scholars who had dedicated their academic lives to the theoretical concept and measurement of food security. As a new scholar, I was intimidated by the senior, non-Indigenous scholars with whom I had been invited to work. I was also daunted by the scientific culture that surrounded the panel's activities. Our meetings took place in posh hotel boardrooms in various Canadian cities, to which we were flown twice a year to carry out our work.

Despite the panel's emphasis on Indigenous food security, only two members were Indigenous scholars – another junior scholar and me. While the number of Indigenous scholars and researchers is growing steadily in Canada – much like in the US, Australia, and Aotearoa – we remain small in numbers. This reality can create additional burdens for Indigenous scholars in academic and research settings (Gaudry and Lorenz, 2018). I have yet to meet an Indigenous scholar who has come into academia purely out of their interest in higher learning, but more frequently they choose their career pathways because of the inequity they witness or endure in their own lives and communities. It is also a rarity for Indigenous scholars to be able to live and work in their own communities. Thus, we almost always work *away* from our families and communities. We do our best to bring our cultures and ways of knowing into our academic practices and research spaces, but most university campuses are largely devoid of Indigenous people, languages or culture. This can make for a lonely and sometimes unfamiliar path, but it is one that many Indigenous scholars have chosen to walk because we love our communities and we are driven to improve their realities (Anderson and Cidro, 2020).

Our first full day of the expert panel began with introductions. Panel members, the Chair and staff members sat around a large U-shaped table and one by one, we introduced ourselves. The single other Indigenous person in the room – also a dear friend – started her introduction in her native language. Given the work we were tasked with, it was a beautiful and important reminder of what should matter most. When she was done, she looked around the room. All of us sat in admiration and awe of her. Despite the calm confidence she exuded, she admitted how nervous she had been to do that introduction. She told us that she had spent the whole summer learning to say those words. Sitting across from my friend, I held back tears, knowing full well the fear and worry that comes along with demonstrating ourselves as Native people. When it was my turn to introduce myself, I followed in the brave footsteps of my friend, and also shared words of introduction in my own language of Anishinabemoen. I stated my name and acknowledged the place I come from, including my River, my Clan and my community. As I spoke, I could hear my mother's voice: '*Chany, you must speak your language at every opportunity.*' But the truth is that in these academic spaces – where Indigenous people and languages have rarely been welcomed and certainly never prioritized – it takes incredible bravery to demonstrate your Native self. I had never been so grateful that my friend went first in the line of introductions. She demonstrated the utmost significance of this practice, and I was so proud to be there with her.

Over the next several meetings, our work turned from conceptualization and data gathering into writing. The process was highly collaborative. As a panel, we recognized that our overall report would be strengthened through the engagement of many perspectives and ideas. During conversations about the chapter on *Indigenous health and well-being*, there was general agreement that it should center on Indigenous connection to the land and its importance for Indigenous Knowledge, the social and cultural practices related to traditional foods and their benefit for food security, nutritional status and overall health and well-being. I was struck when one panel member suggested that this chapter should emphasize not only the health promoting concepts and behaviors related to food security, but also a range of negative health behaviors, including drug and alcohol addictions, violence and abuse and mental health. He suggested this would offer a *balanced approach*. To me, this framing ran counter to the purpose of the report and would contribute to the pathologizing of Indigenous people (O'Neil et al., 1998), a narrative that Indigenous scholars work at great lengths to minimize (McGregor, 2018). While research has been conducted on Indigenous peoples – eliciting a range of genetic, behavioral, health and social data collected about Indigenous people and nations – the use of this data has historically been an act of surveillance, control and ultimately, colonization by the nation state. In Canada, these data have been used – and continue to be used – to affirm ongoing paternalistic, fiduciary relationships between the Crown and Indigenous people. Thus, as that panel member spoke, I grew increasingly agitated. I could feel my face and lips growing flush. My stomach was churning with the realization that I could not remain silent at these suggestions. Looking around the room, not a single other panel member responded to this suggestion, and I could see no obvious dissention. I raised my hand to speak, and everyone turned in my direction. My voice quavering, I simply said: 'But can you explain why this approach will *be useful* for this Report?'

I cannot remember how my colleague responded nor how other panel members reacted in that moment. What mattered most was that I had raised my voice to question the relevance of this approach. Despite my fear of being ostracized or viewed as *hard to work with*, I acted upon that little voice inside my heart. I defended the integrity and beauty of Indigenous Knowledge, and I refused to support the continuation of the narrative that paints our communities as sick and defenseless. Years later, one of the most senior panel members approached me at an academic meeting. He reminded me of that day. With tears in his eyes, he thanked me for having the bravery to address that situation and stated his regret at not speaking up.

The experiences shared here may seem like small acts. And as I look back over my career, they were small. But in a place and time where Indigenous people and ideas remain at the margins of academia, each opportunity to take space offers new confidence that raising my voice and sharing my perspective are acts of decolonization. Over time, these acts will have an impact upon the wider research community, including the Indigenous scholars who come in the generation behind me. On a personal level, these repeated demonstrations of decolonial love (Simpson, 2013) serve to strengthen my sense of identity as an Indigenous scholar. They also sharpen my sense of belonging and purpose in the field of geography, and my ambitions to work every day – through big and small acts – to dismantle the persistence of its colonial structures and associated methodologies.

'Embracing the Gift of Indigenous Intelligence' by Renee Pualani Louis

It is dawn. The sky is lit by a sun sitting just below the eastern horizon. Rays of light brush the tip of Mauna-a-Wākea as the daylight cycle of life on Hawai'i island begins. Inspired by the Kia'i, guardian protector, of Mauna-a-Wākea, I begin a slow rhythmic clapping. Kia'i take on responsibilities of protection, preservation and/or perpetuation of a person, place or thing. It is used here to refer to the people who took on the role and associated responsibilities of protecting an Indigenous Hawai'i Kanaka Maoli perspective of land use on Mauna-a-Wākea, the tallest mountain on earth from sea level. From deep within a vocal accompaniment joins the rhythmic clapping. It is a way to greet Kānehoalani, the celestial entity whose physical form is known as lā, the sun, to set a tone, to situate a person.

We are in this virtual vignette to witness the dawning of a (k)new moment (Meyer, 2013). A moment that re-animates an older perspective of our roles and responsibilities to the people and places where we live, work, and pray. We are here to witness this moment become a mo(ve)ment towards Indigenous research sovereignty/justice. It is a movement responding to those persistent imperialistic tendencies of pursuing Indigenous wisdom to inform a growing appetite of *knowledge is power* seekers while continuing to undermine Indigenous intelligence. There is an important distinction between knowledge and intelligence. Whereas knowledge is generally considered a collection of skills and information one acquires through experiential learning, intelligence is generally considered the ability to apply knowledge to problem solve, and includes the capacity for self-awareness, creativity and moral fortitude. People may share the same knowledge, but each person's intelligence will help them design and create unique solutions to solve real life situations.

Twenty years ago, Dumont articulated Indigenous intelligences as thinking, speaking and acting from the core of Indigenous cultures, traditions and life ways (Dumont, 2002). In those twenty years, many Indigenous community scholars answered his challenge by educating themselves in their Indigenous ways of being, using their traditional practices and actively pushing boundaries to challenge policies, and disrupt processes of colonization. They are acutely aware of geography's colonial heritage of focusing on, gleaning and incorporating Indigenous ancestral knowledge while simultaneously conducting deficit-based research on the same people. This has conveniently allowed for the creation of an irrational false narrative – that Indigenous peoples are not capable of deciding for themselves how best to interpret and implement *their own knowledge* (Coombes et al., 2014).

This is problematic because, with a few exceptions, most geographic research methods continue to incorporate anthropocentric values in their intellectual understandings. A recent ontological shift toward relational approaches, such as in posthumanism, highlights how deeply embedded anthropocentric perspectives are within the discipline (Braidotti, 2013). However, this shift away from anthropocentric perspectives and toward the growth of relational ontologies in geography, is *not* the same as decolonizing the discipline. That is a different and more difficult conversation. Moving away from anthropocentrism allows us to expand our geographic knowledge by engaging in a different, more interconnected understanding of the world which in turn, hopefully, affects how we experience and interact in *all* our spaces. Once geographers and all researchers understand the significance of this ontological shift, it will allow for a more dynamic interchange of knowledge sharing.

One mo(ve)ment that embraces a kincentric ontological approach is Hālau ʻŌhiʻa, a Hawaiʻi stewardship training program founded by Kekuhi Kealiʻikanakaʻoleohaililani. Kekuhi is a highly respected Hawaiʻi Island *Kumu Hula* (Hawaiʻi dance teacher), specializing in protocol. She attentively maintains her responsibility for the eight generations of ancestral knowledge she has been entrusted with and actively imagines appropriate ways to adapt and apply that knowledge tradition to *meet the moment*. She has been adapting and sharing her knowledge traditions for several years as founder of Hālau ʻŌhiʻa. It is in this space where Kanaka Maoli intelligence shines brightly to create more robust learning opportunities that expand our geographic sensibilities.

Hālau, many breaths, is a Hawaiʻi space for learning. ʻŌhiʻa is the most bioculturally important tree in Hawaiʻi. Hālau ʻŌhiʻa is a space for many voices to come together and enhance learning for one another (Kealiikanakaoleohaililani et al., 2019). Conceptually, this is what all places of learning purport. What makes this learning experience different is it is based on Kekuhi's immersive and integrative teachings of Hawaiʻi life ways which are reinforced and manifested in place through experiential engagement with the biocultural landscape (Kealiikanakaoleohaililani et al., 2018). Although specific to Hawaiʻi, learners have applied these teachings to other settings (McMillen et al., 2020), demonstrating how a Kanaka Maoli ontological understanding can be adapted and applied in urban locales such as New York City.

As a learner of Hālau ʻŌhiʻa, I have witnessed and participated in many of the immersive practices and believe it is the optimal form of knowledge exchange because the ontological perspective and methodological approach requires people to recognize and deeply internalize our biocultural connections, respect and honor the specificity of diverse knowledge traditions and actively celebrate learning in the field. As a Kanaka Hawaiʻi cartographer, I have expanded my spatial knowledge framework by engaging in one particular practice, Kuahu. Kuahu is a central practice for Hālau ʻŌhiʻa learners that has its roots in *hālau hula*, space for learning Hawaiʻi dance. Proven outcomes of the practice include:

> …setting spiritual intentions in a community of practice; reframing nature as kin and not commodity; reinforcing that humans and nature are inextricably dependent on one another; providing a venue for the conscious and subconscious integration of conventional, Indigenous, and Local knowledge; creating a safe environment for learners from diverse backgrounds including non-profit organizations, state and federal entities, universities and communities; and enhancing learner capacity to develop intimate connections with place and each other.
>
> *Kealiikanakaoleohaililani et al., 2020, p. B*

These kinds of learning opportunities and mo(ve)ments are found at the margins for a reason. They upset the knowledge hierarchy and supremacy of both colonial and postcolonial structures and systems like those that reward written outputs over community engagement (Kovach, 2019). Indigenous community scholars and practitioners, like Kekuhi, who share their knowledge traditions in a manner consistent with their philosophical approaches and distinct cultural teachings offer more robust spaces for knowledge exchange for all people. Geographers and researchers who are currently engaging in these kinds of learning opportunities struggle to publish their discoveries or find nurturing mentors to help them (Kovach, 2019). Those interested in evolving geographic research paradigms to be more inclusive of Indigenous research paradigms must seriously consider adopting posthuman ontologies. This shift is essential for any geographic research focused on Indigenous cultures and communities as it elevates the Indigenous ingenuity and intellect that best understands the highly localized, generalizable wisdom that Indigenous peoples have maintained over generations. The best way to realize this shift is to accept and respect the dawning of Indigenous research sovereignty.

Indigenous Leadership in Research: ʻwhakaari ma hoiho – Research as Horseplay' by Brad Coombes

Hooves on gravel. I am tired, sore and frustrated. Hooves in mud. I travelled to Te Urewera to complete eight interviews over four days with leaders in local Treaty claims, but this second interview has already consumed three days. Hooves on rock. I consider asking what motivated the interviewee to muster horses then complete this interview on horseback. Hooves in water. I consider reminding him that my last horse ride was 12 years ago but ask instead the first question. Hooves at rest. ʻBecause this is mana motuhake bro.' I am surprised.

The phrase *mana motuhake* was used frequently during my childhood but may since have lost its import with Maori intellectuals. Perhaps *rangatiratanga* (self-determination, chieftainship) now dominates *mana motuhake* in Maori parlance because the former is included in the Treaty of Waitangi, whereas the latter is associated with a short-lived Maori political party. Simpson and colleagues (2020) argue that *mana motuhake* encompasses autonomy *and* identity, so it is not a parochial demand for separatism. It is typically applied to independence in service delivery or to Indigenous governance preferences, but not to research relationships. I was also surprised that in stating his reply, this *kaumatua* (elder) pointed to himself, then to our environs, to himself again, then to *me*. The environmental reference was as clear as the forested beauty of Te Urewera itself, but what in our relationship represented autonomy or self-actualization? After coaxing, he mused further, suggesting that freedom to tell our own stories, in our own ways and in places of cultural significance are core practices of *mana motuhake*. He also noted that I, 'as a Maori researcher,' should appreciate the narrative, the journey, and the narrative more so because of the journey. He was correct. The question about how academic inquiry can foster *mana motuhake* lingered with me much longer than the saddle sores.

Indigenous peoples have multiple reasons for asserting leadership in research. Appropriation of Indigenous knowledge remains a cardinal threat, especially in an era of ecosystem crisis that valorizes local wisdom for 'global' problem-solving. In that context, Indigenous communities demand more than *making space* for Indigenous scholars within conventional research. While knowledge sovereignty may ultimately demand broader forms of Indigenous leadership on Indigenous lands, at the least it requires non-Indigenous researchers to *move over* from their privileged positions in research (Latulippe and Klenk, 2020). Continued dominance of non-Indigenous researchers in Indigenous communities will prolong the deficit understanding of those communities, along with its various social pathologies (McGregor, 2018). After long periods of neo/colonial abuse, our communities seldom function as we desire, or they are characterized by such complexity as to defy exogenous understanding. Consequently, 'non-Indigenous academics' are not 'adequately equipped to understand the multiple and overlapping subjectivities' that emerge and Indigenous research leaders are required to negotiate through plurality (Jenkins et al., 2020, p. 5). As a corollary, any focus on inter-subjective difference, or *White bias* in interpretation, may disregard the continuing dominance of *White privilege* in academic research (Krusz et al., 2020). As a counter to the limited project that seeks to minimize White bias, an emphasis on unsettling academic privilege generates a radically different prescription for research in Indigenous contexts, validating greater Indigenous authority over research practices.

At present, though, academic guilt, introspective remorse about the status of non-Indigenous research allies and self-critique by non-Indigenous researchers have become preoccupations that occlude opportunities for Indigenist scholarship. Because of 'the reluctance-cum-inability of geography to come to grips with the challenge of decolonization,' Barker and Pickerill (2020, p. 641) entreat for non-Indigenous researchers to cease representing the Other and rather to engage in critical self-reflection. While decolonization-of-self may reduce the harm of White privilege in research, the positive impacts of such practices are ambiguous for Indigenous communities (Krusz et al., 2020). Besides, decolonization-of-self as researcher may be unachievable because non-Indigenous researchers cannot step aside from their institutional positionalities (Jenkins et al., 2020). Unreformed benevolence pervades most current scholarship on appropriate research methodologies for application in Indigenous contexts, and that contravenes decolonial ethics.

At its best, the self-analysis of those concerned about past research *on* or *with* Indigenous peoples has extinguished the habitually disappointing remedy that is participatory research. In the everlasting search for the correct preposition (on, with, for or by), it is the relative absence of *by* that is most significant. Mere collaboration is insufficient to disrupt a colonizing presumption of non-Indigenous leadership in research, and it may be co-optive (Jordan and Kapoor, 2016). Notably, the equestrian *kaumatua* did not demand participatory nor even devolved forms of research, but rather he commended a particular style of interviewing – a standard of orthodox social science. Indeed, his position corresponds with human geographers' attempts to *enrich* rather than abandon such conventional research approaches as interviewing through side-line performance, multi-sensory engagement or subtle manipulation of research environs (Dowling et al., 2016). From multiple Indigenous perspectives, interviewing on horseback is culturally authentic and not at all odd. It can be argued, therefore, that rather than participatory approaches that manufacture false consensus, performative action and experiments with non-representational approaches may provide greater space for the creative talents of Indigenous leaders to communicate cultural values of importance (Dowling et al., 2018). Decolonial praxis requires more than critique and must include regeneration of Indigenous ways of knowing and being, affirming the necessary role of Indigenous leaders.

To assert the need for Indigenous leadership in research also requires analysis of the *collective*, flax-roots and rhizomatic foundations of Indigenous leadership itself (Spiller et al., 2020). Indigenous-led research embraces duties to multiple human and more-than-human constituencies, deeply relational connections amongst past and future generations and transcendence of binaries between corporeal and spiritual domains. It is inherently creative and spontaneous, suggesting why such performative, dramatological and expressionist methods as participatory theatre have found favor with Indigenous communities, and why enduring horseback agony is vital (Coombes 2018). When reconfigured as an art form, research is more amenable to Indigenous leadership because, for instance, only one individual can occupy the director's chair at any moment, preserving that space for Indigenous agents of change. Many research projects of recent times value research *with* Indigenous collectives. In contrast, to be led in research requires affective, transcendent moments in which 'the distinction between researchers as separate from the field of their focus is dissolved, as it is no longer a study '…'about' or 'upon,' but from '*within*'' (Spiller et al., 2020, p. 519, our emphasis). Arguably, Indigenous-led pedagogy is the only form of research and learning that can achieve the transcultural consciousness that is required for genuine allyship or for research to contribute to Indigenous transformation. Academic ignorance sustained assimilative policies and cultural genocide in the past, so it is only when 'Indigenous nations…take action to realize our own aspirations' that research will prove emancipatory for Indigenous communities (McGregor, 2018, p. 829).

Conclusion

By connecting current academic debates about the ethics of research to narratives of our becoming and evolution as Indigenous researchers, we sought to encourage consideration of the *necessary* relationships amongst relevance, intelligence, and leadership in research. Geographers have known for some time that their motives, legacies and research practices encounter increasing resistance from Indigenous peoples but, apart from moments of guilty introspection, the extent and impact of their research attention is expanding rather than diminishing. Although that expansion is concomitant with growth in a cohort of Indigenous geographers who may (or may not) choose to perform as cultural intermediaries, more fundamental reforms to geographical practices and personnel are required.

Whether in expert panel discussions, on a mountaintop, or with *hooves on ground*, we must be cognizant that research is reflexive of power, position, and privilege. It is our perspective that non-Indigenous researchers should constantly review their relationship with methods and methodologies. But we do not seek to terminate the interest of allied and non-Indigenous scholars in Indigeneity. Rather, we share our experiences to sharpen understanding about how and why the relational and kincentric ontologies of Indigeneity matter for research praxis.

These considerations are embodied and deeply complicated for us as researchers, even where and sometimes particularly when we work with our own communities. We are therefore both surprised and concerned when non-Indigenous researchers treat Indigenous intelligence and the desire for community relevance as *un*complicated miscellany. We also hope that revealing our own angst in our sometimes-self-critical narratives may validate and empower emerging Indigenous geographers. Your loneliness in this field is understandable, but you can *and should* lean confidently into the associated tensions. It has been our experience that these are the places where both learning and growth will blossom.

As we have shared in the stories of this chapter, our pathways to becoming have been punctuated with moments of doubt, vulnerability, humiliation and a range of other emotions and experiences likely too raw to share in these pages. These moments have challenged us and caused us to wonder if geography is a place of belonging for Indigenous scholarship. Over time, we have come to appreciate that our resilience amidst these moments has supported the formation of our unique identities as Indigenous geographers. Successfully navigating these challenging moments has adorned us with new confidence that it is not we who have to fit into geography, but rather that geography needs to transform to meet Indigenous needs and ways of knowing. If geography as a discipline truly wants to engage with Indigenous peoples and matters, this paradigmatic shift will be a required transformation. We encourage geographers to think in more critical ways about their own practices, behaviors, and ways of knowing as they relate to Indigenous peoples and topics, and we call for practices that center Indigenous sovereignty, intelligence, and leadership in geographic research.

What constitutes relevance for Indigenous communities is ever-changing and embedded in particular contexts. Likewise, Indigenous intelligence is situational, transient and multi-layered – moments of place-based and trans-temporal consciousness that atomize as soon as the research gaze attempts to categorize them. As Indigenous researchers, we have comparably better access to local and familial expressions of those elemental metaphysics, but nonetheless we struggle with their sometimes contrary and always enigmatic complexion. Because it is ever more difficult to access Indigenous precepts of relevance and intelligence from the outside, we contend that research with Indigenous communities must change in at least two ways. First, researchers must defer to greater Indigenous leadership, sharing their resources and using their privilege, but also submitting to the will of Indigenous communities and their intent to address inequities or explore culturally significant trajectories. Second, the incalculable subtlety of Indigenous intelligence and understandings of community relevance should stretch the practices of qualitative research beyond their current orthodoxies. The discovery that Indigenous communities are fatigued by others' attempts to represent them does not necessarily legitimize the burgeoning almanac of non-representational methods and post-human ontologies. Yet, it does suggest that research methods in Indigenous contexts should value freedom of self-expression, creativity and artfulness as means to proclaim, access or experience Indigenous values.

References

Aldred, T.-L., Alderfer-Mumma, C., de Leeuw, S., Farrales, M., Greenwood, M., Hoogeveen, D., O'Toole, R., Parkes, M.W. and Sloan Morgan, V. (2020) Mining sick: Creatively unsettling normative narratives about industry, environment, extraction, and the health geographies of rural, remote, northern, and Indigenous communities in British Columbia. *The Canadian Geographer/Le Géographe canadien*, 65, pp. 82–96. https://doi.org/10.1111/cag.12660.

Anderson, K. and Cidro, J. (2020) Because we love our communities. *Journal of Higher Education Outreach and Engagement*, 24(2), pp. 3–18.

Barker, A.J. and Pickerill, J. (2020) Doings with the land and sea: Decolonising geographies, Indigeneity, and enacting place-agency. *Progress in Human Geography*, 44(4), pp. 640–662.

Braidotti, R. (2013) Posthuman Humanities. *European Educational Research Journal*, 12(1), pp. 1–19.

Coombes, B. (2018) Kaupapa Maori research as participatory enquiry: Where's the action? In: T. Hoskins and A. Jones, eds., *Critical conversations in Kaupapa Maori*, 1st ed. Wellington: Huia Publishers, pp. 29–48.

Coombes, B., Johnson, J.T. and Howitt, R. (2014) Indigenous geographies III: Methodological innovation and the unsettling of participatory research. *Progress in Human Geography*, 38(6), pp. 845–854.

Dowling, R., Lloyd, K. and Suchet-Pearson, S. (2016) Qualitative methods 1: Enriching the interview. *Progress in Human Geography*, 40(5), pp. 679–686.

Dowling R., Lloyd K. and Suchet-Pearson, S. (2018) Qualitative methods III: Experimenting, picturing, sensing. *Progress in Human Geography*, 42(5), pp. 779–788.

Dumont, J. (2002) INDIGENOUS INTELLIGENCE; Have We Lost Our Indigenous Mind? *Native Americas*, XIX (3&4), p. 15.

Gaudry, A. and Lorenz, D. (2018) Indigenization as inclusion, reconciliation, and decolonization: Navigating the different visions for indigenizing the Canadian Academy. *AlterNative: An International Journal of Indigenous Peoples*, 14(3), pp. 218–227.

Howitt, R. and Jackson, S. (1998) Some things do change: Indigenous rights, geographers and geography in Australia. *Australian Geographer*, 29(2), pp. 155–173.

Jenkins, K., Romero Toledo, H. and Videla Oyarzo, A. (2020) Reflections on a failed participatory workshop in Northern Chile: Negotiating boycotts, benefits, and the UN declaration on the rights of Indigenous people. *Emotion, Space and Society*, 37(100721), pp. 1–9.

Jordan, S. and Kapoor, D. (2016) Re-politicizing participatory action research: Unmasking neoliberalism and the illusions of participation. *Educational Action Research*, 24(1), pp. 134–149.

Kealiikanakaoleohaililani, K., Kurashima, N., Francisco, K.S., Giardina, C.P., Louis, R.P., McMillen, H., Asing, C.K., Asing, K., Block, T.A., Browning, M., Camara, K., Camara, L., Dudley, M.L., Frazier, M., Gomes, N., Gordon, A.E., Gordon, M., Heu L., Irvine, A., Kaawa, N., Kirkpatrick, S., Leucht, E., Perry, C.H., Replogle, J., Salbosa, L.-L., Sato, A., Schubert, L., Sterling, A., Uowolo, A.L., Uowolo, J., Walker, B., Whitehead, A.N. and Yogi, D. (2018) Ritual + Sustainability Science? A portal into the science of Aloha. *Sustainability*, 10(10), pp. 3478.

Kealiikanakaoleohaililani, K., McMillen, H., Giardina, C. and Francisco, K. (2019) Cultivating sacred kinship to strengthen resilience. In: L.K. Campbell, E. Svendsen, N.F. Sonti, S.J. Hines and D. Maddox, eds., *Green readiness, response, and recovery: A collaborative synthesis*. (Gen. Tech. Rep. NRS-P-185). Newtown Square, Pennsylvania: U.S. Department of Agriculture, Forest Service, pp. 188–204. https://doi.org/10.2737/NRS-GTR-P-185-paper13.

Kealiikanakaoleohaililani, K., Sato, A., Giardina, C., Litton, C.M., Ramavarapu, S., Hutchins, L., Wight, E.H., Clark, M., Cordell, S., Francisco, K.S., McMillen, H., Pascual, P. and Yogi, D. (2020) Increasing Conservation capacity by embracing ritual: Kuahu as a portal to the sacred. *Pacific Conservation Biology*, 27(4), pp. 327–36. https://doi.org/10.1071/PC20010.

Kovach, M. (2019) Indigenous evaluation frameworks: Can the convention for the safeguarding of the intangible cultural heritage be a guide for recognizing Indigenous scholarship within tenure and promotion standards? *AlterNative: An International Journal of Indigenous Peoples*, 15(4), pp. 299–308.

Krusz, E., Davey, T., Wigginton, B. and Hall, N. (2020) What contributions, if any, can non-Indigenous researchers offer toward decolonizing health research? *Qualitative Health Research*, 30(2), pp. 205–216.

Latulippe, N. and Klenk, N. (2020) Making room and moving over: Knowledge co-production, Indigenous knowledge sovereignty and the politics of global environmental change decision-making. *Current Opinion in Environmental Sustainability*, 42, pp. 7–14.

Leeuw, S. and Hunt, S. (2018) Unsettling decolonizing geographies. *Geography compass*, 12(7), e12376.

McGregor, D. (2018) From 'decolonized' to reconciliation research in Canada: Drawing from Indigenous research paradigms. *International Journal of Critical Geography*, 17, pp. 810–831.

McMillen, H.L., Campbell, L.K., Svendsen, E.S., Kealiikanakaoleohaililani, K., Francisco, K.S. and Giardina, C.P. (2020) Biocultural stewardship, Indigenous and local ecological knowledge, and the urban crucible. *Ecology and Society*, 25(2), p. 9.

Meyer, M. (2013) Holographic epistemology: Native common sense. *China Media Research*, 9(2), pp. 94–101.

O'Neil, J., Reading, J. and Leader, A. (1998) Changing the relations of surveillance: The development of a discourse of resistance in aboriginal epidemiology. *Human Organization*, 57(2), pp. 230–237.

Richmond, C. and Nightingale, E. (2021) Introduction to special section: Geographies of Indigenous health and wellness. *The Canadian Geographer/Le Géographe canadien*, 65(1), pp. 4–7. https://onlinelibr ary.wiley.com/doi/10.1111/cag.12678.

Richmond, C.A.M. (2020) A pathway to "becoming": Stories about indigenization from one Indigenous scholar. In: T. Moeke-Pickering, S. Cote-Meek and A. Pegoraro, eds., *Critical Reflections and Politics on Advancing Women in the Academy*. Hershey, Pennsylvania: IGI Global, pp. 70–86.

Simpson, L.B. (2013) *Islands of decolonial love*. Winnipeg, MB: Arbeiter Ring Publishing.

Simpson, M.L., Greensill, H.M., Nock, S., Meha, P., Harding, T., Shelford, P., Hokowhitu, B., Oetzel, J. and Reddy, R. (2020) Kaumatua mana motuhake in action: Developing a culture-centred peer support programme for managing transitions in later life. *Ageing and Society*, 40, pp. 1822–1845.

Spiller, C., Maunganui Wolfgramm, R., Henry, E. and Pouwhare, R. (2020) Paradigm warriors: Advancing a radical ecosystems view of collective leadership from an Indigenous Māori perspective. *Human Relations*, 73(4), pp. 516–543.

8

GEOHUMANITIES

An Evolving Methodology

Sarah de Leeuw

Introduction

More than 30 years ago, two remarkable (for their time) papers were published in the *Transactions of the Institute of British Geographers*. Following an overview of geography's engagement with humanities, especially literature and creative writing, the slightly earlier of the two publications opined that "[g]eography will deserve to be called an art only when a substantial number of geographers become artists" (Meinig, 1983, p. 25). The paper asserted a pressing need for "a greater openness, a clearing away of pedantic barriers… [and] a toleration of geographical creativity wherever it may lead" (Meinig, 1983, p. 25). A second paper, following just a few months later and written by a Canadian geographer who would go on to be the only geographer-poet to win the country's most prestigious literary prize (The Governor General's Award in Literature), asserted that

> even a poet caught up in passion can point to issues essential for geography. Geographers can't afford to miss out on passion – far less dismiss it. Geography without passion is about alive as a body without blood – ready for the grave-diggers.
>
> *Watson 1983: 391*

Read today, the two papers appear somewhat quaint. Human geography has, since 1983 and in great part due to critical pushback against the discipline's quantitative revolution, fully transformed into a discipline far more sympathetic to the creative expressions being lauded by Meinig and Watson. Stemming from growing tides of Marxist scholarship, and leading into feminist and (more recently) activist-informed queer anti-racist and anti-colonial geographic scholarship, geography is a much different creature today than it was in the early 1980s (Kwan and Schwanen, 2009). Geography has been called to critically account for itself (Harvey, 2001; see also Blomely, 2006) and is increasingly addressing global colonial violence and white-settler heteronormative patriarchal extractive capitalism. Geographers are increasingly producing activist-oriented scholarship and critiquing spatialized social, health, economic, and cultural disparities around the world – including through a turn to poetry (Ward, 2007).

Both papers from 1983 were penned by white-settler cis-gender men using deeply gendered and ethnocentric language to sketch out their arguments. In their reliance on exclusively

DOI: 10.4324/9781003038849-10

male-authored knowledge, in their lack of any critical reflection about positionality or privilege, and in their ideas about where and by whom their arguments might be extended, the papers read as if written for an all-cis-gender male audience of white geographers (which, to be clear, was indeed likely the majority of the audience reading the papers back in 1983). Still, there is something noteworthy about the papers, especially if understood as early harbingers of what, by the early twenty-first century, has been heralded by many as a wholesale sea-change in the discipline of human geography: geography's "creative (re)turn," its re/turn to humanities, and the ascendance of geohumanities as a discreate subdiscipline with associated journals, textbooks, specialty groups, research tools, and philosophical framings (Cresswell and Dixon, 2016; Hawkins, 2013, 2020; Madge, 2014).

What has happened between 1983, when the two papers were published, and today? Why, how, and where have the geohumanities gained so much traction? What is meant by geohumanities? What do geohumanities do, if they are indeed a thing of doing? Are geohumanities important and, if so, why? How do geohumanities fit with the aforementioned moves in human geography toward a discipline invested in feminist, anti-colonial, and anti-racist social justice? Are there weaknesses to geohumanities and, if so, what might those weaknesses be? Where might the geohumanities be in another 30 years? These questions guide this chapter, and their answers form the chapter's content.

Divided into four sections, the chapter offers (first), an overview of geography and the humanities, sketching out some rough and somewhat impartial and incomplete origins of the concept. The following two sections (geohumanities as methodology and geohumanities as method) recognize the impossibility of fully separating the work of thinking about the world (theorizing and conceptual work, or methodology) from the doing and the making material of the world (method). With this in mind, the chapters explore how geohumanities are both a conceptual tool and an applied in-practice tool. Finally, the chapter concludes with some brief contemplations and reflections about how and where geohumanities might go in the future. This conclusion opens space for contemplating some of geohumanities' weaknesses and fallibilities, some the things that – when viewed from a vantage point of three-decades into the future– might result in similar readings about the concept as result when we today read Meinig and Watson's *Transactions of the Institute of British Geographers* papers from 1983.

Geography and (as) the Humanities: The Meanings and Origins of Geohumanities

As many have pointed out, human geography has *always* been a humanities discipline and the humanities have, for a long time, been geographic in nature. In many ways, the very word 'geohumanities' is just a formalization of an idea long in action (Dear et al., 2011). This assertion, however, begs the questions: what are the humanities? The humanities are a large group of disciplines, from history and literature, to anthropology and modern languages, characterized by thinking about and engaging the world (mostly the social and cultural world, as opposed to the ecological or biological world) through critical, speculative, and interpretive means. This is different, for instance, than disciplines in the natural sciences that shy away from subjective interpretation and focus more on empirical evidence and the establishment of theories or laws (Daniels, 2012). Humanities disciplines, from folklore studies to art history, from archeology to poetry and visual arts, have also long engaged the foundations of geographic thought, namely acknowledging the deep importance of space and place as key organizing forces in human and non-human worlds (Cresswell et al., 2016; Hawkins, 2020). Human geographers – as Meinig and Watson pointed out in the early 1980s – have similarly and for a very long time employed the creative arts, undertaken interpretive readings of place and space and, in speculative fashions, read and critically reflected on the world. These overlaps between geographic endeavors, the humanities, and the creative arts (as others have noted) can be found everywhere and across wide swaths of time, from early Greek philosophy's conviction that place was the most foundational force of all things through to artists working with early colonial explores like James Cook in order to put both art and science to work in early map-making efforts (see Casey, 1997; Cresswell et al., 2016; Hawkins, 2020). At the heart of both the European Enlightenment and the Renaissance was a conviction that bringing the sciences and the arts into productive conversation with each other would result in more humane engagements with the world, including in nascent disciplines of the time such as medicine (Cresswell et al., 2016; Hawkins, 2020; Thibault, 2019). Not well considered in the discipline of geography, but certainly long-known to Indigenous peoples around the world, are deep understandings that knowledge of community, land, ecology, and geography has always been inextricable from story, ceremony, song, and visual representation (Richmond, 2018).

Still, with reference to geohumanities specifically, something new does seem to be afoot in the opening decades of the twenty-first century. Indeed, analysis[1] of appearance of the word "geohumanities" in publications indexed by Web of Science shows no use whatsoever of the term prior to 2015, with a systematic growth in the word's use to a high in 2019. Similarly, co-usage of the terms "geography" and "humanities" – while being cited just over 5400 times as indexed by Web of Science – was near zero in 1992 but up to almost 1000 by 2019. The word geohumanities – and the broader concepts of geography as tethered to humanities – is clearly on the rise in geography publications, teachings, and research. As some scholars have warned, the 'something new' represented by geohumanities might be tied to neo-liberal demands for ever-more publishing on what are touted as brand-new and completely innovative concepts, preferably in high-impact journals that boost departments of geography standings as they scramble for global rankings (de Leeuw and Hawkins, 2017). Whatever the impetus, the word and concept "geohumanities" is, however, gaining traction. So, what exactly does geohumanities mean, and where do we find it the discipline of human geography?

Geohumanities, as others have noted, is a somewhat slippery and quickly evolving concept that seems to have become the chosen twenty-first century name for what means, in general and recently, a broadly interconnecting suite of scholarly practices (including teaching, research, activism) found, on the one hand, in geography and, on the other hand, in the humanities and creative arts (Atkinson & Hunt, 2020; Hawkins, 2020). As others have defined it, geohumanities refer to a "zone" of "fortuitous convergence" where there is artful and creative interaction between geography and the humanities (Dear et al., 2011: 1). Early touchstone textbooks in the discipline, including two published a decade ago (see Daniels et al., 2012; Dear et al., 2011), highlighted both the deeply transdisciplinary nature of the geohumanities and the challenge in providing a single unifying or cohesive definition of them. Afterall, the geohumanities can be seen at work when cartographers and GIS specialists produce radical hybrid maps heavily reliant on creative visual expression; when social geographers work with poets to gain new understandings about urban spaces; or when cultural geographers integrate detailed landscape paintings in order to deepen inquiries about place and space (Dear et al., 2011; see also Quoniam 1988). Indeed, it might best be said that trying to clearly define the geohumanities raises more questions than answers; geohumanities are rough-edged and uncomfortable but also radiantly exciting in their lack of clear definition; they live in liminal or adjacent spaces; they exist in a "flux of expectations" and are always in a state of becoming, a beacon of things to come, a new direction (Dear et al., 2011: 3. See also de Leeuw and Magrane, 2019). Perhaps the clearest thing that can be said in efforts to define the geohumanities is that they are generating more and more interest across wider and wider groups of scholars, artists, and activists.

The geohumanities dwell in converging worlds, in worlds of uncertainty and constant transgression which, in part, are constantly breaking down traditional categories of scholarship and thus opening up new directions for practice, activism, creativity, and scholarship (see also Cosgrove, 2012). Given how allusive a definitive conclusive definition of geohumanities is, it becomes worthwhile to look instead at what work is done under the auspice of the concept. What do the geohumanities *do,* as frameworks or conceptual tools for theorizing the world (a methodology) or as practices and sets of tools to answer questions about the world (methods)?

Geohumanities as Methodology, as a Framework for Thinking and Theorizing

A methodology is a framework, an organizing principle, or a lens through which one asks research questions or seeks to understand the world. Very broadly speaking, there are qualitative methodologies and quantitative methodologies: the former, of which geohumanities is a part, are vested in deep, thick textured explorations of ideas and experiences, in open-ended understandings of the world. Qualitative methodologies tend to organize or frame research with questions about "why" as opposed to "what, when, or who," which are often lenses for those using quantitative methodologies. Qualitative methodologies in human geography include a wide expanse of frameworks (such a participatory, phenomenological, and ethnographic frameworks) and philosophical principles (such as determinism, social constructivism, feminism, anti-colonialism, Marxism, and anti-racism). Geohumanities methodologies align with almost all the methodologies common in qualitative human geography research.

As a way of seeing and philosophically orienting to the world (and the questions geographers ask about the world), a geohumanities methodology privileges the creative and the transdisciplinary. Using a geohumanities methodology means accepting slippery open-ended 'unknowings' and a definitive lack of clear closure (de Leeuw and Magrane, 2019). Framing research with geohumanities methodologies will mostly certainly involve: looking for creative materials to work with; believing in the power of text, textuality, narrative and discourse; adapting an orientation to the world that fundamentally tethers place to creativity; turning to the perspectives and voices of artists or 'creatives;' and paying close attention to the arts while at the same time attending to forces of spatiality and place. The use of geohumanities methodologies is in harmony with long standing discussions about experimental geographies and signals a commitment to crossing borders, to thinking about and even making places (and spaces) through clashing, converging, fluctuating, and overlapping means that embrace dialectical understandings of the world.

Adopting geohumanities methodologies and embracing creative experimentation as essential means of seeing, conceptualizing, and theorizing the world should not – and as others have noted– result in a turning away from other methodological frameworks that pay close attention to questions of privilege, power, and inequity (de Leeuw and Hawkins, 2017; Marston and de Leeuw, 2013). Indeed, as writers in experimental and literary geography, along with scholars of visual methodologies, have observed, creative works and worlds to which geohumanities scholars turn are not endpoints unto themselves. Instead, cultural productions must be approached with a keen eye to *how* the work was made: where was it produced; who contributed to its making; and what spatial practices went into it (see also Rose, 2016). In the twenty-first century, twinning geohumanities with critical methodologies like feminism, anti-colonialism, and anti-racism can result in questions about gender, coloniality, and the fictions of biological 'race' as they impact creative production.

As geohumanities methodologies drive more and more work or, put another way, as geohumanities methodologies contribute more and more to the making and theorizing of worlds, it will be important to ask hard questions about how (and by whom) the methodology is enacted. Where are geohumanities methodologies taking place? Who is absent in the worlds being theorized and made through geohumanities methodologies? Geographers taking up geohumanities methodologies are contributing to the production of space and place: as others have noted, geographers do not just *study* space and place, we participate in space- and place-making. In times of Black Lives Matter, Defund Police movements, #MeToo efforts, and Idle No More struggles, and in times of geographers being increasingly called to social justice advocacy and to work that makes the world a better place (Kobayashi et al., 2014), the imperative is strong to engage in critical reflection about the positionalities, privileges, and sociocultural powers underpinning geohumanities methodologies and those who employ them. Much of that critical reflection might be productively focused on the methods, the tools and practices, associated with geohumanities as a methodology. It is to these topics that the chapter now turns.

Geohumanities as Method, as Doing and World Making

If geohumanities as a methodology prompts seeing the world through experimental, open ended, creative, transdisciplinary lenses, geohumanities as a method – as tools and applied practices put to use within rubrics of methodologies – can also sometimes be messy and uncertain. Indeed, how geohumanities get enacted as research method and tools is inventive terrain, constantly shifting and evolving. Quite certainly, just at the moment you are reading this chapter, someone will be experimenting with a new creative geohumanities method. To offer some semblance of coherence, it is, however, helpful to think of geohumanities methods as falling into three broad categories: textual, visual, and performative. These three categories are sometimes enacted through relational practices, a final but not altogether discrete category of geohumanities method. Each of these broad categories of geohumanities methods are being put to use across a spectrum of human geography's subdisciplines, from health geography to feminist geography, queer geography to social geography, from critical GIS geography to cultural geography. Likely because writing is possibly the single strongest commonality running between all geographers and all the discipline's fields of inquiry (de Leeuw, 2017; 2019b), textual geohumanities methods are most prevalent.

Broadly speaking, a textual geohumanities method uses or makes texts comprised of (mostly but not always exclusively) words. Of recent, the method of geopoetics is gaining significant traction in human geography. Geopoetics can be defined as a returning to the geo (earth) in and through acts of poesies – which is the practice of producing poems and poetry. Geopoetics, as opposed to nature poetry, have been defined as inherently activist in nature, a "[resistance] to individuals and systems that are fucking up the planet and consolidating power on the backs of other" (de Leeuw and Magrane, 2019, p. 147). Geopoetics takes seriously the act of turning to poetry and poems as primary sources materials of geographic knowledge but it also embraces writing poetry that is specifically geographic in nature. To this end, a growing number of geographers are now well recognized poets (for instance Acker, 2013, 2009; Cresswell, 2013, 2015, 2020; de Leeuw, 2012, 2015, 2019a; Magrane and Cokinos, 2016) while an increasing number of scholarly papers and texts in human geography are using and including poetry as material for understanding and (re)making different geographies (Cresswell, 2014; de Leeuw, 2017; Madge, 2014; Magrane et al., 2019; Pavia, 2020). Geopoetic methods include: holding and making word-worlds while at the same time embracing silence as a space opened for other not-you word-world; reading, reading, and reading – especially poetry, in order to think, teach and *be* anew in the world; working with poets and working with the natural world to achieve "biocultural" ways of knowing and being (Kristeva et al., 2018); and writing, singing, gesticulating, and being in always new and poetic ways (de Leeuw and Magrane, 2019).

If geopoetics are an important geohumanities method, so too are methods anchored in visual expression. Photovoice, creative community-based mapping, graphix, crafting, mark making, and a wide variety of visual arts from sketching and painting to collaging and printmaking are all and increasingly being called into play by geographers. Many of the creative methods are participatory in nature. They are being used to understand more deeply a community's relationship with place, in order to document place-based transformations, or even in efforts to extend self-reflexive practices in relation to specific spaces and places. Photovoice techniques, alongside a suite of other creative methods have been used to explore health geographies, from asylums to waiting rooms (see Parr and Kearns in de Leeuw et al., 2018) while crafts that include everything from quilting to making tablecloths have been used in community-based inquiries about marginalization, positionality, and geopolitics (Hawkins, 2013; Parr, 2006). On an individual scale, learning to sketch, mark-make, draw and paint have allowed some geographers to see and be in the world a new, with eyes opened more widely to the subtleties of land and landscape (Wiley and Webster, 2019). Deep immersive drawings have been used as a means of accessing topography and ecology in ways impossible through mere narrativization (Quoniam, 1988), while drawing as fieldwork has been documented as an anticolonial feminist practice that can allow researchers to more fully and sympathetically navigate insider/outsider positionalities (Parikh, 2020). In the realm of health geography, graphix and comix are increasingly being turned to, both as source materials and as acts of creation, for accessing and understanding deeper emotional registers of patients, care providers, medical professionals, and illness-subjectivities – including the issues of social (in)justice that surround those emotional registers (Donovan and Ustundag, 2017). Comic and graphic works are also finding themselves in geography journals, as research and spatially creative interventions unto themselves (see Fall, 2021).

Not used as much as writing, literature, poetry (geopoetics) or the creative visual arts, performance and performativity nevertheless hold important places as geohumanities methods. A growing group of human geographers interested in queer/ied and feminist methods are turning to performance (from monologues to drag, from community-based improv to forum theatre) as ways of celebrating the potential of "chaotic" interventions, including as research means in highly diverse and politicized zones of urban spaces (McLean, 2017; 2018). There is growing conversation about dragging as a means of disruption, especially of spatialized neoliberal power regimes (McLean, 2016). Even puppetry has, of recent, been conceptualized as a spatial practice through which to access and understand wide swaths of cultural expressions across both culturally specific and far-reaching global geographies (Banfield, 2021).

Geographers working in relationship with artists or writers or curators or actors and performers or crafters is a practice that crosses textual, visual, and performative lines. While geographers working hand-in-glove with artists of all stripes and descriptions has taken place since the earliest iterations of geography as a discipline (Hawkins, 2013), recent practices of co-curation, assembling, dialogue, co-authorship, or co-production tend to be more intentionally cross- and inter-disciplinary, with an explicit focus on actively doing, and creating something new, as opposed to geographers simply analyzing or drawing from an existing creative archive or expression (Foster and Lorimer, 2007; Hawkins, 2011). Relational or co-production methods need not be limited to two people or to specific places. In efforts to disrupt heteronormative spaces of geography departments, queer and queer-allied geographers have co-curated creative interventions that provide rich insight into how heteronormativity operates quietly, subversively, but in deeply spatial ways (see for instance Kincaid, 2019). These and other relational interventions seek, for instance, to document the potential of queer spaces in making possible new worlds and social relations (Kincaid, no date: https://sites.google.com/view/edenkink aid/coccooning). Many relational geohumanities methods might fruitfully be conceptualized as participatory in nature, where geographers co-create artists books (Hawkins, 2017), work with podcasters to create new media (Kincaid 2020), follow the auditory maps of sound artists in order to more fully understand city-spaces (Butler, 2006), or work in collaboration with visual artists to explore the ethos and values needed in efforts of coming together (Foster and Lorimer, 2007). No matter what form the output, the heart of these efforts is a relationship that, broadly but importantly, gestures towards the geohumanities while simultaneously producing new forms of creative knowledge and material that expand ways of seeing and being in the world. Returning to the two papers that opened this chapter, these relational efforts might, at their most fundamental, be understood to be creating greater openness, refusing barriers, and celebrating geographical creativity "wherever it may lead" (Meinig, 1983, p. 25). What might this mean for geography and geographers in the future?

Geohumanities in the Future: Where Must We Go, What Might be Changed?

Much has transpired in human geography since the early 1980s. The discipline itself has responded to efforts of privileging quantification and scientific objectivity by evolving into a discipline that has growing space, time, and appreciation for arts and humanities. While geography has always been a humanities discipline with deep roots in the creative arts, the relationship is becoming increasingly formalized (while simultaneously staying open ended and evolving) through expanding and solidifying methods and methodologies of geohumanities. The discipline now has a journal dedicated entirely to geohumanities (see www.aag.org/cs/ publications/journals/gh) replete with an online exhibition space featuring clear examples of geographer-artists and artists in relationship to geographers. There are centers of geohumanities expertise in universities around the world (see for instance www.royalholloway.ac.uk/research- and-teaching/research/research-environment/research-institutes-and-centres/centre-for-the- geohumanities/) and geohumanities are being used by health geographers, critical GIS scholars, urban geographers, social geographers and feminist and queer geographers. It would appear that geographers are increasingly being informed by passion; that geographers are becoming artists; and that pedantic disciplinarily barriers are dissolving. Many of the calls to action made by Meinig and Watson in 1983 are being achieved. So, has human geography reached a noble and exalted space of geohumanities perfection? Will the geohumanities of the first two decades of the twenty-first century be looked back upon in 30 years as having achieved all there was to achieve?

While there is certainly much to celebrate, and many new fast-paced innovations continue to happen in geohumanities around the world, it is deeply important – as is the case for so many of geography's subdisciplines– to maintain critical reflexivity. Critical reflectivity allows for innovation, for a trend to be more than just trendy, for transformation to be ongoing and always fresh. To this end, it is worthwhile to note that geohumanities tend to center in and on the Euro-Anglo-global-north, with that vast majority of literatures, research, and makings transpiring in English-speaking academic geographies. Similar critiques have been made about the ascendance of humanities in other disciplines, including in medical- and health-humanities (Naidu 2020). Much geohumanities work is both produced by and focused on racialized-white geographies and geographers. There is a dearth of scholarship and makings produced and led by Black Indigenous People of Colour (BIPOC) or undertaken for expressly anticolonial purposes (de Leeuw and Hunt, 2018; Magrane et al., 2019). This aligns with broader critiques of human geography having dismally low numbers of racialized not-white practitioners and systemic failures of addressing racism in the discipline (Kobayashi, 2006). Almost no work in geohumanities tackles or foregrounds geographies of dis/ability, resulting in the subdiscipline being (not unlike the broader discipline of human geography itself) deeply ableist (Imrie, 1996; see also Parr and Butler, 1999). Similar to critiques about an overly uncritical turn to the wonders of creative cities and creative classes (see work by Florida, 2005), which often ignore issues of gentrification, white-privilege, and entrenchment of neoliberal economic opportunities, there are worries about the geohumanities being a somewhat naïve and nostalgic domain that might, even unconsciously, be reifying hierarchies of power and privilege in geography (de Leeuw and Hawkins, 2017).

The geohumanities are a rapidly expanding and evolving domain of geographic practice, research, activism, making, and thinking. There is a plethora of exciting developments and much general excitement about their almost boundless potentials. For the geohumanities to never reach limits and for them to not become static or reified (re)entrenchments of exclusionary divides in the discipline; however, they likely must be constantly and critically refreshed. Added to. Pushed against. Played with. Re-made. Called into question. Critiqued. To avoid being 'ready for the grave-diggers' (as Watson alluded to almost 40 years ago), geohumanities might well turn to the wisdom of the USA's first Native-American Poet Laureate:

> without poetry, we lose our way. Let there always be poetry – broken lines and rhymes, refusals and gaps, free verse and no hard endings, song and sound, new voices and new spaces – to the geohumanities. Let this call be the beating heart of an always evolving subdiscipline of human geography.

Note

1 Our observation is born out in publication metrics about the use of "decolonizing geographies" or the use of "decolonization" more broadly in geography. A search of geographic publication trends in Web of Science shows a marked increase in the term "decolonizing" from almost no incidences in the 1990s to growth of more than 45 publications per year in 2015. (Search carried out on 11 November 2016, with search terms "decoloni★" and "geograph★." Total 418 records). Further evidence can be seen in the 2017 chair's theme for Annual RGS/IBG conference in London: the theme was "Decolonizing Geographical Knowledges: Opening Geography Out to the World," underscoring the ascendance of decolonization to a central position in the discipline. This observation is born out in publication metrics about the use of "geohumanities" or the use of "geography" twinned in the same journal paper with the word "humanities". A search of geographic publication trends in Web of Science between 1955 and 2021 shows a marked increase over time both in the term "geohumanities" and in the words geography and humanities: from appearing in almost no journals until 2002 to appearing in more than 45 publications by 2015, the words reached all time citation highs of almost 1000 citations in 2019, increased from near zero in 1992 (Search carried out on 10 January 2021, total of 495 records).

References

Acker, M. (2009) *Air-Proof Green*. Pedlar Press. St. John's NL, Canada.

Acker, M. (2013) *The Reflecting Pool*. Pedlar Press. St. John's NL, Canada.

Banfield, J. (2021) 'That's the way to do it!': Establishing the peculiar geographies of puppetry. *Cultural Geographies, 28*(1), 141–156.

Blomley, N. (2006) Uncritical critical geography?. *Progress in Human Geography, 30*(1), 87–94.

Butler, T. (2006) A walk of art: The potential of the sound walk as practice in cultural geography. *Social & Cultural Geography, 7*(6), 889–908.

Casey, E. S. (1997) *The fate of place: A philosophical history*. Berkeley: University of California Press.

Cosgrove, D. (2012) Prologue: Geography with the humanities. In *Envisioning landscapes, making worlds: Geography and the humanities*. Eds. Daniels, S., DeLyser, D., Entrikin, J. N., & Richardson, D. Routledge. xxii–xxvi.

Cresswell, T. (2013) *Soil*. London: Penned in the Margins.

Cresswell, T. (2014) Geographies of poetry/poetries of geography. *Cultural Geographies, 21*(1), 141–146.

Cresswell, T. (2015) *Fence*. London: Penned in the Margins.

Cresswell, T. (2020) *Plastiglomerate*. London: Penned in the Margins.

Cresswell, T., and P. Dixon, D. (2016) GeoHumanities. *International Encyclopedia of Geography: People, the Earth, Environment and Technology: People, the Earth, Environment and Technology*, 1–9.

Donovan, C., and Ustundag, E. (2017) Graphic narratives, trauma and social justice. *Studies in Social Justice, 11*(2), 221–237.

Daniels, S., DeLyser, D., Entrikin, J. N., & Richardson, D. (Eds.). (2012) *Envisioning landscapes, making worlds: Geography and the humanities*. Routledge.

Dear, M., Ketchum, J., Luria, S., & Richardson, D. (Eds.). (2011) *GeoHumanities: Art, history, text at the edge of place*. Routledge.

de Leeuw, S. (2012) *Geographies of a Lover*. Edmonton: NeWest Press.

de Leeuw, S. (2015) *Skeena*. Halfmoon Bay, BC: Caitlin Press.

de Leeuw, S. (2017) Writing as righting: Truth and reconciliation, poetics, and new geo-graphing in colonial Canada. *The Canadian Geographer/Le Géographe Canadien, 61*(3), 306–318.

de Leeuw, S. (2019a) *Outside, America*. Vancouver: Nightwood Editions.

de Leeuw, S. (2019b). Novel poetic protest: What your heart is a muscle the size of a fist offers geographers writing to evoke rather than convey. *Emotion, Space and Society, 30*, 58–61.

de Leeuw, S., Donovan, C., Schafenacker, N., Kearns, R., Neuwelt, P., Squier, S. M., ... & Anderson, J. (2018) Geographies of medical and health humanities: A Cross-Disciplinary conversation. *GeoHumanities, 4*(2), 285–334.

de Leeuw, S., and Hawkins, H. (2017) Critical geographies and geography's creative re/turn: Poetics and practices for new disciplinary spaces. *Gender, Place & Culture, 24*(3), 303–324.

de Leeuw, S., and Hunt, S. (2018) Unsettling decolonizing geographies. *Geography Compass, 12*(7), e12376.

de Leeuw, S., and Magrane, E. (2019) Geopoetics. *Keywords in Radical Geography: Antipode at 50*, 146–150.

Fall, J. (2021). Worlds of Vision. *ACME: An International Journal for Critical Geographies, 20*(1), 17–33. Retrieved from https://acme-journal.org/index.php/acme/article/view/2037

Foster, K., and Lorimer, H. (2007) Cultural geographies in practice: Some reflections on art-geography as collaboration. *Cultural Geographies, 14*(3), 425–432.

Florida, R. L. (2005) *Cities and the creative class*. Psychology Press.

Harvey, D. (2001) *Spaces of capital: Towards a critical geography*. Routledge.

Hawkins, H. (2011) Dialogues and doings: Sketching the relationships between geography and art. *Geography Compass, 5*(7), 464–478.

Hawkins, H. (2013) *For creative geographies: Geography, visual arts and the making of worlds*. Routledge.

Hawkins, H. (2020) Geohumanities. In *The International Encyclopedia of Human Geography*. Ed. Audrey Kobayshi. Elsvier.

Imrie, R. (1996) Ableist geographies, disablist spaces: Towards a reconstruction of golledge's' geography and the disabled'. *Transactions of the Institute of British Geographers*, 397–403.

Kinkaid, E. (2019) Experimenting with creative geographic methods in The Critical Futures Visual Archive. *Cultural Geographies, 26*(2), 245–252.

Kinkaid, E., Emard, K., & Senanayake, N. (2020) The podcast-as-method?: Critical reflections on using podcasts to produce geographic knowledge. *Geographical Review, 110*(1–2), 78–91.

Kobayashi, A. (2006) Why women of colour in geography?. *Gender, Place & Culture, 13*(1), 33–38.

Kobayashi, A., Brooks, M., de Leeuw, S., Cameron, L. and Lewis, N. (2014) "Advocacy." *Sage Progress in Human Geography*. Ed. Roger Lee. Sage Publications. 404–419.

Kristeva, J., Moro, M. R., Ødemark, J., & Engebretsen, E. (2018) Cultural crossings of care: An appeal to the medical humanities. *Medical Humanities, 44*(1), 55–58.

Kwan, M. P., and Schwanen, T. (2009) Quantitative revolution 2: The critical (re) turn. *The Professional Geographer, 61*(3), 283–291.

Madge, C. (2014) On the creative (re) turn to geography: Poetry, politics and passion. *Area, 46*(2), 178–185.

Magrane, E. and Cokinos, C. (2016) *The Sonoran Desert: A Literary Field Guide*. Tucson: The University of Arizona Press.

Marston, S. A., and de Leeuw, S. (2013) Creativity and geography: Toward a politicized intervention. *Geographical Review, 103*(2), iii–xxvi.

Magrane, E., Russo, L., De Leeuw, S., & Perez, C. S. (Eds.). (2019) *Geopoetics in Practice*. Routledge.

McLean, H. (2016) Performing and resisting a drag and drop/plug and play world. *Performance Research, 21*(1), 140–142.

McLean, H. (2017) Hos in the garden: Staging and resisting neoliberal creativity. *Environment and Planning D: Society and Space, 35*(1), 38–56.

McLean, H. (2018) In praise of chaotic research pathways: A feminist response to planetary urbanization. *Environment and Planning D: Society and Space, 36*(3), 547–555.

Meinig, D. W. (1983) Geography as an art. *Transactions of the Institute of British Geographers*, 314–328.

Naidu, T. (2020) Southern exposure: Levelling the Northern tilt in global medical and medical humanities education. *Advances in Health Sciences Education*, 1–14.

Paiva, D. (2020) Poetry as a resonant method for multi-sensory research. *Emotion, Space and Society* 34, 100655.

Parikh, A. (2020) Insider-outsider as process: Drawing as reflexive feminist methodology during fieldwork. *Cultural Geographies, 27*(3), 437–452.

Parr, H. (2006) Mental health, the arts and belongings. *Transactions of the Institute of British Geographers, 31*(2), 150–166.

Parr, H., and Butler, R. (1999) New geographies of illness, impairment and disability. *Mind and Body Spaces: Geographies of Illness, Impairment and Disability*, 1–24.

Quoniam, S. (1988) A painter, geographer of Arizona. *Environment and Planning D: Society and Space, 6*(1), 3–14.

Richmond, C. (2018) The relatedness of people, land, and health: Stories from Anishinabe Elders. *Determinants of Indigenous peoples' health: Beyond the social*, 167–186.

Rose, G. (2016) *Visual methodologies: An introduction to researching with visual materials*. Sage.

Thibault, G. E. (2019) Humanism in medicine: What does it mean and why is it more important than ever?. *Academic Medicine, 94*(8), 1074–1077.

Ward, K. (2007) Geography and public policy: Activist, participatory, and policy geographies. *Progress in Human Geography, 31*(5), 695–705.

Watson, J. W. (1983) The soul of geography. *Transactions of the Institute of British Geographers*, 385–399.

Wylie, J and Webster, C. (2019) Eye-opener: Drawing landscape near and far. *Transactions of the Institute of British Geographers, 44*(1), 32–47.

PART II

Methodologies of Human Geography's Sub-Disciplines

Sarah A. Lovell

In Part II of this volume, contributors address the methodological tensions and emerging questions that dominate many of Human Geography's sub-disciplines and sub-fields. Sub-disciplines are an enduring framework through which human geographers align their work and derive their academic identities. The sub-disciplinary structure provides an anchor for readers, particularly graduate students, seeking context-rich understandings of key methodological developments in their field. For example, the data heavy field of GIS science remains concerned with issues of uncertainty (see Chapter 17), children's geographies address meaningful opportunities for participant engagement (see Chapter 24), while economic geographies tackle research that challenges the capitalist economic imperative (Chapter 22). Underpinning these issues – and indeed the vast majority of chapters in Part II – is a concern with producing research that overcomes normative representations of space and place. Collectively, emerging research practices across the sub-disciplines highlight a growing pluralism in human geography's research methods amidst a surprising degree of ontological unity.

We seek to move beyond dialectical debates dichotomising quantitative and qualitative approaches in human geography, focusing instead on methodological shifts within the discipline. In Part I, Barnes and van Meeteren (Chapter 1) and Rosenberg (Chapter 3) examined the reorientation of human geography toward and beyond logical positivism. As we turn to the sub-disciplines, feminist methodologies emerge as a contemporary influence shaping the development of qualitative research in emotional geographies, children's geographies, and economic geographies (see Foley Chapter 9, Yantzi and Loebach Chapter 24, Dombroski and Roelvink Chapter 22). Feminist researchers emerged as advocates for methodological decisions that gave voice to the "embodied and emotional lived experiences" of women that were otherwise invisible in human geography research (see Chapter 19, Geographies of Disability for further discussion). The debates over representation and power in research that were at the forefront of Feminist methodologies have come to occupy a central position in human geography that expands far beyond a focus on women.

Set against the context of growing debate in human geography, Part II begins with a focus on the methodological opportunities relational ontologies bring to the discipline (see introduction to this volume for a discussion of ontology). In Chapter 9, Ronan Foley reflects on how we access and express the inexpressible in emotional geographies research. Affective

DOI: 10.4324/9781003038849-11

landscapes are considered to be relational assemblages that elicit "fleeting emotional moments" that become the subject of research in emotional geographies. Dittmer, Ghoddousi and Page (Chapter 11) observe that relational ontologies erode distinctions between the researcher and researched and open up a dialogue around the potential to affect (and be affected by) research. The ontological orientation of research thus shifts within the context of posthumanism as it is conceived that "some sense of becoming with the subject of analysis" is unavoidable in the field. Assemblages are described as an agentive "constellation of bodies, objects and material forces that come together," an understanding of reality that echoes Indigenous ways of knowing (Chapter 11, p. 141).

Indigenous ways of knowing underpin the axiologies (values) that inform Indigenous research emphasising mutual respect and a duty of care, argue Parsons and Taylor (Chapter 12). With a grounding in oral tradition, decolonising research methods, such as Talanoa, Sharing Circles and calypso, are subverting the dominance of Western research methods and fostering respectful relationships that align with the values of communities (Parsons and Taylor, Chapter 12; Clarke and Mullings, Chapter 13). Such methodologies are important counterpoints to the growing disenfranchisement Clarke and Mullings recognise in their critique and commentary of Caribbean storytelling (Chapter 13).

Where surveys, interviews, and focus groups once dominated research methods in human geography, posthumanism has upended the human as the sole subject of analysis. This period of ontological renewal has prompted the widespread adoption of ethnographic, archival and participatory methods and the erosion of methodologically siloed sub-disciplines. Within the space of queer geographies, Catungal and Hilt (Chapter 10) argue that such methodological transgressions –including auto-ethnographic theorising and a concern with the ephemeral – has enabled the (hetero)sexualising of space to be challenged. This methodological pivot hinges on an understanding of research as political and enabling 'particular forms of action.' More broadly, the ontological commitments of the sub-discipline are evident in the adoption of emerging research methods and, particularly within the digital space, experimentation with emerging research methods should not, Campbell and Marek (Chapter 18) caution, come at the expense of theoretically "anaemic" human geographies.

Practices of archival research that were once the remit of historical geographies are addressed in chapters on Black geographies (Chapter 15), queer geographies (Chapter 10), and political geographies (Chapter 11) in which traditional sources and renderings of history are challenged. In Chapter 14, Beckingham and Hodder delve beyond a discursive criticism of archival research to call for practices of 'geographical antagonism' in recognition that the geographical histories of a collection – both its form and its omissions –may provide more insight into the stories we have told ourselves as human geographers, than the contents of a collection itself. Beckingham and Hodder (Chapter 14) go beyond the historical to interrogate how emerging technologies – from the algorithms to the collections they draw from –filter the histories we examine as human geographers.

The application of ethnographic approaches in digital geographies presents the opportunity to shed light into the 'black box' of digital systems. Lynch and Farokhi (Chapter 16) argue that ethnographies support an untangling of the enmeshed im/materialities of everyday digital lives. Digital technologies have reshaped our daily lives including our sense of agency and opportunities for methodological innovation Lynch (Chapter 16). Multi-site ethnography is discussed in a chapter dedicated to the geographies of education (see Cheng and De Silva, Chapter 23) as a means of accessing transnational narratives.

In Chapter 20 on animal geographies, Rubio-Ramon and Srinivasan remind us that our choice of research methods is inherently political, silencing some while giving voice to others.

While Rubio-Ramon and Srinivasan consider the representation of human and non-human perspectives as a key challenge for animal geographies, parallel concerns with representation arise in children's geographies. Yantzi and Loebach in Chapter 24, identify the flexibility and playfulness of participatory methods as critical to engaging children and challenging assumptions that they have little to contribute to research. Cheng and De Silva (Chapter 23), however, acknowledge both the value of children's voices and the limitations placed on children's agency by systemic power imbalances. They argue that research methods are needed to capture both the agential and structural influences on the geographies of education.

Concerns with representation in research are underpinned by a political understanding that research has the power to influence actions. Within geographical information science (see Chapter 17) Koo and Chun address the problem of uncertainty within georeferenced data. The authors identify challenges with data that relies on aggregation to extract population-based attributes and the use of social media data sources which omit large portions of the population. Ward (Chapter 21) describes a similarly widening array of data sources posing new opportunities for comparative research in urban geography. They argue the emerging experimentality within the sub-discipline is providing opportunities for creativity previously overlooked. For Wong and Beljaars (Chapter 19) examining the geographies of disability, mixed methods research poses an opportunity to address the "presences, absences, and temporalities of disability" thus destabilizing understandings of disability. In Chapter 15, Emerson dos Santos and Ferreira similarly argue that "creative resistance" requires research that emphasises the spatial agencies seeking to overcome racism. Emerson dos Santos and Ferreira consider Black geographies to be a necessary practice of political resistance in response to the hegemonic use of race.

The role of activism in the academy has been a growing focus of methodological dialogues in human geography with scholarship highlighting the diversity of activist approaches, including the power of 'quiet' activism characterised by a sense of intentionality through small acts (see Pottinger et al., 2017; Chatterton and Pickerill, 2010). This sense of intentionality is fundamental to Dombroski and Roelvink's (Chapter 22) recognition that our methodologies are the lens through which reality is interpreted. They argue for methodological diversity to 'see' beyond the capitalocentric approaches that have dominated economic geographies. Dombroski and Roelvink draw on the work of Sedgwick (2003) to argue that when a researcher documents alternatives to realities dominated by oppression and subjugation, they open themselves up to possibilities of self-nourishment through the affective potential of research. White and Springer (Chapter 25) conclude Part II by giving voice to anarchist geographies supporting interspecies justice and leading to new imaginaries and new research spaces. They make the case for researchers who are activists, agitators and advocates of the ontological values that are increasingly driving research in human geography.

The chapters in Part II have indeed addressed taken for granted research assumptions and methodological directions across the field of human geography. While not all sub-disciplines are captured, we have sought to represent core sub-disciplines of human geography (urban geography, economic geography, political geography) and a selection of sub-fields that have emerged through a critical mass of research centred on methodological developments (historical research, digital geographies, GIS science), social differences (e.g., sexuality, disability, racialisation, age), critical theory (anti-colonialism, anarchism), and subjects of study (health, animals, education). The chapters in this section capture key developments and contemporary methodological debates in these sub-disciplines, which, collectively, are moving forward disciplinary thinking.

Sarah A. Lovell

References

Chatterton, P. and Pickerill, J. (2010), Everyday activism and transitions towards post-capitalist worlds. *Transactions of the Institute of British Geographers*, 35: 475–490. https://doi.org/10.1111/j.1475-5661.2010.00396.x

Pottinger, L. (2017), Planting the seeds of a quiet activism. *Area*, 49: 215–222. https://doi.org/10.1111/area.12318

9

AFFECTIVE LANDSCAPES

Capturing Emotions in Place

Ronan Foley

Introduction

Across human geography, the capturing of human emotions and affective responses to, in and from place, has remained an elusive but ongoing topic of considerable interest (Bazinet and Van Vliet, 2020; Bondi and Davidson 2011). This has stretched from the humanistic geographies of the 1970s, where concerns for emotions in space first emerged, right up to contemporary in-situ and embodied emotional geographies work (Lucherini and Hanks, 2020). Why emotions matter and are a valid subject for human geographical research has been a common theme, though most scholars would subscribe to Anderson and Smith's (2001) observation that to not include emotion within human geography 'makes for anaemic knowledges'. In recognizing that emotions are central to how places/spaces are made and understood, and are individually and collectively felt, they are also shaped by power structures and cultural norms (Barron, 2021). While earlier research suggested that emotions were primarily human in origin, there is increasing recognition of an affective power within natural and built environments (weather/seasons) and subjects (flora & fauna) that also shapes emotional geographies (Williams et al., 2019). This is especially evident in Indigenous geographies and recent more-than-human turns in the subject but also across an increasing diversity of research settings and subjects (Gibbs, 2020). Previously unconsidered landscapes: memorial, imaginative, virtual, symbolic and immersive, are also being brought under an emotional spatial lens (Kinkaid, 2021). Given the complex, fleeting and embodied nature of emotions, an ongoing methodological issue across the decades has been how to effectively excavate affective landscapes in ways that unearth deeper feelings and instincts experienced in place yet stand up as robust and ethically sound evidence (Andrews, 2019; Horton and Kraftl, 2014). In describing some of the different ways this excavation of feeling in place ('place capture') has been approached over time, one can also track wider theoretical and methodological turns in the subject that cross multiple sub-themes within human geography (Hay and Cope, 2021).

From Phenomenology to Full-Blooded Relational Emotional Geographies

To discuss how different methodologies have been applied in emotional geographies research, it is helpful to contextualise how emotions have emerged as a geographical subject (Bondi

DOI: 10.4324/9781003038849-12

and Davidson, 2011; Horton and Kraftl, 2014). In discussing the various turns that have helped embed emotions within geographical research, the 'emotional turn' brought with it a shift from detached observational positionalities to those that were more affective, embodied, engaged, open, and relational. This movement from anaemic to full-blooded methodological engagement also drew from theoretical debates in feminist, post-structuralist, post-materialist and non-representational geographies. In particular, the last of these terms, non-representational geographies, was developed in part to move away from an excessive focus on representation and texts, to consider actions and material objects more fully (see Chapter 30 for further discussion of non-representational Geographies). Representations still matter, but its less a case of either-or, and more a case of 'both-and'. These new lines of thinking were especially important in emotional geographies, which had always struggled with more text-based representation. In particular, Lorimer's (2012) more generous term, *'more-than'-representational*, is helpful for the theoretical grounding of applied work in emotional geographies. In using a term like 'more-than' one gets closer to the nature of affective landscapes and how they are engaged with and understood. Using multiple overlapping methodologies and representations better reflects the complex form, production and experience of emotion in place (Barron, 2021).

As a response to the dominant spatial science paradigm in the mid to late 1970s, early humanistic geographers were forerunners of what we now call emotional geographies. Driven by a specific interest in phenomenological experience, the likes of Tuan, Seamon and others argued for a felt response to more fully understand how place was both made and represented (Casey, 2001). By extension, one's connection to place formed essential proto-emotional geography elements that connected human and non-human worlds. Typical methods included spoken and written accounts of encounters with the natural world but also ethnographic and observational work in both urban and rural settings, though contemporary scholars noted limited empirical explorations within a more explicit theoretical focus (Jackson, 1981).

Through the 1980s and into the 1990s, critical geographical thinking within the 'cultural turn' began to take more seriously how place and emotion interacted and were shaped by wider structural forces in society. An especially influential strand emerged out of feminist geographies on different ways of knowing. This work used emancipatory methods that uncovered and made visible previously hidden emotional voices and narratives, wherein gender, race, sexuality, bodies and experiences of difference were given place and value (Davidson, Bondi and Smith, 2005; Longhurst, 1997; Valentine, 1989). Here specific emotional landscapes became visible in approaches that moved from material space description to complex place-based elements, which addressed the many ways landscapes, and the fleshy bodies within them, could be framed, made and, in particular, felt. Terms such as landscapes of fear and despair aptly described more subaltern voices and collectively made clear how for example, women experienced night-time spaces or how people experienced discrimination and deprivation in ways that were shaped by wider forces (Dear and Wolch, 1987; Davidson, Bondi and Smith, 2005; Valentine, 1989).

The cultural turn also made space for other ways of knowing and seeing, drawing in part from a wider critical focus on representation and how a better understanding of place-making, power and difference could be identified in sharper critical readings between the lines and spaces. In this period, the tracing of emotional landscapes continued to be text-based, but in different formations and assemblages wherein the visual was often prominent (e.g., studies of subjects like graffiti, the documentation of activist events and a wider use of imagery, sound and video that was focused on the politics of injustice and inequalities). In addition, there was a deeper attention to an ethic of care and debates on wider methodological ethics within this research (Kitchin and Tate, 2013). Autoethnographic approaches also became more widely accepted and embedded in human geography, allowing space for researchers to place themselves into research methods as more than just observers (Hay and Cope, 2021).

Definitional debates of how emotions worked in space led to a more 'emotional turn' in the early 2000s. Key papers and texts became explicitly labelled as emotional geographies marking an important space of alignment with new post-structuralist research (Anderson and Smith, 2001; Bondi and Davidson 2011). Through discussions from non-representational theories (NRT), arguments emerged on how terms like emotion and affect inter-acted (Anderson, 2006; Tolia-Kelly, 2006). Some scholars argued a fundamental difference between the concepts, others a more continuum approach, but at heart was a sense that affect in particular was a pre-cognitive and almost unrepresentable form which, unlike emotion, defied any sort of methodological uncovering (Barron, 2021). Yet, despite arguments for affect being non-representable and reso-lutely a-material, it was almost impossible to articulate this without recourse to specific applied settings and practices (Ash and Simpson, 2016). A *more-than* affective approach is helpful here, that can incorporate that continuum of pre- and post-cognitive states associated with emotional geographies work. This link between instinct, cognition and behaviours was also reflected in other branches of emergent emotional work such as psychotherapeutic geographies (Bondi, 2005 & 2014; Philo and Parr, 2003). Methodologically, deep studies of affect remain tricky to operationalise in material settings and a key argument for using broader emotional definitions was to create workable methodologies that meaningfully challenged power and the relevance of emotion within an increasingly neoliberal world (Tolia-Kelly, 2006). In developing more relationally framed geographies of affect/emotion that recognise complex networked spaces and lives, these new methods argued for a fuller sensing-in-the-world (Bondi and Davidson, 2011). Methodologically this translated into more-than-representational geographical research that attempted to somehow measure the immeasurable, to capture the moment, to return to phenomenologically framed co-creative methods that foregrounded more mobile/lived/sensed approaches (Anderson, 2014; Anderson, 2019).

Most recently, it feels as if there is a general acceptance across emotional geographies research for a theoretical positioning around a range such *more-than* approaches that recognise several trends (Latham, 2003; Lorimer, 2012). Concerns with emotional geographies increas-ingly recognise close links with how place affects mental health and wellbeing (at individual and societal levels) and that relational models using connectivity, networking and structures are important in understanding the production and management of emotion in contemporary society (Latham, 2003). Linked to this relational model, we are simultaneously living (before and during the COVID-19 pandemic) in an increasingly digital and online world within which new assemblages of emotion around identity, politics, inequality, climate change and spatial justice are played out in everyday lives. There is a deeper recognition of Indigenous understandings of affective landscapes, such as the Australian Aboriginal version of 'country' as a reciprocal model of caring for the land that specifically articulates a nature-culture affect and emotional connection to place and place loss. Such shifts are reflected in methodological approaches such as citizen social science and artistic practice as ways to capture emotion, often respectfully co-produced within activist research. (Awcock, 2021; Willmot, 2021) Psychotherapeutic geographies additionally identify new landscapes of fear linked to climate anxiety and COVID-19 as existential affective events (Galway et al., 2017; Morrow-Howell et al., 2020). Paying attention to affective landscapes in all these forms shows that a vital and active engagement in places and spaces requires open and empathic methodologies, both for people/place health and wellbeing but also wider social justice, within which multiple more-than methods will be central.

Place Capture: From Words to Bodies to Actions

Shifting ideas on the capturing of place-based emotions is reflected in the breadth of methods used within emotional geographies. While the previous theoretical section was broadly chronological, this section is more thematic but recognises an accretive and complementary element. The themes discussed in this section are very broadly codified as stories, observations and encounters, though good affective landscapes research can often combine all three. All contain a significant embodied component that shifts from single human bodies to others that are more-than-human, collective, transgressive and mobile. The three themes can also be used to consider orthodoxies, technologies and urgencies on how emotional geographies are made and held in the landscape. This includes important aspects of design, collection, analysis and dissemination as well. While the chapter draws from across human geography, the author does acknowledge a positionality as a medical/health geographer. At heart, any study of affective landscapes and emotional geography is always a form of 'place capture' (Britton and Foley, 2020) based on an empathic co-listening to self and place. That place capture must also reflect the ways in which places and intersubjectivities are always active partners; feelings of both 'joy and pain' can emerge from those emplaced encounters (Krinke, 2021). Yet, emotions do not always have to be high and low, and the everyday and everyway of ongoing moods and affections – boredom, satisfaction, ease/unease – act as a balance and are important, if less often discussed.

Stories: Talking and Hearing Affect/Emotion

In using the broad term stories, written and oral accounts are as important now as they were in the 1970s. In the work of Tuan and Seamon place and space were identified as phenomenological settings and lifeworlds in which perceptions and experiences blended to create place bonds (Seamon, 1979; Tuan, 1977). Documenting these in subjective first-hand accounts was given prominence and especially valued felt-responses to place-encounters; Seamon talks about everyday encounters and actions, such as a random meeting in a crowded street or everyday habits within the house as deepening an emotional connection to place (Seamon, 1979). In gathering emplaced stories, questionnaires and interviews, either semi-structured or open, remain essential tools (Hitchings and Latham, 2020a). The documentation of first-hand experience of/in place remains a key foundational element in an emotional geographer's skillset. There remains a skill in these forms of how best to uncover emotions through questions and prompts, something that requires empathy, meaningful listening, flexibility and responsiveness, underpinned by thoughtful ethical considerations on who you are interviewing, why and where and who you are as an interviewer (Askins, 2009).

While early accounts tended to be individualised and did not necessarily challenge norms in society, the shift forward in understandings of narrative and story-making from place as political was a significant element of the cultural turn. In exploring voices of difference, the various ways voices narrated place gave a significant boost to how place and the emotions of place could be captured. These newer forms of narrative – confessional, personal, relational, geo-located, coming from previously excluded/silenced voices - began to show how there were multiple, often competing stories, to be captured from the same place. From initial feminist approaches, these new stories were underpinned by the very specific emotions of being invisible, neglected or ignored. From feminist geographies came a specific concern for embodied voices, not just women's lives but also the lives of other previously silenced groups, such as people with disabilities, gay and lesbian communities, migrants and other racial and ethnic minorities (Bondi and Davidson, 2011). Using research interviews and focus groups, research subjects were given space to tell their stories through traditional forms of qualitative design (Horton and Kraftl, 2014). Topics from feminist research – such as public breast-feeding, public space safety, and body identity – also raised important ethical questions about where, how, why and by who, certain forms of respectful methodologies could be employed in the study of emotional geographies of difference (Datta et al., 2020). Capturing voices from place remains a concern for newer activist work too; more emplaced recent studies seek to consider how stories reflect place and how each place creates its own stories and produce place narratives of value (Hitchings and Latham, 2020a). Often where a story comes from, both literally and metaphorically, is an important part of methodological thinking and directly shapes how that story is told (Dean, 2016; Schwanen et al., 2012). See Chapter 13 for further discussion of storytelling.

For many groups, telling a story from the inside is empowering and methodologies that are co-created and open space for direct rather than mediated stories are essential in contemporary studies, for both ethical and practical reasons. Examples include stories across many geographies, one example of which would be the experience of refugees, migrants and asylum seekers; recent work on solicited diaries with written instructions, proving especially effective in uncovering the emotional lives of Middle-Eastern refugee women both in the home and refugee spaces (Linn, 2021). Social and cultural geographies of racialisation, as discussed in Chapter 15, are lit up by emotional testimonies of discrimination across multiple spaces and settings, from lingering geographies of redlining in relation to the creation of urban ghettos to exclusions from leisure spaces, all of which lead to a deeper unmaking of place-based emotions (Pulido, 2017). Queer geographies also work with complex stories across complex communities that attest to spaces of discrimination and methodologically identify alternative cultural spaces (night-time, online) to gather those stories from (Browne and Brown, 2017, Bonner-Thompson, 2020). Alternative ways of telling stories have developed as emotional responses, documentary evidence, oral histories or even testimonies of often harsh and traumatic experiences in time and space, such as asylum-seeker photos documenting their emotional responses to state-provided housing spaces in Ireland (O'Reilly, 2018). Within medical/health geography, for example, while health service planners generally favour positivist accounts of spaces and networks, critical stories counter these through other stories of lost access, extended waiting times, lack of recognition or community preference (Hopkins and Pain, 2007).

Finally, the very nature of communication and how stories are told and heard are evident in the ongoing development of the web and social media as new spaces and forms for capturing emotion. Here, digital storying and creative work, while maintaining a significant focus on text, has opened even wider access to harder-to-reach respondents, and to collaborative methods seeking to uncover written perspectives on emotion. While traditional historical geography foregrounded written and official texts, the importance of biographical approaches and, in particular, a new openness to the subaltern voice in that archival work has transformed research. Biographical accounts are one core method used, especially in allowing space for previously silenced voices from historical and post-colonial contexts. Such work has a long history for a reason and continues to matter; the hearing, telling and passing on of emplaced stories and experience needs be narrated but also recorded. Within biographical approaches the use of diaries and diarying more widely, such as illness or leisure diaries, for example, have been one tool that sustains by recording an individual story with reference to place and wider relational elements that provide rich accounts (Latham, 2003; Moss and Dyck, 2002). While traditionally written, new technologies in the form of audio podcasting or video diaries make this easier in terms of accessing quiet or silenced voices, both in-person and online, and become additional ways of recording both individual and community/collective biographies (Hawkins, 2019).

Observations: Viewing and Placing Affect/Emotion

Most stories come not just out of perceptions and experiences, but also through observation. While a slightly artificial distinction, it is important to recognise that much valuable empirical material in emotional geographies is captured via observation. This can take several forms of which participant observation (and its corollary, observer participation) have been core components of wider ethnographic methods for decades (Kitchin and Tate, 2013). While giving voice to marginalised individuals and communities matters, there is additional value and skill in using both detached and engaged noticing across different scales, settings and temporalities (Dean, 2016; Barron, 2021). This also requires good researchers taking both ethics and research design seriously, within which piloting and testing remain crucial elements. While traditional ethnographic work draws from sustained and grounded research in one location across a meaningful period (Hitchings and Latham, 2020b), geographers have also set out to capture emotions in place using such methods in shorter time-periods and across heterogenous settings and groups. Such work, in place capture terms, emerges in different ways: as notes, as recordings (visual, sonic etc.), and as shared recollections. The how of that observational place capture is also complex, from detached/engaged observers through to kinetic and vitalist forms of go-along interviews in which the researcher actively participates. Finally, the importance of the place itself, as a core component of place capture, must also make space to consider the changing and mobile places and spaces within which such work is done. There are place-specific elements such as affective atmospheres (in a dance club, at a revivalist church, at a sports stadium or an activist demonstration) that shape how emotion is or can be recorded (Tomalin et al., 2019, Andrews et al., 2014). One can be caught up in and swept up in the emotion within live settings that can turn bad at any moment, while other relational elements such as weather can also affect methodological capacities to carry out effective and robust place capture (Bell et al., 2019).

While not exclusive to the field, there is a particular value in using field notes as a fundamental component of emotional geographies method. In sitting in any environment, and observing any individual or subject group, the simple writing down of what you see, or indeed feel, is a key starting point, but especially so in emotional geographies research, where gesture, tone or body-language are important. Field notes are by definition, more open and contingent than a questionnaire or semi-structured interview. While the technological opportunities for the recording of field notes have changed, with direct sound and video now a possibility, at heart is an intent to pay attention and record some sort of observed emotion. This is not always easy and requires different skills in different settings; wherein emotions can be more evident, whereas affect is harder to observe; the latter especially contingent and fleeting. While a mode of repeated notetaking at fixed times is one route, there is also scope for event or encounter based note taking. In addition, one might also be actively engaged with a group and take notes from that encounter. In particular, writing notes from a focus-group encounter may document emotional responses as a specific theme or perhaps the emotions emergent in the encounter provides some important additional insights. In addition, the space of the encounter – private, outdoor, moving – may have an important role in shaping deeper emotional responses (Foley et al., 2020).

Across human geography more generally, visual methodologies are established approaches within which insights into emotional geographies often emerge, both by design and accident (Rose, 2012). Traditionally, cheap disposable cameras were a central tool, with a research cohort sharing images or meeting for individual or group elicitation, on what their images meant to them. Examples with a specific emotional component included work from children's geographies, where imagery becomes a comfortable shared route to insight (Steger et al, 2021). Myers' (2010) study with people with HIV/AIDS in Auckland, NZ used auto-photography and captioning as methods that allowed participants to record and narrate how the acquisition of HIV affected the experiences of research participants of both literal geographical places as well as their new 'place-in-the-world' linked to their diagnosis and public perceptions around them. Work by Duff (2012) with young people with mental health problems sought to capture how they chose health-enabling spaces in Melbourne. While one assumes a value for green/blue space, what the research captured were more personal choices around grey and interstitial spaces that had resonance to individual emotional responses and wellbeing supports. Other examples include photovoice research with vulnerable groups, like recovering alcoholics, on how they emotionally encountered spaces and used them to make sense of their own recovery (Shortt et al., 2017). Finally, photographs and other forms of imagery can act as emotional triggers or ways of tracing emotions in public space as evidenced in Wylie's (2009) retracing of a walk along a coastal path in the UK and the effect of finding emotional resonance in the plaques listed on benches overlooking the sea, while Lorimer's (2019) visual and spoken encounters at a pet cemetery in Northern Scotland also identify the location of emotion in memorial objects and spaces. Barron's (2021) use of photo-walks with older people in Manchester contains identifiable more-than-representational and relational elements; body language, for example, is identified in-situ as an important place-specific element.

With technological change via the near ubiquitous smartphone and web-based sharing platforms, if anything, access to new forms of image-voice to capture emotion in place have been made easier. Another example from mental health research, Söderström et al., (2016) filmed young people with mental health difficulties tracing their chosen routes through a Swiss city to identify how they negotiated safe spaces for themselves in their everyday mobilities. In addition, mobile phones allow for the capture of aspects of everyday emotional lives across the lifecourse (see Chapter 16 for further discussion of Everyday Digital Geographies). While there can be a focus on more difficult emotions, especially in activist work, it is also important to recognise a place for more positive emotions as well that capture play, joy, relational family and community lives, and wider dimensions of individual and collective wellbeing. With geographical work with older people, the use of imagery has been used both as reminiscence therapy but also to record oral histories of place (Barron, 2021). Here both photo and video imagery work with elicitation with research subjects to provide a literal form of place capture; in which the image acts as an emplaced prompt that captures emotion though there is a skill in managing the power of such imagery to stir and upset too, and a skill on the part of the researcher in managing that (Rose, 2016). Indeed, in much activist and participatory research, there is also a recognisable, even sometimes deliberately provocative form of emotion capture linked to imagery including work from asylum-seeker spaces in Ireland (O'Reilly, 2018), illustrations of traumatised bodies (Philo, 2014), street stickers (Awcock, 2021) and wider publicly shared digital imagery (Leszczynski, 2019). Finally, the use of imagery in psychotherapeutic work uses images, drawing and other sense-based approaches to develop mentalising methods as routes towards understanding trauma and deep emotional histories (Lengen, 2015; Rose, 2012).

Encounters: Sharing and Living Affect/Emotion

While the above sections have been presented as relatively static, this final section considers mobility as a key component in recent emotional geographies research. Within place capture, recording emotional and affective encounters ideally uses live-capture methods that provide embodied routes into emotional geographies (Hitchings and Latham, 2021; Spinney,

2015). Such encounters are always complex, ineffable and contested and the emergence of emotions within active individual or communal spaces takes many living forms (Finlay and Bowman, 2017). These 'encounters-as-practice' or 'representations-in-relation' typically involve immersive place-based recordings in which sound/vision and other senses are brought to bear in literally 'active-ist' ways (Anderson, 2019). Broadly framed as in-situ and go-along methods they have an increasingly important role in emotional geographical method as very specific forms of vital and relational place capture. In theoretical terms, they mirror more-than-representational ideas made evident in kinetic, embodied and communal ways to echo what Andrews (2019) describes broadly as capturing the moment. While there is some overlap with observation and storytelling, studies that work with go-along photo-walks, such as Barron's study of older people in Manchester, link theory and practice together and make a strong case for the added value of such methods in capturing forces of existing, flows and temporalities and fleeting encounters that emerge in the momentary places and spaces (Barron, 2021, 10).

Many examples are found across cultural, health and sport and leisure geographies (Latham and Layton, 2020). From cycling, Spinney's (2006) paper on the kinaesthetic emotions associated with a climb of Mont Ventoux encompass how memory, place reputation, heat and suffering cycling bodies, provides rich explications of affect/emotion, while other accounts from cycling and rambling also attest to how activity spaces help bodies access emotions in both active and activist ways (Doughty, 2013; Waitt et al., 2021 Wylie, 2006). Blue space practices including diving, kayaking, surfing and swimming also engage with such kinaesthetic methods, with more specific immersive components. Examples include video elicitation from diving (Straughan, 2012), GoPro moments from within the surfed wave (Britton & Foley, 2021), a kayak-level perspective moving through water (Foley et. al., 2020) or in differential swimming accounts from within the water (Denton and Aranda, 2020; Foley, 2017). Indeed, these papers are important as documentations of direct and in-the-moment place capture, that relays emotion and affect, but is also methodologically honesty in how tricky these methods can be (Dean, 2016; Denton, Dannreuther and Aranda, 2021; Foley et al., 2020; Hitchings & Latham, 2021). Place capture is ambitious but must also be honest and realistic, especially in in-situ work, and recognise that the more active the method, the less accessible and consistent it might be. As new technologies, GPS, accelerometers, spatial video and action cameras begin to enable such work (Bell et al., 2015; Vannini and Stewart, 2017), one striking example is research that uses wearable stress monitors to literally record emotional responses as bodies move through space. Such wearable technologies, which measure galvanic skin and other physiological markers, can be powerful (if expensive) in providing a specific technical route to capturing emotion (Osborne and Jones 2017). Others technical tools include GPS-enabled work on tracing activity spaces for children or older people and linking these to diaries and other personal accounts that might explain emotionally framed barriers and enablers to fulfilled lives (Ho et al., 2020). Sarah Bell's research on geo-narratives provides some useful insights into the benefits of triangulated go-along research and new perspectives unearthed along such mobile journeys (Bell et al., 2015). Additionally, her work with people with sensory impairments and how these require flexible approaches that provide enabling methods to trace enabling environments is instructive (Bell and Foley, 2021). Indeed, the wider disabilities imperative for all research, 'nothing about us, without us', also emphasises an additional emotional imperative on how research is done as well as on or who with (McPherson, 2011). More broadly the role of bodies as sensors, wherein touch, smell, taste and sound augment sight, are essential components of these types of place capture and act as very specific modes of intimate sensing (Britton and Foley, 2021; Paterson and Dodge, 2012).

In a broader sense, this applies also to vital emotional research methods within everyday community geographies. Beyond individual recordings, more group-based or community-specific research also has in-situ components; for example, studies of walking groups can uncover emotional experiences as they move, across bodies, spaces and lifecourses (Pitt, 2015). There is also a potential scaling up to make such work valuable in providing shared community perspectives that might inform public policy, where emotions might be, for all their apparent elusiveness, made to matter (Anderson and Smith, 2001). Employing mobile emotional methods as a form of citizen social science have been used in studies of community mapping with Indigenous people (Young and Gilmore, 2013), access to gardens for local communities (Pitt, 2015), older communities (Grove, 2021) and more-than-human geographies (Gibbs, 2020). More broadly there is also a growing digital world, within which social media and other digital spaces have become, for all ages, spaces of emotional expression in positive ways for previously hidden or disconnected communities; yet, also new spaces of anxiety and emotional turmoil (Parr et. al., 2015; Twigg et al, 2020). As one of the key effects of the COVID-19 pandemic has been to drive more people online and reduce capacities to carry in-situ research, it also offers new opportunities too around online access to research subjects though issues of robust research design remain.

Valuing Emotional Geographies Methods

In studying affective landscapes, much of the task is in capturing an emotion in place; a task that at times feels like - as affect scholars might attest – making sense of the insensible. At times, it emerges in an embodied response, a shiver or risen hackle, or emergent in body language, but as Lorimer (2012) notes, you know it when you sense it, as a meditative moment on a bicycle or surfboard, or sitting in warm sun with friends and family. Equally it can be a challenging moment of danger and risk, or something older triggered in place as a testimony to psycho-therapeutic geographical thinking (Bennett, 2009; Bondi, 2014). Capturing those moments, both in and out of place is hard to do, record or measure, but worth attempting. Yet, for it to have a value beyond the fleeting, emotional geographers should be active in methods that feed analysis and useful societal outcome, wherein emotion really matters (Anderson and Smith, 2001). In recent in-situ and mobile examples, one is moving and being moved at the same time; where despite the tricky processes, there is a value to uncovering the 'sticky' traces or at least considering how some stick and others don't (Askins, 2009). It is often creative artists who can pin down affect in a painting, photo, artwork or poem, that literally capture the moment. Such creative methods are being employed to open space for emotional expression in alternative and complementary forms, geo-poetic, performative or deeply cartographic (De Leeuw, 2017; Willmott, 2021).

Understanding affective landscapes as relational, as assemblages of fleeting emotional moments in and out of place remains central. Emotions emerge in different modulations in place, wherein they can become both heightened and flattened and it is important to recognise this continuum. Just as, in John Lydon's words, 'anger is an energy', so too is joy, with all the attendant and in-between felt emotions of fear, anxiety, dread, boredom, anticipation, disappointment and ecstasy in between. But energies and energetic spaces are at the heart of embodied and sensory-led methods that also effectively capture emotion in place, reflecting Robin Longhurst's (1997) identification of the body as the place 'closest in'. At the heart of good emotional geographies research is an attempt to get to that closest-in, and often closed-in parts of ourselves and the world around us. How we do that, how we value that, how we record that in ways that are faithful, empathic, realistic, legitimate and leave space for the inexpressable are all part of that challenge. Being relational also means being attentive to temporality and how that too is emergent in place capture methods. Across studies, often, though not exclusively with older people, place triggers, atmospheres, relational accounts and affective memories are all part of intimate sensings of place and how sometimes staying in the moment is essential for both researcher and research partner (Barron, 2021; Dean, 2016).

While mentioned sparingly thus far, ethics play an essential role all. Given emotions can be complex and difficult, researchers have responsibilities, to both themselves and research partners, across the research process, from initial design, to collection, to analysis, reporting and dissemination (kinspainsby hill, 2020). This applies to stories, observations and encounters, each of which may involve exploring sensitive emotions and indeed trauma. Even as such research develops, robust ethical review will also be necessary (Dean, 2016; Foley et al., 2020). Much recent research has been co-produced and affectively co-developed, but it is still ethically important to be sensitive to nuance, with additional challenges, for example, for non-Indigenous researchers to explore emotional geographies of Indigenous and other marginalised groups (Warren, 2017; Wheaton, 2020).

The spaces for emotion to be uncovered within other affective landscapes – memorial, imaginative, virtual, symbolic and immersive – are also considerable. Memorials and symbols, as we have seen recently in the form of statues, flags, buildings and events, have the capacity to stir emotions in very evident and place-based ways and indeed help emplace the importance of emotion more fully in public minds (Horton and Kraftl, 2014). At the heart of these events is a political/activist geography, not necessarily always a progressive one, but one that reflects voices from place. Studying events or indeed longer-term conflicts from a methodological point of view is difficult and complex, where anger, grief, the threat of violence or unpalatable subjects may put researchers off. But often, what lies behind such emotions are longer-term psychotherapeutic dimensions such as post-traumatic stress. Dealing with those deeper affective responses might also be where space for a full-blooded emotional geography, for all its methodological difficulties and discomforts, holds most potential. It is in being immersive that the best forms of place capture take place. Finally, we can use an embodied analogy in methodological terms wherein scientific human geographies acts as the skeleton/bones of place capture; other human geographies act as skin and flesh to add three-dimensional depth, but it is emotional geographies methods that capture the circulatory system, the living/flowing element that makes spatial encounters and affects real (Andrews, 2019). Capturing the flow is never easy; not attempting to do so allows those flows and vital components to dissipate or be lost.

Place capture means also capturing the value of place and the value of emotions in place in ways that are complex and slippery. There is value in capturing the more-than-human, not just in landscape elements but in the different animal and natural inhabitants and rhythms of those same landscapes. Place capture is also about an intimate sensing of place that uncovers meaningful dimensions of affective landscapes, places that touch and move us in their essence. A wide range of sensory geographies remains open and fertile ground for emotional geographies work, especially post-pandemic, with both embodied and technologies tools available to record emotional experiences. Indeed, a range of instruments, psychological, physiological and cartographic, have been developed that seek to directly measure and map affective bodily responses (Ho et. al., 2020; Osborne and Jones, 2017; Willmot, 2021; Young and Gilmore, 2017). Contested examples such as mapping happiness are another example, but more broadly, we must reconcile artificial boundaries between emotional and political work and the insertion of emotion into wider geographies of activism. This is essential to engender meaningful changes in spatial inequalities, something that all emotional geography studies might consider in terms of outputs from their work (Anderson and Smith, 2001).

Finally, it is instructive to write about affective/emotional research at a time of especially heightened emotions. For all the potential reach of a relational and connected world, it does feel as if the flat online world of the recent COVID-19 pandemic has closed off access to affective encounter in affective spaces for many. In particular, the very specific personal and place energies inherent in three dimensional and shared spaces are lost. The value of natural landscapes (green/blue) has been especially identified as spaces of emotional potential, if they are accessible to all (Bell et al., 2015). People in poorer communities have been most affected by COVID-19, living what Wilkinson and Ortega-Alcázar (2018), describe as exhausted geographies with depleted energies and emotional resources. Exhaustion acts as a good analogy for different dimensions of emotion, being felt and expressed but also a hard to represent, yet deeply affective register. Perhaps an additional telling question in emotional geographies research might be to ask what doesn't affect us or what a non-affective landscape might mean. It might be a crueller space that many of us occupy due to an inability or even unwillingness to deal with the inequalities in the world. But this may actually be the sometimes indifferent space that power occupies. Paying attention to affective landscapes in all their forms, and studying them deeply, sustains a vital and active engagement in place and building as a form of empathic therapeutic accretion important for health and wellbeing but also fairness and equality and the ongoing power of place as a trigger to identify important affects and emotions. In the coming decades, the continued exploration and place capture of everyday affects, negative and positive, in personal, community, social and political lives and their stories, images and activities will continue to recognise the embeddedness of affect and emotions in lived worlds and lives.

References

Anderson, B. (2006) Becoming and being hopeful: Towards a theory of affect. *Environment and Planning D* 24:5, 733–752.

Anderson, B. (2019) Cultural geography II: The force of representations. *Progress in Human Geography* 43:6, 1120–32.

Anderson, J. (2014) Exploring the space between words and meaning: Understanding the relational sensibility of surf spaces. *Emotion, Space and Society* 10, 27–34.

Anderson, K. and Smith, S. (2001) Editorial: Emotional geographies. *Transactions of the Institute of British Geographers* 26, 7–10.

Andrews, G.J. (2019) Health geographies II: The posthuman turn. *Progress in Human Geography* 43:6, 1109–1119.

Andrews, G.J., Chen, S. and Myers, S. (2014) The 'taking place' of health and wellbeing: Towards nonrepresentational theory. *Social Science & Medicine* 108, 210–222.

Ash, J. and Simpson, P. (2016) Geography and post-phenomenology. *Progress in Human Geography* 40:1, 48–66.

Askins, K. (2009) 'That's just what I do': Placing emotion in academic activism. *Emotion, Space and Society*, 2:1, 4–13.

Awcock, H. (2021) Stickin' it to the man: The geographies of protest stickers. *Area*, 1–9.

Barron, A. (2021) More-than-representational approaches to the life-course, *Social & Cultural Geography* 22:5, 603–626.

Bazinet, T. and Van Vliet, l. (2020) Affect. In: Kobayashi, A. (Ed.), *International Encyclopedia of Human Geography*, 2nd Ed. London, Elsevier, 47–52.

Bell S.L., Phoenix, C., Lovell, R. and Wheeler, B.W. (2015) Using GPS and geo-narratives: A methodological approach for understanding and situating everyday green space encounters. *Area* 47:1, 88–96.

Bell, S.L., Leyshon, C. and Phoenix, C. (2019) Negotiating nature's weather worlds in the context of life with sight impairment. *Transactions of the Institute of British Geographers* 44:2, 270–283.

Bell, S.L. and Foley, R. (2021) A(nother) time for nature? Situating non-human nature experiences within the emotional transitions of sight loss. *Social Science and Medicine* 276, 113867.

Bennett, K. (2009) Challenging emotions. *Area* 4:3, 244–251.

Bondi, L. (2005) Making connections and thinking through motions: Between geography and psychotherapy. *Transactions of the Institute of British Geographers* 30, 433–448.

Bondi, L., Davidson, J., (2011) Commentary: Lost in translation. *Transactions of the Institute of British Geographers* 36, 595–598.

Bondi, L. (2014) Understanding feelings: Engaging with unconscious communication and embodied knowledge. *Emotion, Space and Society* 10, 44–54.

Bonner-Thompson, C. (2021) Anticipating touch: Haptic geographies of Grindr encounters in Newcastle-upon-Tyne, UK. *Area*, DOI: 10.1111/tran.12417.

Britton, E. and Foley, R. (2021) Sensing Water: Uncovering health and wellbeing in the sea and surf. *Journal of Sports and Social Issues*, 45:1, 60–68.

Browne, K. and Brown, G. (2017) *The Routledge research companion to geographies of sex and sexualities.* Abingdon, Routledge.

Casey, E. (2001) Between Geography and Philosophy: What Does It Mean to Be in the Place-World? *Annals of the Association of American Geographers*, 91:4, 683–693.

Datta, A., Hopkins, P., Johnston, L., Olson, E. and Silva, J.M. (2020) *Routledge Handbook of Gender and Feminist Geographies.* Abingdon, Routledge.

Davidson, J., Bondi, L. and Smith, M. (2005) *Emotional Geographies.* Farnham, Ashgate.

Dean, J. (2016) Walking in their shoes: Utilizing go-along interviews to explore participant engagement with local space, In, Fenton, N. and Baxter, J. (eds), *Practicing Qualitative Methods in Health Geographies,* Farnham, Ashgate, 111–128.

Dear, M. and Wolch, J. (1987) *Landscapes of Despair: From Deinstitutionalization to Homelessness.* Princeton University Press.

de Leeuw, S. (2017) Writing as Righting: Truth and Reconciliation, Poetics, and New GeoGraphing in Colonial Canada. *The Canadian Geographer* 61:3, 306–318.

Denton, H. and Aranda, K. (2019) The wellbeing benefits of sea swimming. Is it time to revisit the sea cure? *Qualitative Research in Sport*, 12:5, 647–663.

Denton, H., Dannreuther, C. and Aranda, K. (2021) Researching at sea: Exploring the 'swim-along' interview method. *Health & Place*, 67, 102466.

Doughty, K., (2013) Walking together: The embodied and mobile production of a therapeutic landscape. *Health & Place* 24, 140–146.

Duff, C. (2012) Exploring the role of 'enabling places' in promoting recovery from mental illness. a qualitative test of a relational model. *Health & Place* 18: 1388–1395.

Finlay, J. M., & Bowman, J. A. (2017) Geographies on the move: A practical and theoretical approach to the mobile interview. *The Professional Geographer*, 69, 263–274.

Foley, R., Bell, S., Gittins, H., Grove, H., Kaley, A., McLauchlan, A, Osborne, T., Power, A., Roberts, E. and Thomas, M., (2020) Disciplined research in undisciplined settings: Critical Explorations of In-Situ & Mobile Methodologies in Geographies of Health and Wellbeing. *Area* 52, 514–522.

Foley, R. (2017) Swimming as an accretive practice in healthy blue space. *Emotion, Space and Society* 22, 43–51.

Galway, L., Beery, T., Jones-Casey, K. and Tasala, K. (2017) Mapping the Solastalgia Literature: A Scoping Review Study, *International Journal of Environmental Research and Public Health*, 16:15, 2662.

Gibbs, L. (2020) Animal geographies I: Hearing the cry and extending beyond, *Progress in Human Geography*, 44:4, 769–777.

Grove, H. (2021) Ageing as well as you can in place: Applying a geographical lens to the capability approach. *Social Science and Medicine*, 288, 10. https://doi.org/10.1016/j.socscimed.2020.113525.

Hawkins, H. (2019) Geography's creative (re)turn: Toward a critical framework. *Progress in Human Geography*, 43:6, 963–984.

Hay, I. and Cope, M. (2021) *Qualitative Research Methods in Human Geography, 5th Edition.* Oxford University Press.

Hitchings, R. and Latham, A. (2020a) Qualitative methods I: On current conventions in interview research. *Progress in Human Geography*, 44:2, 389–398.

Hitchings, R. and Latham, A. (2020b) Qualitative methods II: On the presentation of 'geographical ethnography'. *Progress in Human Geography*, 44:5, 972–980.

Hitchings, R. and Latham, A. (2021) Qualitative methods III: On different ways of describing our work. *Progress in Human Geography*, 45:2, 394–403.

Ho, E., Zhou, G., Liew, J-A, Chiu, T-Y, Huang, S. and Yeoh, B. (2020) Webs of Care: Qualitative GIS Research on Aging, Mobility, and Care Relations in Singapore, *Annals of the American Association of Geographers*, DOI:10.1080/24694452.2020.1807900

Hopkins, P., & Pain, R. (2007) Geographies of age: Thinking relationally. *Area*, 39:3, 287–294.

Horton, J. and Kraftl, P. (2014) *Cultural Geographies: An Introduction*. Abingdon, Routledge.

Jackson, P. (1981) Phenomenology and Social Geography, *Area*, 13:4, 299–305.

Kinkaid, E. (2021) Is post-phenomenology a critical geography? Subjectivity and difference in post-phenomenological geographies. *Progress in Human Geography*, 45:2, 298–316.

kinpaisby-hill, mrs. c. (2020) Participatory Action Research. In: Kobayashi, A. (Ed.), *International Encyclopedia of Human Geography*, 2nd Ed., 9–15.

Kitchin, R. and Tate, N. (2013) *Conducting Research in Human Geography: Theory, methodology and practice*, 2nd edition, Abingdon, Routledge.

Krinke, R. (2021) Maps of Joy and Pain, www.rebeccakrinke.com/Projects/The-, last accessed, June 28th, 2021.

Lapworth, A. (2019) Sensing. *Transactions of the Institute of British Geographers* 44, 657–660.

Latham, A. (2003) Research, performance, and doing human geography: Some reflections on the diary-photograph, diary-interview method. *Environment and Planning A*, 35, 1993–2017.

Latham, A. and Layton, J., (2020) Kinaesthetic cities: Studying the worlds of amateur sports and fitness in contemporary urban environments. *Progress in Human Geography*, 44:5, 852–876.

Lengen, C. (2015) The effects of colours, shapes and boundaries of landscapes on perception, emotion and mentalising processes promoting health and well-being. *Health & Place*, 35, 166–177.

Leszczynski, A. (2019) Digital methods II: Digital-visual methods. *Progress in Human Geography*, 43:6, 1143–1152.

Linn, S. (2021) Solicited diary methods with urban refugee women: Ethical and practical considerations. *Area* 53, 454–463. https://doi.org/10.1111/area.12694

Longhurst, R. (1997) (Dis)embodied geographies. *Progress in Human Geography* 21:4, 486–501.

Lorimer, H. (2012) Cultural geography: non-representational conditions and concerns. *Progress in Human Geography* 32:4, 551–559.

Lorimer, H., (2019) Dear departed: Writing the lifeworlds of place. *Transactions of the Institute of British Geographers*. 44:2, 331–345.

Lucherini and Hanks (2020) Emotional Geographies. In: Kobayashi, A. (Ed.), *International Encyclopedia of Human Geography*, 2nd Ed., 97–102.

Macpherson, H. (2011) Navigating a non-representational research landscape and representing 'under-represented groups': From complexity to strategic essentialism (and back), *Social & Cultural Geography*, 12:6, 544–548.

Morrow-Howell, N., Galucia, N. and Swinford, E. (2020) Recovering from the COVID-19 Pandemic: A focus on older adults, *Journal of Aging & Social Policy*, 32:4–5, 526–535.

Moss, P. and Dyck, I (2002) *Women, Body, Illness: Space and Identity in the Everyday Lives of Women with Chronic Illness*. Oxford, Rowman and Littlefield.

Myers, J. (2010) Moving methods: Constructing emotionally poignant geographies of HIV in Auckland, New Zealand. *Area*, 42:3, 328–338.

O'Reilly, Z. (2018) 'Living Liminality': Everyday experiences of asylum seekers in the 'Direct Provision' system in Ireland, *Gender, Place & Culture*, 25:6, 821–842.

Osborne, T., & Jones, P. (2017) Biosensing and geography: A mixed methods approach. *Applied Geography*, 87, 160–169.

Parr, H., (1998) The politics of methodology in "post-medical geography": Mental health research and the interview. *Health & Place* 4, 341–353.

Parr, H., Stevenson, O., Fyfe, N. and Woolnough, P. (2015) Living absence: The strange geographies of missing people. *Environment and Planning D* 33:2, 191–208.

Paterson, M. and Dodge, M. (eds) (2012) *Touching Space, Placing Touch*. Farnham, Ashgate.

Pearce, P. (2021) Finding one place in another: Post/phenomenology, memory and déjà-vu, *Social & Cultural Geography*, 1–17. DOI: 10.1080/14649365.2021.1922734

Philo, C. (2014) Insecure bodies/selves: Introduction to theme section, *Social & Cultural Geography*, 15:3, 284–290.

Philo, C. and Parr. H. (2003) Introducing psychoanalytic geographies. *Social & Cultural Geography*, 4:3, 283–293.

Pitt, H. (2015) On showing and being shown plants – a guide to methods for more-than-human geography. *Area*, 47:1, 48–55.

Pulido, L. (2017) Geographies of race and ethnicity II: Environmental racism, racial capitalism and state-sanctioned violence, *Progress in Human Geography*, 41, 524–533.

Rose, E. (2012) Encountering place: A psychoanalytic approach for understanding how therapeutic landscapes benefit health and wellbeing. *Health & Place*, 18, 1381–1387.

Rose, G. (2016) *Visual Methodologies: An Introduction to the Interpretation of Visual Materials,* Thousand Oaks: Sage.

Schwanen, T., Hardill, I., & Lucas, S. (2012) Spatialities of ageing: The co-construction and co-evolution of old age and space. *Geoforum*, 43:6, 1291–1295.

Seamon, D. (1979) *A Geography of the Lifeworld*. New York, St. Martin's Press.

Shortt, N., Rhynas, S. and Holloway, A. (2017) Place and recovery from alcohol dependence: A journey through photovoice. *Health & Place* 47, 147–155.

Söderström, O., Abrahamyan-Empson, L., Codeluppi, Z., Söderström, D., Baumann, P.S. and Conus, P (2016) Unpacking 'the City': An experience-based approach to the role of urban living in psychosis, *Health & Place* 42, 104–110.

Spinney, J., (2006) A place of sense: a kinaesthetic ethnography of cyclists on Mont Ventoux. *Environment and Planning D* 24, 709–732.

Spinney J (2015) Close encounters? Mobile methods, (post) phenomenology and affect. *Cultural Geographies* 22:2, 231–246.

Steger, A., Evans, E. and Wee, B. (2021) Emotional cartography as a window into children's wellbeing: Visualizing the felt geographies of place. *Emotion, Space and Society*, 39, 100772.

Straughan, E., (2012) Touched by water: The body in scuba diving. *Emotion, Space and Society* 5, 19–26.

Tolia-Kelly, D.P. (2006) Affect: An ethnocentric encounter? exploring the 'universalist' imperative of emotional/affectual geographies. *Area* 38:2, 213–217.

Tomalin, E., Sadgrovea, J. and Summers, R. (2019) Health, faith and therapeutic landscapes: Places of worship as Black, Asian and Minority Ethnic (BAME) public health settings in the United Kingdom. *Health & Place* 230, 57–65.

Tuan, Y-F. (1977) *Space and Place: The Perspective of Experience*. Minneapolis, University of Minnesota Press.

Twigg, L., Duncan, C. and Weich, S. (2020) Is social media use associated with children's well-being? Results from the UK Household Longitudinal Study. *Journal of Adolescence* 80, 73–83.

Valentine, G. (1989) The geography of women's fear. Area 21:4, 385–390.

Vannini, P., & Stewart, L. M. (2017) The GoPro gaze. *Cultural Geographies*, 24, 149–155. https://doi.org/10.1177/1474474016647369

Waitt, G., Buchanan, I., Lea, T. and Fuller, G. (2021) Embodied spatial mobility (in)justice: Cycling refrains and pedalling geographies of men, masculinities, and love, *Transactions of the Institute of British Geographers* 46(4), 917–28. https://doi.org/10.1111/tran.12464.

Warren, S. (2017) Pluralising the walking interview: Researching (im)mobilities with Muslim women. *Social and Cultural Geography* 18, 786–807.

Wheaton, B., Waiti, J., Cosgriff, M. and Burrows, L. (2020) Coastal blue space and wellbeing research: Looking beyond western tides. *Leisure Studies* 39:1, 83–95.

Wilkinson, E and Ortega-Alcázar, I. (2018) The right to be weary? Endurance and exhaustion in austere times. *Transactions of the Institute of British Geographers* 43, 1–13.

Williams, N., Patchett, M., Lapworth. A., Roberts, T. and Keating, T. (2019) Practising post-humanism in geographical research. *Transactions of the Institute of British Geographers* 44, 637–643.

Wilmott, C. (2021) Affective mapping, In: Kobayashi, A. (Ed.), *International Encyclopedia of Human Geography*, 2nd Ed., 53–60.

Wylie, J. (2009) Landscape, absence and the geographies of love, *Transactions of the Institute of British Geographers* 34, 275–289.

Young, J. and Gilmore, M. (2013) The spatial politics of affect and emotion in participatory GIS. *Annals of the Association of American Geographers* 103:4, 808–823.

10

GEOGRAPHY'S SEXUAL ORIENTATIONS

Queering the Where, the What, and the How

John Paul Catungal and Micah Hilt

Queer Methods and Methodology

What lessons for geographical methods and methodology do queer geographies and queer theorizing offer? This chapter suggests and models a queer approach to geographic knowledge production, identifying specific methods that animate queerness as a mode of scholarly inquiry within and beyond geography. We do this with attention to both methods, the 'how' of research, and methodology, the broader terrain of knowledge concerned with the principles, politics, and procedures that root knowledge production. We argue that queerness and the questions underlying methodology share a common concern with unpacking how power relations shape the designation and governance of certain acts (e.g., of identification, representation and knowledge production) as normal, proper, and appropriate. We highlight the possibilities offered by queer as an ever-evolving and continually contested concept, one that refuses to take for granted the *how*, the *what*, the *where*, and even the *if* of living in and understanding the world. Our main concern and contention in this chapter is to queer geographic inquiry, that is, to ask questions about both the tacit and explicit concerns, approaches, and commitments that are foundational to the conduct of socio-spatial analyses. A queer approach to geographic inquiry attends to Geography's attachments to systems, structures, and foundational sites of power, including sexuality and its intersectional workings. To queer Geography, we argue, is to draw specific spatial attention to the boundaries of Geography's current workings and worlds and to point to other possible, even more expansive, geographies on their way or yet to come (Munoz, 2009).

For us, queer is and must be an opening up, a building on, a deepening, and a troubling, both of ways of being in the world and of how we theorize its workings. Though queerness is sometimes treated as synonymous with homosexuality or other violations of heteronormativity, it is not synonymous with sexuality, though they are entangled with each other, historically, conceptually, and politically (c.f., Cohen, 1997; Eng, Halberstam and Munoz, 2005). To take that intention seriously, queer, we argue, must be deployed as an approach to space not only or primarily to surface and map sexuality, but to understand its *intersectional* spatial production, productions, and contestations. That is, queer as geographical methodology seeks to broaden

DOI: 10.4324/9781003038849-13

the ambit and scope of geographical critique from a demographic focus on sexual minority subjects to a more expansive attention to sexuality's intersections with gender, race, class, ethnicity, location, and other modalities of power.

Methodologically, this more expansive, queer approach has at least three implications for the conduct of geographical research and the methods employed therein. First, it requires that geographers consider what counts as proper objects of study within the field that coalesces under the banner of queer geographies. At stake here are the practices of definition, inclusion, and exclusion that define and produce the bodies of knowledge that constitute this field. Second, a queer approach to the conduct of geographical knowledge production bears upon research decisions, including, for example, how a study population might be defined or sought and what archival, ethnographic or other study sites might be appropriate for research in queer geographies. If "queer subjects and subjectivities are fluid, unstable and perpetually becoming" (Browne and Nash, 2010, p. 1), then research *on* queer subjects and geographies must necessarily contend with the slipperiness of their definition. Third, a queer methodological approach foregrounds context, the 'political' in geographical methodologies and knowledge, including how research itself impacts upon communities, subjects, and spaces on the ground. If methodology is concerned not only with the 'how' of knowing, but also the 'so what?' and 'for what purpose?', then queerness requires an attentiveness to how research itself enables particular forms of action, including interfaces with media, policy, community organizing, and, more broadly, queer everyday lives.

Geographies of Sexuality

In order to chart the emergence of a queer approach to geographical knowledge, it is useful to historicize queer geographies as a field of inquiry, particularly with attention to its tangled relation to geographies of sexualities. Geographies of sexualities are often understood as separate from and a predecessor to queer geographies. Gieseking (2013, p. 15) understands the former as a subdisciplinary formation with interests in "studying LGBTQ spaces and experiences of space" and "studying sexual spaces, or the sexualized experience of space". In contrast, queer geographies is concerned with "much more than the lives of 'queers'" (Oswin, 2008: 90). Instead, it is more usefully conceived as a "bridge and conversation between queer theory and geography" and as "a practice among queer theorists and geographers" (Gieseking, 2013, p. 15).

The emergence of geographies of sexualities was premised on a political insistence that the spatialities of gay (and, later, lesbian) lives are worth studying, which was a response to what Lauria and Knopp (1985) has called a "certain 'squeamishness'" (136) regarding sexuality in Geography. This squeamishness translated into the taken-for-grantedness and normativity of heterosexuality in both space and geographical knowledge production alike, and the resultant exclusion of non-normative sexualities and their spatialities as valid topics of research (Elder, Knopp and Nast, 2003). In response, geographers intervened methodologically by foregrounding gays and lesbians as research participants, the spaces of their lives as key sites of study, and their stories and experiences as valid forms of knowledge and evidence from and with which to produce geographical knowledge. As Gieseking (2013, p. 15) notes, key to the emergence of geographies of sexualities as a subfield in the 1980s and 1990s is the project of visibility: "work on lesbian and gay spaces … began by *counting lesbian and gay people* in order to prove their existence and make them visible in the heterosexual public eye" [our emphasis] . This methodological foregrounding of gay and lesbian subjects and spaces worked to surface and make visible gay and lesbian geographies as an intervention into the "unquestioned heterosexuality of the geographical enterprise" (Elder, Knopp and Nast, 2003, p. 200). In other words, geographers of sexualities addressed absences by methodologically "[making] room for 'sexual deviants" and putting "homosexuality and heterosexuality on the disciplinary map" (Oswin, 2008, p. 90). Bell (1991, p. 328) put the project in this way, with panache: "to bring gay and lesbian geographies out into the open".

Oswin (2008) argues that the work of David Bell, Jon Binnie, and Gill Valentine were especially key to the "coming out" of geographies of sexualities in the 1990s, preceded by work in the 1980s, especially by Larry Knopp and Manuel Castells on gay men's role in urban political and economic change. Oswin (2008) identifies an affinity among these scholars with the work of interdisciplinary queer theorists like Judith Butler, particularly in terms of their insistence that sexual identities and spaces are socially constructed and constantly performed rather than essential, natural, or fixed. Despite this poststructuralist attunement, geographies of sexualities came to settle on some methodological normalizations that reproduced essentialist approaches to sexual subjectivities and spatialities. One example of this is the methodological centering especially of gay men and their urban, commercial gaybourhood spaces, particularly near the beginning of the subdiscipline's history. Writing relatively early on, Bell (1991) already saw this tendency towards a narrow methodological focus on "gay Meccas" as sites of geographical research on sexualities, warning that such a narrow focus could entrench gay men's specifically gendered, racialized, and classed geographies of consumption, leisure, and residence as a stand-in for the pluralities of sexual geographies. Scholars like Gill Valentine and her collaborators did do research that decentered gay men and urban spaces through a focus on lesbians' lived spaces of friendship, intimacy, domesticity, and rural life. Their research took lesbians seriously as research participants whose narratives and experiences constitute important sources of and for geographical theorizing (Johnston and Valentine, 1995; Valentine, 1993a, 1993b; Bell and Valentine, 1995). Nevertheless, despite the influence of the idea of sexual subjectivities and spaces as fluid and performed, research on gay and lesbian geographies sometimes reproduced the fixity of 'gay', 'lesbian' and other sexual categories as identities. Methodologically, this entailed the treatment of 'gay' and 'lesbian' as self-explanatory demographic categories, often without adequate attention given to their intersectionalities with other categories of social difference, including race and class, even while these are acknowledged in passing (c.f., Nast, 1998).

With the establishment and consolidation of geographies of sexualities in the 1990s, scholars "challenged the marginalization and exclusion of lesbian, gay, and bisexual voices" (Binnie, 1997, 223). Methodological interventions were key to their capacity to do so. Geographers of sexualities trained their methods on the processes, systems, and spatialities of discrimination, erasure, and violence against LGBTQ people, both in society more broadly and in the discipline more specifically (see also Binnie and Valentine, 1999). Qualitative methods were especially key to these interventions. For example, through interviews, scholars like Valentine (1993a) elicited stories and experiences as lived modes of knowledge. Similarly, Bell (1991) argues that oral histories are a particularly important method for research on sexual geographies, as written records of the lives of sexual dissidents have often been absent or omitted from formal archives and • sources. Ethnographically attuned engagement with activist and social service organizations and the people involved in them also constituted key methods to understand the agentive capacities of gay and lesbian communities. A key example of the latter is Michael Brown's work on AIDS politics in Vancouver (Brown, 1997). Autobiographical and autoethnographic methods of theorizing, such as Valentine's (1998) analysis of her own experiences of homophobic harassment, also enabled geographers to theorize from their own embodied experiences and thus to insist, as many feminists have before them, on the intertwining of the personal with the methodological (see also Knopp, 1999). Through these methods, geographers of sexualities contributed to the broader task of complicating and dismantling the very intellectual and social processes that shape the lives and spaces of gays and lesbians. In other words, through these methods, geographies of sexualities surfaced the lives and landscapes of sexually minoritized communities and pinpointed how "space is not naturally authentically 'straight' but rather actively produced and (hetero)sexualized" (Binnie, quoted in Oswin, 2008: 90), including through practices of categorization, normalization, erasure, and oppression.

The Queer Turn

In general, work under the rubric of geographies of sexualities asked epistemological and methodological questions about geographical knowledge, including how heteronormativity shaped the conduct of research in the discipline and the kinds of subjects, issues, and spaces that warranted research consideration. Geographies of sexualities thus had a "proto-queer" (Catungal, 2018: 26) quality, as it included "glimpses of latent anti-normative critique" that refused the assumed normalcy of heterosexuality in geography. In pointing to the *sexual* politics of geographical knowledge production, these geographers opened up space for a queer/ er turn in the discipline. Geographers of sexualities did so by "put[ting] the lives of non-heterosexuals on the disciplinary map" (Oswin, 2008, p. 89). Nevertheless, Oswin (2008) argues that geographers of sexualities, even with their deployment of queer as a theoretical concept and political commitment, still tended to synonymize queer with LGBT, with the effect that research on queer space often equated it with gay and lesbian space. In other words, while broader queer theorizing refused identity-focused approaches that assume prefigured sexual subjects and spaces, some research under the rubric of geographies of sexualities still treated 'gay' and 'lesbian' as stable demographic categories, as we noted above. In addition, as Oswin (2008) further notes, 'queer space', when theorized simply as the spatialization of gay and lesbian resistance to heteronormativity, reproduces a single-issue conceptual and methodological approach that then elides sexuality's intersections with race, gender, class, and other categories of difference. As Jasbir Puar writes, "while the claiming of queer space is lauded as the disruption of heterosexual space, rarely is that disruption interrogated also as a disruption of racialized, gendered, and classed spaces" (quoted in Oswin, p. 95). Overall, it is important to note that geographers of sexualities did in fact follow poststructuralist scholars like Foucault and Butler in pointing to systems, processes and acts of power as key to gays' and lesbians' experiences of exclusion and to shine light on sexualities as a set of spatial practices rather than strictly identities. It is also equally important to point out that a significant portion of earlier geographies of sexualities work, with its task of increasing visibility and combating erasure, nevertheless bracketed from conceptual and methodological consideration how queer spaces participated, either directly or indirectly, in projects of normalization and respectability.

We note the proto-queer qualities of geographies of sexualities in part to recognize this work as an active terrain of openings and fissures. These openings and fissures were attentive to queer as a set of investments and provocations focused as much on sexuality itself as it is on the objects, spaces, acts, and modes of inquiry that comprise its study. For example, Knopp (1995) called on geographers of sexualities to broaden the universe of their analytical and methodological consideration and to attend to heterosexualities, bisexualities, other non-normative sexualities, rural and small-town connections to sexualities, and "radical, self-consciously fluid sexualities which reject association with such notions as 'identity' and 'community' altogether" (Knopp, 1995, p. 137). It would take some time for geographers to substantively respond to Knopp's provocation. There were nevertheless resistances, moments, and engagements that moved the conversation more queerly, ones that perhaps anticipated coming queer/er shifts. Perhaps most critically, following Butler, it is worth arguing that queer as a methodological approach is an ongoing project, one that resists fixity and arrival. It is, following Munoz (2009), horizon, rather than event; that which is queer or queered needs to be subjected to further queering (Butler, 1993). Queer is never done. We are never there (Munoz, 2009). The question to us here is how does one render queer as an analytic, methodologically?

Queer Methodologies

Refusal and In/Visibility as Methodologies for Queer Geographies?

If, according to Munoz (2009, p. 72), "queerness is illegible and therefore lost in relation to the straight minds' mapping of space", how might queer geographers look to the methodological – that is, research considerations regarding objects of study, approaches to empirical investigation, forms of evidence, sites of research – for possible lines of flight away from the kinds of identitarian fixations (gay, queer) and liberal binarizations (straight as oppressive / queer as radical) that Oswin (2008) notes haunts the discipline's approach to queerness? How might one methodologically enact the queer goal, following Munoz (2009, p. 73), of "relinquish[ing] one's role (and subsequent privilege) in the heteronormative order" of geographical knowledge production? We ask these questions of ourselves as much as we do of our readers. We draw lessons from queer geographers and allied thinkers in other disciplines to point to some tentative answers to these questions.

One methodological consideration of significance to queer geographers concerns the politics of visibility itself, and the role of research design, data collection and the presentation of knowledge in rendering particular sexual subjects and their spatialities visible. Given Foucault's point about the tight linkage of power and knowledge, and indeed of knowledge being important to acts and systems of governance, it is worth considering how methodological decisions that deliberately hide or make knowledge broadly unavailable might be understood as queer. That is, we think about such refusals as a strategic form of knowledge production and indeed a methodology, informed by ethical and political considerations, for rendering certain things knowable but not others. We draw affinity with theorists of refusal as politics and method, including and especially Mohawk anthropologist Audra Simpson. Reflecting on her ethnographic research for *Mohawk Interruptus*, Simpson (2016) writes of her refusal to be "*that* thick description prosemaster who would reveal in florid detail the ways in which these things were being sorted out" (328) in her community (see also Simpson, 2014). Her refusal worked, at least partly, at the methodological level: it concerned what to do with knowledge that she has gleaned both from her ethnographic research and also from her own embeddedness in Kahnawà:ke community. Her ethnographic refusal is a methodological refusal "in time with" her community's own refusals as assertions of sovereignty (331).

McGranahan (2016) writes: "To refuse is to say no. But, no, it is not just that" (319). Refusal is certainly negative, but it can and should also be regarded as productive insofar as it can work to bring ethical commitments into fruition (ibid.). Here, we draw on an example from earlier work by one of us (Catungal and McCann, 2010), which concerns the role of homophobia in governing the sexual geographies of park spaces in Vancouver. One case study in this work involved the murder of Aaron Webster, a gay park user, near cruising areas or strolls (sites of public sex) that are familiar as such to people 'in the know' (e.g., other users) but not necessarily generally. Eugene McCann and I (Catungal) decided to include a map in the article, to orient our readers to the parts of Vancouver that we examine in our research. We decided, however, not to include finer detail on the specific whereabouts of the stroll, opting instead to include the following text in the map's caption: "Note: We choose not to identify exactly where Webster's body was found since it is near the 'strolls'" (80). This was a refusal informed by the realities of homophobic violence, which was ever more significant given that our work was about the deadly targeting of a gay park user. Simply put, we worried about risking the safety of gay cruisers, and others who might be read as such, by inadvertently publicizing a space that is otherwise strategically invisible except to those with the know-how to read for often subtle or coded signs and practices of cruising in this public green space (c.f., Brown, 2008). In deciding to not map cruising sites in our work, we align with the longstanding practice of anonymizing research sites and subjects in work on sexualities, a methodological and ethical approach that recognizes the durability and force of homophobia and its violence (England, 1994).

This example is admittedly a smaller scale refusal than that theorized and enacted as methodology and ethnographic mode by Audra Simpson, as we described above. Nevertheless, if methodology is about not only how we come to know, but also what to do with knowledge, then the ethical and political commitments that inform the work of queer geographers – in our case, a basic one of minimizing the risk of homophobic violence – is surely key. The politics of non-disclosure in our knowledge production practice is in alignment with – or borrowing from Simpson, as above, "in time with" – the strategic in/visibility that constitutes cruising sites as spaces of sexuality. It is, following Munoz (1996, p. 6), a queer mode of knowing in the face of violence: "Queerness is often transmitted covertly. This has everything to do with the fact that leaving too much of a trace has often meant that the queer subject has left herself open for attack".

Ephemera as Methodological Resource for Queer Geographies

The celebrated musical *Rent* begins with a song that asks "how do we measure a year?" and that contains a litany of ways of answering the question: "in daylights, in sunsets, in midnights and cups of coffee; in inches, in miles and laughter and strife", among other things. *Rent* is about a ragtag group of artists, anarchists, and sexual Others living in New York City during the AIDS crisis in the late 1980s. That the question of measurement starts off the musical is no accident, given that the question of how to determine lives worth living and the politics of measuring up (e.g., to societal norms about class ascendancy and sexual propriety) are key to the narrative and politics of the musical, as it is to the politics of HIV/AIDS itself. We use *Rent* as an entry point here in our discussion of queer methodologies and queer geographies in part because it tackles issues of significance to research on sexualities and spaces, among them the place of marginalized subjects, including queers, in the gentrifying city and the creation of spaces of safety and support for LGBTQ people through art and activism. The topic of measurement in relation to queerness, and its link to the project of visibility that underpins the history of geographies of sexualities approaches, is another such theme.

One of the key methodological moves that enabled geographers to make non-heterosexuals visible is to render them, along with their spaces, measurable, knowable and mappable. Brown and Knopp (2008, p. 42) identify "cartography and mapping as key interventions in disrupting the heteronormativity of space". They laud, for example, *The Gay and Lesbian Atlas* for "its visual affirmation of the axiom that 'we are everywhere'" (ibid.) through mapping the presence of same-sex unmarried partner households across the United States, including in small towns and rural areas where their absence is often assumed. For Brown and Knopp (2008), the *Atlas* worked against the assumption of LGBT metronormativity, mobilizing census data towards the task of "uncloseting" (42), that is making LGBT people visible, via measurable and mappable data, particularly within the national map of the United States (see also Brown, 2000). Brown and Knopp (2008, p. 43) recognize the existence of a paradox between, on the one hand, the conceptual understanding of the fluidity and performativity of sexuality and, on the other hand, the methodological desire to "[fix] sexual subjectivities on a map" through measurement and cartography.

How might queer geographers attend to and understand the queerness of those who elide, by choice or by circumstance, orthodox modes of measurement and mappability? Queer geographers might turn to other modes through which these queer Others mark their presence in the world. Munoz (1996), for example, insists on the utility and value of ephemera as evidence, by which he refers to the kinds of relatively small and sometimes fleeting traces that do not ascend to the level of knowledge in official archives. He notes that it is especially important to attend to "traces, glimmers, residues, and specks of things", as it is often in these forms that evidence of queer lives and acts exist and linger. The attention to the small and minor, the ephemeral that Munoz calls for, is quite a different methodological tack from something like the use of census data. It also enables different kinds of political projects. Not surprisingly, *The Gay and Lesbian Atlas* is tethered to a relatively unproblematized nation-state formation, with the exclusion of gays and lesbians from the polity contested through the mobilization of the nation-state's own regimes of measurement. Munoz recognizes that, in contrast to vaunted data sources such as official archives or state data sets, "the anecdotal, the performative, or … the ephemeral as proof is often undermined by the academy's officiating structures" (p. 7), their solidity as evidence rendered questionable by orthodox measures of what counts as knowledge. Ephemera as queer method, like queerness itself, pushes at the limits of the 'proper', and in so doing, offers us a generatively counter-institutional and more transient, porous, and material approach to knowledge and knowing.

Still though, one critical lesson that Munoz's methodology of the ephemeral offers for geographers is the need to be curious about minor objects and acts, even those traces and evidentiary moments that exist beyond, short of, or in opposition to well-formulated, more official forms of evidence. This might require paying attention during participant observation to seemingly small moments and reading them for their significance, like errant and lingering looks that call up the queer practice of reading the room, for example, for the possibility of desire or the presence of danger. In state archival spaces, this might require reading for traces of queerness between the lines of official state narratives. One example that might be read as employing something of an ephemeral method is Gavin Brown's work on the affective spatial practices of cruising in public washrooms. Through ethnographic participation and observation, he attends to "micro-geographies" of cruising, including, for example, "the exchange of glances" (920), the practice of "mirroring" other cruisers (921) and other acts that comprise "the complex choreography of gestures and glances" (922) in cruising spaces. Quintessentially traces, in that these are the "stuff of fleeting, *ephemeral* moments not intended to be captured" (Turner, quoted in Brown, p. 920, our emphasis), Brown's attentiveness to these moments in order to theorize cruising's queer relationalities and spatialities requires a capacity to *read* the site of cruising for evidence of queer existence and exchange.

Queer Archives, Messy Archives

The use of archival material is a relatively common method for geographers interested in sexualities. Examples include Julie Podmore's (2006) research on the production and politics of lesbian historical geographies in Montreal, which engaged collections from the Archives gaies du Québec, as well as Andrew Tucker's (2009) combination of interviews with archival research for his work on the production of intersectional inequalities among queer men in Cape Town, for which he engaged material located in the Gay and Lesbian Archive (University of Witswatersrand) and the South African National Library (Cape Town).

Archives that specialize in the collection of LGBTQ material are now more common. Most have origins in community efforts to document LGBTQ people's lives and histories. Some have become situated as formal standalone archives with institutional status and resources or as collections in larger institutions. Examples include the ArQuives (formerly the Canadian Lesbian and Gay Archives) in Toronto; the BC Lesbian and Gay Archives, now situated as fonds in the City of Vancouver Archive; the Transgender Archives at the University of Victoria; and the Archives of Lesbian Oral Testimony, a digital repository out of Simon Fraser University. Gieseking (2015) and Catungal (2017) remind us, respectively, that archives are not mere spaces containing 'stuff', but are political spaces of queer knowledge production; embodied and relational sites for negotiating gendered, racialized and sexual affects; and a geography not only of objects but of feelings and histories (see also Cvetkovich, 2003). As with other methods, archival work as queer method requires attending, head on, to how one's intersectional subject positions might shape what it means to engage with archival materials and to be a researcher in archival spaces. It also requires a methodological attunement to questions about which and whose histories are archived (and thus considered worth preserving), in which spaces, and how archives' relationships to funding and the state might shape archival knowledge production and methods (see also Beckingham and Hodder, Chapter 14).

One methodological question concerning archival knowledge of significance to queer geographies concerns not just which official or institutional archives might be useful, but also how an archive might be conceived of or rendered in the first place. Martin Manalansan (2014: 94), for example, turns away from the hegemonic formulation of the archive as a formal "repository or storage of information and documents or a legitimizing instrument of power structures and prevailing authorities". In its place, he draws on feminist scholars to reconceptualize archives as "a space for dwelling and a quotidian site for marginalized subjects as well as gendered and erotically charged energies, meanings, and other bodily processes" (ibid.). In his research on practices and spaces of queer dwelling in New York City, he takes seriously how immigrant domestic spaces and their "hoarder-like household material, symbolic, and emotional conditions" (ibid.) might be considered not only queer (beyond their residents being LGBT), but also archival. Considering, as Manalansan has, the messy and overcrowded shared dwellings of working-class immigrant queers of colour is productive here for thinking about moving beyond the archive's relationship to orderly (or ordered) arrangement of historical knowledges. He asks, via messy queer of colour domestic arrangements and spaces:

> [W]hat happens when disorder and chaos are the elements that make up the archival space? What happens when, instead of orderly catalogs, makeshift arrangements teetering on the brink of anarchy become the "disorder" of things? What kinds of value get attached to persons and things that dwell in mess and disorder, and how can they be dynamically reframed as a way to think more broadly about political acts, aspirations, and stances?
>
> *pp. 102–103*

These questions confront queer geographers who research into and teach about historical and archival knowledge, particularly about the politics of disciplinary commitments to proper archival sources as well as about how to approach mess and disorder as ways of being in the world that, in some senses, refuse (and queer) the order of things.

As with Munoz's focus on the ephemeral, Manalansan's recuperation of the messy as a source of knowledge has profound methodological implications for geographers who study sexualities and spaces and their intersectionalities. How might we find and work with sources of knowledge that escape categorization and are found not in official sites, but in quotidian spaces of living and being? Manalansan's own work, for example, thinks with how working class immigrants of colour curate their domestic spaces through, for example, "finding 'treasures' in the street" (102), a salvaging practice that Manalansan regards as "archiving otherwise" (103). Indeed, the recoding of often mundane found objects as treasures itself points to the capacity of marginalized queers of colour to implement their own systems of categorization and knowledge. Methodological engagement with such otherwise archives (for Manalansan, through intimate and long term ethnographic relationships with residents of these domestic spaces) enables the surfacing of intimate social geographies that would not be possible to glean, in the same way, if the idea of the archive itself is restricted. Thinking with both queer and mess, given their shared vexed relationship to the idea of normalcy, Manalansan (2018, p. 496) argues that "the failure or refusal to domesticate, clean up, temper, and tame the unpredictable and immeasurable lives of queer immigrants also contains a possible escape hatch or alternative from the strictures of oppressive evaluative frames".

What then might it mean for geographers to adopt messy methodologies that take "unpredictable and immeasurable" queerness seriously? We don't have easy answers, in part because to offer some might, paradoxically and ironically, make the concept of mess itself too neat and tidy a methodology. As with queer as a concept, mess forces us to confront the assumed stability and normalcy of our ways of doing research and of being in the academic world. This would entail asking important questions about methodological concepts that we might take for granted, including deeply held, even inflexible, measures of proper scholarship. Indeed, as Munoz (1996, p. 7) wrote some 25 years ago, "[w]ork and thinking that does not employ and subscribe to traditionalist scholarly archives and methodologies are increasingly viewed as being utterly without merit". While not a simple program of methods or a series of "how to's", messiness is nevertheless a methodological reminder that knowledge production is not an innocent act, but one embedded in normative structures, systems, and negotiations. Like queer, messiness refuses the "as-is-ness" of such a world and speaks to the possibility, indeed reality, of other ways of being, doing, and knowing.

Conclusion

Our goal for this chapter was not to produce a ready-made how-to guide to queer geographic methods and methodology. Queer as a concept resists such an approach, in part because its wisdom and utility can be found in its agonistic relationship to processes of normalization, including in knowledge production. Instead, this chapter makes use of queer as a provocative resource for thinking through the politics of knowledge production in Geography. It lays out a genealogy of queer geographies and its tangled relationship to geographies of sexualities, and asks what we hope are generative questions about the approaches and stakes of knowledge production under the umbrella of queer geographies. Along with spotlighting the necessarily political nature of methodological decisions about what constitutes sexualities and their spatialities, we also highlight the paradox that exists between a queer understanding of sexuality's social construction and performativity and the tendency in some sections of geographical scholarship to fix sexuality into a demographic character or identity formation. If nothing else, we hope that this chapter has laid out the utility of queer as a methodological approach that forces us to confront not only the categories we mobilize in our research and what we consider to be proper objects of analysis, but also the political and conceptual directions that these methodological categories and analytical objects plug into, enable and/or foreclose.

We draw substantive attention to three specific methodological concepts that are energized by and in proximity with queerness as an approach: refusal, ephemera, and messiness. Each of these concepts enable questions centrally about methodology in human geography, including: What counts as proper knowledge or evidence? How does the organization of knowledge matter for how it is valued? What do we do with the knowledges we glean from research? For queer geographers and for geographers more broadly, these concepts mess with – and mess up – what we often take for granted as proper or useful approaches to geographical knowledge production. We think with and offer them in this chapter in part as an invitation to our readers: we hope that the political spirit of queer also energizes ongoing revisitations not only of research paradigms, methodological approaches, and conceptual vocabularies, but also of our own respective relationships to and investments in geographical knowledge and its production.

References

Bell, D., and Valentine, G. (1995) Queer country: Rural lesbian and gay lives. *Journal of Rural Studies*, 11(2): 113–122.

Bell, D. J. (1991) Insignificant others: Lesbian and gay geographies. *Area*, 23(4): 323–329.

Binnie J, Valentine G. (1999) Geographies of sexuality – a review of progress. *Progress in Human Geography*, 23(2): 175–187. doi:10.1177/030913259902300202

Brown, G. (2008) Ceramics, clothing and other bodies: Affective geographies of homoerotic cruising encounters. *Social & Cultural Geography*, 9(8): 915–932.

Brown, M., and Knopp, L. (2008) Queering the map: The productive tensions of colliding epistemologies. *Annals of the Association of American Geographers*, 98(1): 40–58.

Brown, M. P. (1997) *Replacing citizenship: AIDS activism and radical democracy.* New York: Guilford Press.

Brown, M. P. (2000) *Closet space: Geographies of metaphor from the body to the globe.* New York: Routledge.

Browne, K., and Nash, C. J. (2010) *Queer methods and methodologies: Intersecting queer theories and social science research.* New York: Routledge.

Butler, J. (1993) 'Critically Queer,' *GLQ* 1(1): 17–32.

Catungal, J. (2018) Toward queer(er) futures: Proliferating the 'sexual' in Filipinx Canadian sexuality studies, in R. Diaz, M. Largo and F. Pino (eds) *Diasporic intimacies: Diasporic Filipinos and Canadian imaginaries*, pp. 23–40.

Catungal, J. P., and McCann, E. J. (2010) Governing sexuality and park space: acts of regulation in Vancouver, BC. *Social & Cultural Geography*, 11(1): 75–94.

Catungal, J. P. (2017) Feeling bodies of knowledge: Situating knowledge production through felt embeddedness. *Tijdschrift voor economische en sociale geografie*, 108(3): 289–301.

Cohen, C. J. (1997) Punks, bulldaggers, and welfare queens: The radical potential of queer politics? *GLQ* 3(4): 437–465.

Cvetkovich, A. (2003) *An archive of feelings: Trauma, sexuality and lesbian public cultures.* Durham: Duke University Press.

Elder, G., Knopp, L., & Nast, H. (2003) Sexuality and space. In Gaile, G. and Willmott, C. (eds.), *Geography in America at the dawn of the 21st century*, pp. 200–208.

Eng, D., Halberstam, J., and Muñoz, J. E. (2005) What's queer about queer studies now? *Social Text*, 23(3-4): 1–17.

England, K. V. (1994) Getting personal: Reflexivity, positionality, and feminist research. *The Professional Geographer*, 46(1): 80–89.

Gieseking, J. J. (2013) A queer geographer's life as an introduction to queer theory, space, and time. In Lau, L., Arsanios, M., Zuniga-Gonzalez, F., Kryder, M. and Mismar, O. (eds.), *Queer geographies: Beirut, Tijuana, Copenhagen*, pp. 14–21.

Gieseking, J. J. (2015) Useful in/stability: The dialectical production of the social and spatial Lesbian Herstory Archives. *Radical History Review*, 2015(122): 25–37.

Johnston, L., and Valentine, G. (1995) Wherever I lay my girlfriend, that's my home: The performance and surveillance of lesbian identities in domestic environments. In Bell, D. and Valentine, G. (*Mapping desire: Geographies of sexualities*, pp. 99–113.

Knopp, L. (1995) Sexuality and space: A framework for anaysis. In Bell, D. and Valentine, G. (eds.), *Mapping desire*, pp. 136–146.

Knopp, L. (1999) Out in academia: The queer politics of one geographer's sexualisation. *Journal of Geography in Higher Education*, 23(1), 116–123.

Lauria, M. & Knopp, L. (1985) Toward an analysis of the role of gay communities in the urban renaissance. *Urban Geography*, 6:2, 152–169. DOI: 10.2747/0272-3638.6.2.152

Manalansan IV, M. F. (2014) The "stuff" of archives: Mess, migration, and queer lives. *Radical History Review*, 2014(120), 94–107.

McGranahan, C. (2016) Theorizing refusal: An introduction. *Cultural Anthropology*, 31(3): 319–325.

Muñoz, J. E. (1996) Ephemera as evidence: Introductory notes to queer acts. *Women and Performance*, 8(2): 5–16.

Muñoz, J. E. (2009) *Cruising utopia: The there and then of queer futurity*. New York: New York University Press.

Nast, H. J. (1998) Unsexy geographies. *Gender, Place and Culture: A Journal of Feminist Geography*, 5(2): 191–206.

Oswin, N. (2008) Critical geographies and the uses of sexuality: Deconstructing queer space. *Progress in Human Geography*, 32(1): 89–103.

Podmore, J. A. (2006) Gone 'underground'? Lesbian visibility and the consolidation of queer space in Montréal. *Social & Cultural Geography*, 7(4): 595–625.

Simpson, A. (2014) *Mohawk interruptus: Political life across the borders of settler states*. Durham: Duke University Press.

Simpson, A. (2016). Consent's revenge. *Cultural Anthropology*, 31(3): 326–333.

Tucker, A. (2009) *Queer visibilities: Space, identity and interaction in Cape Town*. Oxford: Wiley-Blackwell.

Valentine, G. (1993a) (Hetero) sexing space: Lesbian perceptions and experiences of everyday spaces. *Environment and Planning D: Society and Space*, 11(4): 395–413.

Valentine, G. (1993b) Negotiating and managing multiple sexual identities: Lesbian time-space strategies. *Transactions of the Institute of British Geographers*, 18(2): 237–248.

Valentine, G. (1998) "Sticks and stones may break my bones": A personal geography of harassment. *Antipode*, 30(4): 305–332.

11

POLITICAL GEOGRAPHIES

Assemblage Theory as Methodology

Jason Dittmer, Pooya Ghoddousi and Sam Page

Introduction

The recent rise in attention to assemblage theory within political geography has been largely framed as a change in ontology, from the macro- to the micro-, from stability to becoming, and from entities to relations. While this is undoubtedly central to the changes wrought by the theory to the traditional subject matter of the sub-discipline, the focus on the *what* of our scholarly enterprise has necessarily directed some attention away from the *how* of our scholarly enterprise. This chapter seeks to engage with assemblage-as-methodology, focusing on the different ways in which researchers can engage with a world constantly becoming otherwise.

For those still coming to grips with assemblage theory, we offer a brief summary. Assemblages are constellations of bodies, objects, and material forces that come together for a time and exert an emergent agency in the world (Page, 2020). There is a wide diversity of theoretical formulations that travel under the label 'assemblage theory' (Buchanan, 2017; DeLanda, 2006) and related schools of thought (Thrift, 2004; Latour, 2005) and we do not propose to dissect those debates here. Rather, we focus on the common features of them all: an ontology that sees space itself as emergent from constantly shifting force relations that both connect to human politics but also exceed that anthropocentric framing.

Indeed, a refusal to limit analyses to the times and spaces of the human is one of the hallmarks of assemblage theory. Rather, assemblage theory calls into question the boundary between the human and non-human, as well as other bounded notions of the self, instead focusing on the relations that sustain the inequalities associated with these categories (see Gorman and Andrews, Chapter 30, for further discussion). In doing so, it calls our attention to spaces and times that are both 'larger' and 'smaller' than those immediately sensible to us. It is at these scales that things that seem stable to our human senses can most clearly be seen to be in flux, with new elements coming and old ones leaving, or with energies rippling through their constitutive relations and remaking them.

This chapter aims to consider how existing methodologies from the toolbox of the social sciences can speak to these spatio-temporal scales 'above' and 'below' individual human perception, but also how those methodologies are themselves re-worked and affected by being pulled into relation with assemblage theory. In the next section we look to ethnography, the analysis of everyday life's micro-relations, and the myriad, affective circulations that can tip

DOI: 10.4324/9781003038849-14

moments of encounter into full-blown events. The undecidability of the encounter, and the opening up of what may come from it, is heralded by many assemblage theorists as one of the political potentials of the theory (Amin & Thrift, 2002; McFarlane, 2011). In the section after, we turn to the *longue durée* of space-time, looking at how the archives can be reimagined in assemblage theory to not only include the traditional dusty boxes and files of paper but to also include wider heritage conservation efforts and even the earth itself through the concept of the Anthropocene. We conclude by arguing that assemblage-as-methodology opens up new political potentials for researchers through a spatial and temporal extension of the field, in the process breaking down boundaries between the researcher and the researched and the past and the future. By breaking down these boundaries assemblage thinking clears the way for more affirmative, less antagonistic political alternatives.

Ethnography and the Body Becoming

Assemblage theory did not, of course, introduce ethnography to political geography. The eventual impact of the 'new cultural geography' of the 1990s on political geography was relatively slow to unfold but it took into account the critiques of anthropology's engagement with 'the field' by opening it up to a more mobile, relational, and distributed notion of the research site (Crang & Cook, 1995; Marcus, 1998; Burrell, 2009; Elliot & Urry, 2010). But, by the 2000s, a feminist critique of political 'armchair' political geography began to take shape – in which discourse analysis was used to draw conclusions from a distance (Dowler & Sharp, 2001; Marston, 2003). Expanding that critique, Hyndman (2004) and Megoran (2006) highlighted the importance of engaged research that might capture some of the political complexities of everyday life. This has often been expressed as a resistance to state-centric narratives of politics. By proliferating the sites at which politics is studied, ethnography has brought to light new forms of politics that otherwise might not be understood as such, from consumerism (Billo, 2015) to anarchism (Pottinger, 2018) to entrepreneurialism (Wainwright et al., 2018). The focus on the everyday was a crucial point for a more feminist political geography for a range of reasons; of course it was one way to break down the public/private divide that frequently marginalized women, and further it allowed for the inclusion of important dimensions of social life – such as emotions – which had hitherto been deemed outside of the scope of political geography.

Assemblage theory has offered a new engagement with ethnography that pushes political geography in yet new directions. This is largely the result of assemblage theory's challenge to the idea of a coherent, rational subject, which it shares with non-representational theory (see Gorman and Andrews, Chapter 30). Protevi's (2009) re-articulation of the body politic in terms of assemblage theory is useful here.

> First-order bodies politic are individual human subjects, whose bodies are themselves assemblages that shape perception of the social categories through which difference is understood, and that rely on various material flows, such as media, food, water, and so on. Second-order bodies politic include any assemblage in which multiple humans participate, such that there is some degree of collective affective cognition. These, too, depend on material flows, and they have their own metabolisms that stave off dissolution. Crucially, first-order bodies politic are not prior to second-order bodies politic, as in social-contract theory or liberal theories of the state. Rather, first- and second-order bodies politic are mutually engaged in processes of becoming together in synchronic emergence.
>
> *Dittmer, 2017, p. 11*

This 'synchronic emergence' is produced via the ability of these bodies to affect and be affected; that is, these bodies are open systems, with multiple relations of exteriority (Deleuze, 1988; DeLanda, 2006). Affects are the glue that hold these assemblages/bodies politic together, give them their emergent agencies, and drive their processes of becoming.

Fox and Alldred (2015, p. 402, emphasis in original) argue that it is this feature that marks assemblage theory's progressive potential: 'concern is no longer with what bodies or things or social interactions *are*, but with the capacities for action, interaction, feeling and desire produced in bodies or groups by affective flows.' The shift to analysis of 'capacities' rather than properties is crucial because it highlights what is possible should bodies/things be brought together, rather than what defines an entity on its own. That is, the resonance of affects – often unrelated in their causes – can make previously impossible things seem suddenly possible, or even likely. We need no longer be bound by deterministic notions of politics drawing on closed totalities. 'These approaches can provide ways to understand, participate and change the world in a more-than-dialectical way – expanding our conception of politics from antagonistic to agonistic (Amin, 2002; Mouffe, 2000), from negating to affirmative politics' (Ghoddousi & Page, 2020, p. 5).

This focus on agonism, and assemblage theory's focus on territorialization and de-territorialization (or the clarifying and blurring of assemblages), provides a way into ethnographic methods. As per the critique of 'the field' referenced above, ethnographic approaches require us to think in terms of the forces that hold assemblages – such as communities, places, industries, or states – together for a time. These forces are experienced as trans-personal affects, and therefore the ethnographer's body is not only a sensing device that allows the ethnographer – when properly attuned – to feel the 'vibes' in a given collectivity, but also to perceive and to articulate these precognitive and non-individuated affects as emotions (Ghoddousi & Page, 2020, p. 7).

As this indicates, the researcher in an assemblage ethnography is not a subject analyzing an object, or a coherent subject examining the other. Rather, in order to be affected by the assemblage, the researcher must enter into the assemblage. Both are changed – even if minutely – through the relation that is formed. While it has often been noted that the researcher must instill trust in the community to be studied in an ethnography, it is clear also that the attachment must be mutual (Woon, 2013). The researcher and the researched are becoming together, albeit asymmetrically (Ghoddousi & Page, 2020), what Protevi (2009) refers to as synchronic emergence. Therefore, all assemblage ethnography is – to some extent – also a form of autoethnography: 'a form of self-narrative that places the self within a social context' (Reed-Danahay, 1997, p. 9). Such research can easily be caricatured as navel-gazing, but this can be avoided with a few steps, which we detail here.

While being plugged into the assemblage being studied is crucial to experience the affects associated with it, it is possible to modulate those relations so as to maintain a sense of the scholar as a critical analyst. For instance, in their work at the Australian War Memorial, Waterton and Dittmer (2014) both attempted to anonymously observe the other attendees and also used their bodies as 'instruments of research' (Longhurst, Ho, & Johnston, 2008, p. 215) to uncover the affective atmospheres at work in the Memorial. If two bodies both record similar 'readings' of affective flows and their intensities, then there is confirmation of a transpersonal field of affects being experienced in common. If the bodies have conflicting experiences or experiences of differing intensity, that is also a generative result which speaks to *some* dimension of embodied difference within the participating subjects.

To avoid 'navel-gazing' and unreflective experience, Waterton and Dittmer took a series of measures to *disrupt* their immersion. First, they utilized their field note-taking practices not only to literally document their observations but also to heighten their own reflexivity and attentiveness (Nicholson 2012). The constant interruption of notetaking helped to bring them back to their own critical faculties. While this went some way to de-naturalizing the sensory experience of the Memorial, it is possible to over-estimate your own ability to step outside of your own sensory experience and think critically about it. Therefore, Waterton and Dittmer also collected audio, video, and photographic recordings of the museum space to allow for a later focus on affective and sensory dimensions of place, such as ambient sound or lighting design, that often fall below the level of conscious thought in the moment of research. The final technique used by Waterton and Dittmer was to prolong the ethnography beyond the 'normal' duration of a visit; they visited the Memorial for many hours a day, every day, for a week. By remaining in the Memorial for a duration for which the displays and exhibits are not designed, the researchers' attention was drawn from that which the Memorial was engineered to highlight, to the background affects and rhythms of the assemblage that might not be noticed in the temporality of a standard 'visit'. By desensitising the researchers to the intended exhibit.

Of course, an assemblage ethnography must be as outward-looking as the relations of exteriority that define assemblages themselves. Ghoddousi (2019) achieved this through netnography, a technique that blurs the boundaries between researcher and researched by imbricating the researcher within social networks, either digital or fleshy (Kozinets, 2010; Matoes & Durand, 2012). Netnography was first devised for studying digital networks, but as digital networks shape our fleshy modes of socialising, netnographers follow the social into non-digital worlds. Netnography is best approached through the lens of assemblage theory because of the way it necessarily involves non-human actors, such as social media platforms (e.g., WhatsApp groups), their algorithms, and the digital resources that are harnessed by participants in the social network to generate affects within it (e.g., memes, streaming music and video, etc.). These digital artifacts were used to kindle affective states at a distance, which contributed to the increased consistency of the social group, and occasionally to its de-territorialisation.

Crucially, another dimension of the assemblage ethnography/netnography that Ghoddousi undertook was his intentional role within it. This was not some distanced ethnography without political purpose; rather, Ghoddousi intended – *through* his ethnography – to shape the assemblages he was studying (generally related to the Iranian diaspora in London) in ways that changed collective subjectivities and intensified their political potential. As described in Ghoddousi and Page (2020, p. 8) the research method contributed to a body politic,

> [One participant in netnography] added that 'The more distant they keep us [with visa restrictions], the closer we get', thereby implying a political act of resistance against dominant powers emerging out of such practices of everyday life (de Certeau 1984). This example shows how the exchange of affects blurred the boundaries of individual subjects. The collective subjectivity thereby forged exhibited glimpses of an emergent political agency in the *virtual* form that became visible at later stages of research when it fed into new *actual* assemblages of political demonstrations.

One of Ghoddousi's intentions with his research was to shape the diasporic relations of which he was a part so that they might be more progressive and oriented towards amity: avoiding the nasty potentials of identity politics (e.g., exclusivity, essentialism and exceptionalism) and warding off the crystallisation of state-like hierarchies. In this, he was clearly successful, at least for a time.

To conclude this section, assemblage theory blurs the ethnographic field, as well as the boundaries between self/other and subject/object. Rather, they are each composed through their relations of exteriority, and therefore are vulnerable to – and shaped by – the affective circuits in which they are enmeshed. These relations can be manipulated by the researchers' techniques to both experience affective flows but also enable critical reflection on those flows and active intervention in these processes of becoming. Assemblage theory also makes clear the role of the non-human in the production of subjectivity – including that of the researcher and of the researched. Finally, the assemblage approach to ethnography highlights its political potential as a kind of action research. But how can we make sense of assemblage methodology that happens 'above' the scale of individual human perception? What kind of 'action research' is possible at the scale of the more-than-human?

Archives and the Anthropocene: Non-human Agency and Temporality

Of course, ethnography is not the only way to examine the everyday. And, further, the everyday is not always divorced from or opposed to the traditional scale of power: the state (Mitchell 1991). Political geographers in recent years have focused on the idea of the state as an assemblage, with its power the emergent effect of ongoing processes occurring at a range of sites. The banal processes of paper-pushing and bureaucracy are 'everyday' in the experience of the workers populating the state, but to those on the sharp end of a tax audit, the deportation apparatus, or the justice system, it is anything but. Therefore, it is possible to see the state as a body politic (in the Protevian sense), composed of many smaller assemblages, which are themselves populated by a range of objects, bodies, and circulations, all enmeshed with one another in an affective field.

For this reason, it has perhaps always been misleading to refer to theories of 'the state' in the singular. Rather, each state (and each state-like entity – see McConnell, 2016) has evolved in its own way, even if states and polities are shaped by their interactions with one another. Painter (2006, p. 760) highlighted the heterogeneous nature of every state, using the Bakhtinian notion of 'Prosaics' to highlight "the intrinsic heterogeneity and openness of social life and its 'many-voiced' character. It [Prosaics] challenges all authoritative monological master subjects [such as the State] and their efforts to impose authoritative meanings." If states are assemblages, unfolding at a temporality that is both profoundly human and beyond the scale of human lives, then we have to pay attention to the materials that give state effects a more-than-transitory existence; that is, we have to pay attention to state affects (e.g., the bumbled policing of political protest as in Woodward 2014). Further, given the institutionalised diplomatic relations among states, it is possible to conceive of the international system as a web of state apparatuses emergent from everyday practices of diplomacy and shared infrastructures (Dittmer, 2017). That is, not only is the state an assemblage, but so is the political world produced by states. Methodologically, how are we to study this, when the processes of state and inter-state assemblages unfold at temporal and spatial scales beyond individual human experience?

Moving away from the realm of the contemporary everyday and into the longer temporalities of the state requires us to think beyond ethnographic measures, although they may remain useful (the present is, of course, a slice through the longer processes of becoming that every assemblage is undergoing). Accessing recent pasts through interviews is of course possible, but deeper histories are limited by the temporalities of human life expectancy. For this reason, archives become a crucial resource for understanding the changes that have unfolded within states over longer temporalities. Further, as these archives are themselves material formations that require constant processes of assemblage (i.e. curation, organisation, preservation) they themselves can be political – or have political effects. "Just as we need to take into account and render transparent how the ethnographic researcher affects his or her reconstruction of associations, so do we have to grapple with the visibilities and invisibilities produced through the archive" (Müller, 2012, p, 383). The researcher and the archive emerge synchronically via intra-action (Barad, 2007).

Dittmer's research on everyday practices and materialities in the historical UK Foreign Office (2016), for example, was originally stymied by the silences of the archive. This research was originally aimed at the 1782 formation of the Foreign Office, and consequently Dittmer began in the UK National Archives, trawling through early Foreign Office records. However,

> [W]hen the State Papers were compiled at the Public Records Office during the nineteenth century, the correspondence [...] was edited so that things which did not strike the *compilers* as relevant to what *they* held to be foreign policy [...] were left out.
>
> *Neumann, 2012, p. 44*

The absence of everyday institutional documents from the era pushed Dittmer to examine other archives, including the personal papers of Charles James Fox (the first foreign secretary) and the FCO Library (now held by King's College London). None of these archives had meaningful accounts of materiality in the Foreign Office's creation.

This archival black hole, rather than ending Dittmer's inquiry, opened him up to serendipity: it was in the FCO Library that Dittmer happened upon the bound reports of three Parliamentary Select Committees: the 1839 Select Committee on Public Offices (Downing-Street), the 1855 Select Committee on Downing-Street Public Offices Extension Bill, and the 1858 Select Committee on Foreign Office Re-construction. While not in the time period in which Dittmer was originally interested, the testimonies in the reports nevertheless spoke to a period of Foreign Office de- and re-territorialisation, and therefore the everyday practices and materialities within the buildings themselves.

> Where Foreign Office records had been tailored to exclude the everyday routines and materialities of the FCO, Parliamentary records were helpful, because in their discussions over housing the Foreign Office, the everyday practices and materialities of the Foreign Office intersected with the public purse. Foreign Office staff, architects, engineers, and other witnesses to the role of materiality in the Foreign Office spoke before the committees.
>
> *Dittmer, 2016, p. 87*

The serendipity of Dittmer happening upon these reports speaks to assemblage theory's understanding of processes unfolding over different temporalities and occasionally (but not necessarily) resonating with one another. As a sedimentation of histories, the archive is preserved to be a monument to the past, and thus it possesses certain curated properties; the archive requires work by curators and preservationists to remain territorialised over time. The reports' intra-action with Dittmer and his theoretical agenda represent a line of flight that opened up a virtual capacity within the archive, generating insights into everyday life in the Foreign Office that past archivists had decided were too mundane to be preserved. That meaning should emerge from the unexpected and serendipitous alignment of bodies and documents is a fine example of assemblage-as-methodology. The de- and re-territorialisation (Deleuze & Guattari, 1987) of the archival assemblage opened up new potentials within it (see Beckingham and Hodder, Chapter 14, for further discussion).

If we shift temporal scale from that of the state to that of the earth itself, we must also engage with a different type of archive, and therefore with a different type of curatorial practice.

Almeida and Hoyer (2020) offer a methodological approach to the (traditional) archive in light of the onset of the Anthropocene, that is, the role of the sedimented earth as an archive of the non-human – and now human – processes that shape the earth-assemblage. [T]he inscriptive event of the Anthropocene is an *extension* of the archive, where one adds to the readability of books and other texts, the stratifications of the earth. It would not be too radical to claim that at the level of geology, the Earth has a memory. How then do we produce archives that work on behalf of the earth rather than the state?

Recalling assemblage-as-methodology's aim to intervene in, and become with, the world in which researchers are embedded (Fox & Alldred, 2015), Almeida and Hoyer argue for a 'living archive', which opens up new horizons and offers a different future than the doomsday narrative of post-human desolation that is frequently offered. They note that most archives have served to memorialise and foreground the state or other privileged entities under capitalism, and those entities are the structures that have led to the ecological disaster of the Anthropocene. The resources needed to maintain and curate archives, usually provided by wealthy donors or the state, have tended to shape archives' priorities. Instead, Almeida and Hoyer offer the alternative of *community archives*, which seek to pluralise the materials from which narratives of the past, present, and future can be produced, offering new lines of flight that can (potentially) stave off Anthropocene dystopia.

> The living archive is participatory and open, and has a more flexible, adaptive infrastructure that can aid in decolonization. The living archive is referenced in discussions of bodily representation of histories of individuals and communities suffering trauma and is used to counter the concept that published texts should be privileged in archives over living histories, bodily records, and imaginaries.
>
> *Almeida & Hoyer, 2020, p. 18*

Further, each living archive is geographically situated, drawing from local traditions and struggles for definition (their case study is of the Interference Archive in Brooklyn), but is part of the larger constellation, allowing for systemic change via the preservation and amplification of the human and non-human precariat: 'the living archive might provide a path through the ontological murk of the Anthropocene' (Almeida & Hoyer, 2020, p. 20). For example, the Interference Archive collaborated with local Next Epoch Seed Library to map local flora, eschewing native/invasive binaries and imagining a future without humans. Almeda and Hoyer's Living Archive asks us, 'what kind of archive for what kind of future?'

While Almeida and Hoyer situate their praxis in the archive itself, seeing it as productive of new lines of flight, Harrison (2015, p. 27) goes further and reconceptualises the broader category of heritage, defining it as 'collaborative, dialogical and interactive, a material-discursive process in which past and future arise out of dialogue and encounter between multiple embodied subjects in (and with) the present.' In parallel to Almeida and Hoyer, Harrison seeks to pluralise what is considered heritage, seeing these material-discursive processes as local and situated while also seeing their potential to enter into assemblage with one another. By considering heritages as sets of material-discursive codes, Harrison sees – like Almeida and Hoyer – futures as emergent from *practices* of curation: heritage 'has very little to do with the past but actually involves practices that are fundamentally concerned with assembling and designing the *future*' (Harrison, 2015, p. 35). In particular, he draws on the concept of kinship in Aboriginal cultures, which link together the human and natural worlds in an ecological web. This idea is also promoted by Haraway (2016), who provocatively advocates substituting the term 'Chthulucene' for the Anthropocene, because of the tentacular web of connections that embed humanity with the earth. Whereas Almeida and Hoyer argue for more plural practices of archival preservation, Harrison (2015, p. 33) more explicitly engages with the active process of letting heritage go: 'In relation to heritage, [connectivity ontologies] force us to question not only the capacities of various material heritages to persist, but also whether the pasts that they actively create in the present could or should endure into the future.' By turning attention to 'new' forms of heritage curation such as seed banks, endangered languages, and nuclear waste disposal, Harrison effectively calls attention to the things that need preservation in an uncertain world, at least if we are to create a future that is truly sustainable. Indeed, if our heritage is what has brought the earth to the tipping points marked out by the concept of the Anthropocene, then much of what we have been will have to go. Together, Almeida, Hoyer, and Harrison articulate a vision of assemblage-as-methodology that both critically examines and participates in the practices of curation through which unsustainable futures are culled and new possibilities are substituted.

Conclusion

An assemblage approach could range from an onto-epistemological understanding of the world to practical methods for collecting and analysing data in ways that are attuned to their emergent possibilities. In this chapter we have looked at the possibility for assemblage-as-methodology to be deployed in a range of temporal and spatial contexts, from diasporic friend groups to community archives to the state. Central to all these contexts is the necessity of making sensible to human bodies the way in which processes unfold over a wide array of temporalities and spatial frames, occasionally intersecting with one another in ways that produce political resonances. We have emphasised methods that give insights into those various scales, from ethnography and netnography to archival research. Each is suitable for some contexts and unsuitable for others, and we suspect that there are many other methods that will be situationally amenable. Assemblage-as-methodology does not prescribe a single method, although some methods speak better than others to the ontology foregrounded by the theory.

One thread through our review has been the focus on the researcher as an active part of the assemblage being studied. That is, the researcher must open themselves up to that which is being studied, and by extension the opposite is also true. This offers the potential to affect and be affected, enabling auto-ethnographic insights but also political potentials and of course ethical dilemmas. Is it possible to retain a distanced sense of the researcher, with the tropes of 'objectivity' so valued traditionally in the social sciences? While this question may be moot in other areas, such as cultural geography, in political geography some audiences still expect a more-than-subjective analysis, if not quite an objective one. Our experiences indicate that the encounter with affects 'in the field' can be variously modulated and mediated in ways that open researchers up or insulate them. Each might be preferable, depending on the assemblage to be studied or the stage of research (such as when an ethnographer leaves the field to reflect on the experience). What cannot be avoided however, is *some* sense of becoming with the subject of analysis.

Indeed, one common thread through assemblage-as-methodology is an intervention of some kind with the subject matter; this might take the form of coining concepts, converting assemblages, or action research, such as an attempt to understand how practices of curation and processes of assemblage make some futures more or less likely. This intervention, like assemblage itself, can stretch across the human-nonhuman divide and also across multiple temporalities. Indeed, the outcomes may not be apparent in the time scale of the research itself, but they may emerge over time as they are activated by other processes. The objectives of politically-oriented assemblage research are oriented around various lines of flight that can be cultivated (or others that are meant to be preempted): fostering the political potential of a diaspora group, action research with a political party to unseat the government in an election (see Page & Dittmer, 2015; Page, 2019), or promoting sustainable politico-economic arrangements that enable a broader ethics of ecological entanglement. As such, political geography research using assemblage-as-methodology is a flexible set of tools to change the world, drawing together the past, present and future into a single field of analysis.

References

Almeida, N. and Hoyer, J. (2020) The living archive in the Anthropocene. *Journal of Critical Library and Information Studies*, **3**(1), 1–38.

Amin, A. (2002) Ethnicity and the multicultural city: Living with diversity. *Environment and Planning A: Economy and Space*, **34**(6), 959– 980.

Amin, A., and Thrift, N. (2002) *Cities: Reimagining the urban*. Cambridge and Oxford, England: Polity.

Barad, K. (2007) *Meeting the Universe Halfway:Quantum physics and the entanglement of matter and meaning*. Durham, NC: Duke University Press.

Billo, E. (2015) Sovereignty and subterranean resources: An institutional ethnography of Repsol's corporate social responsibility programs in Ecuador. *Geoforum*, **59**, 268–277.

Buchanan, I. (2015) Assemblage theory and its discontents. *Deleuze Studies*, **9**(3), 382–392.

Burrell, J. (2009) The field site as a network: A strategy for locating ethnographic research. *Field Methods*, **21**(2), 181–199.

Crang, M., and Cook, I. (1995) *Doing ethnographies*. Norwich, England: Geobooks.

de Certeau, Michel de (1984) *The practice of everyday life*. Berkeley: University of California Press.

DeLanda, M. (2006) *A new philosophy of society: Assemblage theory and social complexity*. London, England and New York, UK: Continuum.

Deleuze, G. (1988) *Spinoza: Practical philosophy* (R. Hurley, Trans.). San Francisco, CA: City Lights Books.

Deleuze, G. and Guattari, F. (1987) *A Thousand Plateaus: Capitalism and schizophrenia*. New York and London: Continuum.

Dittmer, J. (2016) Theorizing a more-than-human diplomacy: Assembling the British Foreign Office, 1839–1874. *The Hague Journal of Diplomacy* **11**(1), 78–104.

Dittmer, J. (2017) *Diplomatic material: Affect, assemblage, and foreign policy*. Durham, NC: Duke University.

Dowler, L. and Sharp, J. (2001) A feminist geopolitics? *Space and Polity* **5**, 165–176.

Elliot, A., and Urry, J. (2010) *Mobile lives*. New York, NY and Oxon, England: Routledge.

Fox, N. J., and Alldred, P. (2015) New materialist social inquiry: Designs, methods and the research-assemblage. *International Journal of Social Research Methodology*, **18**(4), 399–414.

Ghoddousi, P. (2019) Dimorphic Diasporas: Assembling identity, community belonging, and collective action among Iranians in London. Unpublished thesis, UCL.

Ghoddousi, P. and Page, S. (2020) Using ethnography and assemblage theory in political geography, *Geography Compass* **14**(10), e12533

Haraway, D. (2016) *Staying with the Trouble: Making kin in the Chthulucene*. Durham, NC: Duke University Press.

Harrison, R. (2015) Beyond 'natural' and 'cultural' heritage: Toward an ontological politics of heritage in the age of Anthropocene. *Heritage & Society* **8**(1), 24–42.

Hyndman, J. (2004) Mind the gap: Bridging feminist and political geography through geopolitics. *Political Geography*, **23**(3), 307– 322.

Kozinets, R. V. (2010) Netnography: The marketer's secret weapon. (White paper). Retrieved from www. etnografiadigitale.it/wp-content/uploads/2012/05/NetBase_Netnography_Kozinets_Paper.pdf

Latour, B. (2005) *Re-assembling the social: An introduction to actor-network theory*. Oxford, England: Oxford University.

Longhurst, R., E. Ho, and L. Johnston. (2008) "Using 'the body' as an 'instrument of research': kimch'i and pavlova." *Area* **40**(2), 208–217.

Marcus, G. E. (1995). Ethnography in/of the World System: The Emergence of Multi-Sited Ethnography. *Annual Review of Anthropology*, **24**, 95–117. www.jstor.org/stable/2155931

Marston, S. A. (2003) Political geography in question. *Political Geography*, **22**(6), 633–636.

Matoes, P., and Durand, J. (2012) Residence vs. ancestry in acquisition of Spanish citizenship: A 'netnography' approach. *Migraciones Interncionales*, **6**(4), 9–46.

McConnell, F. (2016) *Rehearsing the State: The political practices of the Tibetan Government-in-Exile*. Oxford: Wiley-Blackwell.

McFarlane, C. (2011) Assemblage and critical urbanism. *City*, **15**(2), 204–224.

Megoran, N. (2006) For ethnography in political geography: Experiencing and re-imagining Ferghana Valley boundary closures. *Political Geography*, **25**, 622– 640.

Mitchell, T. (1991) The limits of the state: Beyond statist approaches and their critics, *American Political Science Review* **85**(1), 77–96.

Mouffe, C. (2000) *The democratic paradox*. London, England: Verso.

Müller, M. (2012) Opening the black box of the organization: Socio-material practices of geopolitical ordering. *Political Geography* **31**(6), 382–383.

Neumann, I. (2012) *At Home With the Diplomats: Inside a European Foreign Ministry*. Ithaca: Cornell University Press.

Nicholson, H. (2012) Attending to sites of learning: London and pedagogies of scale. *Performance Research: A Journal of the Performing Arts* **17**(4), 95–105.

Page, S. (2019)'A machine masquerading as a movement': The 2015 UK General Election Labour campaign investigated through assemblage and affect,' *Political Geography* **70**, 92–102.

Page, S. (2020) Assemblage theory. In A. Kobayashi (Ed.), *International Encyclopedia of Human Geography* (Vol. 1, 2nd ed., pp. 223– 227). New York, NY: Elsevier.

Page, S. and Dittmer, J. (2015) Assembling political parties, *Geography Compass* 9(5), 251–261.

Painter, J. (2006) Prosaic geographies of stateness, *Political Geography* **25**(7), 754–772.

Pottinger, L. (2018) Growing, guarding and generous exchange in an analogue sharing economy. *Geoforum*, **96**, 108–118.

Protevi, J. (2009) *Political affect: Connecting the social and the somatic*. Minneapolis: University of Minnesota.

Reed-Danahay, D. (1997) *Auto/ethnography: Rewriting the self and the social*. London, England: Bloomsbury.

Sharp, J. (2005) Geography and gender: Feminist methodologies in collaboration and in the field. *Progress in Human Geography*, **29**(3), 304–209.

Thrift, N. (2004) Intensities of feeling: Towards a spatial politics of affect. *Geografiska Annaler*, **86B**, 57–78.

Wainwright, T., Kibler, E., Heikkilä, J.-P., and Down, S. (2018) Elite entrepreneurship education: Translating ideas in North Korea. *Environment and Planning A: Economy and Space*, **50**(5), 1008–1026.

Waterton, E. and Dittmer, J. (2014) The museum as assemblage: Bringing forth affect at the Australian War Memorial. *Museum Management and Curatorship*, **29**(2), 122–139.

Wenzel, J. (2018) Past's futures, future's pasts. *Memory Studies* **11**(4), 502–504.

Woodward, K. (2014) Affect, state theory, and the politics of confusion. *Political Geography*, **41**, 21–31.

Woon, C.Y. (2013) For 'emotional fieldwork' in critical geopolitical research on violence and terrorism. *Political Geography*, **33**, 31–41.

12

INDIGENOUS GEOGRAPHIES

Researching and De-colonising Environmental Narratives

Meg Parsons and Lara Taylor

Mā te rongo, ka mōhio Mā te mōhio, ka mārama Mā te mārama, ka mātau Mā te mātau, ka ora.
Through perception comes awareness, through awareness comes understanding, through
understanding comes knowledge, and through knowledge comes well-being

Indigenous Māori proverb

Introduction

A wealth of scholarship now highlights how decolonising methodologies involve the fundamental reconfiguration of how research is conducted with and by Indigenous communities,
so that Indigenous voices and epistemologies are placed at the heart of the research process
(Cannella and Manuelito, 2020; Nakamura, 2010; Smith, 1999). Over the last two decades,
human geographers have engaged in a diversity of projects aimed at decolonising the discipline of geography, research methods and the types of research being conducted (Andrea
Vasquez-Fernandez et al., 2017; Barker and Pickerill, 2019; Clayton and Kumar, 2019). In
particular, decolonial geographical scholarship explores how colonial environmental narratives
were (and still are) used to justify colonial rule as well as the marginalisation of Indigenous
peoples' knowledges, values, governance structures, and environmental management practices
with their ancestral landscapes and waterscapes (Halvorsen, 2019; Parsons et al., 2021). The
vast majority of geographers employing anticolonialism and decolonisation within their work
are non-Indigenous geographers, whose work we commend but do not explicitly focus on
in this chapter (Barker and Pickerill, 2019; Clayton and Kumar, 2019). Instead, we direct
our attention to a small number of Indigenous scholars (including but not limited to human
geographers) who use Indigenous methodologies to investigate geographical issues (Hunt,
2014; Ringham et al., 2016; Simmonds, 2011). First, we summarise the ethics of Indigenous
methods. Second, we explore Indigenous research paradigms and one Indigenous research
methodology (Kaupapa Māori). Next, we highlight and discuss some Indigenous qualitative
methods of interviewing.

DOI: 10.4324/9781003038849-15

Indigenous Research Ethics and Methodologies

Indigenous worldviews take form in traditions, practices, and knowledges, and are articulated in values and beliefs that underpin Indigenous research ethics (Hart, 2010). For instance, Indigenous ontologies and axiologies (nature of values and value judgements) emphasise how humans are connected with more-than-humans through reciprocal relationships that bind all beings together to care for one another (Hart, 2010; Johnson et al., 2016). Louis (2007) contends that the foundation of an Indigenous research paradigm, methodologies, and methods are the four Rs: relational accountability, reciprocal appropriation, respectful representation, and rights and regulations during the research design and practices (Figure 12.1: The Four Rs). Indigenous methodologies stress the need to establish meaningful, mutually beneficial, and ongoing relationships between researchers and Indigenous communities due to Indigenous peoples' experiences of being misled and mistreated by researchers both in historical and contemporary settings (Louis, 2007; Smith, 1999). Therefore, as Indigenous people ourselves we added a sense of responsibility to ourselves and our communities to practice the four Rs, in particular, that we respectfully represent the knowledge, values, tikanga (laws), histories, and experiences of Māori, and ensure that Māori benefit from our research. Along similar lines, Indigenous ontologies and axiologies (philosophies) emphasise how human beings are interconnected with more-than-human beings (biota, biophysical, and supernatural) through reciprocal relationships that bind all parties to duties of care for one another.

Throughout academia over the last three centuries, despite an incredible diversity of different paradigms and methodologies, Indigenous worldviews, values, and assumptions remained

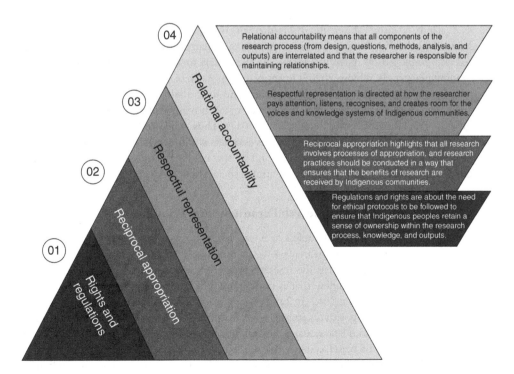

Figure 12.1   The Four R's.
Source: Created by Lead Author based on work of Louis 2007.

excluded. Instead Euro-Western ontologies and epistemologies shaped what methodologies geographers used in their research. Such an exclusion is unsurprising given the violence, marginalisation and discrimination experienced by Indigenous peoples (as well as limited number of Indigenous geographers within the discipline). Yet Indigenous scholars, particularly since the 1980s, have challenged this exclusion and sought to use methodology as a site of resistance and anticolonialism (Cannella and Manuelito, 2020; Chilisa, 2019; Smith, 1999).

Relational accountability is interwoven within a diversity of Indigenous peoples' conceptualisations of sovereignties, belongingness, space and time, and the nature of the world (its reality and futurity), which in turn informs Indigenous methodologies. Methodology is the place where these philosophical assumptions about knowledge, values and the nature of reality, as well as academic theories and practices on a specific research topic converge. Accordingly, efforts to decolonise research methodologies are fundamentally about social justice for Indigenous peoples through empowering and resourcing Indigenous knowledge, authority, and peoples (refer to Figure 12.2).

While emerging as an extension of Indigenous studies, Indigenous research principles and methodologies are now increasingly being adopted by Indigenous researchers both within and outside the academy to recognise and enact Indigenous ways of knowing, within the context of generating research, developing environmental management policies and plans, and re-designing educational programs for Indigenous peoples by and with Indigenous peoples (Kovach, 2015; Smith, 2017). Indeed, Indigenous methodologies are often founded on Indigenous knowledge systems that are always place-specific and culture-specific (Tuck et al., 2014). The same research methods can be employed across a diversity of contexts but are typically adjusted to fit the specific requirements of each Indigenous community, and its systems of knowledge, laws and protocols, and ways of life. Different research frameworks, including Kaupapa Māori (Aotearoa New Zealand), Anishinaabe Mino-Bimaadiziwin (Canada First Nation) and Indigenous Medicine Wheel (USA Native American/American Indian and Canada First Nations), were created and employed by different Indigenous researchers to fit with their knowledges, laws and cultural practices (Chiblow, 2020; Umangay, 2019). As Graham Hingangaroa Smith writes, the adoption of an Indigenous research paradigm and methodology does not mean the complete rejection of Western knowledge, theories or research methods but rather a promotion of Indigenous ways of knowing and being (Smith, 2017). Thus, while Indigenous methodologies may involve the same data collection techniques as others within social sciences, the underlying philosophical assumptions (paradigm), ethics and values (axiology), and ways of knowing (epistemology) that guide Indigenous and conventional social science methodologies differ.

Indigenous Research Paradigms and Frameworks

At its most basic level methodology refers to how a research process will proceed and the theoretical framework on which a research project is founded. It begins with the selection of a research paradigm that will inform the research project. A research paradigm, at its simplest, is a way of describing a worldview that is underpinned by philosophical premises about the nature of social reality (Chilisa, 2019, p. 23). Research paradigms within the social sciences are broadly classified into post-positivist, pragmatic, constructivist, transformative, and Indigenous (see Figure 12.3). Indigenous Botswanan scholar Chilisa describes an Indigenous research paradigm as a "framework of belief systems that emanated from the lived experiences, values, and history of those belittled and marginalised by Euro-Western research paradigms" (Chilisa, 2019, p. 23). In addition to the four Rs, the Indigenous research paradigms are founded on relational ontologies (ways of seeing the world) which emphasis people's connections to other living, non-living, and metaphysical things. Within Indigenous cultures, greater emphasis is placed on an I/We relationship (collectivist) in contrast to Euro-Western societies that focus on an I/you relationship (individualistic). Accordingly, Indigenous knowledge systems are both holistic (shared amongst all living and non-living entities in the cosmos) and relational (about relationships and connections between peoples, places, living and non-living) (Chilisa, 2019, p. 24).

From our cultural backgrounds as two Māori wāhine (women) we, recognise that Te Ao Māori (the Māori world or worldviews) shapes how we understand the world. For instance, in Te Ao Māori human and non-human entities are connected though whakapapa (genealogical connections) so that everything in the world (flora and fauna, mountains, rivers, seas, gods, supernatural beings) is conceptualised as being part of the same extended family (kinship) group. Within Te Ao Māori the whole is always more important than the part; Māori worldviews (of which there are multiple reflecting differences between tribal groups and subjectivities) are holistic, relational, and communitarian (rather than individualistic). The opening whakataukī (Māori proverb) alludes to a Te Ao Māori perspective that situates knowledge, understanding and well-being as interconnected. We decided to include this whakataukī as we, two Māori wāhine (women), identify our personal and professional lives are centred on expanding our knowledges, experiences, and connections with the environment (human and more-than-human) of which we are part. We recognise that our well-being is interconnected with a deeper understanding of ourselves, other living and non-living things and that such holistic understanding is useful and perhaps necessary to address the host of environmental challenges that humanity faces in the twenty-first century. In our research, we drew on Indigenous scholarship as well as Kaupapa Māori, an Indigenous research framework developed by and for Māori to decolonise research conduct (Mahuika, 2019). As Smith wrote in her foundational book *Decolonizing Methodologies*, the fundamental aim of decolonising research processes is "about centring our [Māori] concerns and world views and then coming to know and understand theory and research from our perspectives and for our purposes" (Smith, 1999, p. 41).

Kaupapa Māori is not the same thing as Māori knowledge (mātauranga Māori) or epistemology, and instead, refers to a framework that informs how research is designed and enacted. The approach encourages researchers to ask themselves critical questions before, during and after research with Māori. Kaupapa Māori is prompted as a research agenda that can understand and affect social change. Māori people are no longer only research subjects whose knowledge is extracted, interpreted, and used for the benefit of others (typically non-Māori) but rather the approach positions Māori as active participants in the research process in a way that fosters Māori empowerment and self-determination. Critical components of a Kaupapa Māori approach include the principle of kaupapa (connected to Māori philosophy and values); the centrality of whānau (extended family), hapū (sub-tribe) and iwi (tribe) and whakapapa (genealogy); the recognition of Māori autonomy and rangatiratanga (chiefly authority, self-determination right); the acknowledgement and promotion of mātauranga Māori (Māori knowledge), tikanga (laws), and ako (learning) (Parsons et al., 2021; Reid et al., 2016). Two other principles, recently identified by scholars, are āta (ongoing respectful relationships) and the acknowledgement and active promotion of Te Tiriti o Waitangi (the Treaty of Waitangi). The treaty was signed by representatives of the British Crown and 500 Māori leaders in 1840 and outlined the formation of a partnership between Māori and the Crown to share governance of Aotearoa New Zealand (Orange, 2015). There is a large body of scholarship that employs a Kaupapa Māori framework within education (Smith, 2017; Stewart, 2016) and health (Rolleston et al., 2020). As well as a small but growing number of social science and humanities scholars that are drawing on Kaupapa Māori principles (Jackson, 2015; Mahuika, 2019;

Smith, 2020). A handful of Māori geographers are adopting Kaupapa Māori approaches and the related mana wāhine approach (which focuses on Māori women's experiences) (Barry and Thompson-Fawcett, 2020; Greensill, 2005; Simmonds, 2011). Other Indigenous communities around the globe employ different Indigenous research frameworks and a variety of different research methods. There are efforts to find ways to blend or develop hybrid paradigms too that can allow Indigenous and non-Indigenous researchers to work together (Reid et al., 2020).

Indigenous and Non-Indigenous Frameworks: Exploring The Interface Between Worlds

Held (2019) suggests that for research projects that involve both Indigenous and non-Indigenous scholars they must work together to develop a research paradigm or framework that incorporates both Western and Indigenous worldviews as a way to decolonise research paradigms more broadly (Artaraz and Calestani, 2015; Held, 2019). Indeed, research paradigms do not necessarily exist in isolation from one another; there is often considerable overlap between frameworks (as shown in Figure 12.1), and researchers may not exclusively employ Indigenous research frameworks. In many parts of the globe, new frameworks (as well as government policy) are being developed and enacted that seek to empower Indigenous knowledge and research in transformative and self-determining ways. Diverse actors (Indigenous and non-Indigenous) are increasingly working together to develop a new frameworks and approaches that intend to bridge the divide between different knowledge systems, research paradigms and practices (Davies et al., 2018; Maxwell et al., 2020). An example from Aotearoa New Zealand is the Waka-Taurua framework (see Figure 12.3) highlights how researchers can employ methodological hybridity (drawing on Indigenous and non-Indigenous research paradigms) in a way that not only decolonises environmental research projects but also contributes to decolonising the management and governance of environments. A significant amount of geographical scholarship is emerging as part of such paradigm spanning and disrupting initiatives (also known as inter- or trans-disciplinary collaborations) involving researchers, decision-makers, industry groups, and communities who come together to work on projects due to a shared concerns about a range of issues including climate change, environmental degradation and biodiversity loss (Chilisa, 2019, p. 21). However, the extent of Indigenous involvement in different research initiatives varies considerably (see Figure 12.2).

The positionality of a researcher, the question they are answering, the stage they are at in their career, and the time and resourcing available can all influence the research design, methodologies and methods employed. Significantly, it also reflects the histories, experiences, knowledge systems, customary laws, and preferences of the Indigenous communities who are involved in and/or leading the research, as well as research participants themselves. Some Māori researchers, as well as Māori research participants, report their feelings of discomfort with using Kaupapa Māori methodology due to their lack of fluency in Te Reo Māori (Māori language), unfamiliarity with tikanga (customary laws), and insufficient knowledge of their whakapapa (genealogy) (Fisher, 2014; King et al., 2018). In these instances, Māori scholars (like other Indigenous researchers) often choose to adopt an Indigenous research paradigm but select to use a non-Indigenous research methodology (such as participatory action research) that provides to the most appropriate methodology and methods for their research project and their research participants (keeping in mind the four R's). In many instances the co-development of methodology and methods (and all other aspects of a research process) between researchers and participants is a pre-requisite for authentic and empowering Indigenous research that helps to identify the most appropriate approaches for any particular context.

Empower Indigenous Knowledge

Colonising	Decolonising
◊ Only Western science legitimised ◊ Indigenous knowledge merely acknowledged ◊ Indigenous knowledge extracted from Indigenous experts ◊ Non-Indigenous as primary Indigenous knowledge experts ◊ Cultural expertise of Indigenous researchers overlooked ◊ Scientific expertise of Indigenous Researchers side-lined	◊ Indigenous knowledge & Western science valued ◊ Indigenous activity resourced ◊ By & with Indigenous knowledge experts ◊ Cultural expertise valued ◊ Scientific expertise recognised

Empower Indigenous People

Colonising	Decolonising
◊ Consulted for projects, programmes & organisations ◊ Advice sought to tick the 'Indigenous' box ◊ Informed about the decisions made ◊ Projects about Indigenous ◊ Indigenous rare in the sector ◊ Cultural labour is unpaid or underpaid ◊ Indigenous researchers responsible & isolated	◊ Indigenous-led & co-led projects, programmes & organisations ◊ Advice sought for research values & followed ◊ Involved as decision makers ◊ Projects by & with Indigenous ◊ Many Indigenous in the sector ◊ Additional labour is resourced ◊ Indigenous researchers supported & developed

Empower Indigenous Resources

Colonising	Decolonising
◊ Academic aspirations alone ◊ Academic publication the most important goal ◊ Intellectual property benefit retention by academic institutions ◊ Only Western scientific measures of excellence, impact & success	◊ Indigenous & academic aspirations ◊ Publication & benefit for Indigenous people ◊ Intellectual property benefit sharing or Indigenous ownership ◊ Indigenous worldview of excellence, impact & success is included

Figure 12.2 Decolonising the environmental research sector through empowering Indigenous knowledge, people, and resources.

Figure 12.3 One framework for bridging the divide between different worldviews from Aotearoa New Zealand.

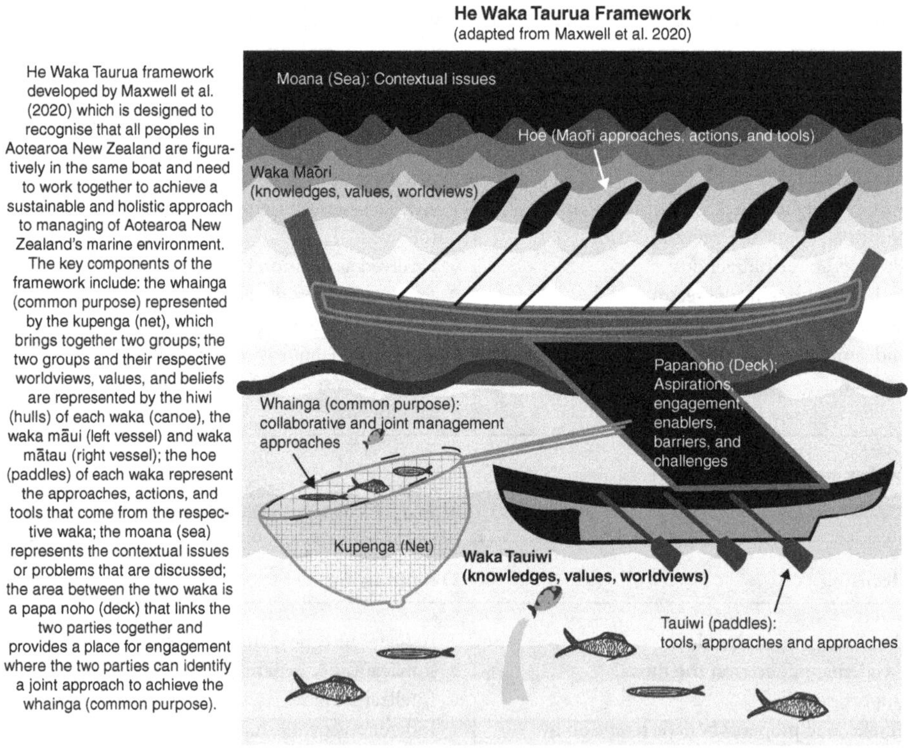

He Waka Taurua Framework
(adapted from Maxwell et al. 2020)

He Waka Taurua framework developed by Maxwell et al. (2020) which is designed to recognise that all peoples in Aotearoa New Zealand are figuratively in the same boat and need to work together to achieve a sustainable and holistic approach to managing of Aotearoa New Zealand's marine environment. The key components of the framework include: the whainga (common purpose) represented by the kupenga (net), which brings together two groups; the two groups and their respective worldviews, values, and beliefs are represented by the hiwi (hulls) of each waka (canoe), the waka māui (left vessel) and waka mātau (right vessel); the hoe (paddles) of each waka represent the approaches, actions, and tools that come from the respective waka; the moana (sea) represents the contextual issues or problems that are discussed; the area between the two waka is a papa noho (deck) that links the two parties together and provides a place for engagement where the two parties can identify a joint approach to achieve the whainga (common purpose).

Moana (Sea): Contextual issues

Hoe (Maori approaches, actions, and tools)

Waka Māori (knowledges, values, worldviews)

Papanoho (Deck): Aspirations, engagement, enablers, barriers, and challenges

Whainga (common purpose): collaborative and joint management approaches

Kupenga (Net)

Waka Tauiwi (knowledges, values, worldviews)

Tauiwi (paddles): tools, approaches and approaches

Figure 12.4 Continuum of Indigenous involvement in research.

Irrespective of what specific research methodology a researcher adopts, steps can be taken to consider how individual researchers and research teams can seek to embrace decolonisation throughout their research process (see Figure 12.2, Figure 12.3 and Figure 12.4). Indigenous researchers and practitioners are producing helpful guidance for researchers to decolonise their work. For example, Figure 12.2, which provides insights on how to empower Indigenous knowledge, people and resources, is adapted from a list of recommendations formed during a two-day gathering of Indigenous Māori researchers in Aotearoa New Zealand, in October 2019, and given to government and research organisations to facilitate a thriving science system. Critically assessing the extent to which Indigenous knowledge, people and resources are empowered enables the researcher(s) to measure how decolonising their research may be (see Figure 12.4). Such analysis requires a high degree of self-reflexivity and flexibility by researchers in the research design and practices (irrespective of whether they are Indigenous or non-Indigenous) (see Figure 12.4). Australian Indigenous scholar, Stuart Barlo records a critical decolonising approach during his PhD research (investigating how to reverse the negative impacts of colonisation on Aboriginal men's dignity), started when he started to explain the human ethics procedures with his participants and how pseudonyms would be used instead of their names in the final thesis (Barlo, 2017). He outlined that this was regular ethics practice to ensure confidentiality and that participants would not receive any backlash from what they said in the study. One of the participants became particularly upset upon hearing that his name would not be included and explained how:

> as a child, he had been placed in a boys' institution where on his arrival he was given a number, and while he remained at the institution, he was referred to by that number and never his name. He said it made him and the other resident feel as if they were nobody and of no value. He went on to tell me that the home had a dog and that the dog had a name.
>
> *Barlo, 2017, p. 10*

While it is standard practice within academic research to de-identify human research participants, this practice is critiqued by many Indigenous scholars and Indigenous communities because of its echoes or repetition of colonial ways of classifying, renaming, extracting, oppressing and removing Indigenous people (from their families, homes, cultures, knowledge systems, and even their names). All Barlo's participants requested that they be named in his thesis, and so he was required to request an amendment to his university's ethics committee. After he informed the committee of the above story, approval to name his participants was given. The act of naming participants (with their permission) was, Barlo's argued, a significant part of decolonising his research project, and "g[a]ve the power of the information back to its owners through allowing their voices to be heard as the knowledge holders" (Barlo, 2017, p. 10). As with Barlo's experience, other researchers (ourselves included) encountered experiences where Indigenous research participants wanting to be named in the research, but researchers faced restrictions on being able to do so by their research institutions, government policies, or academic journal editors (Datta, 2018). Parsons, for instance, was not allowed to name the participants in her PhD research due to Queensland Government regulations and was later challenged by reviewers about the ethical practice of her study when she tried to publish another study in which all the participants were named (at their request) (Crease et al., 2019; Parsons, 2009).

Scholars from across the traditional disciplinary divides are already involved in collaborative projects that enact the use of mixed methodologies in Aotearoa New Zealand on a range of different topics, including climate change, marine governance, and freshwater management. One example is the work of Māori art scholar Huhana Smith who led a transdisciplinary study that investigated the implications of climate change on her hāpu (sub-tribe) and small Māori community of Kuku (located on the western coast of the lower North Island), as well as what adaptation strategies she and her whānau (extended family) adopted to reduce the risks associated with climate change. Smith et al., (2020) employed Western scientific knowledge and mātauranga Māori in a manner that allowed both to co-exist as equally important systems of knowledge, with both quantitative and qualitative methods to collect and analyse the data (Bryant et al., 2017; Smith, 2020). Non-Māori scholars were primarily involved in quantitative data such as the use of climate change modelling to produced downscaling climate change projections for Kuku, the production of local hazard maps, and for projections about the potential loss and damage under different climate change scenarios, as well as the cost/benefit analysis of other climate change adaptation options (Bryant et al., 2017). While Māori scholars were responsible for leading the collection and analysis of qualitative data with Māori community members, which involves running a series of hui (meetings), wānanga (focus groups), and hīkoi (walking) interviews with the community at Tukorehe Marae (the meeting house of the hapū), and creating art exhibitions (which participants co-designed with artists) (Bryant et al., 2017; Smith, 2020). The division of quantitative/qualitative research tasks along apparent cultural lines was an incidental rather than a deliberate effort to establish or maintain binaries, but rather a reflection of different areas of expertise of research team members; non-Māori scholars were physical geographers and economists, while Māori scholars were creative artists and social scientists. However, it does demonstrate a more general trend (discussed later in this chapter) toward a lack of Indigenous geographers using quantatitive research methods and the need to critical consider what decolonial quantitative methodologies are (or could be) (Axelsson and Sköld, 2011).

Indigenous research methodologies, including those that embrace co-design and transdisciplinarity, require a significant investment of time and resources by researchers as well as by Indigenous community members themselves. It only works if there is ample time devoted to build and maintain trust between the different parties and to ensure the four Rs of Indigenous research are followed (Colbourne et al., 2019; Woodward and Marrfurra McTaggart, 2019). For instance, Smith et al.'s (2017) project began with a series of hui, all of which took place within the rohe (tribal land) of the local iwi (tribe) and hapū (sub-tribe) according to their tikanga (customary laws), and involved the research team, iwi and hapū research participants as well as farm representatives coming together to meet, exchange ideas and plan the research project. Meetings were held in Smith's home, at her marae (meeting complex for her hapū) or on the land (one of the dairy farms). The korero (talking) involved the scientists talking to iwi and hapū about their knowledge about the impacts of climate change but also non-scientists (both academics and hapū participants) talking about their knowledge of mātauranga (knowledge) and their experiences of changing environmental conditions. Another critical part of these early discussions was about the different aspirations' iwi, and hapū members held (how the land could or should be developed, where they should live and how they should live) and what values should form the foundation upon which climate adaptation plans and actions should be made. These initial hui discussions were critical in bringing together the members of the research team with community members. The first hui provided scientists to explain to community members their evidence about the potential impacts of climate change on the case study area and issues for further investigations. Iwi and hapū participants were also able to communicate their knowledge, understandings, and direct experiences about local-level environmental changes within the rohe, the diversity of ways climate risks disrupted their daily lives and what they wanted to be improved. Throughout the years of Smith's project hui were held to ensure that community members were always involved in all aspects of the research process (design, data collection, analysis, findings). These hui, therefore, were a pivotal part of the co-design process whereby different scholars worked with one community to identify the issues they were most concerned about, what the critical gaps in scholarly and practical knowledge were (mātauranga

Māori, design, economic and ecological), and how the research project could be tailored to provide significant co-benefits to the iwi (tribe), hapū (sub-tribe), and whānau (family). Such an approach exemplifies not only what an excellent Indigenous-led co-design process should be like, but also how decolonising methodologies can be used by scholars (from across the disciplinary divides) seeking to understand environmental issues in a way that empowers rather than marginalised Indigenous peoples (Bryant et al., 2017; Parsons et al., 2017, 2016). Other similar research projects, also built on the four Rs, are being employed by Indigenous and non-Indigenous scholars investigating marine and freshwater governance, natural resource management, and climate change adaptation (Johnson et al., 2021; Parsons et al., 2021).

Indigenous Methodologies and Methods: Narratives Not Numbers?

We turn our attention now to qualitative methods employed by Indigenous scholars seeking to decolonise environmental narratives. The majority of scholars (ourselves included) use qualitative research methods (the tools that researchers employ to gather data) to investigate Indigenous geographies. There is sometimes considerable overlap between methodology and methods, as outlined by Pacific scholars Suaalii-Sauni and Fulu-Aiolupotea (2014) with regard to Talanoa, due in part to the holistic nature of Indigenous ontologies and epistemologies which means that theory and practice are intertwined. The emphasis on qualitative methods is partly reflective of the extensive oral traditions in many Indigenous societies that lacked a written language prior to colonisation. Thus, oral transmission through stories, songs, poetry, and proverbs (as well through artworks including dances) were the critical way of retaining and transmitting knowledge across generations. Yet some scholars argue that there is a pressing need for Indigenous scholars to develop Indigenous quantitative methodologies because at present the statistics being produced about Indigenous peoples (by non-Indigenous scholars and institutions) are entrenching colonial discourses and policies by marginalising and misrepresenting Indigenous bodies, communities, land rights, and modes of living (Walter and Andersen, 2013). As methods also guide the way the researcher interacts with their participants in a culturally appropriate manner, it is critical that further research is undertaken to explore different ways of decolonising quantitative methodologies, and how non-Indigenous scholars can be engaged in these efforts as well (just as many are working alongside Indigenous scholars to decolonise qualitative methods). It is beyond the scope of interested readers to the work of Indigenous chapter to discuss quantitative methods. However, we recognise its importance and refer to demographers and historians (Coutts et al., 2016; Prout, 2009; Axelsson and Sköld, 2011).

Indigenous Qualitative Methods: Interviews

Over the last two decades, a plethora of researchers has drawn on decolonising and Indigenous perspectives to critique the dominant (Western) interview methods, be it structured, semi-structured, or unstructured (Cannella and Manuelito, 2020; Wilson, 2008). These scholars offer a range of alternative interview approaches that are underpinned by different Indigenous peoples' worldviews, values, and modes of living. The conventional interview is founded on the individual and individuality and positions the researcher as a detached and mostly quiet presence in the interview space revealing little or no information about themselves. In contrast, Indigenous interviews are underpinned by the earlier concept of "relational accountability" emphasising communities' connectedness, togetherness, duty of care and cooperation. Conventional interviews follow typically a set of well-established rules and codes, which draw on Western knowledge systems, values, and social structures. The researcher is positioned as a detached and mostly quiet presence in the interview space revealing little or no information about themselves to the interviewee, in contrast, the researched are expected to be an active and open presence during the interview.

An Indigenous paradigm, however, provides for alternative approaches to interviewing which prioritise Indigenous ways of knowing centred on the interconnectivity of humans and more than humans. Indigenous researchers employed a wide range of interview methods (both Western and Indigenous), however, in the remainder of this chapter we provide a brief overview of four Indigenous interview techniques that are allowing Indigenous scholars to collect qualitative data that challenges, disrupts and decolonises environmental narratives (Absolon, 2020; Datta, 2018; Hunt, 2016; Hutchings, 2004; Mahuika, 2019; Naepi, 2019; Otunuku, 2011; Suaalii-Sauni and Fulu-Aiolupotea, 2014; Tuagalu et al., 2014).

Hikoi (Walking) Method

Mobile methods, specifically walking methods, are increasingly used by social scientists who want to explore ideas of place, identity, and people's relationships with landscapes (Cavanagh and Standley, 2020). Since the mid-2000s, several Māori scholars have begun to develop and employ a distinctly Māori version of a walking interview know as the hikoi method (Gilgen, 2008; Hardy et al., 2015) See also Richmond et al., Chapter 7 in this book. Hikoi (refers to a shared journey, walking and talking) is a practice used by Smith (et al., 2017) and Hardy et al., (2015) in cross-cultural environmental research undertaken in Horowhenua. In both studies, the research teams walked and camped on Māori land alongside members of the community to share their knowledge and to build multiple understandings about the degradation of environmental resources, and to identify future research needs and restoration actions to be taken. Hardy et al., (2015) parallel hikoi captures the concept of a "travelling roadshow" from which strength, kotahitanga (unity), Mātauranga (knowledge and whāinga matua (shared purpose) are created through walking, talking and experiencing things together.

As of November 2020, Māori human geographer Naomi Simmonds is employing hīkoi in her innovative research project (Taku Ara Rā: Ko Māhinaarangai) that explores how retracing the ancestral pathways may affirm what it means to be Raukawa (Māori tribal group) woman in contemporary Aotearoa New Zealand (Hurihanganui, 2020; Simmonds, 2009). Simmonds and a group of other Raukawa wahine (women) are following the journey that their ancestor (Māhinaarangi) took, heavily pregnant and barefoot, from Whakatāne to Otorohanga to reach her lover. The hīkoi is expected to take three weeks, with the group staying at the marae (the meeting house of a sub-tribe/hapū) along the way. The term hīkoi refers to both a goal and a process and is about moving forward, but at all times recognising that the journey itself is significant, and Simmonds' research practice embraces this. At the heart of her hīkoi is a shared kaupapa (purpose): she and the participants are all committed to retracing their ancestor's footsteps to connect to themselves, their ancestors and their landscapes; and mutual trust in each other to share the journey. Sometimes hīkoi are just short walks with a researcher and research participants, and other times the walks are over longer distances and involve multiple researchers and participants. Yet all these journeys are equally important because the sharing of the trip is the more critical aspect of hīkoi. Instead of the action of walking on/through one's rohe (ancestral lands and waters) remains an essential part of tikanga Māori (customary laws), with a Māori person meant to walk on their whenua (land) and visit critical locations regularly to ensure their socio-cultural and spiritual connections to their ancestors (tūpuna) are maintained (Mahuika, 2019; Smith, 2020).

Talaona (Story) Method

To understand Samoan responses to climate risks and the impacts of climate change, Parsons was involved in collaborative research with both Samoan and non-Samoan scholars that employed the Talanoa approach (Farrelly and Nabobo-Baba, 2014; Otunuku, 2011). Coming from the Tongan word for story, Talanoa is an open and unstructured conversational style of interviewing amongst any number of people, reflecting the collectivist culture (Suaalii-Sauni and Fulu-Aiolupotea, 2014). The Talanoa method is now widely employed throughout many Indigenous Pacific Island communities as it provides the participant and the interviewer to share personal stories in a culturally safe way allowing them to develop a sense of trust. Indeed, such informal conversations allow space for deep, meaningful, and sometimes unexpected topics to emerge as interviewees and researchers talk informally. Talanoa interviews (which were initiated and led by the Samoan researcher) allowed the researchers to gain insights into Samoans' embodied and situated knowledge of their environments. Climate change was conceptualised as just one of many factors that impact fa'asamoa (the Samoan way of life) and that Samoans respond to through a wide range of adaptive strategies to reduce risk. The stories we collected effectively destabilised (and decolonised) the dominant global narrative of Samoa (and Samoans) as inherently vulnerable to the impacts of climate change which render the place (and the people) doomed to sink under the rising seas (Parsons et al., 2018).

Parsons and her collaborators chose the Talanoa method as they wanted to employ an Indigenous approach that allowed for Samoan values and cultural protocols to be interwoven with it, most notably that of reciprocity (sharing of knowledge and resources) (Bartlett et al., 2020; Nalau et al., 2018; Parsons et al., 2018). Unlike with Western interview approaches, Talanoa encourages the sharing of stories, experiences, and knowledge between researcher and researched (which also often involved walking, talking, and sharing foods) (Suaalii-Sauni and Fulu-Aiolupotea, 2014). Some of the non-Samoan members of the research team were incredibly anxious about using the Talanoa method because it went against their Western interview training (which encouraged them to only ask questions but not engage. However, Samoan researchers took the lead in the first Talanoa interviews. They sought to teach the non-Samoans some of the critical principles of Samoan culture (which included the giving of gifts of food to participants) (Bartlett et al., 2020). So at times, the participants would ask the researchers questions and the researchers shared their experiences, stories, family histories, and (in some instances) their Indigenous knowledge of extreme weather events and climate change with participants. In this way, each group (researcher and participants) shared their knowledge, stories, and concerns about environmental changes in a way that developed interpersonal and intercultural relationships that continue years after the projects ended (Nalau et al., 2018; Parsons et al., 2018). Such a process is in line with the four Rs of the Indigenous research paradigm, with researchers sharing their stories and experiences with research participants to build relationships centred on trust, respect, and ongoing relationships, which parallels the Australian Indigenous interview method knowing as yarning.

Yarning (Aboriginal Storytelling)

Australian Indigenous scholar Wellington and non-Indigenous researcher Penfold set out to collect qualitative data on Indigenous elders' geographies of home working with the Jerrinja people in New South Wales (Penfold et al., 2019). Initially the researchers intended to conduct semi-structured interviews in participants' homes (in line with conventional Western interview practices) to ensure interviewees felt comfortable, with the house providing useful visual prompts for interviewees (Penfold et al., 2019). However, the researchers soon discovered that many interviewees wanted to sit outdoors in a public space where they could feel the wind and sun on their bodies; which paralleled Aboriginal cultural tradition of sitting outdoors ("being on country") (Burgess et al., 2009; Terare and Rawsthorne, 2020). Accordingly, interviews were moved outdoors to ensure that participants were at ease.

Interviewees were also reluctant to express themselves in the conventional semi-structured interview format, and Penfold et al., (2019) altered their interview approach to one based on the Australian Indigenous storytelling method known as yarning (also known as cup of tea time) (Bessarab and Ng'andu, 2010; Terare and Rawsthorne, 2020). The adoption of yarning (unstructured interviewing technique involving informal sharing of stories between interviewees and interviewers) meant that Jerrinja people were able to recount their embodied experiences of their lives both inside and outside of their physical houses. They noted that homes for them were not the physical dwellings (roof, walls, furniture) but about their relationships with their families and extended kin group, and their capacities to interact with their local environment (which included maintaining cultural and spiritual connections to their ancestral lands). Penfold et al., (2019) and Absolon (2020) note that the process of gathering yarns (stories) allows research projects to be taken in surprising places by participants but it requires researchers to be flexible and responsive to participants' stories.

Sharing Circles

Penfold et al., (2019) also undertook sharing circles (also called talking or storytelling circles) with their participants (Baskin, 2005; Lavallée, 2009; Rothe et al., 2009). Indigenous scholars first modified the approach from Canada and USA for work within the fields of social work and public health to create a communal interview approach situated within Indigenous norms, cultural protocols, and social structures. Amongst First Nations peoples' of Canada the sharing circle begins with a smudging (the burning of sage) to mark the cultural significance of the event and welcome the spirits. Then one elder would ask the group for suggestions about what guidelines the group wanted to follow to ensure everyone felt safe. In Australia, Penfold et al., (2019) adapted the practices of the sharing circle to fit the local context, for example, the smudging ceremony was replaced with an Australian Indigenous Welcome to Country ceremony to similarly recognises ancestors and spirits.

Irrespective of where in the world sharing circles take place, the research technique shares standard features that distinguish from conventional focus groups, specifically the practices around speaking and listening. Sharing processes often involve participants discussing tribal, family and personal trauma (including experiences of disasters, violence, abuse, and racism, including Indigenous children being removed from their families, malnutrition and poverty due to dispossession) and so the rules around how people listen to sensitive and emotionally distressing topics are vitally important (Carriere, 2010; Huambachano, 2019; Kurtz, 2013; Lavallée, 2009). In Australia, Penfold et al., (2019) report that the listening etiquette of a sharing circle ensured that elders were treated with respect worthy of their high status (Penfold et al., 2019); when a Jerrinja elder spoke in a commanding tone, for instance, every other conversation in the room stopped, and everyone would listen without any interruptions. Other times, the sharing circle allowed for more informal discussions. Indeed, the sharing circle requires researchers to remain flexible in their methodological approaches and adapt to the local context and socio-cultural terrain in which they are working.

Concluding Remarks

In this chapter, we sought to provide a review of Indigenous research methodologies and how the development and use of Indigenous research paradigms and methods are assisting in decolonising our understandings of environments as well as human and more-than-human relations. To decolonise research methodologies, we must simultaneously decolonise the disciplines, institutions, sectors and societies in which they reside. Accordingly, as many Indigenous researchers and activists argue, we suggest it is not enough for researchers to write about decolonisation or even enact decolonising research practices within human geography, be it within our research and teaching practices.

References

Absolon, K. (2020) Close to home: An Indigenist project of story gathering. *Journal of Indigenous Social Development* 9, 19–40.

Artaraz, K., Calestani, M. (2015) Suma qamana in Bolivia: Indigenous understandings of well-being and their contribution to a post-neoliberal paradigm. *Latin American Perspectives* 42, 216–233.

Axelsson, P., Sköld, P. (2011) *Indigenous Peoples and Demography: The Complex Relation between Identity and Statistics.* Berghahn Books, New York.

Barker, A.J., Pickerill, J., (2019) Doings with the land and sea: Decolonising geographies, Indigeneity, and enacting place-agency. *Progress in Human Geography* 44, 640–662. https://doi.org/10.1177/03091 32519839863

Barlo, S., (2017) Lessons from the participants in decolonising research. *LCJ* 22, 16–25. https://doi.org/ 10.18793/LCJ2017.22.03

Barry, J., Thompson-Fawcett, M. (2020) Decolonizing the boundaries between the 'planner' and the 'planned': Implications of Indigenous property development. *Planning Theory & Practice* 21, 410–425. https://doi.org/10.1080/14649357.2020.1775874

Bartlett, A., Parsons, M., and Neef, A. (2020) The effects of private household insurance on climate change adaptation strategies in Samoa, in: Neef, A., Pauli, N. (Eds.), *Climate-Induced Disasters in the Asia-Pacific Region: Response, Recovery, Adaptation, Community, Environment and Disaster Risk Management.* Emerald Publishing Limited, London: pp. 167–191. https://doi.org/10.1108/S2040-726220200000022007

Baskin, C. (2005) Storytelling circles: Reflections of aboriginal protocols in research. *Canadian Social Work Review / Revue Canadienne de Service Social,* 22(2), 171–187.

Bessarab, D. and Ng'andu, B. (2010) Yarning about yarning as a legitimate method in Indigenous research. *International Journal of Critical Indigenous Studies* 3, 37–50.

Bryant, M., Allan, P., and Smith, H. (2017) Climate change adaptations for coastal farms: Bridging science and Mātauranga Māori with art and design. *TPJ* 2. https://doi.org/10.15274/tpj.2017.02.02.25

Burgess, C.P., Johnston, F.H., Berry, H.L., McDonnell, J., Yibarbuk, D., Gunabarra, C., Mileran, A., Bailie, R.S. (2009) Healthy country, healthy people: The relationship between Indigenous health status and "caring for country." *The Medical Journal of Australia* 190, 567–572.

Cannella, G.S., and Manuelito, K.D. (2020) *Handbook of Critical and Indigenous Methodologies* (pp. 45–60). SAGE Publications: London and New York. https://doi.org/10.4135/9781483385686

Carriere, J. (2010) Editorial: Gathering, sharing and documenting the wisdom within and across our communities and academic circles. *First Peoples Child & Family Review* 5, 5–7.

Cavanagh, V., Standley, P. (2020) Walking in the landscapes of our ancestors - indigenous perspectives critical in the teaching of geography. *Interaction* 48, 14.

Chiblow, S. (2020) An indigenous research methodology that employs Anishinaabek elders, language speakers and women's knowledge for sustainable water governance. *Water* 12, 3058. https://doi.org/ 10.3390/w12113058

Chilisa, B. (2019) *Indigenous Research Methodologies.* SAGE Publications: London and New York.

Clayton, D. and Kumar, M.S. (2019) Geography and decolonisation. *Journal of Historical Geography* 66, 1–8. https://doi.org/10.1016/j.jhg.2019.10.006

Colbourne, R., Moroz, P., Hall, C., Lendsay, K., Anderson, R.B. (2019) Indigenous works and two eyed seeing: Mapping the case for indigenous-led research. *Qualitative Research in Organizations and Management: An International Journal* 15, 68–86. https://doi.org/10.1108/QROM-04-2019-1754

Coutts, K., Morris, J., and Jones, N. (2016) The Māori statistics framework: A tool for indigenous peoples development. *Statistical Journal of the IAOS* 32, 223–230.

Crease, R.P., Parsons, M., Fisher, K.T. (2019) No climate justice without gender justice": Explorations of the intersections between gender and climate injustices in climate adaptation actions in the Philippines, in: Jafry, T. (Ed.), *Routledge Handbook of Climate Justice.* Routledge, Oxon; New York, pp. 359–377.

Datta, R. (2018) Traditional storytelling: An effective Indigenous research methodology and its implications for environmental research. *AlterNative: An International Journal of Indigenous Peoples* 14, 35–44. https://doi.org/10.1177/1177180117741351

Davies, K., Fisher, K., Foley, M., Greenaway, A., Hewitt, J., Le Heron, R., Mikaere, H., Ratana, K., Spiers, R., Lundquist, C., (2018) Navigating collaborative networks and cumulative effects for Sustainable Seas. *Environmental Science & Policy* 83, 22–32.

Farrelly, T. and Nabobo-Baba, U. (2014) Talanoa as empathic apprenticeship. *Asia Pacific Viewpoint* 55, 319–330.

Fisher, K.T. (2014) Positionality, subjectivity, and race in transnational and transcultural geographical research. *Gender, Place & Culture* 1–18. https://doi.org/10.1080/0966369X.2013.879097

Gilgen, M. (2008) Te whanau o te maungarongo hikoi: Maori practice in motion (3 July 2006 to 9 July 2006). Maori and Psychology Research Unit, University of Waikato.

Greensill, A. (2005) Foreshore and seabed policy: A Māori perspective. *New Zealand Geographer* 61, 158–160. https://doi.org/10.1111/j.1745-7939.2005.00032.x

Halvorsen, S. (2019) Decolonising territory: Dialogues with Latin American knowledges and grassroots strategies. *Progress in Human Geography* 43, 790–814. https://doi.org/10.1177/0309132518777623

Hardy, D., Patterson, M., Smith, H., Taiapa, C., Ruth, M. (2015) Cross-cultural environmental research processes, principles, and methods: Coastal examples from Aotearoa/New Zealand, in: *Handbook of Research Methods and Applications in Environmental Studies.* Edward Elgar Publishing, London, pp. 44–80.

Hart, M.A. (2010) Indigenous worldviews, knowledge, and research: The development of an indigenous research paradigm. *Journal of Indigenous Social Development* 1.

Held, M.B.E. (2019) Decolonizing research paradigms in the context of settler colonialism: An unsettling, mutual, and collaborative effort. *International Journal of Qualitative Methods* 18, 1609406918821574. https://doi.org/10.1177/1609406918821574

Huambachano, M.A. (2019) *Traditional Ecological Knowledge and Indigenous Foodways in the Andes of Peru.* 1,12, 87–110. https://doi.org/10.31261/rias.6866

Hunt, D. (2016) Nikîkîwân 1: Contesting settler colonial archives through indigenous oral history. *Canadian Literature* 25–42.

Hunt, S. (2014) Ontologies of Indigeneity: The politics of embodying a concept. *Cultural Geographies* 21, 27–32. https://doi.org/10.1177/1474474013500226

Hurihanganui, T.A. (2020) Hīkoi of a lifetime: Retracing Māhinaarangi's footsteps [WWW Document]. Radio New Zealand. URL www.rnz.co.nz/news/te-manu-korihi/430610/hikoi-of-a-lifetime-retracing-mahinaarangi-s-footsteps (accessed 11.22.20).

Hutchings, J. (2004) Claiming our ethical space - a mana wahine conceptual framework for discussing genetic modification. He pūkenga kōrero: A Journal of Māori Studies 8, 17–25.

Jackson, A.-M. (2015) Kaupapa Māori theory and critical Discourse Analysis: Transformation and social change. *Alternative: An International Journal of Indigenous Peoples* 11, 256–268. https://doi.org/10.1177/117718011501100304

Johnson, D., Parsons, M., Fisher, K. (2021) Engaging Indigenous perspectives on health, wellbeing and climate change. A new research agenda for holistic climate action in Aotearoa and beyond. *Local Environment* 10, 1–27. https://doi.org/10.1080/13549839.2021.1901266

Johnson, J.T., Howitt, R., Cajete, G., Berkes, F., Louis, R.P., Kliskey, A. (2016) Weaving Indigenous and sustainability sciences to diversify our methods. *Sustainability Science* 11, 1–11.

King, P., Hodgetts, D., Rua, M., Morgan, M. (2018) When the Marae moves into the City: Being Māori in urban Palmerston North. *City & Community* 17, 1189–1208.

Kovach, M. (2015) Emerging from the margins: Indigenous methodologies. In L. Brown & S. Strega (Eds.), *Research as resistance: Revisiting critical, Indigenous, and anti-oppressive approaches* (pp. 19–36). Canadian Scholars' Press.

Kurtz, D.L.M. (2013) Indigenous methodologies: Traversing Indigenous and Western worldviews in research. *AlterNative: An International Journal of Indigenous Peoples* 9, 217–229. https://doi.org/10.1177/117718011300900303

Lavallée, L.F. (2009) Practical application of an Indigenous research framework and two qualitative indigenous research methods: Sharing circles and Anishnaabe symbol-based reflection. *International Journal of Qualitative Methods* 8, 21–40. https://doi.org/10.1177/160940690900800103

Louis, R.P. (2007) Can you hear us now? Voices from the margin: Using indigenous methodologies in geographic research. *Geographical Research* 45, 130–139.

Mahuika, N. (2019) *Rethinking Oral History and Tradition: An Indigenous Perspective*. Oxford University Press: New York and Oxford.

Maxwell, K.H., Ratana, K., Davies, K.K., Taiapa, C., Awatere, S. (2020) Navigating towards marine co-management with Indigenous communities on-board the Waka-Taurua. *Marine Policy* 111, 103722.

Naepi, S. (2019) Masi methodology: Centring Pacific women's voices in research. *AlterNative: An International Journal of Indigenous Peoples* 15, 234–242. https://doi.org/10.1177/1177180119876729

Nakamura, N. (2010) Indigenous methodologies: Suggestions for junior researchers. *Geographical Research* 48, 97–103. https://doi.org/10.1111/j.1745-5871.2009.00625.x

Nalau, J., Becken, S., Schliephack, J., Parsons, M., Brown, C., Mackey, B. (2018) The role of Indigenous and traditional knowledge in ecosystem-based adaptation: A review of the literature and case studies from the Pacific Islands. *Weather, Climate, and Society* 10, 851–865.

Orange, C. (2015) *The Treaty of Waitangi*. Bridget Williams Books, Wellington.

Otunuku, M. (2011) How can talanoa be used effectively an indigenous research methodology with Tongan people? *Pacific-Asian Education* 23, 43–52.

Parsons, M. (2009) Spaces of disease: The creation and management of Aboriginal health and disease in Queensland 1900-1970 (PhD Thesis). University of Sydney.

Parsons, M., Brown, C., Nalau, J., Fisher, K. (2018) Assessing adaptive capacity and adaptation: Insights from Samoan tourism operators. *Climate and Development* 10, 644–663. https://doi.org/10.1080/17565529.2017.1410082

Parsons, M., Fisher, K., and Crease, R.P. (2021) Decolonising blue spaces in the Anthropocene: Freshwater management in Aotearoa New Zealand, Palgrave Studies in Natural Resource Management. Palgrave Macmillan. https://doi.org/10.1007/978-3-030-61071-5

Parsons, M., Fisher, K., and Nalau, J. (2016) Alternative approaches to co-design: Insights from indigenous/academic research collaborations. *Current Opinion in Environmental Sustainability, Sustainability Challenges* 20, 99–105. https://doi.org/10.1016/j.cosust.2016.07.001

Parsons, M., Nalau, J., and Fisher, K. (2017) Alternative perspectives on sustainability: Indigenous knowledge and methodologies. *Challenges in Sustainability* 5, 7–14.

Penfold, H., Waitt, G., McGuirk, P., Wellington, A. (2019) Indigenous relational understandings of the house-as-home: Embodied co-becoming with Jerrinja Country. *Housing Studies* 10, 1–16. https://doi.org/10.1080/02673037.2019.1676399

Prout, S. (2009) Vacuums and veils: Ngaging with statistically 'invisible' Indigenous population dynamics in Yamatji Country, Western Australia. *Geographical Research* 47, 408–421. https://doi.org/10.1111/j.1745-5871.2009.00584.x

Reid, A.J., Eckert, L.E., Lane, J.-F., Young, N., Hinch, S.G., Darimont, C.T., Cooke, S.J., Ban, N.C., Marshall, A. (2020) "Two-eyed seeing": An Indigenous framework to transform fisheries research and management. *Fish and Fisheries* 10, 1–19. https://doi.org/10.1111/faf.12516

Reid, J., Varona, G., Fisher, M., Smith, C. (2016) Understanding Maori 'lived' culture to determine cultural connectedness and wellbeing. *J Pop Research* 33, 31–49. https://doi.org/10.1007/s12546-016-9165-0

Ringham, S., Simmonds, N., and Johnston, L. (2016) Māori tourism geographies: Values, morals and diverse economies. *MAI Journal: A New Zealand Journal of Indigenous Scholarship* 5, 99–112. https://doi.org/10.20507/MAIJournal.2016.5.2.1

Rolleston, A.K., Cassim, S., Kidd, J., Lawrenson, R., Keenan, R., Hokowhitu, B. (2020) Seeing the unseen: evidence of kaupapa Māori health interventions. *AlterNative: An International Journal of Indigenous Peoples* 16, 129–136. https://doi.org/10.1177/1177180120919166

Rothe, J.P., Ozegovic, D., and Carroll, L.J. (2009) Innovation in qualitative interviews: "Sharing Circles" in a First Nations community. *Injury Prevention* 15, 334–340. https://doi.org/10.1136/ip.2008.021261

Simmonds, N. (2011) Mana wahine: Decolonising politics. *Women's Studies Journal* 25, 11–27.

Simmonds, N.B. (2009) Mana Wahine Geographies: Spiritual, Spatial and Embodied Understandings of Papatūānuku (Thesis). The University of Waikato.

Smith, G.H. (2017) Kaupapa Māori theory: Indigenous transforming of education, in: Hoskins, T.K., Jones, A. (Eds.), *Critical Conversations in Kaupapa Maori*. Huia Publishers, Auckland, pp. 70–81.

Smith, H. (2020) Collaborative strategies for re-enhancing Hapū connections to lands and making changes with our climate. *The Contemporary Pacific* 32, 21–46. https://doi.org/10.1353/cp.2020.0002

Smith, L.T. (1999) *Decolonising methodologies*. Zed Books: New York.

Stewart, G. (2016) Indigenous philosophies and education: The case of Kaupapa Māori, in: Peters, M. (Ed.), *Encyclopedia of Educational Philosophy and Theory*. Springer Singapore, Singapore, pp. 1–5. https://doi.org/10.1007/978-981-287-532-7_177-1

Suaalii-Sauni and T., Fulu-Aiolupotea, S.M. (2014) Decolonising Pacific research, building Pacific research communities and developing Pacific research tools: The case of the talanoa and the faafaletui in Samoa. *Asia Pacific Viewpoint* 55, 331–344.

Terare, M. and Rawsthorne, M. (2020) Country is yarning to me: Worldview, health and well-being amongst Australian First Nations People. *Br J Soc Work* 50, 944–960. https://doi.org/10.1093/bjsw/bcz072

Tuagalu, C., Cram, F., Phillips, H., Sauni, P. (2014) *Maori and Pasifika higher education horizons*. Emerald Group Publishing.

Tuck, E., McKenzie, M., and McCoy, K. (2014) Land education: Indigenous, post-colonial, and decolonizing perspectives on place and environmental education research. *Environmental Education Research* 20, 1–23. https://doi.org/10.1080/13504622.2013.877708

Umangay, U. (2019) Palimpsest, contrapuntal and the medicine wheel: An exploration of decolonizing thinking, decolonization and anti-colonial praxis. *Brill Sense*. https://doi.org/10.1163/9789004404588_002

Vasquez-Fernandez, A., Hajjar, R., Sangama, M.S., Lizardo, R.S., Pinedo, M.P., Innes, J., Kozak, R. (2017) Co-creating and decolonizing a methodology using indigenist approaches: Alliance with the Asheninka and Yine-Yami peoples of the Peruvian Amazon. *ACME* 17.

Walter, M. and Andersen, C. (2013) *Indigenous Statistics: A Quantitative Research Methodology*. Left Coast Press: Walnut Creek.

Wilson, S. (2008) *Research is Ceremony: Indigenous Research Methods*. Fernwood Publishing: Falifax.

Woodward, E., and Marrfurra McTaggart, P. (2019) Co-developing Indigenous seasonal calendars to support 'healthy Country, healthy people outcomes. *Global Health Promotion* 26, 26–34.

13

STORYTELLING IN ANTI-COLONIAL GEOGRAPHIES

Caribbean Methodologies with World-Making Possibilities

Shannon Clarke and Beverley Mullings

Introduction: Storytelling as Decolonial/Anticolonial Methodology

Stories and storytelling have re-emerged as powerful methodological tools for unsettling colonialities in the twenty-first century (Tuhiwai Smith, 1999, Wynter and McKittrick, 2015). De-colonial and anti-colonial scholars are recovering storying practices to speak back to essentializing master narratives whilst simultaneously creating autonomous spaces for transformation (Hunt, 2013). Once viewed "as a quaintly humanistic pre-occupation and at worst the building blocks of oppression and inequality" (Cameron, 2012, p. 573), stories can be found across a wide range of academic writing seeking to connect the personal and experiential to broader social structures and relations. Storying makes connections between the storyteller and the listener, and its enactment aims to elicit understanding of the experiential and relational dimensions of everyday life. As Sarah Hunt (2013) observes, 'stories and storytelling are ways of knowing, produced within networks of relational meaning-making', Linda Tuhiwai Smith observes that despite the popular conception that telling and working with stories is a universal experience across all peoples and cultures, most stories reflect the views of the powerful who assume the right to tell the stories of the colonized and oppressed (Archibald et al., 2019). Tracing trends in the discipline of geography, Emilie Cameron (2012) draws attention to what she sees as a renewed interest in the political possibilities of storytelling, after a period of relative abandonment in the 1990s when scholars, drawing on post-structuralist frameworks, became increasingly concerned about the ways that stories had the capacity to reinforce oppressive ideologies and their discursive formations. She argues that to view the feelings and experiences captured in stories as primarily structurally and ideologically determined, can foreclose understandings of the ways that stories "move, emerge and affect in the very act of their telling"(Cameron, 2012, p. 588), and in so doing, produce alternative dimensions of the political. As was clear even in the 1990s, all stories, no matter how mundane or personal, leave a trace. And each trace gestures to a geography of relationships criss-crossed with power, but also possibility. How stories come to be positioned within geometries of power, therefore, has as

DOI: 10.4324/9781003038849-16

much to do with the value attached to the situated knowledge of the storyteller, as it has, the listener. Understanding the political possibilities that stories give rise to – their capacity to rewrite colonizing scripts that place certain peoples outside history and humanity – is a powerful raison d'etre for thinking about storying/story telling methodologically, utilizing the skills of radical inquiry to listen, to hear, and to determine when and how to make other ways of knowing and being, legible.

Jo-anne Archibald (2019) states that storying has always been an important medium for colonized peoples to articulate their worlds, their experiences and their identities, especially in the context of colonial domination. Like many Indigenous scholars, she sees story as an important methodology that: "speaks from and to a deep understanding and philosophy about humans and their relations, [as well as] other entities with whom humans have relationships and responsibilities" (Archibald et al., 2019, p. xi). By rejecting the idea that research can be disassociated from the self, storytelling challenges dominant epistemologies that shape Western definitions of academic rigour and the imperative of neutrality. Sium and Ritskes (2013) argue that stories facilitate the embedding of anti-colonial logics guided by community into the research process. As such, the gathering and telling of stories is inherently political, agentic and participatory.

As scholars turn to stories, and to storytelling as a methodology it is becoming clear that while decolonial and anti-colonial storying practices share common liberatory goals there is great variety in the gathering practices, and processes of making sense of stories. For example, Indigenous scholars in settler colonial contexts see storytelling as integral to recovering previously restricted knowledges and to building ethical relationships between the storyteller and the listener (Tuhiwai-Smith, Hunt, 2013, Kovach, 2009). While scholars writing within Black radical traditions see stories and storytelling as a crucial element in struggles for freedom because of the imaginaries they enable (Wynter and McKittrick, 2015, McKittrick, 2020, Alagraa, 2018). While both scholarly approaches are committed to amplifying the voices of the marginalized, their theoretical frameworks attend to questions that address the harms created by different aspects of coloniality. For example, for many Indigenous scholars, the recovery of worldviews rooted in respect, reciprocity, responsibility, and that connect the whole person to land and to others, is an important aspect of what Jo-Anne Archibald calls 'storywork', while scholars drawing on Black radical traditions emphasize new ways of imagining and creating freedom practices committed to subverting what Sylvia Wynter refers to as the 'overrepresentation of Man'. Critiquing the conflation of Western, biocentric notions of Man (grounded in colonialism and racial slavery) with what it means to be human, Sylvia Wynter reminds us that who we are as a species is a product not only of our biology, but also of our storytelling and myth-making capacities – our ability to imagine and to reshape our worlds through the stories we tell (Wynter and McKittrick, 2015). Her call for us to move towards a definition of the human 'made to the measure of the world' requires us to recognize the importance of narration in the creation of non-Eurocentric genres of the human (Alagraa, 2018, p. 167).

In this chapter, we turn to Caribbean stories and storytelling traditions, to explore the ways in which they index what Anthony Bogues (2012) calls a radical imagination, tied to the political project of freedom and fundamental to speaking back to power. First, we examine how Caribbean experiences of colonialism and racial capitalism produced particular terrains of struggle made accessible through the stories about who we are and who we want to be. In this regard, we examine how the Caribbean quest for freedom shapes the nature of stories and storying practices, across the region and its diaspora. Second, we examine the method of gathering stories – the audiences, the modes of collaboration, the silences and forms of trust that shape how we 'hear' Caribbean voices, how we ask questions, and our ethical responsibility as scholars, as community members and listeners in bearing witness to Caribbean stories. As

Caribbean scholars working within the region and the diaspora, stories have been critical to our scholarship. The stories of our elders, peers, community members and research collaborators have directed not only our research interests but our understanding of core concepts within geography, including the limits and gaps of the canon. We conclude by reflecting on the way that Caribbean storytelling ultimately informs our individual decolonial and anti-colonial practices.

Storytelling in the Caribbean

The history of colonialism in the Americas, and the encounters, patterns of migration and methods of survival that developed in the Caribbean can be read in the oral traditions that endure in this space. These range from creative (commercial) output to social practice, and demonstrate Caribbean orality rooted in liberation, revolution, collaboration, collective subjective recovery, and the invention of new political possibilities. These practices of writing and talking back to dominant structures and social hierarchies through storytelling (Marshall, 2016; Hua, 2013) contribute to what Bedour Alagraa (2018) calls the "Caribbean radical imagination." Alagraa's analysis considers how Caribbean writers – through fiction and other narrative forms – create new subject positions that challenge colonially imposed ideas about humanity. Narratives, or what writer Sylvia Wynter terms "autopoeisis," is the act of creating oneself; in this way, it is also an act of freedom (Alagraa, 2018, p. 2). In her analysis of Black life in science, Katherine McKittrick (2020) also turns to Wynter and autopoeisis to consider how stories help us to know what it means to be alive. She writes: "Autopoeisis is the process through which we repeat the conditions of our mode of existence in order to keep the living system – our environmental and existential world, our humanness – living" (McKittrick, 2020, p. 133).

This takes on greater significance in the Caribbean where African enslavement, Asian indentureship, attempted Indigenous disappearance, emancipation and ongoing forms of racial capitalism have created a specific terrain of struggle. Although geographers have documented the storying turn (Cameron, 2012) few have explicitly engaged the work of racialized scholars, including those from the Caribbean. We consider what storytelling and storying practices offer the Caribbean and its diaspora in the post-independence era of neoliberal restructuring, labour migration, and environmental precarity (Noxolo and Featherstone, 2014). Whether through folklore, calypso, carnival, liming or ole talk, Caribbean orality subverts racial, colonial and capitalist power structures. Key to the traditions examined in this chapter are improvisation and performance, memory and collaboration. While the following examples cannot capture every oral tradition in the region, nor the full history of Caribbean orality, they do represent some ways in which racial subjectivities (among others) are communicated, disrupted and recreated.

Honor Ford-Smith (2019) writes of Caribbean performance as cultural maroonage that produces the social space for embodied and relational recognition. For Smith, as with Wynter, creative expression and storytelling – specifically, stories about the past – challenge dominant narratives. McKittrick (2020) argues that Black intellectual and physiological struggle is a labour of disobedience and a refusal of post-plantation logics. She writes:

> This disobedience sparks a stream of black texts and practices that continually imagine a way of being black as linked to landlessness and activity-based collaborative intellectual labor. In this, we share how we arrive at struggle; how we get there, what stories allow us to get there, what stories disrupt what we know and what stories enable new ways of thinking about freedom. This is black life and its affective expression. Maybe this is what possibility feels like.
>
> *McKittrick, 2020, p. 72*

Challenging the normative understanding of libraries as physical archives, Cherry-Ann Smart (2019) writes that the forced separation of millions of people from their homes represents a "library in crisis" and enslaved Africans as embodying the survival and preservation of knowledge from the continent. If the physicality of libraries has become less important in the twenty-first century, as Smart (2019) suggests, it is easy to understand how the production and transmission of intangible forms of knowledge constitute a moving library in the African diaspora. The displacement and enslavement of African peoples facilitated the creation of new forms of knowledge production. Smart writes: "This knowledge was not only traditional cultures, language or performance…Instead, they were wilful applications of forms or processes for human advancement, the re-engineering of knowledge based on circumstances, available resources and time" (Smart, 2019, p. 18). This includes the formation of language itself, or the assemblage of speech into Caribbean Creole. Caribbean orality in this chapter includes the dialects that developed out of displacement and which are reproduced in daily life and in creative spaces, sometimes strategically (McLaren, 2009; Murdoch, 2009, De Souza, 2003).

Further, these dialects and languages exemplify another feature of Caribbean oral tradition and performance: improvisation. The slave ship and the plantation are spaces of violence, but they are also, McKittrick (2020, p. 153) argues, spaces of livingness, and the work of Black geographies is to honour the creation and intellectual labour that emerge from them (Bledsoe, 2015; Harney & Morton, 2013; McKittrick and Woods, 2007). Survival manifests in performance, too. Limbo, drumming and dance served not only to entertain, but also to share information (Smart, 2019; Williams 2017; Vété-Congolo, 2007). Williams (2017) takes the position that enslaved people and their descendants embody an undervalued but ubiquitous "cultural repository" that is prominent in Caribbean (and we would add global) popular culture. That repository of historical memory, through performance, is a way of making and negating social meaning, codes, and boundaries. Further, if bodies are embodied archives, then performance can be a way of telling stories and sharing knowledge when other ways of knowing are prohibited, devalued and marginalized (Ford-Smith, 2019). In his examination of Caribbean aesthetics in theatre and dance, Williams (2017) argues that "style here, is not arbitrary or artistic viewing reality, but the incarnation/embodiment of cultural and political validation" (220). Thus, what might be experienced, to outsiders, as vulgar or rowdy is a political act: "At play, here, is the oily, slick, shape-shifting, trickster-like quality of the masquerader, reminiscent of the tragicomic folk hero Anancy" (Williams, 2017, p. 221).

Anancy, as he is called in the Anglophone Caribbean, appears in folk stories as a clever, shape-shifting spider that occasionally takes human form with a gift for "spinning" tales to outwit foes. This figure of deception and indeterminacy makes it, at different points, "a hero, villain [or] dupe" (DeSouza, 2003, p. 340). In a close examination of Anancy, De Souza traces its African roots and finds variations throughout the Caribbean and its diaspora, which maintain certain themes or "African continuities." Besides trickery, Anancy is often socially marginalized; seeking short-term economic gain to cope with scarce resources; control of social situations where rigid social hierarchies seem to be working against it; and "keen awareness of his victims' expectations, which enables him to manipulate them at will." (De Souza, 2003, p. 342). One key shift in the way these stories have travelled lies in the need for a folk hero that triumphs over power. Whereas in African versions these tales contain a moral message, De Souza (2003) notes that success for the Caribbean Anancy figure signified resistance, not disobedience. Lying is celebrated because cunning is necessary for survival (Vété-Congolo, 2007).

Hanétha Vété-Congolo (2007) turns Anancy, Compère Lapin, and other Caribbean folk figures to understand how African folktales adapted to the Americas served to help enslaved and subjugated peoples imagine new futures while surviving in the present. Echoing Smart (2019), Vété-Congolo argues that these tales served the practical purpose of communicating successful resistance, or *maroonage*. *Maroonage* refers to acts of resistance by enslaved people in the Caribbean that ranged from sabotage on the plantation (procrastination, periodically running away) to the abandonment of plantations entirely to form autonomous communities of collective resistance, notably, the maroon communities of Jamaica and the *quilombos* of Brazil (Knight, 2012, pp. 62–67, see also: Chapter 15, Emerson dos Santos and Ferreira, this volume). These acts of resistance created new geographies and led to resistant ways of knowing in the Caribbean, what Vété-Congolo (citing Edouard Glissant) refers to as an aesthetic "illustrated by interioriality" (Vété-Congolo, 2007, p. 4). These stories and other folktales are powerful for their ability to conceal and communicate concealment.

> Storytelling demonstrates the method and furnishes the pedagogy for resistance. The characters that are best suited for that mission, who also present a true representation of the daily trials of life, are Compère Lapin and Anancy... his methodology is meaningful only if everyone is aware of the circumstances that the tale reflects, and the experience being described. The storyteller and the listener must have the same understanding of the tale.
>
> *Vété-Congolo, 2007, p. 5*

Calypso and carnival are both practices that "talk back" to colonial power, past and present (Funk, 2008). Murdoch (2009) writes that Caribbean musical practices, including calypso, can trace their roots back to the plantation, where the same forms that animate calypso – satire, social commentary, protest, parody – are found in the narrative practices of dissent among enslaved peoples. In the decolonial and (post-)independence era, calypso continues to serve as a space of contestation, and an archive for future audiences. The calypsonian, characterized by colourful storytelling and commentary on social, political and economic processes, has been read as a descendant of the West African griot. Griots were hired to flatter royalty and "sing their praises, celebrate their lives, and document their achievements" (Funk, 2008, p. 1). The Calypsonian of the colonial era functions in a similar capacity, which Funk (2008) notes in the songs about the British Crown ("Coronation of Queen Elizabeth" by Roaring Lion; "I Was There at the Coronation" by Young Tiger) and many other pieces written to commemorate the visiting dignitaries or celebrate the war effort (Funk, 2008).

In the decolonial and post-independence period, calypsonians adopted the same narrative techniques of storytelling, double meaning, lyricism, critique and satire to articulate the frustrations of the young nation with its elite leadership, and imperial collaborators. Funk (2008) points to Lord Kitchener (of Trinidad and Tobago) and his lyric "They Got It," about Ghana's independence to highlight the development of anti-colonial attitudes in this music. These songs capture key moments in Caribbean history and, we argue, serve as a critical migrating archive of Caribbean subjectivities, for example, the failed attempt at a West Indian federation (see Barbadian Mighty Sparrow's, "Federation") or the region's relationship with a new imperial power in the form of the United States (Trinidadian Lord Invader's, "Rum and Coca Cola") (Funk, 2008).

Calypsonians, like their literary counterparts throughout the diaspora, also chronicle the experience of being from and living within Caribbean spaces around the world. This allows for a reading of history and political consciousness through Caribbean oral tradition: "If one reviews the broad sweep of [Lord Kitchener's] calypsos over the prior decade and a half, one of the themes they reflect is a gradual disenchantment with life in England, a growing allegiance to his African roots, and a support for the decolonization of the British Empire" (Funk, 2008: 5). The tradition of performance, and the critiques and commentary it aims to communicate are important for thinking through Caribbean stories methodologically, and particularly for thinking about the ways in which race and racial subjectivities can be theorized. The reading of these performances, and other forms of oral tradition must keep in mind their subversive possibilities and goals. Further, we must grapple with silences, ambiguity and things left unsaid not as "gaps" in data but as commentary of its own. In considering silences in her own research, Deborah Thomas (2019, p. 213) argues that "silences can become strategic technologies through which we are invited to confront collectively held public secrets, to name and recognize the sociohistorical entanglements that complexly create our relational present."

A final tradition considered here is "liming" or the often informal, unstructured and improvised (Anglo-)Caribbean art of socializing. While the practice is recognizable in different forms throughout the region (*bemberria*, *jangueo*, or *compartir*), like Anancy tales, it shares some key traits no matter where it lands: the flexibility of location; the exchange of food, drink and, importantly, stories. It is not limited by topic, timeframe, or social hierarchies, and participants are free to come and go as they please (Santana et al., 2019, p. 107). Liming is "a testing ground to measure verbal improvisation and ingenuity, woven with a healthy dose of humor" (Clarke, 2012, p. 303). Humor − along with the presence and sharing of food − is perhaps the most notable aspect of liming. Writes Santana et al., (2019, p. 108): "Caribbean humor is seen as a way to address difficult topics or situations, an important coping mechanism, and a means of collective negotiation of meaning in the Caribbean."

The authors observe rhythms of liming that distinguish it from gatherings elsewhere: it crosses and reinforces social lines and revolves around tacit rules of the social and emotional division of labour. There is the storyteller; the person(s) providing food and drink; the musician, the host; and the various supporting narrative roles: "the punctuator of the tall tales, the elaborator of the tall tales, the fact checker …" No matter the role, participants are expected to perform and perform well: "An individual who contributes positively to a good lime is likely to be perceived as someone who can also be a positive contributor in general both in the workplace and in the political realm" (Clarke and Charles, 2012, p. 305).

Santana et al., (2019) turn to Samuel Selvon's novels *The Lonely Londoners* and *The Housing Lark* to illustrate the prevalence of liming in Caribbean literature. Much of these novels are spent with characters in conversation with each other, often assembled in spontaneous circumstances and in multiple places (at Waterloo Station, a bus stop, at a rooming house, someone's basement apartment). There are often multiple voices talking over one another, without hierarchy, and without linearity, and this is juxtaposed to the way outsiders communicate (in this case, the white British observers in these novels). To the uninitiated, this talk seems disorganized, unruly and unproductive (Santana et al., 2019, p. 109); in the Caribbean and throughout the diaspora, liming is instructive, cathartic and potentially materially rewarding.

Liming is a storying practice through which researchers can also read resistance, and through which the embodied library that Smart (2009) proposes can be found. Taken together, they offer a challenge to Eurocentric epistemologies and approaches to qualitative research that dominate even heterogeneous spaces like the Caribbean (Santana et al., 2019, p. 105).

Methods for Gathering Caribbean Stories

Jamaican poet and novelist Olive Senior (1996) writes of storytelling as a collaborative effort between listener and teller. She argues that to think about one's audience is to consider one's co-conspiracy in the project of recreation, and that attention to audience is what makes storywork by Caribbean writers and Caribbean people an anti-colonial practice. As Senior observes "What has been handed down to us in written form ... are descriptions of cultures by observers in the superior position – even the missionaries – with virtually no participants of the events being described; in other words, what has been meaningful to the observer, not the observed" (Senior, 1996, p.43). There is a difference, then, between what members of Caribbean diasporic communities tell others, and what is shared within our communities. Engaging in Caribbean storywork therefore, demands that the questions scholars ask, the ways they listen, and how scholars engage with the stories they hear, be attuned to the ethics and terms upon which the knowledges embedded in stories have been articulated or shared. Methods for interrogating stories such as interviews, oral histories, textual analysis, photo elicitation are simply that: tools. How they are used to gain knowledge of the worlds that racialized Caribbeans negotiate, challenge and imagine differently, speaks to the specific challenges of meaning-making that Caribbean stories and storytelling present.

Listening

Careful attention to the voices made audible in Caribbean storytelling is an important practice that reveals not only the situated knowledges of storytellers, but also the intentionality and embodied politics of those who speak for others. Sherene Razack (1998) cautions readers not to ignore the role of the listener, particularly when there is an imbalance of power between them and the storyteller. For, as she points out, the stories we hear are filtered through our own experiences and moral codes, which affects how they are interpreted. Without attending to our own subject positions as listeners, empathy is not enough for honest analysis.

For example, Aljoe (2004) considers slave narratives and the way in which these voices have been incorporated into different literatures. This includes abolitionist papers, but also legal records, diaries and travel narratives as well as separate texts, such as *The History of Mary Prince, a West Indian Slave Related by Herself* and many others. "Although the majority of these documents exist in the colonial archive and as such are entangled with the politics of colonialism, when read against the grain of singular totalizing history, these narratives provide an important resource for understanding the experience of slavery and its aftermath in the African diaspora" (Aljoe, 2004, p. 1).

Thus, Caribbean oral histories can also be understood not as autobiographies but as *testimonios*. The distinction recognizes that these personal narratives are articulations of a collective experience – for example, slave narratives as testimony of the experience of enslavement in the Caribbean as much as the personal recollections of dispossession and unfreedom. Of Mary Prince's narrative, Aljoe writes:

> For example, one important feature of the *testimonio* genre is the floating 'I'. By this we mean, the 'I' has the grammatical status of what linguists call a "shifter," a linguistic function that can be assumed indiscriminately by anyone; it is not just the uniqueness of herself or her experience but its ability to stand for the experience of her community as a whole.
>
> *Aljoe, 2004, p. 8*

This is not to reduce these stories to an essential reality. Rather, it acknowledges the multiple subjectivities at work in creating these texts (Aljoe, 2004).

Translation

Because stories like slave narratives require collaboration for their readability – that is, the final texts are always the work of narrator, transcriber and editor – there is no single subjectivity within them. Editors make decisions about remaining faithful to Caribbean dialect in their transcriptions and maintaining the readability of the final text. While the previous example referred to narratives of the enslaved, the question of power and agency in the translation of stories told in Caribbean dialect remains a vital consideration in the process of honoring those who rarely had the opportunity to make their voices public, with the full range of their expression. Creole, as argued above, competes with standardized European languages for dominance throughout the region; further, Aljoe (2004, p. 6) writes that Creole represents an "alternative worldview." Because it operates as a form of anti-colonial resistance, decisions about how and whether to translate this speech are questions of power. Working with narrative histories and Caribbean orality, then, also forces researchers to navigate their insider/outsider status. Even those familiar with these storytelling practices, speech and narrative processes must engage with narratives as researchers with significant control over how personal histories are received and interpreted.

Interpretation, Responsibility and Ethics

Perhaps the most challenging aspect of Caribbean storying practices is the process of making meaning from the ways that Caribbean cultures use stories. Narrative research must contend with missing details, the ethical implications of asking participants to recover and engage with memories of loss, dispossession and the violence of racism, whilst simultaneously figuring out how to work with narratives that are non-linear and potentially contradictory. Dore (2016) considers questions of interpretation while reflecting on her research in Cuba. Alongside a large team collecting oral histories on the island, Dore discusses the realities of interpreting this data when the listeners seemed to be hearing different things in the identical recordings: "Our discussions reminded me that listening to interviews is as subjective as reading historical documents, frequently more so. What each of us heard, or said we heard, was tempered by our politics, experience, and expectations, by the ubiquitous presence of the Cuban state, and sometimes by what we hoped the narrators would say" (Dore, 2016, p. 240).

Dore (2016) recounts the analysis of interviews with Juan, an Afro-Cuban from a township outside of Havana. Reviewing his interviews with her British-Cuban research team proved difficult, specifically because what some interpreted as satire others took literally (at times both sexist and racist). Juan clearly enjoyed performing: "His lines about Dalmatians, *indeados* for breeding, and *chorongo* braids were, I felt pretty sure, part of an act designed to shock his audience. And what audience could be better than academic interviewers from the Academy of Sciences? Juan knew, of course, that he was taking a risk, yet he simply couldn't resist titillating us squares" (Dore, 2016, p. 246). Further, the racial and gendered identities of her team might have also influenced the way Juan approached these interviews and his decisions about what to withhold, share, embellish or downplay.

The researchers also diverged in their interpretation of the data when Juan spoke about race. Not understanding the taboo of talking about racial discrimination in Cuba, what some took as unsurprising or unremarkable, Cuban researchers flagged as subversive and worthy of further examination. A disparaging comment about dreadlocks drew competing interpretations: while some read this as performative, and others – such as Dore – as regressive, one researcher argued that for people of African descent "hypercleanliness is a conscious or unconscious strategy to combat whites' stereotype that they, blacks, are unclean" (Dore, 2016, p. 245). This reflection highlights the challenges and opportunities of working with Caribbean narratives and oral traditions. Specifically, the historical and material context in which these expressions of resistance, subjectivity and racial consciousness develop are crucial to the data collection process and analysis.

As Dore (2016) demonstrates, even when these narratives can be taken as "fact" in the scientific sense (that is, a participant's statements are on-record and can be verified by multiple listeners) how that data is interpreted poses a challenge. Recognizing the non-linear, spiritual and insurgent nature of Caribbean stories is a crucial first step in the process of reading and interpretation where articulations between the situated knowledges of storytellers and those who gather stories are laid bare.

Community Co-production of Knowledge

Liming is an enriching and definitively Caribbean form of knowledge production that flattens the hierarchies that ordinarily define the relationship between researchers and storytelling participants. As Santana et al., (2019, p. 118) observe:

> Liming and Ole Talk generate knowledge in the form of an unstructured array of ideas, stories, opinions, jokes, highly performative allocutions and profound reflections... As a fluid, multi-layered framework for interaction, liming enables ideas, narratives, opinions and reflections to be discussed in a non-linear way.

Santana et al., (2019) configured limes across the Caribbean diaspora and audio recorded the events, arguing that the practice allows for knowledge production that more accurately reflects Caribbean ways of being and thinking. The aim of this research was to understand the characteristics of liming and Ole Talk in order to demonstrate how it might be employed as a methodology. They describe truly improvised settings and events in line with the character of these gatherings and thus, a significant lack of control on their part, regarding the process. Largely dependent on their hosts for recruitment and subject to the unstructured nature of these gatherings, the researchers were not present as facilitators. Participants and researchers asked and answered questions, and determined the direction of the lime. Everyone was free to come and go as they pleased, away from the microphones and the purview of the researchers, if the topic at hand did not appeal. The unstructured nature of the research gathering, and the unpredictability of the lime itself allowed for ideas to be explored or abandoned in ways that point to the significance of those topics, for the group, and unsettled at least one researcher, who reflects their dual position as participant and observer. On their silence during a conversation about homophobia in the Caribbean and its relationship to colonialism, they write: "My silence stemmed from a concern that expressing my disagreement could cause the participants [to] say what they thought the researcher wanted to hear" (Santana et al., 2019, p. 114).

In this way, liming as cultural practice, the authors argue, is not incompatible with the research process of gathering data, recording and analysis (Santana et al., 2019, p. 111). In these settings the researcher-participant hierarchy dissolves: researchers are included, ignored and talked over just as much as any participant at the gathering. Their identities as Caribbean academics also played a role in their position as researchers and participants: "Nevertheless, there was a prevailing conflict between the cultural competencies and habitus we held as Caribbean islanders and were expected to use in a liming context (openness, sharing, humor, teasing) and our training in Western academia on best research practices" (Santana et al., 2019, p. 113).

Santana et al.'s (2019) use of liming allows us to consider questions of listening, translation, ethics and responsibility. While the configuration of these limes might not stray far from standard research practice (for example, snowballing recruitment methods), the marginality of the researchers' position in tension with their lived experience as Caribbean people, highlights the way in which Caribbean practices are often at odds with hegemonic methods of data collection. In the above example, the researchers are required to give up a degree of control over not only who is present, but the direction of the research itself. As participants come and go, the researchers must contend with sudden gaps and silences. When the lime ends – a collaborative production, containing multiple points of view and subjectivities, both audible and inaudible – it is up to the researchers to translate these conversations and performances, and interpret the contributions (including moments of disengagement) for their peers. Thus, it is the responsibility of the researcher to consider the lens through which they interpret each moment of the storytelling process and to think deeply about how they are translated (or not) for non-participants. The untranslatability of these spaces is then also part of working with Caribbean stories and storytelling practices. As demonstrated by Dore's (2016) reflection above, an ethical engagement with these practices may mean sitting with the possibility that what has been observed is beyond one's full capacity to know, even when it has been faithfully represented.

Conclusion

Anti-colonial geographies have turned to storytelling to challenge dominant epistemologies, colonial violence, and the marginalization of diverse ways of knowing. Caribbean scholarship and writing in Black geography (see also Emerson dos Santos and Ferreira, Chapter 15, this volume) have provided critical interventions in this regard. Caribbean storytelling practices often involve themes of flight (marronage), movement (migration), and the embrace of newness (self-fashioning) that reflect the material and symbolic struggles against racial violence and dispossession that remain dominant experiences in the lives of many in the region and its diaspora. Stories here function as maps that draw our attention to the specific ways that coloniality is experienced, negotiated and resisted. Stories and forms of storytelling that address the ways that individuals and communities reconcile themselves with the weight of history while imagining and seizing opportunities to make themselves anew, are important reminders of the existence of alternative genres of the human. And for that reason, stories are an indispensable part of struggles to build emancipatory futures. The challenge for scholars who gather Caribbean stories, is to learn how to hear, how to ask and how to interpret stories, whilst understanding that in gathering stories one is already a collaborator and participant in the process through which, as McKittrick (2020) observes, Black people come to know what they know about how to live freely. Stories and storytelling, then, are crucial to this work. Reflecting on her work with Sylvia Wynter, McKittrick (2020) writes: "I learned from her that sharing stories *is* creative rigorous radical theory. The act of sharing stories *is* the theory and the methodology" (p. 73, emphasis in original).

References

Alagraa, B. (2018) Homo Narrans and the Science of the Word: Toward a Caribbean Radical Imagination *Journal of the Critical Ethnic Studies Association, 4*(2). 164–181.

Aljoe, N. N. (2004) Caribbean Slave Narratives: Creole in Form and Genre *Anthurium: A Caribbean Studies Journal, 2*(1).

Archibald, J., Q. Xiiem, J. Lee-Morgan and J. De Santolo, Eds. (2019) *Decolonizing Research: Indigenous Storywork as Methodology*. London, Zed Books.

Barnwell, A. (2020) "Keeping the Nation's Secrets: "Colonial Storytelling" within Australian Families." *Journal of Family History*, *46*(1), 46–61.

Bledsoe, A. (2015) The Negation and Reassertion of Black Geographies in Brazil. *ACME: An International E-Journal for Critical Geographies*, *14*(1), 324–345.

Bogues, A. (2012) And What about the Human? Freedom, Human Emancipation, and the Radical Imagination. *Boundary*, *2*(39), 36.

Cameron, E. (2012) "New geographies of story and storytelling." *Progress in Human Geography*, *36*(5), 573–592.

Clarke, R. and Charles, R. N. (2012) Caribbean Liming: A Metaphor for Building Social Capital. *International Journal of Cross Cultural Management*, *12*(3), 299–313.

De Souza, P. (2003) Creolizing Anancy: Signifyin(g) Processes in New World Spider Tales. *Matatu*, *27*(1), 339–363.

Dore, E. (2016) Hearing Voices: Cuban Oral History *Hispanic American Historical Review*, *96*(2), 239–247.

Ford-Smith, H. (2019) The Body and Performance in 1970s Jamaica: Toward a Decolonial Method. *Small Axe: A Journal of Criticism*, *23*(1), 150–168.

Funk, R. (2005) In the Battle for Emergent Independence: Calypsos of Decolonization. *Anthurium: A Caribbean Studies Journal*, *3*(2).

Gibson-Graham, J.K. (2006) *A Postcapitalist Politics*. Minneapolis: University of Minnesota Press.

Harney, S. and Moten, F. (2013) *The Undercommons: Fugitive Planning & Black Study*. Brooklyn, NY: Minor Compositions Press.

Hua, A. (2013) Black diaspora feminism and writing: Memories, storytelling, and the narrative world as sites of resistance. Transnational feminist diaspora literary imaginary *African and Black Diaspora: An International Journal*, *6*(1), 30–42.

Hunt, S. (2013) Ontologies of Indigeneity: The Politics of Embodying a Concept. *Cultural Geographies*, 21(1): 27–32.

Jackson, S. N. (2012) *Creole Indigeneity: Between Myth and Nation in the Caribbean* [Kobo].

Knight, F. W. (2012) *The Caribbean: The Genesis of a Fragmented Nationalism* (Third ed.). New York: Oxford University Press.

Kovach, M. (2009) *Indigenous methodologies: Characteristics, conversations, and contexts*. Toronto: University of Toronto Press.

Marshall, E. Z. (2016) Resistance through 'Robber-Talk': Storytelling Strategies and the Carnival Trickster. *The University of the West Indies*, 210–226. doi:10.1080/00086495.2016.1203178

McKittrick, K. (ed.) (2014) *Sylvia Wynter: On Being Human as Praxis*. Durham, North Carolina, Duke University Press.

McKittrick, K. (2020) *Dear Science and Other Stories*. Duke University Press.

McKittrick, K., & Woods, C. (Eds.). (2007) *Black Geographies and the Politics of Place*. Toronto: Between the Lines.

McLaren, J. (2009) African Diaspora Vernacular Traditions and the Dilemma of Identity. *Research in the African Literatures*, *40*(1), 97–111.

Murdoch, H. (2009) A Legacy of Trauma: Caribbean Slavery, Race, Class, and Contemporary Identity in "Abeng'. *Research in the African Literatures*, *40*(4), 65–88.

Noxolo, P. and Featherstone, D. (2014) Commentary: Co-producing Caribbean Geographies of In/security. *Transactions of the Institute of British Geographers*, *39*(39), 603–607.

Razack, S. E. (1998) *Looking White People in the Eye: Gender, Race and Culture in Courtrooms and Classrooms*. Toronto: University of Toronto Press.

Rice, C. and I. Mündel (2018) Story-Making as Methodology: Disrupting Dominant Stories through Multimedia Storytelling. *Canadian Review of Sociology/Revue canadienne de sociologie*, 55(2): 211–231.

Santana, A. F., Nakhid, C., Nakhid-Chatoor, M., & Wilson-Scott, S. (2019) Liming and Ole Talk: Foundations for and Characteristics of a Culturally Relevant Caribbean Methodology. *Caribbean Studies*, *47*(1), 99–125.

Senior, O. (1996) Lessons from the Fruit Stand: Or, Writing for the Listener. *Journal of Modern Literature*, *20*(1), 39–44.

Sium, A. and E. Ritskes (2013) Speaking Truth to Power: Indigenous Storytelling as an Act of Living Resistance. *Decolonization: Indigeneity, Education & Society* **2**(1): 1–X.

Smart, C. A. (2019) African Oral Tradition, Cultural Retentions and the Transmission of Knowledge in the West Indies *International Federation of Library Associations and Institutions*, *45*(1), 16–25.

Smith, L. T. (2013) *Decolonizing Methodologies: Research and Indigenous Peoples*. London, Zed Books.

Smith, L. T., Q. Q. Xiiem, J. B. J. Lee-Morgan and J. De Santolo (2019) *Decolonizing Research: Indigenous Storywork as Methodology*, Bloomsbury Publishing.

Thomas, D. A. (2019) *Political Life in the Wake of the Plantation: Sovereignty, Witnessing, Repair* (Kobo ed.).

Tuhiwai Smith, Linda 1999. *Decolonizing Methodologies: Research and Indigenous Peoples*. London: Zed Books.

Vete-Congolo, H. (2007) Caribbean Storytales: A Methodology for Resistance. *Anthurium: A Caribbean Studies Journal*, *5*(1).

Williams, E. (2017) The Anancy Technique: A Gateway to Postcolonial Performance. *Caribbean Quarterly*, 63, 215–233..

Wynter, S. (2003) Unsettling the Coloniality of Being/Power/Truth/Freedom Towards the Human. After Man, Its Overrepresentation-An Argument. *CR: The New Centennial Review*, 3(3), 257–337.

Wynter, S. and K. McKittrick (2015) Unparalleled Catastrophe for Our Species? Or, to Give Humanness a Different Future: Conversations. *Sylvia Wynter: On Being Human as Praxis*. K. McKittrick (ed). Durham, NC, Duke University Press: 9–89.

14

HISTORICAL GEOGRAPHIES

Geographical Antagonism and Archives

David Beckingham and Jake Hodder

Introduction

Archives are not simply containers of sources, little more than the places we go to answer our research questions. Over recent decades, the archive has come to be seen as a political space whose materials and systems of organisation have ordered the world in particular and uneven ways. In this reading, the archive is both a physical place and an 'instituting imaginary' that reflects and reinforces the political worlds of its creation (Mbembe, 2002). Methods texts stress that researchers need to be aware of the partiality both of archival sources and the knowledge claims of those who have created and curated them (see Craggs, 2016). They usefully establish that we must not take the stories of archives at face value, and should examine the possibilities for tracing themes and even voices marginalised or silenced by archives' organising practices. Such reflections are at the centre of historian Antoinette Burton's call for a '[s]trategic antagonism toward sources'. This, Burton argues, 'is, or should be, the hallmark of all historians interested in a critical engagement with the past, rather than in its reproduction' (Burton, 2011, pp. 103–4). In this chapter we lay out a distinctive *geographical antagonism* of archives. Underpinning the assumptions on which they were founded, their collections organised, and their power employed or challenged, is a series of geographical relations that reveal the cultural, social and political work of archives.

Our intention in this chapter is not to provide a comprehensive 'how to' guide for historical research – exemplary versions are offered elsewhere (Black, 2010; Craggs, 2016). Theoretically, however, research proceeds from an understanding that archives, particularly in state and colonial settings, are partial: their structure reflects and reproduces power asymmetries that might marginalise or silence particular groups and themes. And even where such groups might be present, they might be represented in records through the voices and recording practices of those in power (Craggs, 2016: 112). In imperial and colonial settings, researchers have therefore debated the merits of reading these archives 'along' or 'against' that organisational grain (Stoler 2009). Working against the grain, for example, historical geographers have sought to uncover alternative voices and actions that unsettled the power claims embedded in surviving archival records (Moore, 2018, p. 3). Such approaches reflect a methodological conviction that texts might reveal more than simply the 'unmediated will of the dominant power', that archives can be read to tell different stories (Lambert, 2009, p. 56). Reading 'along the grain', however,

DOI: 10.4324/9781003038849-17

Stoler influentially (2009, p. 50) warned researchers not to assume 'we know the scripts' (also see de Leeuw, 2012). Such debates flag a possible tension between the theoretical – or even ethical – ambitions and the practical possibilities of archival work.

In studies of class, gender and race, theorising absence in archives has prompted a practical consideration of the diversification of source materials (Bressey, 2020; Mills, 2013; Hodder, 2017). We might also note here a critique of archiving practices that privilege the written over oral (Ogborn, 2019) and material (Slatter, 2019) sources, and official over informal and personal collections (Ashmore et al., 2012) (see Chapter 13 in this collection on storytelling, for example). More broadly, methodological work has examined the relationship of visual evidence to disciplinary practices (Hayes, 2018; Piana and Watkins, 2020; Rose, 2003[1]). Geographers have also confronted archival exclusions with a range of creative and participatory methods, looking to, and even co-producing, alternative sources of historical geographical knowledge (Dwyer, and Davies, 2010; Lorimer, 2009).

Reflecting on the constitution of geographical knowledge is a useful way to confront Geography's own disciplinary history and the stories it has told about itself. Feminist scholars have used alternative sources such as travel writing, for example, to critique a double discounting of female geographical knowledge at the time and in later disciplinary accounts (for relevant reviews see: Ferretti, 2019; Moore, 2018). In a similar manner, authorised accounts of exploration often wrote out 'indigenous intermediaries' whose role as guides and translators would otherwise undermine contemporary heroic ideals (Martin, 2020, p. 26). Martin warns that writing such figures back in should not simply reinforce the epistemological frameworks that accounted for their erasure in the first place. This work is a reminder that writing can itself be a powerful asymmetrical archival practice, and that the purpose of writing should be considered (Griffin, 2018; Haines, 2017; Moore, 2018). So should its place. Craggs and Neate (2020) raise a fundamental question of orientation: where are we writing from? They ask, 'What happens if we start from Nigeria?'. Here, the Anglo-American dominance of disciplinary accounts, indeed their constitution of what Geography is, can be unsettled by archival evidence and new oral history interviews from Nigerian geographers.

Our argument for a distinctively geographical antagonism builds on the claim that archives – understood as both physical spaces and collections of distinctive materials – are containers and producers of specific *geographical* imaginaries. After Andrew Hill Clark (1971, p. 14), we might still say that one of the most notable differences between geographers and other students of the past 'digging in the same files', is a concern for the 'quantity, quality, and integrity of the locational provenance of the information sought'. Attending to provenance, we argue, is a way to interrogate not only how archives came into being and where they are located now, but also to connect the diverse spatial stories in those records to the work that such materials have been asked to support. In the remainder of this chapter, using examples from our research, we interrogate what geographical antagonism looks like in traditional, physical – that is to say, for convenience, 'analogue' – archives, and then apply this to digital and digitised collections.

Geographical Antagonism in Analogue Archives considers how the real-world topography of 'analogue' archival records mirrored the emerging geographies of state governance and rule. Historically, the location, ordering and display of collections naturalised state power at various scales: imperial, national and local. As Putnam (2016, p. 381) writes, just as 'governance was structured in nested geographic units, so too is the information generated by governing'. However, by considering categories which cut across these boundaries – such as internationalism and race – we explore how the scalar assumptions of archives can be unsettled or disrupted. In the second section, *Geographical Antagonism in Digital Archives*, we consider geographical antagonism and digital archives. Digitisation, which has increased significantly in

174

recent years, has made the historical and geographical claims of archives harder to see, even as it has made it easier for researchers to make connections across disparate collections. Whereas the first wave of digitisation focused primarily on published sources, such as books and newspapers, in recent years a more diverse range of materials have been involved (Hahn, 2006), including the archives of the Royal Geographical Society. The digitisation of one of the discipline's pre-eminent collections offers an important opportunity to interrogate the distinct challenges posed for critical historical scholarship by the shift from finding aid to search algorithm, and from physical building to digital platform (Hodder and Beckingham 2022).

Geographical Antagonism in Analogue Archives

Historical geographical research is an exercise in spatial and temporal connection. At its most basic, geographical antagonism requires us to take seriously the locational geography of archive collections: how materials have ended up located where they are, how they were organised and displayed, and what historical and geographical work their assembly did. Emblematically, the establishment of The National Archives in France in 1790 or the Public Record Office in England in 1838 shows how the development of archives supported the emergence of the nation state (Steedman, 1998). Archives have also played a similar role in imperial imaginaries, with complex geographies. In his monumental assessment of mapping and empire, Matthew Edney uses the metaphor of archiving to explain how imperial knowledge collection and projection over *British* India assumed a kind of coherence and objectivity: 'the purpose of the archive is to relate knowledge about particular places to a larger conception of space' (Edney, 1997, p. 41). Archiving was a process by which often disparate and fragmentary information could be organised, in ways which helped centre the imperial state and its metropole London (Richards, 1993). An interesting example was provided by geographer Clements Markham, who submitted a proposal to reorganise the archive of the India Office as it subsumed the materials of the East India Company (Mitchell et al., 2019, p. 16). The organisational principles of Markham's proposed archive – where older plans, maps and surveys were to be combined with incoming reports and arranged in 'territorialised' units – would determine the organisation of the world 'out there' simultaneously within the bounds of the archive. This example, though never fully realised, clearly shows the importance of interrogating how archives are organised across scales in ways that claim to order space and time.

It is possible to consider the ordering effects of physical locations. The architecture and atmosphere of archive buildings can combine to give them 'something of the nature of a temple and a cemetery' (Mbembe, 2002, p. 19). Robert Adam's Register House in Edinburgh was designed as 'a proper repository for the Records of Scotland', to quote the minute recording its commission in 1772. Today it is the home of the National Records of Scotland; and it continues to display to its users and those who pass by outside the spatial and symbolic power of information that it had established in the expanding physical infrastructure of Edinburgh's New Town (Tait, 1974, pp. 117 and 120). Also in Edinburgh, we can see the entwining of physical and imaginative organisation in Patrick Geddes's 1892 refurbishment of the Outlook Tower on the Royal Mile. This was a 'temple of geography', according to one contemporary, where spatial classification was raised vertically from the world of the ground floor to a top-floor Edinburgh gallery complete with camera obscura. It was 'a symbolic writing of space' argues David Livingstone (2003, pp. 37–8), in which Edinburgh was presented as the result 'of global evolutionary forces' (35). Ruth Craggs (2008) has demonstrated how the Royal Empire Society library in London similarly made an ordering argument through the management of space and experience. By zoning the collection, the library worked alongside practices of archiving to produce a moral geography of empire that offered a political and economic mandate for colonial intervention. Both the 'back-stage' organisational system and the 'front-stage' spaces of display were central archival claims to order knowledge about the world.

Archives are not simply places where the past is buried, where records come to reside only when they have ceased to be useful. It is important to conceive of archives as shaping the future, as Markham's example shows. If some more modern archive locations, such as warehouse units, feel spatially and imaginatively separated from the governmental priorities of the present, there is a clear injunction to recover the politics and pasts of the other locations in which their records operated. Analysing the movement of materials between locations can reveal the otherwise hidden scalar sorting that Putnam (2016) argues made archives instruments of state authority. In a study of reformatory treatment for drunkenness, for example, Beckingham (2019) uses a range of civil service files to theorise the movement of information and inmates. The case geographies of patients who were moved between institutions disrupted the geographical imagination of government into neat scales of local and national responsibility. Files, importantly, were 'spaces of action and not simply record' (Beckingham 2019, p. 1448).

A geographical story of scale and location can be theorised from collections of political projects whose imagination sat above or against national and imperial interests. The dispersed geography of records of internationalism is a case in point (Hodder et al., 2021). Benjamin Thorpe (2018) has explored how one such archive, of the Pan-European Union, became a repeated target of thefts by national powers wanting to thwart its 'dangerous' internationalist ideals. This is an important example of sociologist Mike Featherstone's (2006, p. 592) point that archives have a 'spatial history', that they have been 'destroyed, stolen, purchased and relocated'. Indeed, to combat archival seizure the League of Nations' Institute for Intellectual Co-operation established a 'Permanent Consultative Committee on Archives' and, in 1948, UNESCO supported the creation of The International Council on Archives to offer a normative and legal framework for disputed archival holdings (Kecskeméti, 2017). This was given impetus by decolonisation (Shepard, 2015), as newly independent states disputed the ownership and truth claims of collections and sought their own 'instituting imaginaries'. We can see then the locational power of collections, and the ways in which the deliberate *displacement* of archives, might work to disrupt alternative political imaginations (Lowry, 2019a; 2019b).

Transnational themes such as race and diaspora also disrupt the scale of archival organising. Paying attention to location, especially with collections built from below, so to speak, offers a way to antagonise the political work of state record-keeping practices. Reflecting on the experience of Afro-Caribbean and Asian diasporas, Stuart Hall (2001, p. 90) notes that: 'An "archive" does have something specific in terms of its boundaries – though what has been said about location and "diaspora" means that what constitutes its [an archive's] boundaries in any specific case is not a natural "given" but remains to be critically specified.' Against the boundaries of many conventional archives, collections built by Black activist and community groups are embedded in local political concerns and broader-scale Pan-African and Black internationalist politics (Ishmael, 2018). Natalie Hyacinth's (2019, p. 24) report on Black-led archives in the UK reveals a dispersed geography that reflects the political struggles which the collections endeavour to document: there is a strikingly clear relationship 'between the precarity of Black archives [and] the precarity of Black lives'. It is therefore necessary to situate individual collections within a wider topography of Black-led archives (see also Chapter 15 on Black geographies, this volume). Here, the national collections of the Black Cultural Archive in Brixton sit alongside community ones such as the Nottingham Black Archive or Ahmed Iqbal Ullah RACE Centre in Manchester. Such archiving work remains a profoundly spatial practice and the locational provenance of these records is as central to the work they do as the more firmly locatable imperial collections we considered at the start of this section.

The infrastructure and organisation of archives then, even separate from their contents, have the capacity to reveal geographical stories. So too do their physical locations. The way that materials are organised and displayed does not merely reflect pre-existing political scales or social and cultural categories; our plea for antagonism is to investigate how record-keeping practices can actively bring them into being. To scholars of the imperial archive this is a familiar claim: it marks the grain that we can read along or against. But it merits repeating that these imaginaries naturalise other units of governance. We suggest that considering different aspects of organisation and location can reveal the political work behind that organisation, and by so doing help in the historical identification of policy challenges or alternative imaginaries. Archives of internationalism and race explicitly antagonise simple stories of location and authority especially in cases where records were created to move '*between* scales' and have per-haps settled quite serendipitously (Legg, 2016, p. 247, original emphasis). The most seemingly national of archives can contain remarkably personal experiences, and likewise local archives are containers of global histories.

Geographical Antagonism in Digital Archives

In this section, we examine the importance of geographical antagonism in a world of digital archives. Mass digitisation has brought many benefits: it has preserved collections for future generations and, putting to one side questions of access rights and subscription costs, it has the potential to democratise access to research materials. But there are also reasons to be cautious. The pace and nature of digitisation are uneven: well-financed institutions – as well as popular or easily digitisable material – tend to get prioritised. In a world where researchers can reach their evidentiary threshold through digital sources alone, it becomes increasingly difficult to justify the time and cost of visiting harder-to-reach undigitised collections. As such digitisa-tion might reinforce existing archival hierarchies, and research topics, rather than radically reshaping them.

Despite the scale of changes underway, there is a significant gap in the literature. To follow Caroline Bressey (2020, p. 2), 'there has been little formal reflection on the concerns, challenges or methodological opportunities presented by the formation of digital archives'. This is strange given the volume of reflections on conventional archive work and, relatedly, on archives from the perspective of digital humanities. Bressey's work examines the potential of aggregated digitised sources, such as newspaper collections, to support research into groups and themes marginalised within conventional archives. It may be tempting to believe that the effect of aggregation is to scale scarcity into abundance, and so digitisation becomes a tool for emancipatory research. Citing the work of feminist historians, however, Bressey offers caution: digitisation 'doesn't necessarily transform the nature of the sources we are searching' (Hunter, 2017, p. 210, cited in Bressey, 2020). The silencing and erasure intrinsic to histor-ical recording practices remains an obstacle to researchers despite the new possibilities for connection offered by digital platforms.

Here we strike a different note of caution. We suggest that the locational antagonism demanded of analogue sources is no less essential in the digital archive. On the contrary, digital collections often work to obfuscate the kinds of geographies we have outlined above and thereby demand more concertedly antagonistic treatment. In order to develop our case, we need first to interrogate how digitisation is reshaping the research *process* in ways which remain largely untheorised. It is precisely because historical geographers have so strongly encouraged critical engagement with the constitution and sorting functions of archives, and because digitisation is imprecisely and selectively transforming *how* those functions work, that it demands our attention.

Here we might consider the recent digitisation of the archives of the Royal Geographical Society by Wiley Digital Archives.[2] In its most straightforward sense, digitisation obscures the physical space of the archive itself, which, as we have shown, is intimately entangled with the archive's role as a technology of rule. This is a really simple point. If we consider the display of the RGS collection's materials in its historic space on Kensington Gore to be central to understanding the work that those records performed (see Griffiths and Baker, 2020), how do we remain attentive to the impress of that historical setting when accessing such collections remotely? The answer is almost certainly to interrogate the platforms that host and display archive materials and the algorithms we use to search them. Such an approach is to be mindful of the differences between digital and analogue archives, even when the documents themselves may be the same (Jeurgens, 2013). That is to say, that digitisation is not simply the duplication of archive materials or transference of archival research practices from one space to another, but always necessarily involves transformation. This includes the loss of some forms of information and addition of others.

Here it is worth noting that digitisation refers to two related but distinct processes: first, the scanning of documents to produce a visual image and, second, the digitisation of their contents into ASCII or Unicode, often with the addition of metadata. Advances in AI algorithms and Optical Character Recognition (OCR) mean that digitised materials are often searchable by the minutiae of their content (words on the page, or what is termed 'full-text search') as well as their context (date, location, keywords). The transformative nature of digitisation resides in this latter form of digitisation. In the case of the RGS collection, for example, enhanced research features include exportable datasets for charts, tables and statistics; the geo-tagging of maps on current coordinates displayed through a GIS suite; and the ability to translate into 105 languages.

Where previously we had to access the archive in a very particular physical location, digitisation seems to promise the geographical release of visiting the archive from anywhere. But digital space is not synonymous with anywhere. On the contrary, the digital archive exists in a very specific digital space: that of the, often commercial, research platform. These platforms have built-in, proprietary research tools that publishers keenly identify as the added value they bring. For example, Wiley Digital Archives is the only programme in the library market to offer Automated Text Recognition (ATR), an AI that turns handwriting into typeset allowing for full-text search. If publishers assert copyright through such platforms, and citational practices change as a result, then something about the research process has been transformed. Nicholas Dirks described how researchers establish their credibility through referencing and footnotes, in ways which construct archives as spaces 'free of context, argument, ideology – indeed history itself' (Dirks, 2002, p. 48). In a digital era, platforms might seem to be free of context, but like analogue archives they reorder, extract, tabulate and map results in ways that profoundly shape the research process and questions we ask of materials. This is not necessarily prescriptive, of course. As with analogue archives, we can consciously choose to read digital sources along or against the grain; but to do so demands a consciously antagonistic approach and a clear theorisation of technology.

If, as we have asserted, the real-world geography (organisation, location, and mobility) of archives matters, how can we critically recover the provenance of archival materials when those same documents seem to be geographically unbound by digitisation? In many senses, the promise of digitisation is precisely this ability to work against geographical barriers and to pull together disparate, dislocated collections. It is for this reason that digitisation has coincided with broader intellectual trends such as the rise of transnational history. As Lara Putnam (2016: 380) has argued, transnational approaches are driven by the frictionless, disintermediated discovery offered by text-searchable digital archives. And yet she cautions us to take seriously the 'real-world geography of textual sources' which is itself evidence of the complexity of transnational interactions in the past and which is often hidden by remote access. Here we are particularly interested in analogue collections which have been digitised but similar arguments could be made of so-called 'born-digital' materials. Roy Rosenzweig (2003, p. 743) notes that born-digital sources are 'so easily altered and copied, lacks physical marks of its origins, and, indeed, even the clear notion of an "original"' that their 'real-world geography' (such as information about who created them and where) can be erased entirely.

Digitisation obfuscates these 'real-world' geographies by disassociating materials from the traditional archival practice of *respect de fonds*, which is designed to recognise both provenance and the relationship of documents to one another in the arrangement of collections (Bartlett, 1992). Understanding and interrogating these organisational principles, whether we read with or against them, has been a vital part of critical historical scholarship. It is also the basis of our citational practices. Our references do not simply show where something is from, which is to say its physical location; they point to the collection in which a document sits and suggest a relationship to other material. Digitisation and full-text search change these research practices in fundamental ways. In the digital archive we arrive at materials not through the arrangement of the archive, but through search tools that prioritise individual documents based on their relevance to our search terms. These are then often displayed to us in ways which show little fidelity to the provenance or organisational framework by which the analogue documents were originally arranged. Indeed, this is the 'efficiency' and 'empowerment' so frequently promised to the researcher by digital platforms: the ability to surgically extract individual sources which speak directly to our research questions.

The reordering of power in the digital archive is not from the archivist to the researcher, we might suggest, but from archivists to algorithms. Scholars rarely know the relevance metrics which an algorithm is using to collate and display their results; and, in the case of proprietary algorithms, it might not even be possible to know them. Ted Underwood (2014) argues that 'search' is a deceptively simple term for what is happening when we use digital collections. Search implies something comparable to using a finding aid to retrieve a document, but in many ways the task being completed has more in common with data mining and, even then, data mining of 'a strangely focused form that only shows you what you already know to expect' (Underwood, 2014: 66). This is particularly important when we consider that the evidentiary threshold of qualitative work is still closely tied to the labour-intensive world of analogue research. As Underwood (2014: 66) notes, in a digital database of 'millions of sentences, full-text search can turn up twenty examples of anything'. But the problem here is less what we find than what we do not; search algorithms arrange by relevance in ways which filter out alternative ideas. At the heart of geographical antagonism is the recognition that archive materials do not exist in a vacuum. They are products of the times and spaces of their creation and use. Their significance can often only be fully understood when read against that broader context, a key principle of traditional archival arrangement. Today, 'we are using search algorithms we have never theorised, and arguably using them in a strongly projective way that is at odds with historicism' (Underwood, 2014, p. 70). If, to return to Antoinette Burton (2011), archives mask as much as they reveal, this is no less true of digital collections.

Conclusion

In this chapter we have used the provocation of Antoinette Burton to call for a geographical antagonism of archives and collections. We have presented the different ways of understanding how locational provenance remains central to the concerns of historical geography. The organisation and display of archival records were tools of statecraft. They shaped political-geographical imaginaries. These simple points about organisation and display continue to matter. In the examples of 'traditional' archives, we have shown how examining the physical space of the archive and the diverse locational geography of archival records can reveal important understandings of power. This well-established concern has been central to the work of geographers, reading along or against the grain of archives. But we have also called attention to the mobility and visibility of collections as a further way to emphasise how archives are, and enable, a series of spatial and scalar connections.

The familiarity of some of the earlier discussion should serve to highlight why these concerns matter in an era of digitisation. We have highlighted the distinction between simply scanning original documents and posting those online, and the development of new complex platforms that can more fundamentally change how we interrogate sources. When finding aids dictate what can be said, as historical research has shown, it matters. And just as historical modes of cataloguing construct documents in spatial and temporal relations, so too do digital platforms and search algorithms. Here, we do not criticise the fact of digitisation, even if calling out its partiality is a way to celebrate the kind of time and labour-intensive scholarship that exists outside the digital frame. Digitisation is revolutionary and remarkable. And it is for this reason, that we propose geographical antagonism as foundational for new digital methodologies.

Notes

1 This is part of a 'roundtable' special issue, with contributions from Mike Crang, Felix Driver, David Matless and James Ryan.
2 The RGS archive contains the world's largest private collection of maps, charts, and atlases totalling more than 150,000, as well as manuscripts, field notes, expedition reports, scrapbooks, correspondence, diaries, sketches, and lantern slides.

References

Ashmore, P., Craggs, R., and Neate, H. (2012) Working-with: Talking and sorting in personal archives. *Journal of Historical Geography*, 38(1), 81–89.
Bartlett, N. (1992) Respect des fonds: The origins of the modern archival principle of provenance. *Primary Sources & Original Works*, 1(1-2), 107–115.
Beckingham, D. (2019) Bureaucracy, case geography and the governance of the inebriate in Scotland (1898–1918). *Environment and Planning C: Politics and Space*, 37(8), 1434–1451.
Black, I.S. (2010) Analysing historical and archival sources. In N. Clifford, S. French and G. Valentine (eds.) *Key Methods in Geography* (London: SAGE Publications) 466–484.
Bressey, C. (2020) Surfacing black and brown bodies in the digital archive: Domestic workers in late nineteenth-century Australia. *Journal of Historical Geography*, 70, 1–11.
Burton, A. (2011) *Empire in Question: Reading, Writing and Teaching British Imperialism* (Durham, NC: Duke University Press).
Clark, A.H. (1971) First things first. In R. Ehrenberg (ed) *Pattern and Process: Research in Historical Geography* (Washington DC: Howard University Press) 9–21.
Craggs, R. (2008) Situating the imperial archive: The Royal Empire Society Library, 1868-1945. *Journal of Historical Geography*, 34(1), 48–67.

Craggs, R. (2016) Historical and archival research. In N. Clifford, S. French and G. Valentine (eds.) *Key Methods in Geography* (London: SAGE Publications) 111–128.

Craggs, R. and Neate, H. (2020) What happens if we start from Nigeria? Diversifying histories of geography. *Annals of the American Association of Geographers*, 110(3), 899–916.

de Leeuw, R. (2012) Alice through the looking glass: Emotion, personal connection, and reading colonial archives along the grain, *Journal of Historical Geography*, 39(3), 273–281.

Dwyer, C. and Davies, G. (2010) Qualitative Methods III: Animating archives, artful interventions and online environments. *Progress in Human Geography*, 34(1), 88–97.

Edney, M.H. (1997) *Mapping an Empire: The Geographical Construction of British India, 1765-1843* (Chicago, IL: The University of Chicago Press).

Featherstone, M. (2006) Archive. *Theory, Culture & Society*, 23(2-3), 591–596.

Ferretti, F. (2019) History and philosophy of geography I: Decolonising the discipline, diversifying archives and historicising radicalism. *Progress in Human Geography*, 44(6) 1161–1171.

Griffin, P. (2018) Making usable pasts: Collaboration, labour and activism in the archive. *Area*, 50(4), 501–508.

Griffiths, M. and Baker, K. (2020) Decolonising the spaces of geographical knowledge production: The RGS-IBG at Kensington Gore, *Area* 52(2), 455–458.

Hahn, T.B. (2006) Impacts of mass digitization projects on libraries and information policy. *Bulletin of the American Society for Information Science and Technology*, 33(1), 20–24.

Haines, E. (2017) Visions from behind a desk? Archival performance and the re-enactment of colonial bureaucracy. *Area*, 51(1), 25–34.

Hall, S. (2001) Constituting an archive. *Third text*, 15(54), 89–92.

Hayes, E. (2018) Geographical light: The magic lantern, the reform of the Royal Geographical Society and the professionalization of geography c.1885–1894. *Journal of Historical Geography* 62, 24–36.

Hodder, J. and Beckingham, D. (2022) Digital archives and recombinant historical geographies. *Progress in Human Geography*, in press.

Hodder, J. (2017) On absence and abundance: Biography as method in archival research. *Area*, 49(4), 452–459.

Hodder, J., Heffernan, M., and Legg, S. (2021) The archival geographies of twentieth-century internationalism: Nation, empire and race. *Journal of Historical Geography*, 71, 1–11.

Hunter, K.M. (2017) Silence in noisy archives: Reflections on Judith Allen's "evidence and silence – feminism and the limits of history" (1986) in the Era of Mass Digitisation. *Australian Feminist Studies*, 32(91-92), 202–212.

Hyacinth, N. (2019) Black archives in the UK report: Opportunities, challenges and moving forward. *Race, Culture and Equality Working Group Report, Royal Geographical Society*. [available online: https://raceingeographydotorg.files.wordpr, accessed 19th February 2021].

Ishmael, H.J. (2018) Reclaiming history: Arthur Schomburg. *Archives and Manuscripts*, 46(3), 269–288.

Jeurgens, C. (2013) The scent of the digital archive: Dilemmas with archive digitisation. *BMGN-Low Countries Historical Review*, 128(4), 30–54.

Kecskeméti, C. (2017) Archives seizures: The evolution of international law in displaced archives, in J. Lowry (ed.) *Displaced Archives* (Abingdon and New York, NY: Routledge).

Lambert, D. (2009) "Taken captive by the mystery of the Great River": Towards an historical geography of British geography and Atlantic slavery. *Journal of Historical Geography*, 35, 44–65.

Legg, S. (2016) Anti-vice lives: Peopling the archives of prostitution in interwar India. In H. Fischer-Tiné, J.R. Pliley, R. Kramm (eds.) *Global Anti-Vice Activism, 1890-1950: Fighting Drinks, Drugs, and "Immorality"* (Cambridge: Cambridge University Press).

Livingstone, D. (2003) *Putting Science in its Place: Geographies of Scientific Knowledge* (Chicago and London: University of Chicago Press).

Lorimer, H. (2009) Caught in the nick of time: Archives and fieldwork. In D. DeLyser, S. Aitken, M.A. Crang, S. Herbert and L. McDowell (eds.) *The SAGE Handbook of Qualitative Geography* (London: SAGE) 248–273.

Lowry, J. (2019a) "Displaced archives": Proposing a research agenda. *Archival Science*, 19(4), 349–358.

Lowry, J. (2019b) Radical empathy, the imaginary and affect in (post) colonial records: How to break out of international stalemates on displaced archives. *Archival Science*, 19(2), 185–203.

Martin, P.R. (2020) Indigenous tales of the Beaufort Sea: Arctic exploration and the circulation of geographical knowledge. *Journal of Historical Geography*, 67, 24–35.

Mbembe, A. (2002) The power of the archive and its limits. In *Refiguring the Archive* (Springer, Dordrecht) 19–27.

Mills, S. (2013) Cultural–historical geographies of the archive: Fragments, objects and ghosts. *Geography Compass*, 7, 701–713.

Mitchell, P., Lester, A., and Boehme, K. (2019) "The centre of the muniment": Archival order and reverential historiography in the India Office, 1875. *Journal of Historical Geography*, 63, 12–22.

Moore, F. (2018) Historical geography, feminist research and the gender politics of the present. *Geography Compass*, 12 (e12398).

Moore, F. (2010) Tales from the archive: Methodological and ethical issues in historical geography research. *Area*, 42(3), 262–270.

Ogborn, M. (2019) *Freedom of Speech: Talk and Slavery in the Anglo-Caribbean World* (Chicago: The University of Chicago Press).

Piana, P. and Watkins, C. (2020) Questioning the view: Historical geography and topographical art. *Geography Compass*, 14 (e12483).

Putnam, L. (2016) The transnational and the text-searchable: Digitized sources and the shadows they cast. *The American Historical Review*, 121(2), 377–402.

Richards, T. (1993) *The Imperial Archive: Knowledge and the Fantasy of Empire* (London: Verso).

Rose, G. (2003) On the need to ask how, exactly, is geography "visual"? *Antipode*, 35(2), 212–221.

Rosenzweig, R. (2003) Scarcity or abundance? Preserving the past in a digital era. *The American Historical Review*, 108(3), 735–762.

Shepard, T. (2015) "Of Sovereignty": Disputed archives, "wholly modern" archives, and the post-decolonization French and Algerian Republics, 1962–2012. *The American Historical Review*, 120(3), 869–883.

Slatter, R. (2019) Materiality and the extended geographies of religion: The institutional design and everyday experiences of London's Wesleyan Methodist circuits, 1851–1932. *Journal of Historical Geography*, 64, 60–71.

Steedman, C. (1998) The space of memory: In an archive. *History of the Human Sciences*, 11(4), 65–83.

Stoler, A.L. (2009) *Along the Archival Grain: Epistemic Anxieties and Colonial Common Sense* (Princeton: Princeton University Press).

Tait, A.A. (1974) The Register House: The Adam building. *The Scottish Historical Review*, 53(156), 115–123.

Thorpe, B.J. (2018) *The time and space of Richard Coudenhove-Kalergi's Pan-Europe, 1923-1939* (PhD thesis, University of Nottingham).

Underwood, T. (2014) Theorizing research practices we forgot to theorize twenty years ago. *Representations*, 127(1), 64–72.

15

BLACK GEOGRAPHIES

Methodological Reflections

Renato Emerson dos Santos and Priscilla Ferreira

Introduction

This chapter is a methodological reflection on contributions of the growing field of Black geographies to decenter traditional approaches in the discipline of geography that have neglected, marginalized and made invisible the spatial experiences, spatial agency and epistemic contributions of Black populations. In this chapter, we will not engage with exhaustive comparative analysis of these different regional or national schools of thought or intellectual traditions in geographies of race. Rather, we will discuss some key features of distinct projects in terms of their approaches to think through their methodological challenges. How are the associations between different spatial imaginations and understandings of race – from "race relations" to "processes of racialization" (Jackson, 1987; Bonnet, 1996; Sansone, 1996) – mobilized in different social and intellectual contexts in the production of Black geographies? What are Black geographies, and what are their current methodological challenges and pathways?

The Construction of "Race" and Racist Geography

The subject of "race" is long-standing in geography, with approaches, understandings, theoretical, methodological, and epistemological bases varying in time and space. From the nineteenth century to the middle of the twentieth century, in various nations there developed what Bonnet (1996) dubs "Racial Geography." Based in an analytic co-articulation of race and climate, notions like "civilization," and hierarchies among cultures, this geography contributed to the establishment of racist social configurations. Aside from European colonialism and imperialism in Africa and Asia, such racist geographies contributed to projects of whitening populations and constructing racial hierarchies in countries of ancient colonial formation in the Americas. These forces helped to establish essentialist understandings of the social category of race that have remained entrenched in these societies even after changes to the social orders that engendered them. In order to understand the methodological potential of contemporary Black geographies, we need to reckon with the racist roots of geographical thought and practice. These racialized geographies were part of racist hegemonic projects of such societies.

The association between spatial classifications such as the regionalization of the globe in continents that divide Europe and Asia, which are in fact a contiguous extension of land,

DOI: 10.4324/9781003038849-18

and classifications of racial groups gave rise to a set of hierarchies mutually reinforced in a Eurocentric global order (Quijano, 2010). Such order deploys racialized hierarchies of class and the organization of forms and strategies of exploitation of labor power and the generation of surplus value (Balibar and Wallerstein, 1988); geopolitical hierarchies between centers and peripheries (Grosfoguel, 2003; Lander, 2005); and associations with the imposition of patriarchal hierarchies of gender and sexualities (Anzaldua, 1987; Crenshaw, 1991; Collins, 2016). The construction and diffusion of such racialized epistemic constructs, and the positioning of social subjects in relations of power, domination, exploitation, and oppression, find validation in the pretense of scientific geographical thinking which endow themselves with attributes such as rigorous, neutral, and objective to legitimize while obscuring the very political situatedness of colonial legacies embedded in modern geographical formations. Bonnet's (1996) critique of the formation of whiteness as race-less and non-relational to geographical projects of white supremacy speaks to the point we are raising here. When discussing the assumption of non-relationality ascribed to the emergence of whiteness as a social construct, Bonnet critiques the premise that race and racism is about sentiments and actions around Blackness or non-white social locations, as if whiteness as a social identity and positionality had not itself been forged through interdependent relations of uneven power; that is, racism is co-constitutive with whiteness.

Likewise, in the Brazilian context, racialized hierarchies were present in works of French geographers that inaugurated the so-called "scientific geography" in the country as evidenced in Monbeig (1984) and Deffontaines (1940; 1944). A Brazilian school textbook of 1944 analyzed by Santos (2009, p. 38) taught a racial classification which, after assigning each racial group a regional origin and describing their differentiating physical traits (skin color, hair, facial features), credited the white race, or Caucasian, with "well-developed intelligence and superior civilization," and the Black race, or Ethiopian, with "rudimentary civilization." These essentialist understandings of race and racial classification, and their relationship with space were to subsequently undergo profound critique along with their societal projects, the spaces constructed upon the bases of these perspectives, and the social relations and experiences of racialized space that they helped to shape.

Since the end of the twentieth century, geographies derived from and/or part of a growing struggle against racism have begun to emerge marked by differences according with their contexts. In the United States, England, or in Latin American countries like Brazil and Colombia, these geographies emerge striated by differences of social context in terms of national (Omi & Winant, 1986) or regional racial formations (Lao-Montes, 2020). They are influenced by geopolitical positions of epistemological disputes (Alatas, 2003) and appear in projects engaged with the critique and transformation of the racial relations – which means that they are also influenced by different formations of anti-racist struggle. When such geopolitical projects are centered Black lives, Black traditions, and seek to overcome anti-Black racism (Moore, 2007), the spatial dimensions of these social experiences and political struggles are increasingly named as "Black geographies. Such geographies arise across different places and regions, with roots in distinct historicities, temporalities, and dialogues, as exemplified in the British school of study of race relations and geographies of race since the 1980s which sought to align "debates over the social construction of 'race' and a discussion of the territorial expression of various forms of racism" (Jackson, 1987, p. 4). Bledsoe and Wright (2019) make the case for the plurality of Black geographies in the US and globally (2019) and, in Brazil, Santos (2012) discusses the multi-layered dimensions of spatial expressions of racial relations, of racism and of anti-racism.

Black geographies, therefore, is not just about the study of disadvantageous experiences of segregation, dispossession, impoverishment of Black communities and how colonialism and racial capitalism have framed deleterious spatial patterns of majority-Black places across the globe and the African Diaspora in particular, as we have noted was the focus of earlier studies of geographies of race. Black geographies as a field of study and everyday praxis also entail and reveal the spatial agency of Black populations in defining spatial conditions of existence for their collective survival, thriving and imaginative futures. Moreover, it highlights how Blackness relationally extends proactive spatial agency upon majority white places and spaces when Black populations refuse confinement, segregation, exclusion, dispossession, cultural appropriation, and, consequently, triggers reactive patterns of attempted control, violence, and punishment by white supremacist spatial projects. Black geographies exceed scholarly investigation of how race relations are mapped onto spatial formations. It implies a political project that both conceptually and through everyday enactments not only expose the multiplicity of geographical imprints of the fatal coupling of power and race, as Gilmore (2002) has asserted, but also calls for and articulates with intention and methods the transformation of dominant racialized geographical projects designed to thwart Black life and Black freedom.

Situating Black Geographies

Black geographies as a field of study within academia has emerged, moreover, as a broad array of work engaged in a political project and an open, diverse episteme whose unity is given through commitment to the overcoming of racism and the forms of power with which it is involved, through contributions based in Geography. In this search, traditions of political struggle for liberation against colonialism, imperialism, the exploitation of labor, patriarchy, and different forms of oppression both align and diverge. Different traditions of Black struggles are drawn upon as references of the political roots of theoretical and epistemological influences and are often both objects and tools for rereading social dynamics. These multiple framings of Black struggle as object of study, as epistemological lenses or ideological inspiration have mobilized struggles and resistance against slavery and for freedom from colonization, colonialism, imperialism, white supremacy, discrimination, segregation, racial violence, and against all narratives of subalternity, as seen in the praxis of Black feminisms, Pan-Africanism, and Black movements in various countries. Both individual and collective legacies of struggles are re-centered as examples of the potency of Black agency in the making of Black geographies. In what follows, we will also attend to how non-conflictual dimensions of movement building, such as care, are resignified and valorized as keys to understanding these agencies.

The point of departure, then, is a theoretical, political, and epistemic commitment to overcoming racism, understood as a system of domination and planetary relations of power as the modus operandi of modernity/coloniality and their historical processes that function as the social logic inherent in the classification, hierarchization, and social subalternity of individuals and groups. Racism thus permeates all levels of social relations, engendering "geographicities" of race relations – that is, the spatialization of oppressive race relations. However, racism is met with the creative resistance from those individuals and groups that develop and refine spatial agency actualizing spatial projects through their living and livelihoods.

The commitment of Black Geographies to social transformation and the overcoming of racism involves the revelation of relations of power in their spatial dimensions, reconstruction of submerged historical narratives, and reckoning with negation of their spatialities in dominant approaches. Toward this end, Black geographies also seek to critique and reverse the effects of the power of scientific geographical knowledge and the incidence of the hegemonic white, heteropatriarchal, imperial gaze upon spatial imaginations historically produced by Geography. In this sense, Black geographies grapples with the spatializations of racism by shedding light on struggles against racism and by validating social experiences silenced by traditional disciplinary canons. The theory and practice of Black geographies, therefore, disrupts and undoes the epistemicide established by the scientific enterprise aimed at anti-Black erasure.

The political and epistemological challenges and commitments inherent to Black Geographies imply the exercise of theoretical and methodological creativity. To seek spatialized understandings of race relations or racialized analysis of structural everyday spatial experiences requires and invites new pathways for spatial imagination. If a geography produced based on projects of social transformation is to continue articulating itself as geography, while maintaining its disciplinary commitments with spatial thinking, it must seek to re-imagine methodological approaches aligned with ways of knowing, apprehending, enacting Black spatial experiences and multifaceted Black sense of place.

What are the methodological implications brought to light by Black geographies? What strategies can be seen to emerge nurtured by these projects? In the sections that follow, we seek to contribute an understanding of how these questions are being confronted in the domain of Black geographies, and reemphasize the degree to which they pressure the forms of scientific practice in the field of geography. Firstly, we present a systematization (neither exhaustive, nor totalizing) of how Black geographies have developed spatialized readings of race relations. We identify three exercises of racial imagination and of race relations founded on "rationales based in space". Then, we examine strategies for reimagining the methodologies in line with what has come to be called Black geographies.

Reimagining Spaces, Spatialities and Spatializations of Race, Racialization, Racism and Anti-Racism

An initial exercise in the analytic movement of the production of Black Geographies is the understanding of *different forms of the racialization of spaces.* This involves readings of how race relations produce graphic marks in space in the form of spatial structures, as well as in the association of symbolic references and racialized identifications with spaces. These associations influence patterns of group behavior and state institutions. Thus, instead of fixed conceptions of races, as classes of human beings distinguished by corporeal and/or cultural criteria, it is pertinent to think of racialization as dynamic processes of classification that take place in specific social contexts. Racialization appears as a process of attributing changing meanings of race to the bodies of individuals, but also to collectivities and cultural practices. On one hand, racialization is the imputation of a classification based on race and racism, which connects race (as classification) and racism (as a system of oppression), sustaining practices of hierarchy and subordination, prejudice, discrimination, segregation, and violence. On the other hand, racialization involves and triggers agency of individual and collective subjects. Moreover, racialization also undergirds place-making in that, by extension, racially-informed agencies, conflicts, values, views, and interventions are also ascribed to space. That means, that understanding space also requires a relational purview that captures the dynamic nature of race relations-racialization. Such framing calls up a move away from apprehending space as fixed, but rather as a field of possibilities, open to transformation.

One case in point of such theoretical shifts posited by Black geographies regards the way that analytical approaches to the issue of segregation have been undergoing some change. Since the 1980s the focus has shifted toward a more complex understanding of race and ethnicity, and of the reading of social, cultural, and economic interactions of different groups whose identities are interpreted as historically contingent. Segregation thus becomes central in the constitution of "geographies of racism" (Jackson, 1987). Although Bonnett (1996) shows that such studies still run the risk of essentializing racial matters, this register traces an emerging concern with "racialization of spaces," a trend that profoundly marks Black geographies in their quests for socially historicized understandings of the relationship between race and space.

We approach the displacement of the framing that relates race and space by way of the example of segregation because this term can be read as a noun – a "fact"/"thing," a pattern of distribution/concentration/localization of groups in space – but also as a verb a "process," a network of actions and phenomena that produce such a distribution, that which involves the search to mitigate the practices that produce such a fact. In this sense, racialized practices attain visibility just as processes of urban transformation are demonstrated in their racial and anti-Black dimensions in terms of gentrification (Kent-Stoll, 2020), urbicide (McKitrick, 2011), ethnic/racial exile (Sanchez, 2012), or the whitening of territory (Santos et al., 2018). Racialization emerges here as a component of rationalities and behaviors of social, political, and economic agents in critique of urban readings that marginalize and erase the anti-Black character of urban history and urban planning (Sandercock, 1998), as the underside of colonial narratives of a modernity that hides its racist dimensions and African Diasporic resistances (Lao, 2019).

Symbolic dimensions and identities of race and space appear as well in the readings of those Black communities reminiscent of the resistances to slavery, as in the *quilombos* of Brazil (Anjos 2009) and *palenques* of Colombia (Freire 2020). Traditionally Black communities and their forms of relation with nature inform a Black sense of place constituted through spatio-temporal relations. Oslender (2002), for example, speaks of "aquatic space" in reference to the ways that Black communities on the Colombian Pacific coast organize everyday life according to the different aspect of water cycles, tides, rain, and floods. De Almeida (1989) notes how multiple forms of land tenure in traditionally Black communities are self-defined as "*Terras de Preto*" (lands that belong to Blacks folk) or "*Terras de Santo*" (land that belong to saints, orishas or devotional practice), while Gomes (2012) documents repositories of ethno-botanical knowledges found in backyards and temples that are used for medicinal or religious purposes. Sodré (1988) further points out a myriad of ways that groups claim belonging and re-establish a sense of connection to Africa.

A Black sense of place also reveals itself in how places are named, an intervention that entails an act of power, as Bordieu (1989) has suggested. Similarly, Santos (2009) shows how, in Brazil, several majority-Black spaces have received toponomies based in racial discrimination, but also many others articulate Africanisms. It is also shown how, in recent periods, places such as neighborhoods and occupations of landless movements in rural spaces and of homeless movements in urban spaces have been named with reference to the *Quilombo dos Palmares* and its most prominent leaders, such as Zumbi and Dandara. *Quilombo dos Palmares* was the largest and most important community of Africans who freed themselves from slavery in the colonial period and that resisted for more than 100 years in Brazil. These marks of the conscious memory of Black struggles in space function as tools for the resignification of the past and appearance of contemporary subjectivities. This is particularly evident in Brazil due to the important mobilization of leaders such as Abdias do Nascimento, one the most important Black Brazilian leaders of the twentieth century, who proposed that *quilombo* be thought of as each and every instance of Black struggle (1980). A growing literature in Black Geographies has been drawing on *quilombismo* as an analytical framework.

The case of those reminiscent communities of *quilombos* leads to a second analytic movement in Black Geographies, which is the **mobilization of a relational sense of place in Black struggles.** These communities, once having secured the right to the title of landed tenure by the Brazilian Constitution of 1988, have reconstructed their political struggle, shifting their claim to the struggle for land to a struggle for territory (Arruti, 1999). A plethora of social movements in Latin America, including the well-known *Processo das Culturas Negras* in the Colombian Pacific region (Escobar, 2004), have influenced a turn from understanding territory merely as a category of analysis towards addressing it as a political category of active struggle (Haesbaert, 2020). They claim that these territories belong to subjects whose identities are grounded on "cosmovisions" centered on deep relationships with nature and reverence to ancestrality, as well as ties of reciprocity and solidarity constructed over generations of struggles to ensure the right to belonging. In posing such claims they call our attention to the ways in which plural understandings of racialization engender open and relational interrogations of geographic categories of analysis. For instance, territory, territoriality, and territorialization are given new meaning as conceptual tools to the extent that they are mobilized as categories of collective social practice and struggle, and as such could be dubbed a politics of territory. Likewise, in their work, Santos and Soeterik (2015) analyze the action of the Black Movement in Brazil in its struggles for anti-racist education, mobilizing the notion of "politics of scale" (Swinguedow, 1997) to demonstrate how this social movement plays local, national and international political games to achieve its objectives.

Forms of what Jackson (1987) has dubbed "geographies of [Black] resistance" have emerged with a focus directed toward the understanding of articulations between space and race situated in specific locations of Black protest. This relation between Black protest and place, together with these studies that speak of "Black territories" (Offen 2003 and "politics of scale of the Black Movement" (Santos and Soeterik, 2015), address spatial concepts and categories of analysis of Geography to understand (and effectuate) anti-racist struggle to bring into focus "spatial repertoires of action" of the Black Movement (Santos, 2019).

The reading of "spatial repertoires of action" connects with an important third aspect of analysis in Black Geographies, that is *Black agency as key to the imaginative exercise of examining spatialities traditionally rendered invisible by disciplinary canons.* The reading of the "Black sense of place" (Mckittrick, 2011) reveals the way in which Black social and spatial positions create specific forms of agency. Connecting themselves with political critiques and proposals, and with epistemologies that explore situated knowledges (Haraway, 2009) particularly at the margins (Crenshaw, 1991), Black geographies work out the constitution of spaces and spatialities by way of difference, an analytic movement also noted by Soja (1996). Thus, such spatial imaginations depart from singular racialized social experiences and temporalities. In tandem with the necessity of opening up spatial imaginations in search of radical projects of social transformation (Massey, 2004), Black spatial imagination appears thus as a tool for looking at the aspirations, suffering, conquests, desires, and demands that appear in individual and collective histories, cultural practices, aesthetic heritages and identities, archives of knowledge and social, economic, and political repertoires. Black life trajectories allow us to reconstruct spatial narratives, interpretations and hermeneutics of modernity (Lao-Montes, 2019), opening up new connections and translocal spatial imaginations, as well as forms of periodization – which is to say, questioning Eurocentric narratives that erase the agency of subjects by means of hegemonic conceptions of space and time.

Lao-Montes uses the example of the historian James Sweet's research on Domingos Álvares, a subject born during the eighteenth century in the region of what is today Benin from a family of Vodun priests. He was kidnapped and trafficked to Brazil, where his knowledge of healing and religion allowed and compelled him to travel through different regions and the cities of Recôncavo Baiano, Rio de Janeiro, Recife, and later, Lisbon. He was a respected individual who was sought after by enslavers for his medical capacity. However, he was also feared due to his political upsurging against the social order of slavery, and was thus quickly sold forward multiple times. Domingos Álvares appears as a subject (an agent) and vehicle of diffusion of African knowledge of medicine, spirituality, and political action, as an agent in the construction of the African Diaspora in the "Black Atlantic" (Gilroy 2002) by way of circulation of African knowledges and purposeful dedication to rebuilding self-reliant intentional afro-descent communities. Functioning as a kind of key for understanding the formation of "Afromodernities" as a form of "subaltern cosmopolitanism," after Lao-Montes, this example is based in a spatialized Black biography/trajectory that allows us to overcome the idea of the African Diaspora as a "backwards ghetto" within modernity or as a "peripheral modernity," readings that proceed from hegemonic spatial imaginations. This resignification departs from a critique and theoretical, political, and epistemological displacement of hegemonic views on Black experiences and spaces which, according to McKittrick (2011), are seen through the prism of – and only as – dispossession. This turn permits us to perceive them as potency, insofar as we seek to valorize solidarities within, between, and beyond spaces (Lipsitz, 2007).

Therefore, taking as point of departure the theoretical, political, and epistemic commitment to social transformation of anti-racist struggles, Black geographies seek to reveal the obscured political values of cultural practices, aesthetic heritages, traditional African religions, networks of affective, family relations, and other forms of collectivization of strategies, as associative forms typically outside the public sphere and the most visible civil society. It draws on and highlights multiple ways of producing, documenting, and circulating knowledge that stem from different intellectual traditions and encompass subjectivities thus far largely illegible to white-centric geographical epistemes. As Diana Taylor (2013, p. 20) argues, "the repertoire allows for an alternative perspective on historical processes of transnational encounters and invites a remapping of the Americas, this time by following traditions of embodied practice." These repertoires are seen as different sources of resistances and the reinvention of existences, constructive of Black agencies in modernity/coloniality, and thus enable and enjoin other imaginations of space and race to be revealed and mobilized in demands, struggles, and processes of social transformation that seek to confront racism. All of these theoretical movements to engage different empirical elements as sources for spatial studies impel us to think through the methodological challenges of Black geographies we will discuss in the next section.

Methodological Challenges and Pathways of Black Geographies

The panorama of developments of Black geographies discussed above, though not exhaustive, leads us toward an open field of inquiry to understand how race relations are articulated, and how processes of racialization, Black struggles and agencies are mobilized as geographic rationales, as well as geographic and spatial tools. We contend that what delineates the field of study of Black geographies are the demonstrated ethical and political commitments to social transformation, to ending racism and its intersecting forms of oppressions. These commitments stem from the necessity of revealing oppressions and relations of power and the construction of other spatial imaginations as pathways for overcoming the current social order in its racial dimensions. Social critique thus connects with a critique of hegemonic scientific knowledge in its dominant relations with other forms and genealogies of knowledge. Such geographies therefore imply methodological challenges and displacements. In these, by means of transformative political commitments, the epistemological and methodological dimensions of the production of knowledge appear as political attitudes under the scrutiny of the impacts and effects of power. Methodologies are not seen here, however, as a form of proceeding, as a sequence of "disciplined" acts of research obediently directed by ties to approaches considered academically legitimate. We contend that methodologies are not merely a form of investigating, gathering, and processing information, but rather one component of a political orientation and position in the production of engaged knowledge. Reimagining race and space requires re-imagining methodological habits. We need methodological experimentation that troubles scientific assumptions which affirm that rigid methods ensures validity, while neglecting the political consequences of how these dominant methodologies police thought, imagination, and outcomes, and try to keep divergent thinking on its track. Accordingly, we highlight four trends that have emerged as expanding agendas in the field of Black geographies.

One aspect of reimagining engrained methodological practices concerns epistemological critique. Black geographies engage knowledge production that takes as its inspiration and antecedents an ample and diverse set of political and epistemological movements or currents in whose nexus both the project of society and the project of science are indissociable. The set of references of struggle and the production of knowledge engenders robust repertoires of Black geographies, from Black anti-racist organizing to anti-colonial struggles to Black feminisms, including recent decolonial currents and the dialogue with more dated fields of critical thought, such as strains of Marxism. From this amalgamation, which is one of the prime attributes of Black geographies, the search to break with Eurocentrism is not seen as having reference to a specific geographic location, but to all attitudes and thinking trampled in the generalization and universalization of particular social experiences of subaltern groups, leading the way to the epistemicide concurrent with such silencing. Countering such wasteful, dismissive and domineering geopolitics of knowledge, the very praxis of an epistemic project of knowledge *otherwise* valorizes the experience of social differences that subvert the consistent attempted erasure of the geographical agency of subalternized groups. Thus, more than an "epistemic vigilance" (Bourdieu et al., 1999), it is Mignolo's proposition of "epistemic disobedience" (2004), Crenshaw's call to "think from the margins" (1991), and Haraway's emphasis on "localized knowledges" (2009) that invite us to reconsider epistemological, theoretical, and political premises with profound methodological consequences. The critique of the epistemological displacements to which we have referred requires not merely the validating of new subjects, but also the centering of new sources of knowledge production as demonstrated in the previous section.

A second aspect we notice in recent methodological changes in the subfield of Black geographies places renewed emphasis on the ensemble of sources and methodologies in the domain of Black geographies and calls for drawing on archives historically disqualified by traditional methodological canons (see also Chapter 14 by Beckingham and Hodder on archives, this volume). In Black geographies, traditional sources and repositories, such as maps and censuses are resignified, repositioned, confronted, and "decentered" – challenging the monopoly of the State as producer of these objects. One example in point, is the popularizing experiences of "community-driven mapping", co-elaborated *with* the subjects of communities and context being mapped, as well as counter-mapping carried out *by* non-professional cartographers, ordinary people, who claim the tools of cartographic production. Used as tools in territorial disputes, such cartographic representation are instruments of identity and group cohesion in intentional communities that equip ordinary people to fight for rights and recognition in "cartographic activisms". Such cartographic activism goes beyond the production of maps and become the means to contest norms of the production and validation of knowledges within the fields of cartography and geography (Santos, 2011). In Brazil, religious communities of African descent who have historically been victim of state-led racism, religious intolerance and political repression, have produced mappings that resist erasure and challenge spatial imaginations. These include mappings of temples, violations, and attacks (Fonseca and Giacomini, 2013), but also mappings that seek to identify the spatiality of other groups as associative practices, shelter spaces for movement organizing, locations for the enactment of public rituals, areas for the collection of medicinal flora and worship in the forest, locations for rituals in the waters (rivers, streams, and beaches used by religious groups), and occupied territories, among others (INTECAB/PNCSA, 2005).

Beyond the displacement of the protagonism in the production of knowledge, these mappings converge with others that rupture with cartographic conventions in the search for cultural representations informed by the codes and heritage of these groups. These examples reveal a few methodological elements worth highlighting. First, beyond epistemological displacement to dissolve the barrier between subject and object in the production of knowledge through the protagonism and authority over the maps, the representation of religious, associative, and symbolic practices drive us to access the relational imagination of the spatialities of these groups. Thus, the valorization of group subjectivities, such as spiritual value ascribed to natural elements and places, necessarily translates from a methodological standpoint into the acknowledgement of their archives and forms of transmission of knowledge: orality, storytelling, photographs, memories, and performances. Chapter 13 by Clarke and Mullings provides a detailed discussion of Caribbean storytelling as anti-colonial methodology. Embodied practices and knowledge emerge as key lenses to read spatial genealogies, trajectories, memories, ancestralities, material and immaterial flows of knowledges, political influences, and more. Breaking with the distinction between repertoire and archive (Taylor, 2013), Black geographies seek rather to identify and validate an ensemble of repertoires that foster the apprehension and valuing of the social, political, economic, cultural, and epistemic agency of individuals and collectives as constituting narratives that juxtapose themselves to Eurocentric hegemonies.

Appreciation of sources, repertoires and archives leads us to a third methodological aspect in Black geographies, which is multidisciplinarity and interdisciplinarity. Disciplinary boundaries delineate a field of possibilities for in-depth development of spatial thinking, and at the same time, they impose barriers for the creation of other ways of thinking and producing other narratives. The very search for non-essentialist meanings of race and for dynamic understandings of processes of racialization already pushes geographies to establish dialogues with theoretical contributions from other disciplines. Discussions about power and epistemology demands extrapolating such boundaries. The exploration of diverse sources and archives reaffirms the relevance of interdisciplinary dialogue in forging different stances toward analysis and interpretation of the different ways in which subalternized subjects produce, store, share and transmit information in a plurality of ways – orally, through imagery, aesthetic and performative forms, among others. Interdisciplinarity is then key to the necessary methodological creativity that allows us to knit webs of knowledge, put ideas together and associate them in a relational way, as we unleash our imagination to envision anew methodological perspectives and tool, something so needed and urgent that has been denied by the hegemonic Eurocentric epistemes.

Finally, we find that the political-intellectual project of Black geographies pushes disciplinary boundaries, opens up the discipline, and invites multidisciplinary expansion while critically re-elaborating conceptual frameworks central to the field. Black geography scholars strongly emphasize spatial dimensions and spatial "rationale" to analyze race, racialization, race relations, and Black agencies. For instance, Black geographies critiques traditional ways of conceptualizing space as a totality. It troubles the idea of a "metonymic reason" – that is, a rationale that takes the parts for the whole, an analytical approach that Boaventura Santos (2004) argues is a cornerstone of supremacist Eurocentric thinking. As discussed earlier in the text, it was foundational to the discipline of geography to conflate the idea of geographical space with terrestrial space as a fixed totality. Such totality – Planet Earth – was taken to be the sum of parts – continents – and each continent was associated with a racial group in definitive, fixed, essentialist ways (Europe equals white; Africa equals Black, and so on). Therefore, traditionally, spatial thinking has fixated on the dualistic assumption of spatial separation. Simultaneously, this same political-analytical move has fixated a distinction and separation between racial groups, thus conflating each world region with supposedly stark hierarchical values attributed to each racial group.

The manipulation of the conceptual tools of geographic science has relied on applying methodologies oriented by abstraction – or rather reduction – by means of establishing selective valid sources, defining parameters to legitimize heteropatriarchal white geographical subjects and projects in detriment of non-white, non-gender conforming ones. By developing the infra-structure of dominance upon epistemic premises of distinction, separation, non-relationality, and hierarchy, traditional disciplinary paradigms have disqualified, subalternized, and silenced the racialized Black geographical subjects we seek to acknowledge and give voice to. Re-imagining space requires an epistemological endeavor to expand, envision, and create other societal projects, as Massey (2004) suggested.

We must understand the political quest and the analytical and methodological work that Black geographies and Black geographers have been doing to articulate the relational meanings of space in the domain of Black geographies. It is an invitation for all to approach spatial thinking in relational ways, which calls for understanding race in non-dualistic but in relational ways. In other words, it demands a radical shift of resource and epistemic power dynamics that subvert white privilege within the discipline and allows for multiple positionalities to reveal themselves. Thus, in our view, the goal of Black geographies is to make space for geograph-ical subjects to reflect the diversity of spatial subjectivity, relationships, and agencies to address theoretical oversights, methodological limitations, and epistemic reparation. That requires an epistemic regime that incorporates creative and expansive methodologies aimed at troubling fixed meanings, revealing interdependence, inviting self-reflexivity, and fostering epistemic co-responsibilities. Such methodologies will necessarily have to re-design methods and re-invent metrics capacious enough to comprehend and embody the rich multiplicity of geo-graphical knowledge and experiences of non-white, non-heteropatriarchal, non-conforming subjects.

Conclusion

In this chapter, we have argued that Black geographies comprise a broad and diverse field of inquiry that share a commitment with epistemic interventions to reveal and counter racist sources and effects of power-knowledge inequality. We also contended that the expanding field of Black geographies finds cohesion as a political-intellectual praxis by drawing on everyday enactments of Black sociality, on legacies of Black movements and Black intellectual traditions to articulate anti-racism narratives, vision, and projects.

Therefore, one of the central challenges at hand involves deepening a critique of dominant and fixed meanings of race while seeking a more relational understanding of race, power, and space. Such work entails methodological approaches that consider the vast plurality of epistemic subjects and grant legitimacy to sources of knowledge and information, archives, and social repertoires in their material and immaterial dimensions and worth. This rupture with the excluding methodological mandates of traditional science implies disrupting discip-linary boundaries whenever they do not serve the advancement of the field of geography. Methodological disruptions should forge spatial thinking that asserts the spatial agency, geo-graphical experience, and sense of place of Black populations vis-à-vis processes of racialization and spatialization of racism and anti-racism.

References

Alatas, S. F. (2003) Academic Dependency and the Global Division of Labour in the Social Science. *Current Sociology*, 51(6), pp. 599–613.

Almeida, A. W. B. de. (1989) Terras de preto, terras de santo, terras de índio – uso comum e conflito. *Cadernos do Naea*, 10, pp. 163–196.

Anderson, B. (2008) Comunidades imaginadas: Reflexões sobre a origem e a difusão do nacionalismo. São Paulo: Companhia das Letras.

Anjos, R. S.A. dos. (2009) Quilombos – Geografia Africana, Cartografia Étnica, Territórios Tradicionais. Brasília: Mapas Editora.

Anzaldúa, G. (1987) Borderlands: La Frontera. San Francisco: Spinters/Aunt Lute Press.

Arruti, J. M. (1999) Propriedade ou território. *Tempo e Presença,* 21(307).

Balibar, É. and Wallerstein, I. (1988) Raza, Nación y Clase. Madri: Iapala.

Bledsoe, A. and Wright, W.J. (2019) The pluralities of black geographies. *Antipode, 51*(2), pp.419–437.

Bonnet, A. (1996) Constructions of 'race', place and discipline: Geographies of 'racial' identity and racism. *Ethnic and Racial Studies* 19, pp.864–883.

Bourdieu, P. (1989) O poder simbólico. Lisboa/Rio de Janeiro: DIFEL/Bertrand Brasil.

Bourdieu, P. and Chamboredon, J.C., Passeron, J.C (1999) A profissão de sociólogo: Preliminaries epistemológicas. Petrópolis, RJ: Vozes.

Collins, P. H. (2016) Aprendendo com a outsider within: A significação sociológica do pensamento feminista negro. *Revista Sociedade e Estado*, 31 (1), 99–127.

Crenshaw, K. (1991) Mapeando as margens: Interseccionalidade, políticas de identidade e violência contra mulheres não-brancas. *Stanford Law Review*, 43 (6), 1241–1299.

Deffontaines, P. (1940) Geografia Humana do Brasil. Rio de Janeiro.

Deffontaines, P. (1944) Ensaios de divisões regionais e estudo de uma civilização pioneira: O Estado do Espírito Santo. *Boletim Geográfico*, 2 (19).

Escobar, A. (2004) 'Actores, redes e novos produtores de conhecimento: Os movimentos sociaise a transição paradigmática nas ciências', in B. S. Santos (ed.) *Conhecimento prudente para uma vida decente: Um discurso sobre as ciências revisitado*. Cortez, São Paulo.

Fonseca, D. P. R da; Giacomini, S. M. (2013) Presença do axé: mapeando terreiros no Rio de Janeiro. Pallas, Rio de Janeiro.

Fortune, A. (2018) Los Negros Cimarrones En Tierra Firme Y Su Lucha Por La Libertad, in M. A. Gandásegui, F. D. Castillo, and H. A. Carrera. *Antología Del Pensamiento Crítico Panameño Contemporáneo*, CLACSO, Argentina, pp. 309–378.

Freire, M. S. de L. (2020) Peddling Sweets and Pioneering Territory: Black women and work in Colombia's Caribbean Region. *Vibrant: Virtual Brazilian Anthropology*, 17. Retrieved on May, 2021 from https://doi.org/10.1590/1809-43412020v17d501.

Gilmore, R. W. (2002) Fatal couplings of power and difference: Notes on racism and geography. *The Professional Geographer*, 54(1), 15–24.

Gilroy, P. (2002) O Atlântico Negro: modernidade e dupla consciência. Rio de Janeiro: 34/Universidade Cândido Mendes.

Gomes, Â. (2012) Etnobotânica e territorialidades negras urbanas da Grande Belo Horizonte: Terreiros e quintais, in R. E. dos Santos (Ed). *Questões urbanas e racismo*, 1ed, DP et Alii, Petropolis, pp. 276–308.

Gothan, K. F. (2000) Urban Space, restrictive covenants and the origins of racial segregation in a US City, 1900-50. *International Journal of Urban and Regional Research*, 24 (3).

Grosfoguel, R. (2010) Para descolonizar os estudos de economia política e os estudos pós-coloniais: Transmodernidade, pensamento de fronteira e colonialidade global, in B. S. Santos and M. P. Meneses (Ed). *Epistemologias do Sul*, Cortez, São Paulo.

Guimarães, A. S. A. (2018) Formações nacionais de classe e raça, in A. Baroni and F. Rios (Eds). *Negros nas cidades brasileiras (1890-1950)*, Intermeios, Fapesp, São Paulo.

Haesbaert, R. Território(s) numa perspectiva latino-americana. *Journal of Latin American Geography*, 19 (1), 141–151.

Haraway, D. (2009) Saberes localizados: A questão da ciência para o feminismo e o privilégio da perspectiva parcial. *Cadernos Pagu* 5, pp. 7–41. Retreived from https://periodicos.sbu.unicamp.br/ojs/index.php/cadpagu/article/view/1773 on June 5, 2021.

Jackson, P. (1987) 'The idea of 'race' and the geography of racism', in Peter Jackson (Ed). *Race & Racism: Essays in Social Geography*. London: Allen & Unwin.

Lander, E. (2005) 'Ciências sociais: Saberes coloniais e eurocêntricos', in Edgardo Lander (Ed.), A colonialidade do saber: eurocentrismo e ciências sociais, perspectivas latino-americanas. Buenos Aires: Consejo Latinoamericano de Ciencias Sociales – CLACSO.

Lao-Montes, A. (2019) Metrópolis Negras de Benin a Río de Janeiro y de Harlem a La Habana: Modernidades Afroamericanas y Cosmopolitismos Subalternos. Conferencia Magistral, Casa de las Americas, La Habana, Cuba, June.

Láo-Montes. A. (2020) Hacia una Analítica de Formaciones Étnico-Raciales, Racismos y Política Racial. Contrapunteos diaspóricos, cartografías políticas de nuestra Afroamérica. Universidad Externado de Colombia, March, 3rd.

Lipsitz, G. (2007) The racialization of space and the spatialization of race: Theorizing the hidden architecture of landscape. *Landscape Journal*, 26(1), 10–23.

Masssey, D. (2004) Filosofia e política da espacialidade: Algumas considerações. Revista *GEOgraphia*, 6 (12).

McKittrick, K. (2011) On plantations, prisons, and a black sense of place. *Journal of Social and Cultural Geography*, 12(8), 947–963.

Mignolo, W. (2003) Histórias locais / projetos globais: Colonialidade, saberes subalternos e pensamento liminar. Belo Horizonte: Editora UFMG.

Mignolo, W. (2004) 'Os esplendores e as misérias da 'ciência': Colonialidade, geopolítica do conhecimento e pluri-versalidade epistémica', in B.S. Boaventura (Ed.), *Conhecimento prudente para uma vida decente: um discurso sobre as ciências revisitado*, Cortez, São Paulo.

Monbeig, P. (1984) Pioneiros e fazendeiros de São Paulo. Tradução Ary França e Raul de Andrade e Silva. São Paulo: Hucitec/Polis.

Moore, C. (2007) Racismo & Sociedade: novas bases epistemológicas para entender o racismo. Belo Horizonte: Mazza.

Nascimento, A. do. (1980) O quilombismo: documentos de uma militância pan-africanista / Abdias do Nascimento. Petrópolis: Vozes.

Nascimento, A. do. (2016) O genocídio do negro brasileiro: processo de um racismo mascarado. São Paulo: Editora Perspectiva.

Offen, K. (2003) The territorial turn: Making black territories in Pacific Colombia. *Journal of Latin American Geography* 2(1), 43–73.

Omi, M. and Howard, W. (2014) [1986, 1st ed.]. Racial Formation in the United States. Routledge, New York.

Oslender, U. (2002) Espacio, lugar y movimientos sociales: Hacia una espacialidad de resistencia. Scripta Nova. *Revista electrónica de geografía y ciencias sociales*, Universidad de Barcelona, 6 (115) www.ub.es/geocrit/sn/sn-115.htm [ISSN: 1138-9788]

Quijano, A. (2010) Colonialidade do poder e classificação social, in B. de S. Santos and M. P. Meneses (Eds.), *Epistemologias do Sul*, Cortez, São Paulo.

Sánchez, A. G. (2012) Espacialidades del destierro y la re-existencia: Afrodescendientes desterrados en Medellín, Colombia. Medellín: La carreta Editores, Instiuto de Estudios Regionales, INER, Universidad de Antioquia.

Sandercock, L. (1998) *Introduction: Framing Insurgent Historiographies for Planning. Making the Invisible Visible. A Multicultural Planning History.* Los Angeles: University of California Press.

Sansone, L. (1996) Nem somente preto ou negro: O sistema de classificação racial no Brasil que muda. *Afro-Ásia,* 18, 165–188.

Santos, S. B. (Ed.), (2004) Conhecimento prudente para uma vida decente: Um discurso sobre as ciências revisitado. Cortez: São Paulo.

Santos, R. E dos. (2009) Rediscutindo o ensino de Geografia: Temas da Lei 10.639. Rio de Janeiro: CEAP.

Santos, R. E. dos. (2011) Ativismos cartográficos: Notas sobre formas e usos da representação espacial e jogos de poder. Revista Geografica de America Central (online), v. 2, pp. 1–17.

Santos, R. E. dos. (2012) Sobre espacialidades das relações raciais: Raça, racialidade e racismo no espaço urbano. In: Renato Emerson dos Santos. (Org.). Questões urbanas e racismo. 1ed. Petrópolis: DP et Alii. v.1, pp. 36–66.

Santos, R. E. dos (2015) Soeterik, Inti Maya. Scales of political action and social movements in education: the case of the Brazilian Black Movement and Law 10.639. *Globalisation, Societies and Education,* v. 13, pp. 1–19.

Santos, R. E. dos. (2019) 'Repertórios espaciais de ação na luta anti-racismo: O caso da Pequena África no Rio de Janeiro', in F. Sánchez and P. C. Moreira (Eds.), *Cartografias do conflito no Rio de Janeiro*, 1ed, Letra Capital, Rio de Janeiro, pp. 12–27.

Santos, R. E. dos. (2018) 'Uma leitura sobre espacialidades das relações raciais: Raça, racialidade e racismo no espaço urbano', in F. L. de Oliveira; J. R. Lima, P. Novais de. (Eds.), *Território e planejamento: perspectivas transdisciplinares*, 1ed: Letra Capital, Rio de Janeiro, pp. 345–366.

Santos, R. E. dos; Silva K. S.; Ribeiro, L. P.; Silva, N. C. (2018) 'Disputas de lugar e a Pequena África no Centro do Rio de Janeiro: Reação ou ação? Resistência ou r-existência e protagonismo?', in N. Rena; D. Freitas; A. I. Sá and M. Brandão (Eds) *Seminário Internacional Urbanismo Biopolítico*, 1st. ed, Fluxos, Belo Horizonte, pp. 464–491.

Scherer-Warren, I. (1993) Redes de movimentos sociais. São Paulo: Edições Loyola.

Sodré, M. (1988) O terreiro e a cidade: A forma social negro-brasileira. Petrópolis: Vozes.

Soja, E. (1996) *Thirdspace: Journeys to Los Angeles and Other Real-And-Imagined Places*. Massachusetts: Blackwell Publishers.

Swynguedouw, E. (1997) 'Neither global nor local: Glocalization and the politics of scale', in K. Cox (Ed.), *Spaces of globalization: reasserting the power of the local*, The Guilford Press, New York/London, pp. 137–166.

Taylor, D. (2013) O arquivo e o repertório: Performance e memória cultural nas Américas. Editora UFMG.

Wieviorka, M. (2007) O racismo, uma introdução. São Paulo: Perspectiva.

16

DIGITAL GEOGRAPHIES AND EVERYDAY LIFE

Space, Materiality, Agency

Casey R. Lynch and Bahareh Farrokhi

Introduction

Geographers have long examined the emergence and evolution of digital technologies and their roles in broader socio-spatial relations, from early reflections on the nature of "cyberspace" (Batty, 1993) and debates about the role of Geographic Information Systems (GIS) in society (Sheppard, 1995) to the production of gendered relationships in information and communication technologies (ICT) (Holloway et al., 2000). More recently, the growth of research in this area has led some to highlight a 'digital turn' in the discipline characterized by geographies produced through, produced by, and of the digital (Ash et al., 2018). Digital geographies has emerged as a broad current of inquiry encompassing a range of research on 'smart' spaces, Big Data, algorithms, platforms, social media, digital infrastructures, and robots, and their various roles in shaping spatial relations, the production of economies and labor practices, exercises of power and resistance, affective experiences of space, and processes of subject formation, among other topics. The breadth of this literature reflects the multiplicity of ways digital technologies are permeating and transforming everyday life (Duggan, 2017). For Henri Lefebvre (2014), a focus on 'everyday life' grounds critique in concrete, material reality—in the rhythms and spaces where individuals and collectives encounter and negotiate the conditions of life. A focus on the 'everyday' in digital geographies thus aims to understand how extensive processes of digitalization and technological change are produced, experienced, negotiated, and potentially contested unevenly by differentially-situated subjects in place.

This chapter examines some of the major theoretical and methodological developments and challenges associated with this rapidly expanding literature in human geography. Questioning the evolving role of digital systems in everyday life requires rethinking ontological and epistemological assumptions around space and place, materiality and immateriality, and technological and posthuman forms of agency. Geographers are increasingly exploring local, personal, and embodied socio-technical practices in their entanglement with vast and complex infrastructures and systems. As they do so, geographers face the dual challenge of conducting situated, engaged research with a diversity of subjects while also striving to make sense of increasingly complex, opaque, and spatially-extensive technological systems controlled largely by private companies operating across the globe. This chapter explores how the re-theorization of concepts like

DOI: 10.4324/9781003038849-19

space, materiality, and agency implies new challenges and opportunities for methodological practice around the digital geographies of everyday life.

Socio-Technics of Everyday Life

Early commentators on networked technologies and their impacts argued that their development spelled the 'death of geography.' As Morgan (2004, p. 4), explains: "it is often assumed that space-time relations have been so radically compressed that it is possible to completely annihilate space with time." Others sought to theorize the Internet as a new kind of space— cyberspace—"which is layered on top of, within, and between the fabric of traditional geographical space" (Batty, 1993, p. 616), or to delineate among the various "virtual geographies" emerging in the computer age. Batty (1997) distinguished among cspace (or computer space) as the internal space of computers and computer networks, cyberspace as the space of networked digital communication, and cyberplace as "the impact of the infrastructure of cyberspace on the infrastructure of traditional place" (p. 340). Cyberspace was seen to produce unique methodological challenges as it is "largely invisible to conventional methods of observation and measurement, and this poses enormous challenges to the very way we might engage in developing a science of this phenomenon, not to mention its design" (Batty, 1993). These early commentaries highlight how digitalization has long prompted retheorizations of space. Does the trend toward faster and more extensive forms of networked communication spell the annihilation of space? Does cyberspace constitute a unique kind of space? If so, what is its relation to "geographical space," traditional conceptions of place, or the internal spaces of computers? And how might these relationships be studied in practice?

Over the past several decades, geographers have approached these questions from numerous angles and in some cases challenged their underlying premises. Geographers have critiqued the spatial metaphors commonly used to describe computer networking and challenged the distinction between cyberspace and geographical space, instead exploring their ongoing and evolving entanglement or co-constitution (Graham, 2013). Thrift and French (2002), for instance, explore the ways computer code increasingly helps determine how the spaces and rhythms of life are produced and maintained, reshaping "what can show up in the everyday world" (p. 310) and what cannot with clear implications for how we conceive of the possibilities for political action. Similarly, Kitchin and Dodge (2014) "detail how software produces new ways of doing things, speeds up and automates existing practices, reshapes information exchange, transforms social and economic relations and formations, and creates new horizons for cultural activity" (p. 3). Increasingly, geographers cannot begin to understand how space is produced and reproduced without understanding the extensive infrastructures of data production and algorithmic assemblages that continually (re)direct flows of information, people, capital, and resources across space.

Further, while most everyone would agree that early pronouncements of the annihilation of space were overly simplistic, geographers have paid close attention to the ways space-time relations are reordered through everyday socio-technical processes and practices. Innovations in computing, including the emergence of the smart phone and the rise of new digital platforms, for instance, have reordered the space-time relations of work, consumption, and socialization, among other realms of life (Barns, 2019). Reflecting on these practices, Leszczynski (2018, p. 18) proposes "mediation as a conceptual framework for understanding the multiple yet contingent comings-together of technology, people, and place and space that are productive of our quotidian lived realities." In other words, the theoretical lens has shifted from an early focus on the relationship between an assumed 'virtual' space and physical space to the ways networked technologies increasingly produce or mediate spatio-temporalities in their complex entanglements in everyday life.

At the same time, others have examined how digital technologies do not just produce and mediate everyday spatial processes, but themselves are produced by and dependent upon the specific spatio-materialities of digital infrastructures (Pickren, 2018; Plantin & Punathambekar, 2019). In their discussion of data center development in the Pacific Northwest of the United States, Levenda and Mahmoudi (2019) examine how contemporary digital capitalism is dependent upon the exploitation of constructed "natures"—including cheap land, cheap water, and cheap energy. They explain how "although data are intangible, data require storage and communication, which necessitates material and tangible infrastructures. These infrastructures bring rare-earth minerals, electricity infrastructures, laboring bodies, fiber optic cables, knowledge, code, water, and fossil fuel together in a disturbing whole" (Levenda & Mahmoudi, 2019, p. 5). The assembling of these infrastructures differentially reshapes everyday life for communities in the sites of data center development, mineral extraction, or energy production, at the same time that those data centers are key to the functioning of digital systems designed to intervene in the everyday practices of people around the globe.

This work raises important questions not just about the relationship between virtual and physical space, but also about the complex im/materialities of digital life. Scholars have sought to challenge common narratives about the supposed immateriality of digital information or processes, calling for increased attention to the infrastructures on which digitality depends, the forms of control exercised through them, and the inequalities they produce through their development and deployment (Kinsley, 2014; Lynch, 2021). For instance, Kinsley (2014) argues: "It is important to see the agency of code and data, evoked as 'virtual' technologies, as inextricably tied to the appliances and infrastructures through which they are materially expressed" (p. 366). Along this same vein, Pink and Fors (2017b) question "the emergent and processual nature of the relationality of the digital and material as entangled within the same 'process of becoming.'" These scholars and others shift attention away from analyses of individual technologies and toward the processes of technological development and deployment. In doing so, geographers have been sought to highlight the ways 'the digital' materializes differently across space.

As geographers have reconceived space to account for the evolving role of digital technologies in shaping it, and as they have increasingly recognized the complex im/materialities through which 'the digital' manifests, they have also reconsidered assumptions about the nature of agency and the different capacities among humans and technologies. In particular, many geographers have come to stress the agential abilities of digital systems, drawing insights and inspiration from a broad range of theorists and concepts, from Bernard Steigler's (1998) view of the co-evolution of humanity and technics, to Bruno Latour's (2005) Actor Network Theory, Deleuze and Guattari's (1987) assemblage theory, and Donna Haraway's (1987) notion of the cyborg, among others. This work also draws on longer histories of posthuman, vitalist, and object-oriented theories across the discipline (Meehan et al., 2014; Panelli, 2010). These theoretical developments have shifted the subjects and objects of digital geographic research, as scholars have sought new ways to engage non-human entities that recognize their agential capacities to shape the world beyond and in conjunction with the intentions and actions of 'human' subjects. For instance, Ash's (2015: 89) exploration of inorganically organized objects theorizes "technical objects as having a homeostatic autonomy, generated by a series of material thresholds, which in turn frame their capacity to affect and be affected." Further, a recognition of the agency of technical objects is clear in the language used to describe networked and software-enabled spaces as 'smart,' 'sentient,' or 'thinking'—despite a general lack of deeper theorizing about what these terms might imply (Lynch & Del Casino, 2020).

Others have questioned the focus on the agency of digital technologies at the expense of exploring what Gillian Rose (2017) calls "posthuman agency." Rose explains how theories of agential objects and technical systems rightly aim to correct for centuries of writing centered on a purportedly sovereign 'Human' subject. Yet, in the move to examine the agency of nonhuman, technical objects, geographers have largely failed to reconceptualize the possibilities for always-differentiated forms of (post)*human* agency. Rose highlights how human agency is presented as a kind of excess, remainder, or resistance to the agency of digital systems in much of the work in digital geographies. In contrast, a fully posthuman account of agency must account for the co-production of differentially human and technical forms of agency. Drawing on the work of Bernard Stiegler, Rose (2017) understands post*human* agency as "necessarily coproduced with digital technics" through practices of reinvention that also actively (re)produce differentiations among posthumans. These practices of reinvention may reinforce existing differentiations based on race, gender, or class, while also giving rise to new, emergent differentiations based on the algorithmic logics and the spatialities and temporalities they help produce.

This concern with differentiated agency and inequality in relation to digital systems, however, is not new. Scholars have long been interested in questions of the "digital divide" understood as inequality in access, understanding, and effective usage of emerging digital technologies. This work has highlighted the need to recognize the unequal and differentiated nature of human-technological relations. The causes of inequality highlighted in work on the digital divide range from the high cost of technology or lack of availability in particular places, to a lack of skills and education needed to employ those technologies. In the United States, scholars have highlighted people with low incomes, racialized communities, women, inner city and rural residents and the elderly as groups that have been historically disadvantaged in relation to ICT (Gilbert et al., 2008; Mossberger et al., 2003). Others have examined how relationships to ICT and notions of technological knowledge and expertise have been produced along gendered and racialized lines, as well as how those relationships might be remade (Gilbert, 2010; Holloway et al., 2000; Lynch, 2020b). Thus, while geographers have sought to explore relationships among human and technological agents, there is no single, universal 'Human' to which they can look. Digital geographies of everyday life must always be aware of the existing and emergent inequalities that mediate and are mediated by entanglements with digital technologies and systems.

Methodological Practices

The theoretical developments in digital geographies discussed above have highlighted the evolving nature of space and place as increasingly mediated by digital systems, the complex im/materialities of digital life, and the distributed and unequal forms of posthuman and techno-logical agency. These developments raise major questions for methodological practice as geographers work to make sense of such relationships in the spaces of everyday life. Faced with these challenges, geographers are developing innovative methodological approaches allowing them to critically investigate the production and function of particular technological systems, everyday socio-technical practices and experiences, and the possibilities for these relations to be remade through place-based digital activism and organizing.

Scholars of digital geographies have called for critical, qualitative, and ethnographic studies focused on specific aspects of broader socio-technical assemblages, including software code (Kitchin & Dodge, 2014), algorithms (Christin, 2020), platforms (Fields et al., 2020), and databases (Burns & Wark, 2019)—recognizing their agential capacities in producing the spatialities, temporalities, and exclusions of everyday life. For instance, in their presentation of software studies, Kitchin and Dodge (2014) call for "techniques that focus on the processes and practices of code, and which reveal its discursive and material constitution and effects," which might include "genealogies, ethnographies, observant participation, and envisioning using mapping and spatialization techniques" (p. 255). In a similar way, Burns and Wark (2019) call for database ethnographies "observing the ways such infrastructures frame knowledges and behavior, and the ways they translate phenomena into data" (p. 600). These research approaches call on scholars to spend significant time exploring and analyzing the inner workings of technical apparatuses in order to elucidate how they operate and come to have an impact on the world.

Much of this research aims to "open the black box of processors and arcane algorithms to understand how software … does work in the world by instructing various technologies how to act" (Kitchin & Dodge, 2014, p. 13). Yet, more recently some have questioned the ability of critical social scientists and others to realistically "open the black box" as well as the utility of such an approach for offering an account of socio-technical processes. Fields et al., (2020), for instance, argue that "being unable to see inside the black box is not necessarily such 'a profound epistemic problem', because opacity is 'a basic condition of human life' (Bucher, 2016, pp. 86–87)" (p. 463). Geographers and others critically studying phenomena like "platform urbanism"—a recognition of the increasingly central role of social media and platforms like Airbnb, Deliveroo, or Uber in urban sociality, governance, and economics (Barns, 2019)—should then reflect on this opacity and how it operates in the world. Fields et al., (2020), for instance, outline three methods for studying platforms and their effects in everyday life: narratives, counter-mapping, and proxying. Narrative methods, for instance, focus on telling multiple stories from different perspectives about the processes of technological change and their differential impacts across space and time, "offering potential for evaluating the multiplicity of political sites through which platform urbanism takes place" (Fields et al., 2020, p. 464). Meanwhile, counter-mapping, such as in the case of the Anti-Eviction Mapping Project (Maharawal & McElroy, 2018), "offers potentials for subversion and transgression of the workings of platform urbanism by situating digital platforms in the experiences of those who both help comprise platform urbanism and are its potentially unwilling subjects" (Fields et al., 2020, p. 464). Finally, proxying aims to trace "data as it moves (or finds itself obstructed) between social actors, institutions and sites" and "emphasizes looking at the material conditions and implications of data deployment through contingent and contested social practices, settings, and institutional arrangements." In this approach, the opacity of the inner workings of platforms is not a problem, as researchers find opportunities to observe, interact with, and follow digital platforms in practice.

In a similar way, Louise Amoore's extensive work on algorithms has involved detailed analyses of the internal logics and processes of specific algorithmic assemblages, though she also stresses that the desire for transparency or to open the black box is inherently limiting (Amoore, 2018, 2020). Instead, researchers need "to be attentive to the accounts algorithms are giving of themselves" which necessarily begins "with the intractably partial and ungrounded accounts of humans *and* algorithms" (Amoore, 2020, p. 20). That is to say, researchers need to follow the deployment of algorithms in practice to understand how they come to shape the world. This includes moments when algorithmic assemblages may not operate as expected or prompt emergent phenomena in their deployment in practice. Leszczynski (2020) refers to these moments as examples of "glitch," such as when Uber's surge pricing for rides from New York City's John Fitzgerald Kennedy Airport during 2017 protests against discriminatory immigration policies led to the social media campaign #deleteUber. Regarding such moments, Amoore (2020) argues that "the appearance of a moment of madness [sic] is a valuable instant for an ethicopolitics of algorithms because this is a moment when algorithms give accounts of themselves" (p. 23). These are moments when we might gain a better understanding of the logics of the algorithm, its affordances and its limitations. Thus, while Amoore examines the internal logics and functions of digital systems, this approach is coupled with insights from interviews and participant observation with engineers and decision-makers and documentary evidence of institutional contexts and processes to understand how they are deployed in practice. In many ways this approach resembles the ethnographies of Science and Technology Studies (STS) scholars, employing a broad range of methods in order to trace out the production of scientific and socio-technical apparatuses and the ways they come to exercise authority and produce the world in particular ways (Hess, 2001; Seaver, 2017).

Others have sought to examine key digital assemblages by tracing genealogies of their production and deployment in particular social, political, and institutional contexts. For instance, Jefferson (2020) adopts a genealogical approach to trace the historical development of contemporary forms of hi-tech racialized policing in the United States, including the rise of predictive policing systems, electronic monitoring, and crime data centers. Using traditional archival methods, Jefferson traces the emergence of standardized crime reporting systems beginning in the nineteenth century; the emergence of relational databases, crime mapping and GIS in the 1960s; the beginning of the War on Crime and urban de-industrialization; and the development and evolution of digital, data-driven police apparatuses in New York and Chicago since the 1990s. In doing so, Jefferson highlights the evolution of data-driven logics in policing, the social and political production of policing data, and the shifting constellation of actors involved in the development of these apparatuses, from local, state, and national officials, to private industry and academic researchers. This genealogical approach destabilizes teleological and techno-determinist narratives about the production of the digital assemblages that work to (re)produce everyday life—in this case, for the millions of individuals caught up in the vast apparatuses of digitally-enabled racial criminalization.

Jefferson's genealogical approach is situated within a longer history of critical research around GIS that has sought to develop new methodological practices that account for the social and political nature of all data production and analyze the role of GIS in (re)shaping everyday realities. In this vein, since the 1990s participatory and public-participatory GIS (PGIS or P/PGIS) have explored uses of GIS technology in place-based community projects and research. PGIS aims to integrate "local and indigenous knowledge with 'expert data'" (Dunn, 2007, p. 619) to make GIS applications more useable for a greater diversity of people and purposes. For instance, PGIS approaches may integrate local knowledge such as stories, experiences and cultural values with conventional maps. Such approaches allow researchers to follow the ways communities might make use of digital information and platforms, though they can also be used to trace the impacts of technological change and digital economies on local communities themselves. For instance, in the Anti-Eviction Mapping Project, Maharawal & McElroy (2018) utilized PGIS to map the sites of evictions in San Francisco's Mission District in relation to private bus stops used to transport workers to tech company campuses outside the city. In recent years, the possibilities for this kind of research have rapidly expanded. As Kar et al., (2016, p. 296) explain, "dramatic changes in Information and Communication Technologies (ICT) and terminology have obliged P/PGIS to situate itself in the era of Web 2.0 and prove its continued value." New geospatial software and platforms such as GoogleMaps, Google Earth, free and open-source GIS, and location-enabled social networking sites like Foursquare or Twitter offer new opportunities to engage a diversity of individuals and perspectives in research around geospatial data in their everyday lives (Kar et al., 2016).

Expanding on the methods discussed above, several geographers have conducted ethnographies that recognize and reflect upon the co-constitution of the 'human' and the technical, the im/materialities of digital life, and the entanglement of virtual and physical space. Duggan

(2017) argues that despite its legacy of anthropocentrism, "ethnography is an inherently processual way of thinking about and studying the sociocultural practices of the world" (p. 2) and that it "is adapting to the multifarious ways in which 'the digital' is now intertwined with everyday life" (p. 8). Such an approach combines 'traditional' ethnographic methods, such as long-term participant observation and interviews, with methods that aim to understand the inner logics of particular technologies or apparatuses. For instance, studying smart home technology, Maalsen (2020) employs "smart devices as co-ethnographers, suggesting that their ability to be in places and at times not available to us as researchers, can give us an important insight into the smart home" (p. 1545). This is similar to Pink and Fors (2017a) research on individuals employing self-tracking technologies like FitBits and exercise or sleep-tracking mobile platforms. They employed a combination of 'sensory ethnography interviews,' video-reenactment methods, and auto-ethnographic self-tracking in order to "collaborate with participants to co-produce understandings of tacit or unspoken ways of knowing" (Pink & Fors, 2017a, p. 385). This ethnographic approach "sees the digital-material as constituted within and as part of the wider configurations of things and processes, including bodies" (p. 379), reflecting broader trends toward posthuman ethnographies (Kipnis, 2015; Whitehead, 2009) and demonstrating how geographers creatively adapt ethnographic methods in exploring the roles of digital technologies in everyday life.

Digital Geographies in Action: Researching Technological Sovereignty in Barcelona

To demonstrate the way researchers combine methods to consider the multiple agencies, im/materialities, and complex spatialities of digital life, this final section discusses the experiences of one of the authors, Casey Lynch, in conducting ethnographic research on community groups organizing for "technological sovereignty" in Barcelona. The work examined the goals and practices of a loose network of community technology projects focused on experimenting with decentralized, democratic control over processes of digital development through the use of open-source technologies in place-based alternative economies (Lynch, 2020a, 2020b, 2021). This included platform cooperatives, public 3D printing labs, and community-managed internet infrastructures, among others. The goal of the research was to understand the possibilities for counter-hegemonic models of digital development that contest the narrow logics and imaginaries of the smart city. The project employed traditional ethnographic methods like participant observation and interviews but combined these with direct experimentation with specific technologies and systems. In doing so, he aimed to both "open the black box" to understand their capacities and affordances and to trace their effects and entanglements in practice.

While much of the research involved 'traditional' ethnographic methods, it is important to recognize that even these practices are increasingly transformed by digital technologies. In Lynch's research in Barcelona, many of the groups organizing for technological sovereignty met regularly in-person (hosting meetings or community events) while also maintaining multiple channels of digital communication and collaboration. These included group chats or channels on messaging apps like WhatsApp, Telegram, or Signal, email listservs, online message boards, shared documents for collaboration on specific projects, video tutorials on YouTube, and a variety of project websites. Conducting ethnographic research thus involved navigating these hybrid spaces in order to fully understand the complexity of the groups' practices.

This mix of in-person and digital communication and coordination may have been somewhat more pronounced in such tech-centered projects but is no way unique to this case. While traditional forms of ethnography have long privileged face-to-face encounters as somehow more 'authentic' or 'real,' or virtual worlds as separate and unique in relation to the 'real world,' researchers are increasingly recognizing the need to rethink "the field" as hybrid and unbounded (Przybylski, 2021). As so many aspects of everyday life become mediated through digital devices and systems, ethnographic practice regardless of research topic needs to adapt to this reality and reflect on its significance for social life. In Lynch's research in Barcelona, recognizing this reality allowed him to engage communities in multiple ways, trace the emerging practices through which different groups and initiatives organized themselves, and reflect on the significance of different actors' or groups' digital practices in relation to their broader project and goals. Lynch was able to see how different groups' ideals and visions of alternative digital futures impacted their uses of messaging apps based on concerns of privacy and accessibility, their use of open-source tools for collaborative work, or of wikis to co-produce and share information among the community.

In addition to these more traditional methods, Lynch aimed to understand the capacities and affordances of particular technologies through direct engagement and experimentation—exploring how particular technologies work and the logics on which they are based through a critical ethnographic lens. In one example of this, his research into the community-managed internet infrastructure, Guifi.net, involved independently studying the basics of wireless communication, the capacities of specific models of wireless antennas, and the different approaches to organizing wireless networks. He read technical reports and manuals as well as guides written specifically by and for Guifi.net participants. This experience helped him reconcile the expressed social, political, and economic goals of the project—to create a commons-based, decentralized broadband network—with the specific capacities of the technological systems in question.

Lynch also set up an internet connection through the community infrastructure in his own home, purchasing the required hardware, reviewing installation guides, coordinating with neighbors to access the network, and experimenting with different physical arrangements. Over the course of two years, he worked to maintain this connection—reorienting the antenna after strong winds, replacing damaged chords, and coordinating with neighbors to resolve internet outages. This experience allowed him to directly explore the complex im/materialities and spatiality of internet connectivity, while also recognizing how his own agency was entangled with and co-constituted by the agency of a range of non-human actors—from the assemblage of hardware and software on which the network relies, to the physical layout of the city and individual buildings, to the wind. This experience was key to understanding the broader practices, opportunities, and challenges of the Guifi.net project, such as the difficulty for the network to expand to the Gothic Quarter of Barcelona. When approached to install a node (internet connection point) in a community center in the neighborhood, Guifi.net organizers quickly recognized that the physical characteristics of the neighborhood, the limitations of wireless antennas, and the decline of long-term residents in the area due to the tourist economy and Airbnb made the project unfeasible (Lynch, 2021). Lynch's own experimentation with the technology in question allowed him to make sense of these and other moments encountered in his ethnographic research as specific technologies became entangled in the everyday lives and practices of community projects in Barcelona.

Conclusion

While not contained within a defined sub-discipline, research on digital geographies—broadly conceived—has seen remarkable growth over the past several years prompted by the unprecedented speed of socio-technical change brought on by the "digital revolution" and its effects across nearly all aspects of everyday life. In order to make sense of these changes, geographers have been forced to rethink previous epistemological and ontological assumptions about the nature of space, the im/materiality of digital life, posthuman and technological forms of agency, and the inequalities with which they are entangled.

Recognizing the complexity of digitally-mediated life has led geographers to experiment with a broad range of research methods and practices. Some have sought methods for "opening the black box" of particular technologies to better understand their internal logics and thus how they produce certain outcomes. Others have rather examined how certain technologies come to work in the world—aiming to better understand particular technical assemblages through their actions and effects in practice. Genealogical approaches allow geographers to question hegemonic narratives of linear, teleological digital development and highlight the unequal social, political, and economic contexts in which particular technologies come to be. Finally, ethnographic and participatory research practices, often coupled with critical analyses of digital systems, allow geographers to explore the experiential, more-than-human, spatially-uneven, and occasionally contested nature of digital change in particular places or among particular communities.

Overall, the broad and rapidly evolving nature of research on digital geographies and everyday life means that researchers are likely to continually face new and difficult challenges in their research practices. Geographers will continue to grapple with new and emerging technologies as they come to reshape everyday processes, practices, and spaces through their entanglement of differentiated forms of posthuman agency. As new questions arise, geographers should maintain an openness to experimentation, hybridity, creativity, and transdisciplinary collaborations, as well as an appreciation for multiple ways of knowing. As Kinsley et al., (2020) argue, the emergence and development of digital geographies as an area of inquiry across the discipline "can and should be an opening out, not a closing down."

References

Amoore, L. (2018) Cloud geographies: Computing, data, sovereignty. *Progress in Human Geography*, *42*(1), 4–24.

Amoore, L. (2020) *Cloud Ethics: Algorithms and the Attributes of Ourselves and Others.* Duke University Press Books.

Ash, J. (2015) Technology and affect: Towards a theory of inorganically organised objects. *Emotion, Space and Society*, *14*, 84–90.

Ash, J., Kitchin, R., and Leszczynski, A. (2018) Digital turn, digital geographies? *Progress in Human Geography*, *42*(1), 25–43.

Barns, S. (2019) Negotiating the platform pivot: From participatory digital ecosystems to infrastructures of everyday life. *Geography Compass*, *13*(9), e12464.

Batty, M. (1993) The geography of cyberspace. *Environment and Planning B: Planning and Design*, *20*(6), 615–616.

Batty, M. (1997) Virtual geography. *Futures*, *29*(4), 337–352.

Burns, R., and Wark, G. (2019) Where's the database in digital ethnography? Exploring database ethnography for open data research. *Qualitative Research*, 1468794119885040.

Christin, A. (2020) The ethnographer and the algorithm: Beyond the black box. *Theory and Society*. https://doi.org/10.1007/s11186-020-09411-3

Deleuze, G., and Guattari, F. (1987) *A Thousand Plateaus: Capitalism and Schizophrenia.* University of Minnesota Press.

Dunn, C. E. (2007) Participatory GIS — a people's GIS? *Progress in Human Geography*, *31*(5), 616–637.

Duggan, M. (2017) Questioning "digital ethnography" in an era of ubiquitous computing. *Geography Compass*, *11*(5), https://doi.org/10.1111/gec3.12313

Fields, D., Bissell, D., and Macrorie, R. (2020) Platform methods: Studying platform urbanism outside the black box. *Urban Geography*, *41*(3), 462–468.

Gilbert, M. (2010) Theorizing digital and urban inequalities: Critical geographies of "race", gender, and technological capital. *Information, Communication & Society*, *13*(7), 1000–1018.

Gilbert, M. R., Masucci, M., Homko, C., & Bove, A. A. (2008) Theorizing the digital divide: Information and communication technology use frameworks among poor women using a telemedicine system. *Geoforum*, *39*(2), 912–925.

Graham, M. (2013) Geography/internet: Ethereal alternate dimensions of cyberspace or grounded augmented realities? *The Geographical Journal*, *179*(2), 177–182.

Haraway, D. (1987) A manifesto for Cyborgs: Science, technology, and socialist feminism in the 1980s. *Australian Feminist Studies*, *2*(4), 1–42.

Hess, D. (2001) Ethnography and the development of science and technology studies. In P. Atkinson, A. Coffey, S. Delamont, J. Lofland, & L. Lofland, *Handbook of Ethnography* (pp. 234–245). SAGE Publications Ltd.

Holloway, S. L., Valentine, G., and Bingham, N. (2000) Institutionalising technologies: Masculinities, femininities, and the heterosexual economy of the IT classroom. *Environment and Planning A*, *32*(4), 617–633.

Jefferson, B. (2020) *Digitize and Punish: Racial Criminalization in the Digital Age*. University of Minnesota Press.

Kar, B., Sieber, R., Haklay, M., & Ghose, R. (2016) Public participation GIS and participatory GIS in the era of GeoWeb. *The Cartographic Journal*, *53*(4), 296–299.

Kinsley, S. (2014) The matter of "virtual" geographies. *Progress in Human Geography*, *38*(3), 364–384.

Kinsley, S., McLean, J., and Maalsen, S. (2020) Editorial. *Digital Geography and Society*, *1*, 100002.

Kipnis, A.B. (2015) Agency between humanism and posthumanism: Latour and his opponents. *HAU: Journal of Ethnographic Theory*, *5*(2), 43–58.

Kitchin, R., and Dodge, M. (2014) *Code/Space: Software and Everyday Life*. The MIT Press.

Latour, B. (2005). *Reassembling the Social: An Introduction to Actor-Network-Theory*. Oxford University Press.

Lefebvre, H. (2014) *Critique of Everyday Life: The Three-Volume Text*. Verso.

Leszczynski, A. (2018) Spatialities. In *Digital Geographies* (pp. 13–23). SAGE.

Leszczynski, A. (2020) Glitchy vignettes of platform urbanism. *Environment and Planning D: Society and Space*, *38*(2), 189–208.

Levenda, A.M., and Mahmoudi, D. (2019) Silicon forest and server farms: The (urban) nature of digital capitalism in the Pacific Northwest. *Culture Machine*, 1–14.

Lynch, C.R. (2020a) Contesting digital futures: Urban politics, alternative economies, and the movement for technological sovereignty in Barcelona. *Antipode*, *52*(3), 660–680.

Lynch, C.R. (2020b) Unruly digital subjects: Social entanglements, identity, and the politics of technological expertise. *Digital Geography and Society*, *1*, 100001.

Lynch, C.R. (2021) Internet infrastructure and the commons: Grassroots knowledge sharing in Barcelona. *Regional Studies*. Online before print. DOI: 10.1080/00343404.2020.1869200

Lynch, C.R., and Del Casino, V. J. (2020) Smart spaces, information processing, and the question of intelligence. *Annals of the American Association of Geographers*, *110*(2), 382–390.

Maalsen, S. (2020) Revising the smart home as assemblage. *Housing Studies*, *35*(9), 1534–1549.

Maharawal, M.M., and McElroy, E. (2018) The anti-eviction mapping project: Counter mapping and oral history toward Bay Area housing justice. *Annals of the American Association of Geographers*, *108*(2), 380–389.

Meehan, K.M., Shaw, I. G. R., and Marston, S. A. (2014) The state of objects. *Political Geography*, *39*, 60–62.

Morgan, K. (2004) The exaggerated death of geography: Learning, proximity and territorial innovation systems. *Journal of Economic Geography*, *4*(1), 3–21.

Mossberger, K., Tolbert, C. J., and Stansbury, M. (2003) *Virtual Inequality: Beyond the Digital Divide*. Georgetown University Press.

Panelli, R. (2010) More-than-human social geographies: Posthuman and other possibilities. *Progress in Human Geography*, *34*(1), 79–87.

Pickren, G. (2018) 'The global assemblage of digital flow': Critical data studies and the infrastructures of computing. *Progress in Human Geography*, *42*(2), 225–243.

Pink, S., and Fors, V. (2017) Being in a mediated world: Self-tracking and the mind–body–environment. *Cultural Geographies*, *24*(3), 375–388.

Plantin, J.-C., and Punathambekar, A. (2019) Digital media infrastructures: Pipes, platforms, and politics. *Media, Culture & Society*, *41*(2), 163–174.

Przybylski, L. (2021) *Hybrid Ethnography: Online, Offline, and In Between.* SAGE Publications.

Rose, G. (2017) Posthuman agency in the digitally mediated city: Exteriorization, individuation, reinvention. *Annals of the American Association of Geographers, 107*(4), 779–793.

Seaver, N. (2017) Algorithms as culture: Some tactics for the ethnography of algorithmic systems. *Big Data & Society, 4*(2), 2053951717738104.

Sheppard, E. (1995) GIS and society: Towards a research agenda. *Cartography and Geographic Information Systems, 22*(1), 5–16.

Stiegler, B. (1998) *Technics and Time, 1: The Fault of Epimetheus.* Stanford University Press.

Thrift, N., and French, S. (n.d.). The automatic production of space. *Transactions of the Institute of British Geographer s, 27*(3), 309–335.

Whitehead, T. L. (2004). What is Ethnography? Methodological, Ontological, and Epistemological Attributes (online). In EICCARS working paper). Available at: www.cusag.umd.edu/documents/workingpapers/epiontattrib.pdf

Whitehead, N. L. (2009) Post-human anthropology. *Identities, 16*(1), 1–32.

17

GISCIENCE

Addressing Aggregation and Uncertainty

Hyeongmo Koo and Yongwan Chun

Introduction

Uncertainty, which is inevitably embedded in geographic data, appears in two components of geographic data: location and attribute information. Location information describes locations of geographic features and/or phenomena on the Earth surface. In a vector representation, locational (or positional) uncertainty concerns how accurate the x, y coordinates of points, lines, or polygons are, and similarly it relates to the location of cells in a raster representation. That is, it concerns deviations of observed coordinates from their true coordinates. In human geography research, locational uncertainty is of less concern for geographical data published by authorized organizations such as census boundaries by U.S. Census Bureau. However, this issue is well recognized for other data sources including open source data (Delmelle et al., 2019), geocoding (Bonner et al., 2003; Koo et al., 2018a), GPS (Zandbergen, 2009), and public participation GIS (PPGIS; Brown, 2017).

Attribute information concerns non-locational characteristics of geographic features and/or phenomena. It is generally obtained from a census such as the US Census or collected through a survey. Major sources of attribute uncertainty include sampling and measurement errors. Because attributes are estimated with samples, sampling error can occur even for a national level census, such as the US decennial census data or American Community Survey (ACS) data. This is related to how well samples can represent the population. Measurement error occurs when observed values are different from their true values. In a sample survey, the primary sources of measurement errors are the effects of questionnaire design, interviewers, respondents, and data-collection methods (Kasprzyk, 2005).

Specification error commonly occurs with inappropriate model specifications such as an unsuitable distribution assumption (e.g., a use of normal distribution for a count variable) and a non-spatial regression model for a dataset that clearly shows a level of spatial autocorrelation (e.g., Fingleton and Le Gallo, 2008; Hu et al., 2020). All sources of uncertainty can interact with each other and have an influence on geographic data analysis. Hence, it is crucial to recognize uncertainty in geographic data, to understand propagation of uncertainty in processing data, and to properly incorporate uncertainty in geographic data analysis. From the perspective of data analysis and/or modeling, specification error also occurs. Griffith et al., (2015) present a research agenda for uncertainty in geographic data analysis, which include uncertainty visualization, incorporation of uncertainty in spatial pattern analysis, and incorporation of uncertainty in spatial data modeling. This chapter discusses how the three aspects of uncertainty have been addressed in the literature in the following sections.

Uncertainty in GIS Data

This section discusses uncertainty in attributes mainly focusing on the ACS. It begins with a brief description of ACS and uncertainty embedded within it. Then, regionalization is described as an approach to reduce impacts of highly unreliable attributes by aggregating areal units. Finally, the modifiable areal unit problem (MAUP) that is caused by aggregation processes is discussed.

American Community Survey

The ACS inherently includes uncertainty, mainly due to sampling error (Sun and Wong, 2010). Sampling error has two main causes, which are a small sample size and heterogeneity in the population (Spielman et al., 2014). The ACS collects small samples that is about 2–3 percent of housing addresses based on an annual basis. This sampling process yields greater uncertainty in geographical areas with small populations rather than those with large populations (Wong and Sun, 2013), because a limited number of samples can be collected in small areas. The ACS overcomes the limitation of small samples by pooling samples in small geographical areas over multiple years (i.e., aggregating temporal resolution of variables). Thus, 1-year ACS products are only available for the areas having a least 65,000 people, and 3-year products for 20,000 people (US Census Bureau, 2009). Estimates of small geographical areas (e.g., census tracts and census block groups), can be obtained only from 5-year ACS products, but still uncertainty in a small area is generally greater than that in a large area. In addition, uncertainty cannot be fully explained by a sample size due to heterogeneity of the population, and, hence, a proper use of ACS requires understanding uncertainties in individual estimates (Citro and Kalton, 2007).

The ACS offers individual estimates with margins of errors (MOEs). ACS MOEs denote a 90 percent confidence bound for corresponding estimates, and thus, a larger MOE means the associated estimate has greater uncertainty. The US Census Bureau recommends considering MOEs for comparing and analyzing estimates (US Census Bureau, 2009). For instance, in a simple comparison of estimates between two different geographical areas, the difference between the estimates is determined based on a statistical significance test (e.g., *t*-test for mean difference). Similarly, ACS MOEs should be considered when comparing estimates based on a choropleth map (Koo et al., 2017; Sun et al., 2015) and analyzing difference between estimates related to their location (i.e., measuring spatial autocorrelation) (Jung et al., 2019a; Koo et al., 2019).

American Community Survey MOEs do not directly indicate the quality of corresponding estimates because they cannot be compared between different variables. Specifically, MOEs are provided on the same scale as their associated estimates. For example, MOEs for median age and income variables are expressed in years and dollars, respectively, and the amount of difference between years and dollars should not be comparable. Also, MOEs are generally associated with the scale of each estimate (i.e., estimates with large values are likely to have large MOEs).

A coefficient of variation (CV) is often utilized for comparing uncertainty among different variables. A CV is a measure of the relative amount of uncertainty associated with estimates (i.e., the ratio of standard error derived from an MOE to an estimate) (Spielman and Folch, 2015). Thus, a small value for a CV level denotes a low level of uncertainty in an estimate, but no universal criteria of CV values exist to determine the reliability of an estimate. Empirically, the National Research Council suggests that a CV level of 10 to 12 percent or less indicates a reasonable level of reliability (Citro and Kalton, 2007). Another source suggests CV levels of less than 12 percent, 12 to 40 percent, and over 40 percent indicate high, moderate, and low levels of reliability respectively (ESRI, 2011).

The US Census Bureau introduces two strategies for dealing with highly unreliable ACS estimates by reducing CVs (that is, increasing sample sizes) (US Census Bureau, 2009). The first approach is collapsing characteristic detail of estimates. For instance, if the CV level of median income estimates for Hispanic households in census tracts is critically high, median income estimates of all household types can be alternatively utilized by collapsing detail (Spielman and Folch, 2015). However, this strategy cannot be applied to a variable having no coarser level of detail (e.g., there is no level of detail below the estimate of median house values). The second strategy is grouping estimates across geographical areas, which is one type of 'regionalization'. Regionalization is a geographical strategy for improving the reliability of ACS estimates, but it involves a complex combinatorial problem when one groups areas (Spielman et al., 2014).

Regionalization: Aggregations of Area Units

Regionalization generally refers to a process of aggregating geographical areas into a certain number of regions (Openshaw and Rao, 1995), and has been widely utilized in human geography for various purposes, including political (or electoral) districting (e.g., Kim et al., 2017) and housing market delineating (e.g., Helbich et al., 2013). Technically, regionalization is a type of clustering method, where observations are grouped into a smaller number of clusters for maximizing homogeneity within clusters or minimizing heterogeneity among clusters based on one or multiple variables. However, regionalization is often subject to a contiguity constraint, which restricts a geographical area within a cluster so that it shares a common border with at least one neighbor in the same cluster to create a region (Folch and Spielman, 2014).

Folch and Spielman (2014) and Spielman and Folch (2015) suggest a regionalization algorithm as the strategy for reducing the uncertainty of ACS estimates. Their regionalization algorithm has three goals: 1) reducing the CV below a predetermined threshold in each region, 2) minimizing heterogeneity of attributes within each region, and 3) maximizing the number of regions (i.e., minimizing aggregation). The goals can be achieved through spatial optimization algorithms with an objective function and constraints. Specifically, the first goal is accomplished through a constraint to have the CV below a specific threshold for every estimate in each region. The threshold can apply to all variables or a specific variable in each region. To achieve the second goal, they use an objective function to minimize the sum of the squared deviations from the means of regions. The third goal intends to preserve the geographic detail of the original input dataset by maximizing the number of regions, and can be achieved by applying the max-p algorithm (Duque et al., 2012). Folch and Spielman (2014) provide a heuristic solution of the max-p algorithm with a two-phase approach. Their regionalization algorithm can assist in creating reliable ACS datasets but also leads to the MAUP.

Modifiable Areal Unit Problem

A quantitative analysis of geographic data is commonly conducted with attributes reported based on areal units. Its analytical results are affected by a choice of areal units and can vary considerably. This problem is referred to as the MAUP. Openshaw and Taylor (1979), in one of the early studies of MAUP, show that the correlation between two variables can vary when areal units are aggregated differently. For a simple example, a correlation between two variables based on census block groups and census tracts is often different from each other. Fundamentally, the MAUP is associated to a level of 'artifact' or 'arbitrariness' in demarcations of areas, even for areal units such as census units, administrative regions, or voting districts that are defined by authorized organizations.

The MAUP can be explained by aggregation of areal units. Given a set of small areal units, aggregating them into a smaller number of large areal units leads to a new spatial tessellation. Numerous possible ways of aggregation can lead to different analytical results. The MAUP can have two effects: a scale effect or a zoning effect. A scale effect is related to a level of aggregation, or spatial resolution of aggregated units. A high level of aggregation results in a small number of aggregated areal units in a resulting spatial tessellation, and each aggregated areal unit tends to cover a large area. In contrast, a low level of aggregation results in a tessellation with a relatively large number of aggregated units. A scale effect in empirical data often has a nested structure. For example, census tracts are constructed through aggregation of census block groups in the US Census. A zoning effect is associated with different zonation results for a same number of aggregated units. Aggregated areal units in a resulting tessellation can vary in size and shape (e.g., congressional districts vary in size and shape).

In the literature, the MAUP has been widely recognized. Fotheringham and Wong (1991) discuss the effects of the MAUP in multivariate analysis, specifically, multivariate linear regression. Arbia and Petrarca (2011) extend the discussion to spatial econometric models focusing on estimation bias for regression coefficients. Lee et al., (2019) analyze the effects on the level of spatial autocorrelation by examining variations of Moran's *I* values in their simulation experiments.

Uncertainty Visualization

As data visualization has been recognized as an effective approach to convey information, uncertainty visualization has been examined in the literature in cartography and GIS. Specifically, a visual representation of uncertainty can furnish supplementary information for corresponding attributes that are represented in a map (MacEachren, 1992).

Bivariate Mapping for Visualizing Uncertainty Attributes

Research has investigated how well uncertainty as well as corresponding attributes can be simultaneously represented in conventional mapping. Specifically, two approaches have been compared: side-by-side maps (or adjacent display) or bivariate mapping. Figure 17.1 illustrates these two approaches using median household income for counties in Texas from the 5-year (2009-2014) ACS data. While uncertainty is presented in a separate map in the side-by-side maps (Figure 17.1a), both attribute and uncertainty are combined in a single map in the bivariate mapping (Figure 17.1b). Studies including MacEachren et al., (1998), Kubíček and Šašinka (2011), and Francis et al., (2015) conclude that bivariate mapping is more effective than side-by-side mapping because users do not need to change their focus between two maps.

One critical issue, especially for uncertainty visualization with bivariate mapping, is how to select effective symbols for an attribute and its uncertainty. In cartographic design, an effective symbol includes the consideration of shape, size, texture, color, and transparency, among other characteristics (Roth 2016). Among visual variables, three have been widely investigated for uncertainty visualization in the literature: color components (hue, value, and saturation), transparency, and texture. For example, David and Keller (1997) and Slingsby et al., (2011) recommend color hue and color value (or lightness) for uncertainty symbols, while color saturation is

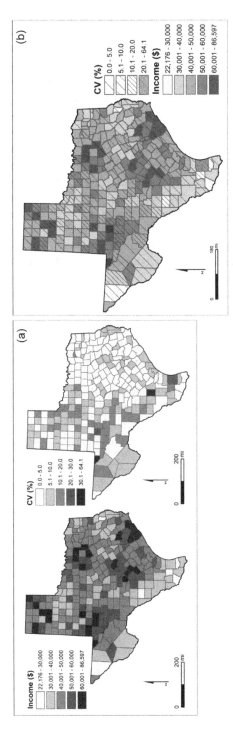

Figure 17.1 Visualization of uncertainty using (a) side-by-side mapping and (b) bivariate mapping.

also commonly considered (e.g., MacEachren, 1992, Deitrick and Wentz, 2015). In addition, transparency (or opacity) is recognized as an effective visual variable (Dreki, 2002; MacEachren et al., 2012), and texture is also widely recommended (Goodchild et al., 1994; Xiao et al., 2007; Sun and Wong, 2010).

More recently, Koo et al., (2018b) discuss a structured outline for effective attribute uncertainty visualization using bivariate mapping. They present three approaches: (1) a choropleth map with overlaid uncertainty symbols, (2) a proportional symbol map with color properties for uncertainty, and (3) composite symbols in which both attribute and uncertainty are represented. Figure 17.2a illustrates a choropleth map for an attribute with overlaid symbols for uncertainty, and Figure 17.2b presents a proportional symbol map in which the sizes of the symbols represent attributes and the color components (i.e., saturation or value) represent uncertainty. They argue that the former with a choropleth map is more appropriate for standardized attributes (e.g., population density), the latter with a proportional symbol map is more appropriated for raw counts (e.g., raw population size). In Figure 17.2c, the level of uncertainty is represented with the thickness of circle outlines or the height of the colored portion of each bar (i.e., upper and lower bounds around attributes).

From a practical perspective, generally uncertainty visualization is not available in popular GIS environments. For example, while a choropleth map with overlaid symbols (e.g., Figure 17.2a) can be generated using a layer structure (e.g., a layer for the choropleth map and another layer for the uncertainty symbols) in GIS, the other two types of uncertainty mapping are not readily available in GIS. A lack of readily available tools has prevented uncertainty mapping from being popularly utilized.[1]

Map Classification with an Incorporation of Uncertainty

Choropleth maps are popularly utilized in thematic mapping, and GIS has made it easy to make a choropleth map. Map classification methods have been widely discussed in cartography (e.g., Slocum et al., 2008) and are readily available in GIS. Recently, studies have investigated a potential shortcoming of map classification that considers only attributes. It is important because a visual inspection can be affected by a map pattern, which is an outcome of map classification. Xiao et al., (2007) discussed robustness of maps classification and developed a robustness measure with attributes and their uncertainty. Their robustness measure is defined based on a probability that each observation falls into the class to which the observation is assigned in a map classification result. A percentage of observations whose probabilities are larger than a threshold value is used for the robust measure.

Sun and Wong (2010) focused on whether or not classes are statistically different from its neighboring classes. They measured it based on two-sample z-tests between the largest value in a lower class and the smallest value in an upper class under an assumption that ACS observations follow a norm distribution with mean of its attributes (i.e., survey estimates) and standard errors computed from its MOE. They define 'confidence levels' (CL) as ($1 - p$ value of a two-sample z-test), which can be expressed as:

$$CL_{ij} = \phi\left(\frac{|\bar{x}_i - \bar{x}_j|}{\sqrt{\sigma_i^2 + \sigma_j^2}} \right) \tag{1}$$

where $\bar{x}_i \, \& \, \bar{x}_j$ denotes the means of observations i and j, $\sigma_i \, \& \, \sigma_j$ denote corresponding standard errors, and $\phi(\cdot)$ denotes the probability function. For example, two distributions can

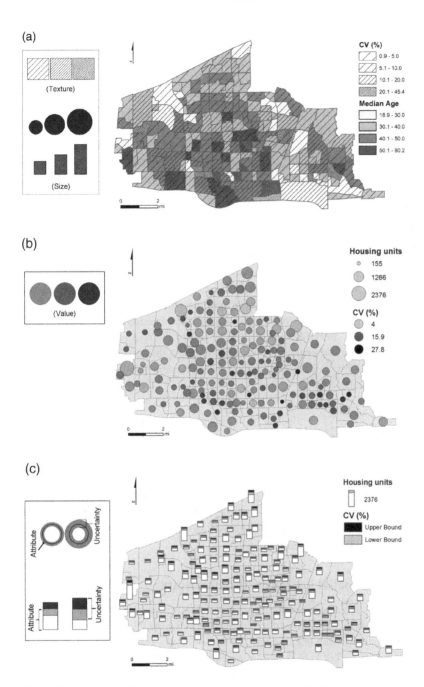

Figure 17.2 Three configurations for uncertainty mapping with bivariate mapping.

overlap with each other when standard errors are large as illustrated in Figure 17.3a. Then, their means may not be statistically different and the corresponding *p*-value becomes large. Accordingly, the CL is low. In contrast, Figure 17.3b exemplifies a high CL with a little overlap between the distributions.

While a CL is measured only between the largest and the small values of two neighboring map classes, Sun et al., (2015) extend it to compare all possible pairs among observations in two classes and use the minimum CL as a 'class separability index' (CS). It can be expressed as $CS_{A,B} = CL_{ij}$ where $CS_{A,B}$ denotes class separability between class A and class B. Figure 17.4a presents a choropleth map based on class separability index. The legend presents SP values between two consecutive classes as well as class intervals. That is, it shows how well the classes are separated when the unreliability information of the attributes is incorporated. Figure 17.4b presents a plot of the classification result with the attributes (blue dots) and their 95 percent confidence intervals (horizontal bars). A potential issue of a CS-based choropleth map is that a majority of observations can fall into one class and the others have a small number of observations.

Koo et al., (2017) present an optimal map classification with attributes and their uncertainty information extending the shortest path problem. Cromley and Campbell (1991) show that an optimal map classification has a similar form as the shortest path problem in an acyclic network with observations being ordered. Figure 17.5a illustrates a network with four observations. The arcs represent groupings of observations and have attached costs (or impedance) in the shortest path problem. For map classification, costs can be calculated as the sum of deviations in a group. An optimal solution minimizes the sum of costs. While only estimates are used to calculate the sum of deviations in the conventional map classification (e.g., Cromley and Campbell, 1991), Koo et al., (2017) utilize the class separability index to incorporate uncertainty. They also propose to use Bhattacharyya distance, which is a dissimilarity measure between two distributions (e.g., Coleman and Andrews, 1979). For two normal distributions, it can be calculated as:

$$BD(i, j) = \frac{1}{4}\left(\frac{\left(\bar{x}_i - \bar{x}_j\right)^2}{\sigma_i^2 + \sigma_j^2} \right) + \frac{1}{4} ln\left(\frac{1}{4}\left(\frac{\sigma_i^2}{\sigma_j^2} + \frac{\sigma_j^2}{\sigma_i^2} + 2 \right) \right). \tag{2}$$

Bhattacharyya distance compares not only the mean difference (the first term) but also variability (the second term). Figures 17.5b and 17.5c present optimal classification results for median household income from the 5-year (2009-2014) ACS for the counties in Texas. These two maps are more visually balanced compared to the map in Figure 17.4a. Note that this optimal map classification is implemented in SAAR.

Spatial Patterns with Uncertainty

Spatial data analysis often requires a researcher to find substantial spatial patterns of a phenomenon. For example, an identification of spatial clusters can help understand human geographical phenomena such as crime (e.g., Levine, 2006; Koo et al., 2020) or disease incidence (e.g. Goovaerts and Jacquez, 2005). Such a phenomenon in which observations tend to be similar (or dissimilar) with their spatially neighboring observations is called spatial autocorrelation (Griffith 1987). When a significant level of spatial autocorrelation exists, observations are not random in space. A conventional statistical model such as linear regression may result in biased estimates and/or unstable statistical decisions because spatial autocorrelation can violate the independence assumption. Spatial regression models can furnish an appropriate modeling

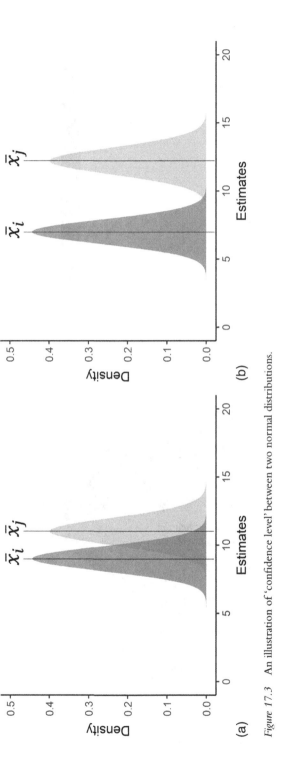

Figure 17.3   An illustration of 'confidence level' between two normal distributions.

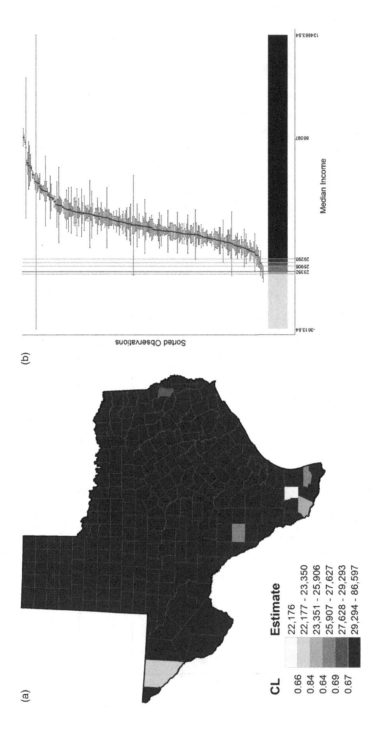

Figure 17.4 An illustration of the separability index mapping using the median household income in Texas from the 5-year (2009–2014) ACS data. (a) a choropleth map and (b) a plot of the classification result with the attributes and their 95 percent confidence intervals (horizontal bars).

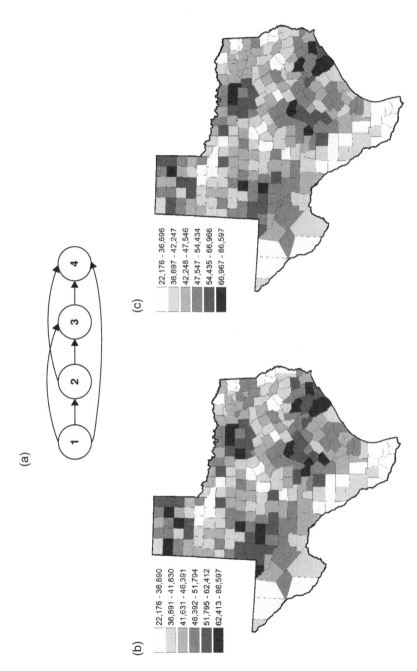

Figure 17.5 Optimal map classification with uncertainty for the median household income in Texas from the 5-year (2009–2014) ACS data. (a) an illustration of a network structure for map classification, (b) an optimal map classification result with the class separability index, and (c) an optimal map classification result with Bhattacharyya distance.

approach (e.g., Anselin 2009). Hence, it is important to examine whether significant spatial autocorrelation exists.

The most popular statistical measure for spatial autocorrelation is Moran's *I*. Moran's *I*, as well as other popular spatial autocorrelation measures such as Geary's *C*, do not consider uncertainty of attributes in their formulations. That is, attributes are treated as 'error free'. One potential problem is that a statistical decision can be unstable when uncertainty of attributes is not considered. Figure 17.6a shows a simulation result of Moran's *I* for 1,000 random number sets drawn with the median household income for Texas counties from the 5-year (2009-2014) ACS data. Under an assumption that each median household income follows a normal distribution with mean of the attribute and standard deviation calculated from its MOE, a random value for each county is drawn from the corresponding normal distribution. One set consists of such random values for the 258 counties, and its Moran's *I* value is calculated. The histogram in Figure 17.6 shows the distribution of Moran's *I* values for 1,000 random sets with the Moran's *I* value for the attributes only (represented with a vertical bar). This simulation result shows that Moran's *I* can be potentially affected by attribute uncertainty. The vertical bar is on the right side of the Moran's *I* distribution for the simulated random sets. This pattern is more noticeable in Figure 17.6b, which presents a result of the same simulation experiment for census tracts in Dallas county using median household income of Hispanic households. The vertical bar is located at the far-right side of the distribution. This also demonstrates that Moran's *I* can be potentially biased when uncertainty is not considered.

This potential bias problem is also reported in Koo et al., (2019). They conducted a similar simulation experiment with random estimates that covers a range of spatial autocorrelation levels and random standard errors. They also considered three different regular square tessellations ($n = 100, 250, 500$). Random values for attributes were generated using a spatial autoregressive progress: that is, $\mathbf{X} = (\mathbf{I} - \rho\mathbf{W})^{-1}\boldsymbol{\varepsilon}$, where $\mathbf{I}$ denotes an identity matrix, ρ denotes a spatial autocorrelation parameter, $\mathbf{W}$ denotes a spatial weights matrix, and $\boldsymbol{\varepsilon}$ denotes random numbers generated from the standard normal distribution $N(0,1)$. On the other hand, random values for standard errors were generated using truncated normal distributions. They report that Moran's *I* with attributes tends to indicate a stronger level of spatial autocorrelation, as can be seen in Figure 17.6. Specifically, they discuss three observations from their experiment. First, the extreme spatial autocorrelation pattern of Moran's *I* is stronger for ρ value is far from 0 (that

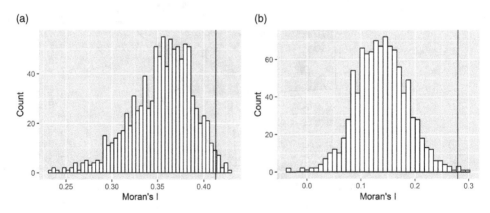

Figure 17.6 A simulation result of Moran's *I* values using attributes from the 5-year (2009–2014) ACS data. (a) median household income for Texas counties and (b) median household income of Hispanic households for census tracts in Dallas county.

is, closer to 1 or −1). When ρ is close to 0, Moran's I is around the center of the corresponding distribution. Second, this potential bias issue gets severe as the number of observation (n) gets larger and/or the level of uncertainty gets lager. Third, when ρ is positive, Moran's I gets stronger where there is positive spatial autocorrelation. When ρ is negative, Moran's I gets stronger where there is negative spatial autocorrelation.

Jung et al., (2019a) also discuss the bias of Moran's I when uncertainty is not incorporated using ACS data. They present a modification of Moran's I based on an errors-in-variable context. Here, observed values ($\mathbf{X^*}$) are composed of true values ($\mathbf{X}$) and errors ($\boldsymbol{\varepsilon}$); that is, ($\mathbf{X^*}$) = $\mathbf{X}$ + $\boldsymbol{\varepsilon}$. The adjusted Moran's I can be expressed in a matrix form as:

$$I^* = \frac{\mathbf{X^{*T}WX^*}}{\mathbf{X^{*T}X^*}} = \frac{(\mathbf{X}+\boldsymbol{\varepsilon})^T \mathbf{W}(\mathbf{X}+\boldsymbol{\varepsilon})}{(\mathbf{X}+\boldsymbol{\varepsilon})^T (\mathbf{X}+\boldsymbol{\varepsilon})} \tag{3}$$

where T denotes matrix transpose operator. Utilizing the independence assumption on the errors ($\boldsymbol{\varepsilon}$) and asymptotic properties, they further modify the formula as $I^* = \frac{Var(X)}{Var(X^*)} \times I$ where I denotes Moran's I. They also present a modification of local Moran's I in a similar derivation. They conclude that a spatial autocorrelation measure without ignoring uncertainty is likely to be biased and may fail to detect a true spatially autocorrelated pattern.

Incorporation of uncertainty in measuring spatial autocorrelation is still under-investigated despite its potential problem. One recent development is presented in Koo et al., (2019). They propose a spatial autocorrelation measure using Bhattacharyya coefficient (BC), which is derived from Bhattacharyya distance. The relationship can be expressed as:

$$BC(i,j) = exp(-BD(i,j)) \tag{4}$$

BC indicates the amount of overlap between two distributions. It ranges from 0 to 1, where 0 means no overlap and 1 means an identical distribution (i.e., a full overlap). BC considers the difference between two means and their variances simultaneously. One underlying idea is that two attributes with large uncertainties may not be considerably different when the uncertainties are considered. Hence, a similarity of two observations needs to be considered with both attributes and their uncertainty. Koo et al., (2019) formulate spatial BC (SBC) as:

$$SBC = \frac{\sum_i \sum_j w_{ij}BC(i,j)}{\sum_i w_{ij}} = \frac{n}{\sum_i w_{ij}} \cdot \frac{\sum_i \sum_j w_{ij}BC(i,j)}{\sum_i BC(i,i)} \tag{5}$$

Note that $BC(i,i) = 1$ and, hence, $\sum_i BC(i,i) = n$. This formula shows that SBC has a similar formulation as Moran's I. SBC ranges from 0 to 1 and a high SBC value indicates BC values between spatially neighboring units are similar. SBC increases when spatial autocorrelation for the attributes increases conditioned on a constant uncertainty level. Also, an increase of uncertainty with the same estimates leads to an increase of SBC. They demonstrate how SBC behaves with a simulation experiment as well as an application with ACS data.[2]

Jung et al., (2019b) present another approach to account for uncertainty in rate values. They are concerned about uncertainty in denominator values in the calculation of rate values.

For example, uncertainty in population sizes that are used in calculation of disease rates can have an impact on a spatial autocorrelation measure. Extending the empirical Byes smoothing approach by Assunção and Reis (1999) that considers different population sizes in the denominator of rate measures, they propose the heteroscedasticity-consistent empirical Bayes (HC-EB) measure that accounts for uncertainty of estimates in the denominator. Then, they measure spatial autocorrelation with a modified Moran's *I* using HC-EB values. Similarly, they present a local version of the modified spatial autocorrelation measure.[3]

Although new approaches are proposed, an incorporation of uncertainty in spatial autocorrelation measures needs further investigations. An incorporation of errors can make spatial autocorrelation measures complex. Also, a new measure with uncertainty may not be interpreted in the same way as a conventional spatial autocorrelation measure.

Uncertainty in GIS Analysis

A GIS analysis is often conducted with statistical models including regression. This section discusses first specification error, which is a major source of uncertainty in statistical modeling. It is often related to an inappropriate functional form (e.g., statistical equations or assumptions). Then this section describes measurement error models that incorporate errors in independent variables in regression. Measurement error models can furnish an appropriate tool to incorporate impacts of data quality (e.g., MOEs of ACS) in spatial data modeling.

Specification Error

An analysis of a human geographical phenomenon often involves describing relationships or associations between a dependent (or response) variable and independent (or explanatory) variables using statistical models including regression. Uncertainty in a functional formulation can cause specification error (or misspecification). While major sources of specification error have been discussed including using improper functional forms and/or improper assumptions (e.g., Griffith, 2018), three types of sources are discussed in this section.

First, one common specification error is to use a linear model when an underlying relationship is non-linear. For example, using linear regression that assumes a linear relationship can lead to specification error, when a non-linear relationship is more appropriate. In addition, a utilization of normal approximation is a source of specification error for count or percent observations. For example, the logarithm values of observations may not follow a normal distribution. An excessive number of zero observations can deepen this issue. Rather, Poisson regression can furnish a more suitable model specification (Chun, 2008; Flowerdew 2010).

Second, a failure of incorporating spatial autocorrelation can cause specification error when an underling level of spatial autocorrelation is observed. A presence of spatial autocorrelation among observations can violate the independence assumption and can lead to estimation bias in linear regression (Cordy and Griffith 1993). Spatial regression models, including spatial lag model and spatial error models, can be utilized for normally distributed data (Anselin 2009). Also, Moran eigenvector spatial filtering (MESF) can furnish a useful methodology to accommodate spatial components in GLM as well as linear regression (Griffith et al., 2019).[4]

Third, another common specification error is that relevant independent variables are missing. The explanatory power of a model decreases with relevant variables missing (i.e., determinants of the dependent variable). This is referred to omitted-variable bias (Wooldridge 2013). In addition, if missing variables have an underlying spatial pattern, regression residuals are likely to show a significant level of spatial autocorrelation and, hence, estimation bias can occur.

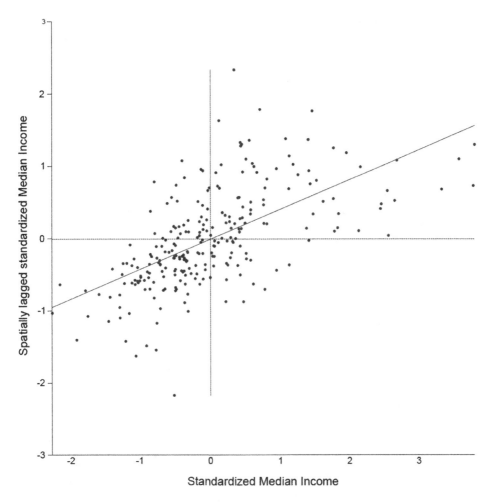

Figure 17.7 Moran scatterplot of the median household income in Texas from the 5-year (2009–2014) ACS data.

A spatial model specification can reduce the impacts of missing relevant variables (Griffith and Chun, 2015).

In addition, Hu et al., (2020) discuss another specification issue in the context of spatial autocorrelation. They emphasize a mixture of positive and negative spatial autocorrelation in observations. A Moran scatterplot can illustrate a mixed pattern of positive and negative spatial autocorrelation. In Figure 17.7, the global Moran's I value for the median household incomes of Texas counties is 0.4130 (p-value < 0.001), which indicates positive spatial autocorrelation. However, from a local spatial autocorrelation perspective, it is a mixture of the two types of spatial autocorrelation. While the observations in the first and third quadrants represent positive local spatial autocorrelation, the observations in the second and fourth quadrants show negative spatial autocorrelation. Since global Moran's I represents the level of spatial autocorrelation with a single number, the components of these two types of spatial autocorrelation can cancel out each other in its calculation. Here, the global Moran's I value only present positive spatial auto-correlation and does not indicate negative autocorrelation components. Similarly, a mixture of positive and negative spatial autocorrelation may not be properly represented in popular spatial regression models that have a single parameter for spatial autocorrelation. Hu et al., (2020) present a decomposition of positive and negative spatial autocorrelation components using MESF and include both components in a regression model. With an application of modeling breast cancer incidences in Broward county, FL, they show that having both components improves a model performance and would lead to a proper model specification.

Measurement Error Model

A regression analysis is conducted to find a relationship between a dependent variable and independent variables under an assumption that observations are true values. It is assumed that variables are correctly measured without errors. Linear regression can be expressed as $\mathbf{Y} = \mathbf{X}\beta + \epsilon$ with true X values. When observations have error, that is, $\mathbf{X^*} = \mathbf{X} + \epsilon$, linear regression equation becomes

$$\mathbf{Y} = \mathbf{X}\beta + \xi = (\mathbf{X^*} - \epsilon)\beta + \xi = \mathbf{X^*}\beta + (\xi - \epsilon\beta) = \mathbf{X^*}\beta + \mathbf{u} \tag{6}$$

where ξ denotes random errors. This model is referred to as a measurement error model or error-in-variable model (Fuller, 1987). The regression of $\mathbf{Y}$ on $\mathbf{X^*}$ has an error term $\mathbf{u}$. That is, when measurement error is ignored, $\mathbf{X^*}$ and $\mathbf{u}$ are correlated to each other (both of them have ϵ). This correlation between independent variables and error term leads to endogeneity bias and, hence, regression results become unreliable. Specifically, estimated coefficients become biased toward to 0, which is called attenuation bias (Stefanski, 2000). A number of methods to estimate measurement error models have been developed in the literature. Carroll et al., (2006) discuss a range of estimation methods including an instrument variables approach, simulation approaches, and Bayesian methods.[5]

Measurement error models are utilized in various research fields including statistics and econometrics, but they are still rarely utilized in geography research. Nevertheless, measurement error models in a spatial context can be found, especially in public health research. Measurement-based variables such as air pollution and heavy metal levels are analyzed with measurement models. For example, Carroll et al., (1997) analyzed ozone levels in Harris county, Texas, and Xia and Carlin (1998) incorporated measurement errors in spatio-temporal models of lung cancer mortalities in Ohio. With the availability of MOEs in ACS, geographical research recognizes a potential impact of data quality on results of statistical models. For example, Donegan et al., (2021) present a Bayesian hierarchical model to incorporate MOEs in spatial data modeling. However, measurement errors, specifically, MOEs of socio-economic variables from ACS are still largely ignored in geographic research. This topic should be investigated further in future research.

Conclusions

Uncertainty has been constantly recognized in the geography literature. This chapter reviews uncertainty issues in georeferenced data (attributes) and then discusses recent developments to explore and incorporate uncertainty in geographical research. Specifically, three challenging areas are discussed: uncertainty visualization, measurement of spatial patterns with uncertainty, and spatial data analysis with uncertainty. The chapter discusses the importance of incorporating attribute uncertainty in geographical data analysis because results can render a different decision outcome from conventional modeling approaches in which uncertainty is not appropriately addressed.

GIScientists have faced two key uncertainty related research challenges in recent years. First, the uncertainty geographic context problem (UGCoP) has been discussed in the literature. It indicates that while the effects of socio-economic and/or environmental factors are investigated in geographical research, area-based attributes may not be relevant to individuals (Kwan, 2012). For example, an exposure to air pollution will be different between drivers who are in a car and runners who are directly exposed to polluted air when they are in the same geographical space. That is, they are in a different geographic context. This can help researchers to investigate associations of factors with more relevant measures for individuals who are in a different geographical environment. Advances in personal sensing technology with location-aware devices such as GPS and mobile phones can help to measure surrounding environments of individuals. Second, while new emerging data sources including social media and volunteered geographic information (VGI) become increasingly available, an awareness of uncertainty in new data sources and a proper use of them are critical. For example, social media data with georeferenced information can provide opportunities to investigate geographic issues including even near-real time responses to natural disasters. However, there are representative issues because social media data may not represent all parts of the population appropriately with a dominance of a specific user groups (e.g., more young people). Also, VGI data are prone to have uncertainty or incorrect information more than conventional data that are produced with a more thorough data quality control process.

As geospatial Big Data have been increasingly available, methodological advancements to deal with geographic Big Data are required. For example, social media data contain generally millions of observations with location tags, a huge number of satellite images are captured on a daily basis, population surveys produce and maintain high quality and a large volume of data for the fast-changing world, and individual level observations become available instead of aggregated observations based on areal units. Geographic research has faced challenges to process, analyze, and visualize such data effectively and efficiently, and this pattern is likely to be more prevailing in future research. Methodological advancements can assist researchers to investigate geographical phenomena and to construct knowledge.

Notes

1 Koo et al., (2018c) provide readily available functions for the three uncertainty visualization approaches. They developed a software package called SAAR using ArcGIS Engine by Esri and R. SAAR also supports visualization tools for exploratory spatial data analysis including Moran scatterplot and conditional mapping, and confirmatory analysis tools including regression, spatial regression, and Moran eigenvector spatial filtering (Griffith et al., 2019). The installation file of SAAR can be can be found at https://thesaar.github.io/. Alternatively, ESF Tool (https://github.com/esftool/esftool), which can be considered as a light version of SAAR, also provides uncertainty mapping tools. ESF Tool does not need any license because it is developed with DotSpatial (https://github.com/DotSpatial/DotSpatial), an open source GIS library, while SAAR requires an ArcGIS desktop license.
2 R code for SBC is available on GitHub (https://github.com/hyeongmokoo/SBC). In addition, they present a local version of SBC (Koo et al., 2021).
3 An implementation of the HC-EB measure in R is available (https://github.com/phjung/HCEB).
4 Implementation of MESF is available in SAAR (Koo et al., 2018c) with a graphical user interface.
5 These estimation methods are available in popular statistical packages such as SAS and R (e.g., ivreg, simex, and eivtools packages in R). They extend the discussion for nonlinear measurement error models.

References

Anselin, L. (2009) Spatial regression. *The SAGE handbook of spatial analysis*, 1, 255–276.

Arbia, G., and Petrarca, F. (2011) Effects of MAUP on spatial econometric models. *Letters in Spatial and Resource Sciences*, 4(3), 173.

Assunção, R. M., and Reis, E. A. (1999) A new proposal to adjust Moran's *I* for population density. *Statistics in medicine*, 18(16), 2147–2162.

Bonner, M. R., Han, D., Nie, J., Rogerson, P., Vena, J. E., & Freudenheim, J. L. (2003) Positional accuracy of geocoded addresses in epidemiologic research. *Epidemiology*, 14(4), 408–412.

Brown, G. (2017) A review of sampling effects and response bias in internet participatory mapping (PPGIS/PGIS/VGI). *Transactions in GIS*, 21(1), 39–56.

Campbell, G. M., and Cromley, R. G. (1991) Optimal simplification of cartographic lines using shortest-path formulations. *Journal of the Operational Research Society*, 42(9), 793–802.

Carroll, R. J., Chen, R., George, E. I., Li, T. H., Newton, H. J., Schmiediche, H., & Wang, N. (1997) Ozone exposure and population density in Harris County, Texas. *Journal of the American Statistical Association*, 92(438), 392–404.

Carroll, R. J., Ruppert, D., Stefanski, L. A., & Crainiceanu, C. M. (2006) *Measurement error in nonlinear models: A modern perspective*. CRC Press.

Chun, Y. (2008) Modeling network autocorrelation within migration flows by eigenvector spatial filtering. *Journal of Geographical Systems*, 10(4), 317–344.

Citro, C.F., and Kalton, G. (2007) *Using the American Community Survey: Benefits and challenges*. National Academy Press.

Cordy, C. B., and Griffith, D. A. (1993) Efficiency of least squares estimators in the presence of spatial autocorrelation. *Communications in Statistics-Simulation and Computation*, 22(4), 1161–1179.

Coleman, G. B., and Andrews, H. C. (1979) Image segmentation by clustering. *Proceedings of the IEEE*, 67(5), 773–785.

Davis, T. J., and Keller, C. P. (1997) Modelling and visualizing multiple spatial uncertainties. *Computers & Geosciences*, 23(4), 397–408.

Delmelle, E. M., Marsh, D. M., Dony, C., & Delamater, P. L. (2019) Travel impedance agreement among online road network data providers. *International Journal of Geographical Information Science*, 33(6), 1251–1269.

Drecki, I. (2002) Visualisation of uncertainty in geographical data, in: W. Shi, P. Fisher, M. Goodchild (Eds.), *Spatial Data Quality*, Taylor & Francis, London, pp. 140–159.

Deitrick, S., and Wentz, E. A. (2015) Developing implicit uncertainty visualization methods motivated by theories in decision science. *Annals of the Association of American Geographers*, 105(3), 531–551.

Donegan, C., Chun, Y., and Griffith, D. A., (2021) Modeling community health with areal data: Bayesian inference with survey standard errors and spatial structure, (working paper).

Duque, J. C., Anselin, L., and Rey, S. J. (2012). The max-*p*-regions problem. *Journal of Regional Science*, 52, 397–419.

Environmental Systems Research Institute (Esri). (2011) *The American Community Survey*. An ESRI white paper.

Folch, D. C., and Spielman, S. E. (2014) Identifying regions based on flexible user-defined constraints. *International Journal of Geographical Information Science*, 28, 164–184.

Fotheringham, A. S., and Wong, D. W. (1991) The modifiable areal unit problem in multivariate statistical analysis. *Environment and planning A*, 23(7), 1025–1044.

Fingleton, B., and Le Gallo, J. (2008) Estimating spatial models with endogenous variables, a spatial lag and spatially dependent disturbances: Finite sample properties. *Papers in Regional Science*, 87(3), 319–339.

Flowerdew, R. (2010) Modelling migration with Poisson regression. In *Technologies for migration and commuting analysis: Spatial interaction data applications* (pp. 261–279). IGI Global.

Francis, J., Tontisirin, N., Anantsuksomsri, S., Vink, J., & Zhong, V. (2015) Alternative strategies for mapping ACS estimates and error of estimation. In *Emerging techniques in applied demography* (pp. 247–273). Springer, Dordrecht.

Fuller, W. A. (1987) *Measurement Error Models*, New York: John Wiley.

Goodchild, M., Buttenfield, B., and Wood, J. (1994) Introduction to Visualizing Data Validity in Hearnshaw H and Unwin, D.(eds) *Visualization in Geographic Information Systems*, pp. 141–149.

Goovaerts, P., and Jacquez, G. M. (2005) Detection of temporal changes in the spatial distribution of cancer rates using local Moran's I and geostatistically simulated spatial neutral models. *Journal of Geographical Systems*, 7(1), 137–159.

Griffith, D. A. (1987) *Spatial autocorrelation: A primer*. Washington DC: Association of American Geographers.

Griffith, D. A. (2018) Uncertainty and context in geography and GIScience: Reflections on spatial autocorrelation, spatial sampling, and health data. *Annals of the American Association of Geographers*, 108(6), 1499–1505.

Griffith, D. A., and Chun, Y. (2016) Evaluating eigenvector spatial filter corrections for omitted georeferenced variables. *Econometrics*, 4(2), 29.

Griffith, D., Chun, Y., and Li, B. (2019) *Spatial regression analysis using eigenvector spatial filtering*. Academic Press.

Griffith, D. A., David Wong, and Y. Chun, (2015) Uncertainty related research issues in spatial analysis, In J. Shi, B. Wu, and A. Stein (eds.), *Uncertainty Modelling and Quality Control for Spatial Data*, London: Taylor & Francis Group/CRC Press, pp. 3–11.

Helbich, M., Brunauer, W., Hagenauer, J., & Leitner, M. (2013) Data-driven regionalization of housing markets. *Annals of the Association of American Geographers*, 103, 871–889.

Hu, L., Chun, Y., and Griffith, D. A. (2020) Uncovering a positive and negative spatial autocorrelation mixture pattern: A spatial analysis of breast cancer incidences in Broward County, Florida, 2000–2010. *Journal of Geographical Systems*, 1–18.

Jung, P. H., Thill, J. C., and Issel, M. (2019a) Spatial autocorrelation and data uncertainty in the American Community Survey: A critique. *International Journal of Geographical Information Science*, 33(6), 1155–1211.

Jung, P. H., Thill, J. C., and Issel, M. (2019b) Spatial autocorrelation statistics of areal prevalence rates under high uncertainty in denominator data. *Geographical Analysis*, 51(3), 354–380.

Kasprzyk, D. (2005) Chapter IX, Measurement error in household surveys: Sources and measurement. *Household sample surveys in developing and transition countries*. New York: United Nations. pp. 171–198.

Kim, K., Chun, Y., and Kim, H. (2017) A robust heuristic approach for regionalization problems. In Thill, J. C., & Dragicevic, S. (eds) *GeoComputational Analysis and Modeling of Regional Systems*. Springer, Cham. pp305–324.

Koo, H., Chun, Y., and Griffith, D. A. (2017) Optimal map classification incorporating uncertainty information. *Annals of the American Association of Geographers*, 107(3), 575–590.

Koo, H., Chun, Y., and Griffith, D. A. (2018a) Modeling positional uncertainty acquired through street geocoding, *International Journal of Applied Geospatial Research*, 9(4), pp. 1–22.

Koo, H., Chun, Y., and Griffith, D. A. (2018b) Geovisualizing attribute uncertainty of interval and ratio variables: A framework and an implementation for vector data. *Journal of Visual Languages & Computing*, 44, 89–96.

Koo, H., Chun, Y., and Griffith, D. A. (2018c) Integrating spatial data analysis functionalities in a GIS environment: Spatial Analysis using ArcGIS Engine and R (SAAR). *Transactions in GIS*, 22(3), 721–736.

Koo, H., Chun, Y. and Wong, D. W. (2021) Measuring local spatial autocorrelation with data reliability information. *The Professional Geographer*, accepted.

Koo, H., Lee, M., Chun, Y., and Griffith, D. A. (2020) Space-time cluster detection with cross-space-time relative risk functions. *Cartography and Geographic Information Science*, 47(1), 67–78.

Koo, H., Wong, D. W., and Chun, Y. (2019) Measuring global spatial autocorrelation with data reliability information. *The Professional Geographer*, 71(3), 551–565.

Kubíček, P., and Šašinka, Č. (2011) Thematic uncertainty visualization usability–comparison of basic methods. *Annals of GIS*, 17(4), 253–263.

Kwan, M. P. (2012) The uncertain geographic context problem. *Annals of the Association of American Geographers*, 102(5): 958–968.

Lee, S. I., Lee, M., Chun, Y., and Griffith, D. A. (2019) Uncertainty in the effects of the modifiable areal unit problem under different levels of spatial autocorrelation: A simulation study. *International Journal of Geographical Information Science*, 33(6), 1135–1154.

Levine, N. (2006) Crime mapping and the Crimestat program. *Geographical Analysis*, 38(1), 41–56.

MacEachren, A. M. (1992) Visualizing uncertain information. *Cartographic Perspectives*, (13), 10–19.

MacEachren, A. M., Brewer, C. A., and Pickle, L. W. (1998) Visualizing georeferenced data: Representing reliability of health statistics. *Environment and Planning A*, 30(9), 1547–1561.

MacEachren, A. M., Roth, R. E., O'Brien, J., Li, B., Swingley, D., & Gahegan, M. (2012) Visual semiotics & uncertainty visualization: An empirical study. *IEEE Transactions on Visualization and Computer Graphics*, 18(12), 2496–2505.

Openshaw, S. and Taylor, P. J. (1979) A million or so correlation coefficients: Three experiments on the modifiable areal unit problem. In N. Wrigley, ed. *Statistical Applications in the Spatial Sciences*, London: Pion, pp. 127–144.

Openshaw, S, Rao, L. (1995) Algorithms for re-engineering 1991 Census geography. *Environment and Planning A*, 27, 425–446.

Roth, R. E. (2016) Visual variables. *International Encyclopedia of Geography: People, the Earth, Environment and Technology*, 1–11.

Slingsby, A., Dykes, J., and Wood, J. (2011) Exploring uncertainty in geodemographics with interactive graphics. *IEEE Transactions on Visualization and Computer Graphics*, 17(12), 2545–2554.

Slocum, T. A, McMaster, R. B., Kessler, F. C., and Howard, H. H. (2008) *Thematic cartography and geovisualization*. 3rd edition, Prentice Hall: Upper Saddle River, New Jersey.

Spielman, S. E., and Folch, D. C. (2015) Reducing uncertainty in the American Community Survey through data-driven regionalization. *PLoS ONE*, 10, 1–21.

Spielman, S. E., Folch, D., and Nagle, N. (2014) Patterns and causes of uncertainty in the American Community Survey. *Applied Geography*, 46, 147–157.

Stillwell, J., Daras, K., and Bell, M. (2018) Spatial aggregation methods for investigating the MAUP effects in migration analysis. *Applied Spatial Analysis and Policy*, 11(4), 693–711.

Sun, M., and Wong, D. W. (2010) Incorporating data quality information in mapping American Community Survey data. *Cartography and Geographic Information Science*, 37(4), 285–299.

Sun, M., Wong, D. W., and Kronenfeld, B. J. (2015) A classification method for choropleth maps incorporating data reliability information. *The Professional Geographer*, 67(1), 72–83.

Stefanski, L. A. (2000) Measurement error models. *Journal of the American Statistical Association*, 95(452), 1353–1358.

US Census Bureau. (2009) *A Compass for Understanding and Using American Community Survey Data: What Researchers Need to Know*. Washington, DC: U.S. Government Printing Office.

Wong, D. W., and Sun, M. (2013) Handling data quality information of survey data in GIS: A case of using the American Community Survey data. *Spatial Demography*, 1, 3–16.

Wooldridge, J. M. (2013) Omitted variable bias: The simple case. *Introductory Econometrics: A Modern Approach*. 5th edition, Mason, OH: Cengage Learning, 91–94.

Xia, H., and Carlin, B. P. (1998) Spatio-temporal models with errors in covariates: Mapping Ohio lung cancer mortality. *Statistics in medicine*, 17(18), 2025–2043.

Xiao, N., Calder, C. A., and Armstrong, M. P. (2007) Assessing the effect of attribute uncertainty on the robustness of choropleth map classification. *International Journal of Geographical Information Science*, 21(2), 121–144.

Zandbergen, P. A. (2009) Accuracy of iPhone locations: A comparison of assisted GPS, WiFi and cellular positioning. *Transactions in GIS*, 13, 5–25.

18

HEALTH GEOGRAPHY AND BIG DATA ADVENTURES

Methodological Innovations, Opportunities and Challenges

Malcolm Campbell and Lukas Marek

Introduction: Moving People, Evolving Places

This chapter continues with the theme of understanding some of the myriads of methodological and theoretical challenges within the discipline of human geography. We focus on the particular challenges that arise from the sub-discipline of health and medical geography, or the geography of health (Earickson, 2009, Moon, 2009). Within this sub-disciplinary field, there is a rich history, as is often the case in other sub-disciplines of geography, with interdisciplinary reach into adjoining disciplines such as epidemiology, sociology, economics, or anthropology, to name just a few. This means that health and medical geographers are often in the position of attempting to reconcile epistemological and methodological debates in aligned fields as well as within their own sub-disciplines. This is a challenge, but it is also an opportunity. We argue, both robustly and optimistically, that there is an opportunity in the inherent tensions of spanning multiple bodies of epistemological and methodological endeavour, to innovate and discover new ways of doing things that can be not only helpful but transformational in the production of health and medical knowledge. We take the position of critical scholars who have been involved in applied research, carefully reflecting on the various lessons learned in the praxis of research, affording opportunities for methodological innovation as well as better understanding the tensions and the trends that are driving the research agenda at large.

A geographical commentary (Cupples, 2020) aptly stated that 'geography is a promiscuous discipline' that crosses physical and social sciences, arts and humanities. It is also a magpie discipline, collecting and curating from closely aligned disciplines. This is what attracted us (perhaps you too) to geography in the first instance, the ability to curate and collect. The cautionary note is that other birds may not appreciate the geographical community 'borrowing' its intellectual treasure. As such, the discipline of geography has this key strength, which is also a weakness. Geographers are often attracted towards the interface with other disciplinary perspectives and methods. This interdisciplinarity essence permeates into the sub-disciplines of geography, of which health and medical geography is a component part. The authors are situated within this field and so frame many of our observations and critiques from within this context.

DOI: 10.4324/9781003038849-21

The chapter is structured to help guide us along the journey towards a hopeful future of continuing evolution of health and medical geography. We begin by discussing several important 'pillars' of knowledge, and points of tension, to begin painting a rich background to situate the reader; the role of exposure, time, context, activity space, big data, smart cities, privacy and ethics. Each of these 'pillars' add colour to the canvas. Next, we move to explore the key opportunities and challenges of using big data in geographical research, with a brief summary of some research in action. We conclude by weaving together the salient lessons learned and positing a future research agenda, optimistically expecting further methodological innovation.

From Aggregate to Individual: A Wealth of Individualised Big Data

A change in the ability to collect individualised location data has led to methodological innovation occurring within health and medical geography, with a move from almost solely aggregate level data (e.g., Census, Address based), which focuses on fixed places, towards individual (or ego-centric) level mobile data, such as GPS location data from smartphones or other devices. This work has been ongoing for some time (Perchoux et al., 2013, Birenboim and Shoval, 2016) with the initial promise giving rise to a host of emerging challenges. The key issues we have seen in our experience of attempting and completing several research projects using individual GPS data are discussed in more detail later in the chapter in the 'research in action' section. An important distinction to recognise is the difference in the quantum of data collected using a mobile (dynamic) in comparison to a fixed (static) approach. Figure 18.1 aptly demonstrates the difference in terms of the volume of data created minute by minute, compared to a once in five (or ten) year single Census data point. This is important as a point of departure. There is a very substantial difference between using a single data point over several years (perhaps ten years) as a sole representation of place for individuals compared to data for several hundred people and the half a million points that can be created in a single year, for a single person.

Exposure: Interactions Between People and Place

An important concept within health and medical geography, and indeed epidemiology, is the concept of exposure. Has someone, something, or a group of people been 'exposed', or not, to a particular phenomenon? Amongst the easier exposures to grasp would be the physical and natural environmental exposures, for example air pollution (Zou et al., 2019) as measured by instruments that capture Particulate Matter (PM_{10} or $PM_{2.5}$). The physical environment exposures usually represent a reasonably obvious and measurable change in the physical environmental conditions; air pollution is better or worse on a scale. Noise can be measured in a similar manner (decibels, dB). A commonly used natural environmental exposure could be to either greenspaces or bluespaces, which can have important impacts on health and wellbeing (Nutsford et al., 2016, Mitchell and Popham, 2007, Foley and Kistemann, 2015), and whether these spaces are usable or visible makes a difference to exposure(s). We can also be exposed to features in the built environment that potentially have a health impact (Renalds et al., 2010); for example Fast Food Outlets (FFO), (Wiki et al., 2020, Pearce et al., 2009) or Alcohol Outlets (AO) (Hobbs et al., 2020, Rowland et al., 2016, Young et al., 2013, Lu et al., 2018) which are tangible examples of the built environment that are measurable and observable. However, we also need to consider an important temporal difference between these categories of exposure: air pollution and other similar phenomena can change second by second. In other words, these exposures are highly dynamic in certain places and times; there is spatio-temporal variation. The equivalent exposure to a statically situated FFO or an AO is not spatially variable,

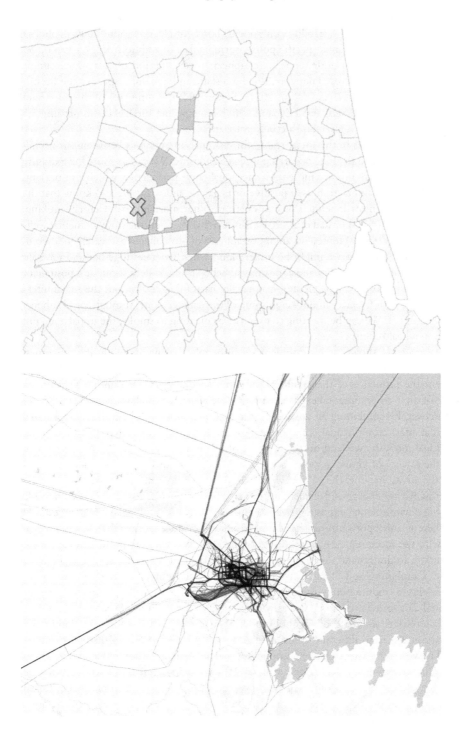

Figure 18.1 Fixed (static) and mobile (dynamic) exposure spaces: Authors' analysis and data.

but has an important temporal variability component. For example, one cannot buy alcohol or fast food if an outlet is not open at specific points during a day. In addition, access to an outlet improves (decreases) when an outlet appears (disappears) over a period of years, for example as the built environment evolves. Therefore, there is a static and dynamic aspect to the health-environment exposures themselves (Kwan, 2021). Figure 18.2 is a summary of the important difference between a static (fixed) and a dynamic (mobile) exposure surface, demonstrating the difference between a discrete place and a continuous surface.

There are also exposures in the social realm, for example, interactions between our friends and family, co-workers and the like, occurring in space and time. We can measure the exposure to others (Yes/No), but this misses a more subjective, qualitative perspective, which is important. In other words, was the interaction itself positive or negative, and why? We also know that the lack of social interaction, or loneliness, is equivalent to other mortality risk factors; smoking, alcoholism or obesity (Holt-Lunstad et al., 2010, Tiwari, 2013). Thus, the study of mental well-being and social interaction(s) represents an important area of our collective research endeavour. It also matters what activities we might be undertaking when we meet together and interact. Each social interaction has a quality and a meaning attached to it; crudely, it can be a positive or negative interaction. For example, alcohol is normally harmful to health, but the social inter-action of drinking with friends can outweigh some of the damage done, through social support (Dunbar et al., 2017). We can be meeting to exercise, or meeting to smoke, the impacts are not the same from a health perspective.

While the evidence shows the environment affects people's behaviour and health outcomes, it is often a combination of exposures rather than a single aspect of the environment. We can represent the influence of the environment more accurately by accounting for the co-occurrence of these environmental exposures in multiple environmental domains (Marek et al., 2021). Moreover, by combining all possible (social, natural, built and physical) environmental exposures that affect an individual, i.e. those exposures that are not inherited, in combination with individual mobility, we land on the concept of exposome (Wild, 2012). The exposome represents the totality of known exposures we face throughout our lives and includes the food we ingest, the air we breathe, the objects we touch, the psychological stresses we face, and the activities in which we engage and as such, it can be viewed as the totality of all known possible (non-genetic) environmental exposures over a lifetime.

Thus, there is an opportunity in the ability to combine the smaller and richer individual level data with the more representative administrative data sets, for example a national census or health survey, to understand the representativeness of the patterns we are seeing at the individual scale. Combining and blending datasets from multiple sources on multiple exposures maintains fidelity to existing epistemologies, whilst adding a flirtatious leap into the novel or crowdsourced data sources, curating data as wide ranging as Air Pollution (Marek et al., 2018), Airbnb's (Campbell et al., 2019) or Alcohol Outlets (Daras et al., 2019). Moreover, combining multiple exposure(s) from different domains (e.g., social, physical, built and natural environ-ments) will move a step closer to an 'exposomic' approach. Further, we argue that by slicing the lifecourse (Christensen et al., 2011, Bayat et al., 2021) into 'typologies' of movement it may be possible to understand the exposome (e.g., early years, or later life as an example area of focus) for specific groups or periods in the lifecourse (Falkingham et al., 2016). We also urge consid-eration of carefully 'blending' quantitative or positivist approaches with theoretically informed, qualitative approaches (Paez et al., 2007); avoiding the pitfalls of the last quantitative revolution perhaps. Further, regularly updated (big) research databases of linked population records based on authoritative datasets and surveys, such as the Integrated Data Infrastructure (IDI) in New

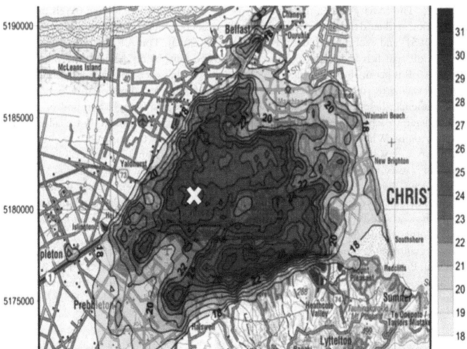

Figure 18.2   Static (fixed) and Dynamic (mobile) exposures.

Zealand, have already been used to generate new knowledge and inform the policy about long-term population behaviours and mobility (Marek et al., 2021).

Introducing Time: Or (Re)Introducing Time

Something that is surprising, but perhaps should not be, is that in thinking through the role of time, others have previously undertaken the heavy intellectual and theoretical lifting. Time geography (Lenntorp, 1976, Hägerstraand, 1970) has been around for half a century. The more recent (re)discovery of time geography is partially a function of ever increasing computing power which now allows many of the concepts to be operationalised, such as more precise space-time paths in the environment, in ways not previously possible (Miller, 1991, Andrews, 2021). In tandem, the reams of data that are being created also allows us to combine data and computing power in a way that offers us new insights into the ways in which health and well-being may be impacted by location or our context (Figure 18.3). The temporal components of both static and dynamic exposure have often been overlooked or summarised; is exposure to a specific location or environment possible at that point in time.

When we wish to disaggregate (or aggregate) spatio-temporal data into meaningful categories for further analysis, to summarise, generalise or so forth, there are two concepts worth a discussion at this point; the activity space and the Uncertain Geographic Context Problem (UGCoP). The activity spaces, in simple terms, are the places that someone travels or moves through during their daily lives. Again, this ontological contribution has existed for some time (Lewin, 1951) and evolved over time (Perchoux et al., 2013). This is linked, but distinct from the UGCoP, which is about the appropriate area-based units being delineated (Kwan, 2012). The UGCoP is about the truly relevant geographic area that exerts behavioural influence on people as well as the temporal uncertainty in both the timing and duration in which individuals experience these contexts (places). What is the appropriate neighbourhood or context to consider for each exposure(s), for each individual? This is perhaps both a challenge and an opportunity. When using individual level data we are able to construct a context from which exposure is derived. There are trade-offs between the utility of any analysis and protection of privacy of both people and places. A more recent review paper (Smith et al., 2019) demonstrated, thoughtfully and thoroughly, how the myriad of different ways in which activity space can be conceptualised may have important consequences for our research. There is not necessarily a standardised approach on the specific method or conceptualisation that is the 'best', in other words, the most appropriate and accurate.

In geography more broadly, there are well-rehearsed considerations related to the Modifiable Areal Unit Problem (MAUP) and ecological fallacy that also feature, but which are not discussed here. If we are able to use individual spatio-temporal data at fine resolution, there is the possibility of mitigating some of the nefarious effects of aggregated data, such as the MAUP, but, new approaches introduce new challenges for the researcher.

Big Data: 'Buzzword Bingo' Or Epistemological Advancement?

During our research, a key observation, and a point of debate in the academy, has been the reasonably obvious disconnect between rhetoric and real world reality of big data in action. We started with this disconnect in our introduction and now return to this theme in the context of 'big data' in research. This disconnect, perhaps, represents one of the fundamental tensions and trends evident in using big data for health and medical geography research as well as the associated epistemological disruptions (Kitchin, 2014, Gruebner et al., 2017). Other chapters

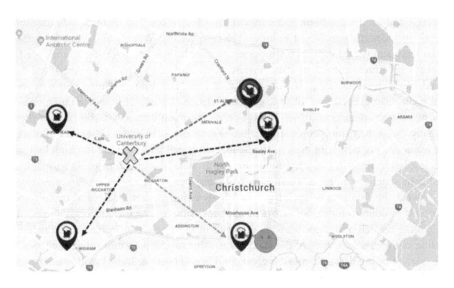

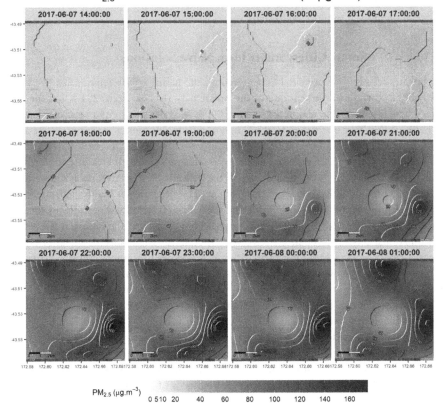

Figure 18.3 Static (fixed) and Dynamic (mobile) exposures with temporal constraints.

(see Chapters 4 and 17) in this book also consider the issue of big data. A brief definition of big data often includes the five 'V's of Volume, Velocity, Variety, Veracity and Value. Volume meaning the large amount of data, perhaps petabytes[1] worth; Velocity referring to the speed, in other words real-time creation of data; Variety, with respect to data being structured or unstructured; Veracity representing data reliability and trust; and Value meaning that data are only useful if they can generate useful (valuable) knowledge. There are of course much fuller definitions of big data (Kitchin, 2014). Thus, one of the main critiques of big data is that it means many things to many people, much like the term 'Smart City'. There are distinct challenges when dealing with 'big data' in fine spatio-temporal detail and the methodological approaches needed in order to understand the data collected. The quantum of data to clean, analyse and comprehend is an issue when using truly 'big data'. The researcher can often lose any sense of what the data means, applying increasingly complex methodological techniques to the data. It is possible to propagate errors by applying some of the various 'black boxes' of machine learning models and algorithms to our data in the pursuit of insight, without full understanding of the precise nature of the algorithms. What to do with missing data, particularly in the context of big data becomes a big(ger) problem. There are some broad frameworks available to deal with missing data, dependent on how the missing data arose, either by removal of data, interpolation or imputation. Again, this can introduce a systematic bias in any resultant analysis and inference. We argue that big data has some big challenges to overcome not least of which is a firm theoretical foundation to stand on.

Smart Cities: Smart Idea, Naïve Solutions

Linked to debates on big data, are debates centered on the nature of the Smart City. We have been reasonably forthright in our critique of the Smart City (Marek et al., 2017) and the use of big data. There is a justifiable reason for stalking the prey of the smart city in the intellectual savannah before unleashing our savage attack. There is a strong flavour, a veneer of intellectual hubris put forth, which is counterproductive (or even dangerous) as 'smart cities' have increasingly hijacked and prostituted terminology in order to sell a product with little attention to the ethical and privacy challenges. It can be argued that the definition of 'Smart City' is strategically ambiguous, rarely presented in a clear, concise manner. The definition is highly dependent on the storyteller. While corporate rhetoric from entities such as IBM, Cisco or Siemens focuses mostly on the technological solutions that can be provided to clients (the cities), it is hard to imagine a functional city without its citizens (Marek et al., 2018, Marek et al., 2017). It is rather easy for city bureaucrats utilising smart cities technologies to slip into using big data for improved control and optimisation of processes; organising traffic flow, management of waste and infrastructure, to facial recognition technologies in the urban landscape. However, if the aim of local government is to be successful in building a smart city, the efforts should be focused on citizens in the first instance.

Only on rare occasions do cities get a chance to fully incorporate smart city technologies into their built environments and infrastructure. After significant earthquakes in 2010 and 2011, Christchurch, New Zealand was afforded an opportunity to rebuild and re-create the city, a smarter city. However, due to the more pressing needs of citizens and property owners (Matthewman and Goode, 2020), only a limited number of mainly infrastructure projects (e.g., smart bins) were realised with limited success (Marek et al., 2018, McNair and Arnold, 2016). While this is an example from a midsize city, the processes and tensions are seen globally in the deployment of smart cities and big data. This is perhaps why the utilisation of smart city technologies is operated through piloting a limited number of small(er) scale projects (Hollands,

2015). We will return to pilotism subsequently. In essence, much is promised but nothing appears to fundamentally change, except the transfer of large volumes of money.

Privacy and Ethics: Just Because We Can, Doesn't Meant We Should

Privacy and ethics are interlinked, others may disagree and wish for a separate discussion of each. We argue that one must consider these in tandem. Many of the more recent arguments around privacy relate to the role of 'technology companies' and 'social media', particularly in the context of data breaches, re-selling and general misuse of data. In essence, there is a Faustian bargain; we provide our data voluntarily for the 'service' of being connected on social media (Kramer et al., 2014). We search for items and products online freely and get tailored results. Along the way, location and preference data are captured alongside important and sensitive information about our social networks and behaviour. In the hands of a health and medical geography researcher we can utilise this data in order to understand important aspects of exposure, mental wellbeing or social connection. However, as researchers, we undertake an ethics process before collecting any such data. This is a critical point of departure from the commercial use of such data. When teaching students in an undergraduate health geography course, we often expose the underlying data collected by some of these companies to demonstrate both the inherent usefulness of this data to a research-minded individual, whilst highlighting the disclosive information that can be exposed (read 'sold') to other companies. A key point of critique is the idea that this process is legal but in our experience, a vanishingly small subset of people read and fully understand the terms and conditions of services they are using. It has been demonstrated that it is possible to use 'anonymised' data, with 15 demographic attributes, to re-identify the vast majority of people (Rocher et al., 2019). In stark contrast, in studies we have conducted collecting such sensitive information, it is imperative that we have gained ethical approval and informed consent from the participants of any study before proceeding. Hopefully, the contrast is clear. The most pernicious challenges relate to the (big) data collected, which is inherently sensitive. The data will often contain home and work addresses as well as other time sensitive daily routine information (e.g., what time you left for work). Analysing this data requires deep and thoughtful consideration about how to present any data from analysis of raw data.

An example of the usefulness of such data came to the fore during the COVID-19 pandemic when location information and information on social networks was vital in understanding the transmission of the SARS-CoV-2 virus (and disease). Many of us are now willingly complicit in tracking ourselves day to day for a public health good. This neatly encapsulates the privacy/control nexus; who owns the data and what will it be used for? Indeed, during the pandemic, the authors used national aggregated mobile phone data (Campbell et al., 2021b) for the purpose of modelling disease spread and evaluation of mobility restrictions aligned to the New Zealand COVID-19 Alert System,[2] Figure 18.4 shows an example of the type of data (for now) freely available. Therefore, in some respects the genie is already out of the bottle, and arguably there is increasing public cognisance of the benefits and drawbacks of the privacy/public health trade-offs.

Challenges and Opportunities for Methodological Innovation

We conclude this chapter with a summary of some of the key opportunities and challenges traversed, ending on an optimistic note about the potential to innovate methodologically, whilst maintaining key aspects of a health and medical geography heritage.

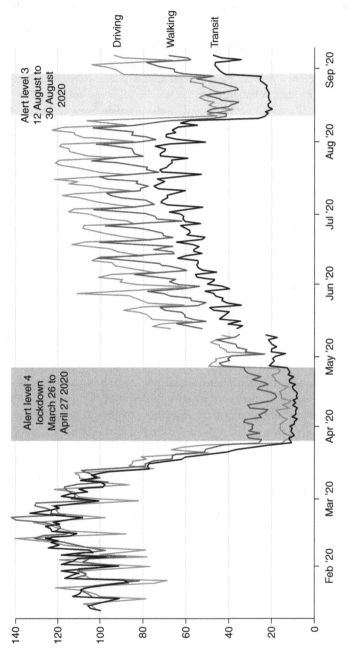

Figure 18.4   Apple mobility data, Auckland, New Zealand during COVID-19 restrictions, 2020.

Challenges: Pilotism and Reproducibility

We see two key areas remaining that will continue to be a challenge within the current paradigm of geographic research. First, 'pilotism', the use of small(er) studies that perhaps never emerge beyond a case study or a pilot. Second, reproducibility; the challenge of how to independently verify scientific findings.

An important critique is that there are often quantitative studies with minimal sample sizes (we ourselves plead guilty), which end up being pilots. There are justifiable reasons for the 'pilot' approach including limited resources and time, but this is inherently problematic in building a generalisable framework within quantitative human geography. It is also very difficult to recruit and sustain a sufficient sample (greater than n=30) with statistical representativeness of more than one group (e.g., comparing within a sample by gender, age, ethnic compositions). So, larger samples are required to be able to make any inference about groups within any sample.

A further key 'thorny' issue is how to reproduce scientific findings from studies that have collected and analysed detailed and personal spatio-temporal data. One can argue that simulation methods offer a way to attempt to observe the general, but not specific, trends. As such, many studies are not truly replicable, but are reproducible on a different sample of people. Therefore, offering the code used for analysis is a way of ensuring best practice, at least in the analysis. Other chapters in this book (see Chapter 17) also contain particularly relevant examples of this issue.

Opportunities: Interventions in the Neighbourhood

In contrast to some of the challenges outlined above, opportunities either already being realised, or on the horizon ready to seize, do exist. Some aspects discussed in the previous section (challenges) also afford an opportunity to evolve our understandings as we tackle the challenges outlined above. We also urge qualitative and quantitative health geographers to draw on the relative strengths of combined methodological perspectives.

Stepping towards a more complete understanding of neighbourhood and context can be seen as an opportunity in that we can now create a 'true' spatio-temporal context for exposure(s). Returning to a recent review paper (Smith et al., 2019), understanding the consequences of how we determine context, as well as conceptualise and model it, could be a step forward in better assessing the exposures related to health behaviours and outcomes. This is not a novel observation, but one that affords an opportunity to move towards tackling a 'wicked problem' in health geography. We argue it is now possible, and essential, that we better understand the extent to which these assumptions about context hold (Thornton et al., 2017).

An understanding of peoples 'actual' spatial-temporal contexts may well allow for personalised interventions in (close to) real time, in specific places. In other words, interventions in the neighbourhood. This is a most exciting prospect, but proves to be a reasonable technological challenge coupled with a wicked ethical dilemma. Even if we could intervene in real time to help 'nudge' people towards healthier environments for example, should we do so if there is a possibility of being wrong and doing harm? This may be low(er) risk in some situations, for example a change to alcohol licencing laws, but perhaps high(er) risk when dealing with a serious respiratory illness and the potential for hospitalisations occurring.

Research in Action: The Sensing City, #Well ConnectedNZ and mGeoHealth

This section aims to address a series of research projects that operationalised the key points of content above in the real world in a suite of live research projects. Over several years the authors have collaborated locally, regionally and internationally with a range of medical and academic colleagues in interdisciplinary research. The common thread connecting the projects discussed here, is the use of smartphone or GPS location data in order to address a health focused research question, an mGeoHealth. We embarked full of enthusiasm on a journey (with our technological 'solutions') to discover new knowledge. The lived experience was somewhat less liberating and resulted in a series of critical reflections before emerging with a more refined approach. Our projects included tracking whole samples of students as part of a pedagogical approach to highlighting the inherent value of location to health and medical Geographers as well as the less well understood importance of privacy and ethics. We also undertook a similar exercise with respiratory patients in order to understand their exposure to the environment, specifically air pollutants that can exacerbate their health condition (Marek et al., 2018). Finally, we used a relatively rudimentary web mapping tool[3] in #WellconnectedNZ which aimed to increase social connections amongst patients (Vannier et al., 2021).

Perhaps our most important observation from conducting this series of research projects with 'big data' in a health geography context is that there is an important and non-trivial disconnect between the rhetoric and the real world reality. A summary of the situation is as simple as this; technological problems are solvable, but people need a tailored approach. Measuring air quality was possible in great detail (see Figure 18.3) and afforded a new opportunity to understand exposures beyond a fixed (static) address. Problems with sensors were possible to overcome, though expensive. An unforeseen issue was the extent to which study participants (who were often seriously ill and so a trip outside their home can be a challenge) did not move and thus static exposure and dynamic (movement based) exposures were often largely similar. This meant that the static (home) exposures were more important for this group of relatively immobile people. We often uncovered smaller differences in movement, rather than larger differences in dynamic exposure. This was in stark contrast to a group of students who were frequently (temporally) and widely (spatially) mobile. We therefore observed differences in 'typology' of movement between those who were ill and the generally younger cohort of students. This also led to an emerging research question about time-weighted exposures – an area of research that seems to have started blooming in the last several years (Campbell et al., 2021a).

Another critical problem was that despite care and attention being paid to ensuring the technology for collecting the data on location was the same, the response from individuals was not the same to the technology. In other words, individual interaction with technology was non-uniform and thus required much more careful exploration. This is not meant as a trite observation, despite perhaps appearing so. It is inevitable that people have a variety of patterns of behaviours with the technology that is used to collect research data. This means that there is potential bias introduced in the data if specific sub-groups of the population share patterns of behaviour with respect to technology which may result in more or less missing data. This can take the form of some participants forgetting to carry a GPS recording device or smartphone or forgetting to charge a battery or even not interacting with the phone at all. A summary of this issue was that often the simplest solutions were preferable to the more complex. This was part of the Genesis of the #WellConnectedNZ project. At its core, the contribution from a geographical perspective was to produce a web mapping application of possible locations of social connection as well as providing context on the digital divide. A key finding that was not obvious at the point of departure is the important impact(s) of the digital divide, or the inequality in access to technology (broadly defined). Indeed, as researchers, perhaps the most humbling of research experiences, was recognising that a key assumption about the ubiquity of smartphones and knowledge of this technology being easy to use as well as being the 'best' solution was not a wise assumption to have made. Moreover, the medical professionals whom we were working alongside, were also stunned by the disconnect between the assumption of technical competence and the reality, which was rather different. The patients often did not have, or did not understand the technology being used. Therefore, it was often not the case that participants in our study understood, desired or wanted access to smartphone technology. The myriad of ways in which our technological 'solution' was 'broken' proved to be quite illuminating. This tangent has an important lesson. As the push towards digital and online solutions accelerates, we may see a new 'digital divide' related to health and wellbeing exacerbated by the exclusion of a certain proportion of our populations. We are increasingly convinced that there is an important role for health and medical geographers in providing critique about the acceleration towards the 'smart'/'digital'/'online' solutions to provide health (and other government) services. There are important and unaddressed questions about those who are excluded, whether intentionally or not. Our approach, a simple map, proved much more impactful and effective than the original desire: a combination of tracking and monitoring. We (the research team) opted for a mixed methods approach in order to more fully understand and co-create the approaches needed. In New Zealand, the Te Reo (Māori language) phrase *He aha te mea nui o te ao. He tāngata, he tāngata, he tāngata* is often used. Translated this means, *What is the most important thing in the world? It is people, it is people, it is people.* The lesson here is not to forget both the wealth of indigenous knowledge that exists alongside western science as well as not forgetting that the data points we collect as researchers are deeply related to lived experiences and daily lives of actual people. Being quantitative (health) geographers, we often tend to see the data points when analysing the data; however, in the context of human and health geography, one should be cognisant that the data represents people's lived experience and requires sensitivity.

Future Research Directions: Where Next for Health Geography, Big Data, Time and Space?

As new methods and terminology (e.g., analytics) evolve and are utilised, there is a genuine concern amongst some Geographers about the 'hollowing out' of disciplines by a pseudo-rebranding of some existing techniques into new fields which are essentially reorganisations of existing disciplines with 'analytics', 'data science' or 'geospatial' appended. The temptation to 'Make Geography Great Again' is a whimsical summary of the problem, without meaning to score political points. Thinking back to the 'quantitative revolution' (see Rosenberg, Chapter 3 of this volume), there is potential for reinvention of the wheel. It is important to be intellectually honest about the reality of (re) discovering older ideas and then applying these to current research projects. We argue, the risk to the academy is that there is theoretical anaemia emerging in our attempts to apply any and every method, because we can. We confess we have been drawn towards this hypnotic rabbit hole, and it is hard to escape when the counterpoint is to do the intellectual heavy lifting of thinking through a theoretical lens; carefully considering ontology, methodology and epistemology of a discipline.

We close this chapter on a hopeful note by highlighting some key avenues for research in the future that look to emerging questions or areas where research could be advanced but is perhaps still nascent. First, we would argue that much of the work on understanding social exposures or social interactions is still emerging, particularly concerning the potential to mix both the quantitative methods for dynamic data discussed in this chapter alongside qualitative information about the nature, quality or purpose of social interactions. Some excellent early examples exist (Alexandre et al., 2020). At the time of writing, a salient example is the emergence of Bluetooth technology in COVID-19 contact tracing smartphone applications, which could be turned on its head and used to capture social connection (rather than infectious disease spread) advancing our understanding of social interactions.

Second, a richer and more accurate understanding of the actual, rather than proximal, nature of exposures to aspects of the built environment is also now possible. Specifically, does an individual go to the nearest outlet (supermarket, alcohol outlet, etc.) and if not, why not? As above, advancing our understanding from the n-nearest built or natural exposures to the actual exposure may help us to understand more precisely the mechanisms through which people choose to move towards or away from, particular aspects of the built or natural environment. This moves us beyond a population weighted, or aggregated exposure, to an understanding of the 'true' individual exposure(s).

There is also an opportunity to (re) use the richer personal data from smaller samples with simulation models to more accurately simulate health outcomes and behaviours for populations or sub-groups of the population using established techniques. Agent-based models (Yang et al., 2011, Badland et al., 2013, Badham et al., 2018) or Spatial Microsimulation methods (Campbell, 2011, Campbell and Ballas, 2016) are potentially useful in this regard. These estimation techniques have previously afforded an opportunity to upscale existing data, when that data did not previously exist in the correct form and provide policy scenarios that can be useful.

A final 'big hairy goal' to conclude is better understanding our individual movements, and all known non-genetic environmental exposures across the lifecourse (the exposome). To what extent do our patterns of movement (and exposure) change as we move from infant, to child, to teenage, to adult, to retirement for example? In understanding the extent of any such differences, we can tailor policy interventions to people at each life stage to have the maximum positive impact. This requires data-intensive research on both people and places longitudinally.

We close by arguing that evolving our understanding of the health and environment interactions using big data whilst drawing on the rich heritage, whether epistemological or methodological of health and medical geography, situated within the discipline of geography will be a fruitful way to proceed.

Notes

1 1 Petabyte = 1000000000000000 Bytes, a Million Billion bytes.
2 https://covid19.govt.nz/alert-system/
3 www.wellconnectednz.org

References

Alexandre, N., Cédric S., Chaix, B. and Yan, K. (2020) Combining social network and activity space data for health research: Tools and methods. *Health & Place,* 66, 102454.
Andrews, G. J. (2021) Bios and arrows: On time in health geographies. *Geography Compass,* 15, e12559.
Badham, J., Chattoe-Brown, E., Gilbert, N., Chalabi, Z., Kee, F. and Hunter, R. F. (2018) Developing agent-based models of complex health behaviour. *Health & Place,* 54, 170–177.
Badland, H., White, M., Macaulay, G., Eagleson, S., Mavoa, S., Pettit C. and Giles-Corti, B. (2013) Using simple agent-based modeling to inform and enhance neighborhood walkability. *International Journal of Health Geographics,* 12, 58–58.

Bayat, S., Naglie, G., Rapoport, M. J., Stasiulis, E., Widener, M. J. and Mihailidis, A. (2021) A GPS-based framework for understanding outdoor mobility patterns of older adults with dementia: An exploratory study. *Gerontology.*

Birenboim, A. and Shoval, N. (2016) Mobility research in the age of the Smartphone. *Annals of the American Association of Geographers,* 106, 283–291.

Campbell, M. and Ballas, D. (2016) SimAlba: A spatial microsimulation approach to the analysis of health inequalities. *Frontiers in Public Health,* 4.

Campbell, M., Marek, L. and Hobbs, M. (2021a) Reconsidering movement and exposure: Towards a more dynamic health geography. *Geography Compass,* 15, e12566.

Campbell, M., Marek, L., Wiki, J., Hobbs, M., Sabel, C. E., Mccarthy, J. and Kingham, S. (2021b) National movement patterns during the COVID-19 pandemic in New Zealand: The unexplored role of neighbourhood deprivation. *Journal of Epidemiology and Community Health,* jech-2020–216108.

Campbell, M., McNair, H., Mackay M. and Perkins, H. C. (2019) Disrupting the regional housing market: Airbnb in New Zealand. *Regional Studies, Regional Science,* 6, 139–142.

Campbell, M. H. (2011) *Exploring the social and spatial inequalities of ill-health in Scotland: A spatial microsimulation approach.* University of Sheffield.

Christensen, P., Mikkelsen, M. R., Nielsen, T. A. S. and Harder, H. (2011) Children, mobility, and space: Using GPS and mobile phone technologies in ethnographic research. *Journal of Mixed Methods Research,* 5, 227–246.

Cupples, J. (2020) No sense of place: Geoscientisation and the epistemic erasure of geography. *New Zealand Geographer,* 76, 3–13.

Daras, K., Green, M. A., Davies, A., Barr, B. and Singleton, A. (2019) Open data on health-related neighbourhood features in Great Britain. *Scientific Data,* 6, 107.

Dunbar, R. I. M., Launay, J., Wlodarski, R., Robertson, C., Pearce, E., Carney, J. and Maccarron, P. 2017. Functional benefits of (modest) alcohol consumption. *Adaptive Human Behavior and Physiology,* 3, 118–133.

Earickson, R. (2009) Medical Geography. In: Kitchin, R. and Thrift, N. (eds.) *International Encyclopedia of Human Geography.* Oxford: Elsevier.

Falkingham, J., Sage, J., Stone, J. and Vlachantoni, A. (2016) Residential mobility across the life course: Continuity and change across three cohorts in Britain. *Advances in Life Course Research,* 30, 111–123.

Foley, R. and Kistemann, T. (2015) Blue space geographies: Enabling health in place. *Health & Place,* 35, 157–165.

Gruebner, O., Sykora, M., Lowe, S. R., Shankardass, K., Galea, S. and Subramanian, S. V. (2017) Big data opportunities for social behavioral and mental health research. *Social Science & Medicine,* 189, 167–169.

Hägerstraand, T. (1970) What about people in regional science? *Papers in Regional Science,* 24, 7–24.

Hobbs, M., Marek, L., Wiki, J., Campbell, M., Deng, B. Y., Sharpe, H., McCarthy, J. and Kingham, S. (2020) Close proximity to alcohol outlets is associated with increased crime and hazardous drinking: Pooled nationally representative data from New Zealand. *Health & Place,* 65, 102397.

Hollands, R. G. (2015) Critical interventions into the corporate smart city. *Cambridge Journal of Regions, Economy and Society,* 8, 61–77.

Holt-Lunstad, J., Smith, T. B. and Layton, J. B. (2010) Social Relationships and Mortality Risk: A Meta-analytic Review. *PLOS Medicine,* 7, e1000316.

Kitchin, R. (2014) Big Data, new epistemologies and paradigm shifts. *Big Data & Society,* 1, 2053951714528481.

Kramer, A. D. I., Guillory, J. E. and Hancock, J. T. (2014) Experimental evidence of massive-scale emotional contagion through social networks. *Proceedings of the National Academy of Sciences,* 111, 8788–8790.

Kwan, M.-P. (2012) The uncertain geographic context problem. *Annals of the Association of American Geographers,* 102, 958–968.

Kwan, M.-P. (2021) The stationarity bias in research on the environmental determinants of health. *Health & Place,* 70, 102609.

Lenntorp, B. (1976) *Paths in Space-Time Environments: A Time-Geographic Study of Movement Possibilities of Individuals,* Lund, Sweden, Royal University of Lund.

Lewin, K. (1951) *Field Theory in Social Science,* New York, Harper and Row.

Lu, H., Zhang, X., Holt, J. B., Kanny, D. and Croft, J. B. (2018) Quantifying spatial accessibility in public health practice and research: An application to on-premise alcohol outlets, United States, 2013. *International Journal of Health Geographics*, 17, 23.

Marek, L., Campbell, M. and Bui, L. (2017) Shaking for innovation: The (re)building of a (smart) city in a post disaster environment. *Cities*, 63, 41–50.

Marek, L., Campbell, M., Epton, M., Kingham, S. and Storer, M. (2018) Winter Is Coming: A Socio-Environmental Monitoring and Spatiotemporal Modelling Approach for Better Understanding a Respiratory Disease. *ISPRS International Journal of Geo-Information*, 7, 432.

Marek, L., Hobbs, M., Wiki, J., Kingham, S. and Campbell, M. (2021) The good, the bad, and the environment: Developing an area-based measure of access to health-promoting and health-constraining environments in New Zealand. *International Journal of Health Geographics*, 20, 16–16.

Marek, L., Greenwell, J., Hobbs, M., McCarthy, J., Wiki, J., Campbell, M., Kingham, S., Tomintz, M. (2021) Combining large linked social service microdata and geospatial data to identify vulnerable populations in New Zealand. *Big Data Applications in Geography and Planning*. Edward Elgar Publishing.

Matthewman, S. and Goode, L. (2020) *City of quakes: Excavating the future in Christchurch*, Sociological Association of Aotearoa New Zealand.

McNair, H. D. and Arnold, L. M. (2016) Crowd-sorting: Reducing bias in decision making through consensus generated crowdsourced spatial information. International Conference on GIScience Short Paper Proceedings 2016.

Miller, H. J. (1991) Modelling accessibility using space-time prism concepts within Geographical Information Systems. *International Journal of Geographical Information Systems*, 5, 287–301.

Mitchell, R. and Popham, F. (2007) Greenspace, urbanity and health: Relationships in England. *Journal of Epidemiology and Community Health*, 61, 681–683.

Moon, G. (2009) Health geography. *In:* KITCHIN, R. and THRIFT, N. (eds.) *International Encyclopedia of Human Geography*. Oxford: Elsevier.

Nutsford, D., Pearson, A. L., Kingham, S. and Reitsma, F. (2016) Residential exposure to visible blue space (but not green space) associated with lower psychological distress in a capital city. *Health & Place*, 39, 70–78.

Paez, A., Scott, D., Potoglou, D., Kanaroglou, P. and Newbold, K.B. (2007) Elderly mobility: demographic and spatial analysis of trip making in the Hamilton CMA, Canada. *Urban Studies*, 44, 123–146.

Pearce, J., Hiscock, R., Blakely, T. and Witten, K. (2009) A national study of the association between neighbourhood access to fast-food outlets and the diet and weight of local residents. *Health & Place*, 15, 193–197.

Perchoux, C., Chaix, B., Cummins, S. and Kestens, Y. (2013) Conceptualization and measurement of environmental exposure in epidemiology: Accounting for activity space related to daily mobility. *Health & Place*, 21, 86–93.

Renalds, A., Smith, T. H. and Hale P. J. (2010) A systematic review of built environment and health. *Fam Community Health*, 33, 68–78.

Rocher, L., Hendrickx, J. M. and De Montjoye, Y.-A. (2019) Estimating the success of re-identifications in incomplete datasets using generative models. *Nature Communications*, 10, 3069.

Rowland, B., Evans-Whipp, T., Hemphill, S., Leung, R., Livingston, M. and Toumbourou, J. W. (2016) The density of alcohol outlets and adolescent alcohol consumption: An Australian longitudinal analysis. *Health & Place*, 37, 43–49.

Smith, L., Foley, L. and Panter, J. (2019) Activity spaces in studies of the environment and physical activity: A review and synthesis of implications for causality. *Health & Place*, 58, 102113.

Thornton, L. E., Crawford, D. A., Lamb, K. E. and Ball, K. (2017) Where do people purchase food? A novel approach to investigating food purchasing locations. *International Journal of Health Geographics*, 16, 9.

Tiwari, S. C. (2013) Loneliness: A disease? *Indian Journal of Psychiatry*, 55, 320–322.

Vannier, C., Mulligan, H., Wilkinson, A., Elder, S., Malik, A., Morrish, D., Campbell, M., Kingham, S. and Epton, M. (2021) Strengthening community connection and personal well-being through volunteering in New Zealand. *Health & Social Care in the Community*.

Wiki, J., Kingham, S. and Campbell, M. (2020) A geospatial analysis of Type 2 Diabetes Mellitus and the food environment in urban New Zealand. *Social Science & Medicine*, 113231.

Wild, C. P. (2012) The exposome: From concept to utility. *International Journal of Epidemiology*, 41, 24–32.

Yang, Y., Diez Roux, A. V., Auchincloss, A. H., Rodriguez, D. A. and Brown, D. G. (2011) A spatial agent-based model for the simulation of adults' daily walking within a city. *American Journal of Preventive Medicine,* 40, 353–361.

Young, R., MacDonald, L. and Ellaway, A. (2013) Associations between proximity and density of local alcohol outlets and alcohol use among Scottish adolescents. *Health & Place,* 19, 124–130.

Zou, B., You, J., Lin, Y., Duan, X., Zhao, X., Fang, X., Campen, M. J. and Li, S. (2019) Air pollution intervention and life-saving effect in China. *Environ Int,* 125, 529–541.

19

GEOGRAPHIES OF DISABILITY

On the Potential of Mixed Methods

Sandy Wong and Diana Beljaars

Introduction

Since the 1970s, geographers have made important theoretical and methodological contributions to understanding the socio-spatial circumstances affecting the lives of people with disabilities, which include physical and mental conditions. The history of disability geography reveals a sharp departure from positivism and quantitative approaches to robust engagements with critical social theories and qualitative methodologies. Quantitative research in disability geography is relatively uncommon today, resulting in partial knowledges of the spatial distributions of disabled people and the environmental attributes of where they live, work, and play. Furthermore, there is rare cross-fertilization between quantitative and qualitative approaches, mirroring the split in geography more broadly between spatial scientific and humanistic perspectives.

This chapter traces the trajectory of disability geography to consider the potential of mixing quantitative and qualitative approaches for generating new geographic insights into the everyday experiences of disabled individuals. We begin with an overview of disability concepts and definitions, followed by a summary of major historical and intellectual developments in disability geography. Then, we explore the possibilities for mixing methods, using illustrative studies on: the socio-spatial mobilities of individuals living with visual impairment, the socio-spatial educational environments of young pupils with disabilities, and the socio-spatial practice of critical accessibility mapping. The chapter ends with reflections on researcher positionality and directions for future scholarship.

Defining Disability

How disability is defined affects how we relate to disability, who identifies as being disabled, how we understand the everyday geographies of people living with disability, and how we formulate policies to combat ableism. In this chapter, when we refer to disability, we consider it to encompass a diversity of physical and mental conditions that are fluid and contingent rather than permanent or fixed; and as a range of mind and body differences rather than as a 'loss' in capacity or a deviation from an 'ideal' state that is mirrored in the normatively conceived image of an 'able' body. Physical conditions include bodily differences that influence people's corporeal functionality and physical mobility, such as limb loss, spinal cord injury, and visual

DOI: 10.4324/9781003038849-22

and hearing impairments. Mental conditions include those that shape individuals' thoughts, emotions, and behaviors, such as cognitive disabilities and anxiety disorders. These examples are by no means exhaustive and are subject to change over time and by varying (social) contexts.

Ableism – the structural and societal devaluation of and discrimination against disabled people – influences the politics of disability identification. The claiming or rejecting of disability as an identity stems from people's perceptions of and relationship to disability, which are in turn rooted in beliefs around what bodies ought to be 'capable' of. Many individuals may not identify as being disabled. As an example, there is ambiguity as to when vision loss becomes disabling. Indeed, most people wearing vision aids, such as glasses and contact lenses, are unlikely to identify as disabled. Others actively reject the identity due to the stigma of disability and fears of social exclusion. In some cases, people's disability may be invisible or not well understood in the context of ableism, making it difficult for both individuals themselves and others to recognize their condition as a disability; for instance, the taboos surrounding clinical depression. For others, their disability may be temporary rather than permanent (e.g., migraines), so while they recognize their condition as disabling, they may not identify as being disabled due to the short duration of their condition. At the same time, there are many people who do claim disability as their identity and actively challenge the ableism enmeshed in disability identification. Disabled activists and advocates have debated the use of identity-first versus people-first language (i.e., disabled person versus person with disabilities) and rejected disability euphemisms (e.g., special needs, differently abled). People recognize that disability is an important part of their identity and that identity politics centered around disability are necessary for dismantling ableism. To quote Alice Wong, a disabled activist in the U.S.,

> To me, disability is not a monolith, nor is it a clear-cut binary of disabled and nondisabled. Disability is mutable and ever-evolving. Disability is both apparent and nonapparent. Disability is pain, struggle, brilliance, abundance, and joy. Disability is sociopolitical, cultural, and biological. Being visible and claiming a disabled identity brings risks as much as it brings pride
>
> *Wong, 2020, p. xxii*

In geography, 'disability' as a keyword has expanded from a medical condition to a sociopolitical construction, reflecting the changing models used to understand the experience of disability (Crooks, Dorn, and Wilton, 2008, Wilton and Evans, 2020). Until the 1980s, the medical model was primarily employed to characterize disability. The medical model considers disability to be an individualized medical condition such that only medical solutions are conceivable. Medical interventions are intended to restore the body or mind to a 'normal' condition, one that allows people to have 'complete' functional capacity. A person is considered disabled when medical interventions are unable to return them to 'normal' (Wilton and Evans, 2020). While the medical model represents a narrow framework for understanding disability because it is asocial and ahistorical, it continues to pervade the way we think about disability. Even we (the authors) cannot avoid it –earlier when we listed examples of conditions that we considered to fall under the category of disability, we used medical terminology to do so. This is because these conditions are widely recognized and we want our readers to be able to consider tangible examples while thinking through various theoretical and methodological issues.

Starting in the 1980s and 1990s, the social model emerged. This model critiques the medical model's emphasis on body/mind impairments and functional limitations for defining disability as a purely biological reality (Wilton and Evans, 2020). Instead, disability is conceived through the social model as restrictions to community participation due to inaccessible physical environments and discriminatory attitudinal and social milieus. In other words, a person is regarded as having a disability when they are unable to fully participate in daily activities. Since the causes of disability are environmental, the solutions involve collective efforts among communities and societies towards improving accessibility and inclusion, and addressing ableism. The social model brings to light the multitude of ways in which disability is socially constructed and part of a larger social context rather than being simply located and confined within individual bodies. It refers to the ways in which disability enters the economic realm, digresses from the norms that govern a society, and reflects what society considers as valuable (Sontag 2001). Here, the 'abnormality' of disability is unveiled as differing from certain societal standards of what people ought to be able to do that are largely derived from people's capacities to contribute to the economy. In effect, it lays bare how bodies are commodified and lose value under capitalism (Gleeson 1999). Casting this wider net, it unsettles the stability of the disability concept and draws attention to its existence as a critical reflection of the society from which it is mobilized. In an attempt to bring these models together, Shakespeare (2006) argues for understanding disablement to emerge in the interactions between body and society.

In the late 1990s and early 2000s, relational models were utilized to call attention to the ways in which social relations and cultural representations configure our understandings of disability (Wilton and Evans, 2020, Hall and Wilton, 2017). Deconstructing the notion of disability, relational models disrupt the materialist emphasis of the social model (i.e., disability as produced by capitalism and other 'disabling' socio-political structures) to consider how performativity shapes the relations between disabled identities and socio-spatial environments. Rather than treating disability as taking place 'in space', relational models start from the assertion that disability should be understood as actively producing space. They draw attention to how medical categories of impairment are socially produced as opposed to being 'natural' and unquestioned classifications. In doing so, relational models further challenge the tenets of the medical model by questioning the authority of the medical gaze to define and categorize individual impairment. A relational model also brings human agency into view. Rather than being passive victims of ableist contexts, disabled people actively and dynamically engage with, contest, and transform social spaces (Wilton and Evans, 2020). As a whole, relational approaches further dismantle the medical status quo, and establish disability as inherently fluid by bringing into focus the ways in which everyday practices and discourses influence it.

Historical and Intellectual Trajectory of Disability Geography

The historical and intellectual trajectory of disability geography developed in parallel with societal shifts from defining disability as a medical label to a socio-political construction. Its history also reflects changing theorizations about health and well-being within geography, which over time expanded its framings of health from biomedical to socio-ecological models. Disability geography was initially positivistic. Geographers in the 1970s began using quantitative and spatial analytic methods to investigate the spatial distributions and experiences of disabled people, with an emphasis on physical disabilities. The 1990s marked a transition point when some geographers recognized that positivist frameworks were limited, in addition to the humanist turn in the 1970s and 1980s, noting the apparent lack of interest in medicalized lived experience in human geography. Following radical and feminist turns in human geography and reflecting these turns in the social sciences more broadly, research approaches consequently shifted to engage with critical social theories and qualitative methodologies, in addition to increased recognition of and work on learning and intellectual disabilities. The summary that follows provides a brief overview of the contributions of the two main intellectual traditions and the challenges of engagement between them. (More comprehensive resources on disability geography can be found elsewhere, e.g., Park, Radford, and Vickers, 1998, Gleeson, 1999, Crooks, Dorn, and Wilton, 2008, Chouinard, Hall, and Wilton, 2010, Hall and Wilton, 2017 and Wilton and Evans, 2020).

The earliest work on disability geography emerged in the 1980s, when disability was largely conceived of as an individualized biomedical condition and geographers utilized positivist approaches to investigate disability (Chouinard, Hall, and Wilton, 2010). Some used regression analysis to examine the incidence and spatial distribution of disabling conditions (Lovett and Gattrell, 1988, Mayer, 1981). Others used surveys, descriptive statistics, and regression analysis to examine access and mobility issues in urban and rural environments for disabled people (Gant and Smith 1988, Wolch 1980). Another group of geographers applied behavioral approaches to assess how individuals with visual impairment cognitively map spatial information to navigate the built environment (Golledge, 1993, Kitchin, Blades, and Golledge, 1997, Marston, Golledge, and Costanzo, 1997). The former research stream was important for understanding the spatial distributions and mobility of people with disabilities, and the latter was vital to the development of assistive technologies such as tactile maps, haptic tools, and GPS navigational devices (Wilton and Evans, 2020).

The earliest engagements with critical social theory also emerged in the 1980s, but it was not until the 1990s that a vigorous body of scholarship developed (Chouinard, 2018, Wilton and Evans, 2020). In the late 1980s, geographers traced the historical geography of asylums for mentally ill people in the UK (Philo, 1987) and how deinstitutionalization contributed to the development of service-dependent 'ghettos' in North American cities (Dear and Wolch, 2014). Starting in the 1990s, geographers highlighted materialist approaches and sought to analyze how socio-political and built environments produced disability. Deriving disability as a function of capitalism as the first social critical comprehensive theory of disability geography, Gleeson (1999) outlined an 'embodied historical-geographical materialism' as an analytical approach to examine how overlapping spheres of social relations generated environments that excluded and devalued disabled people. Imrie (2013) called attention to the disabling effects of inaccessible built environments in the UK as they obstruct the mobility and social participation of visually impaired pedestrians. In the 2000s, disability geographers engaged with other theories and tools to generate new insights on embodied experiences of disablement, expanded the meaning of disability to include more bodies and perspectives, considered new spaces as sites of inquiry, and explored spatial metaphors in the experiences of disabled people (Chouinard, Hall, and Wilton, 2010, Crooks, Dorn, and Wilton, 2008).

Indeed, it is the foregrounding of the multiple forms of spatiality through which geographers have examined disability, in comparison to disability study scholars, that has led to the broad variety of critical understandings of disablement. Moss and Dyck (2003) used a feminist perspective to understand how the embodied identities of women with chronic illness were a product of both the material and the discursive. Feminist perspectives, and in particular feminist methodologies, also highlight how the embodied and emotional lived experience of disability are vital in how spaces are encountered and negotiated (Davidson, 2017, Davidson, Bondi, and Smith, 2005). Posthumanist inspired approaches also called on emotion and affect to understand disability to bring to light how disability emerges and is negated within the socio-material situation of the body (Anderson and Harrison, 2010). In particular, non-representational theory has been engaged with to understand how disabled subjectivities transpire from encounters between individuals' embodied practices and other bodies, spaces, and objects (Hall and Wilton, 2017, Macpherson, 2009).

Despite the advancements made in expanding the meaning of disability and in broadening the boundaries of disability geography scholarship, there continues to be a split between those who employ approaches rooted in humanism or positivism. Researchers operating from each paradigm have continued to develop their research independently, producing knowledges that are rarely in conversation with one another (Dorn, Keirns, and Del Casino, 2010). The early heated exchanges between Golledge on the one side and Gleeson and Imrie on the other regarding 'correct' conceptualizations of disability represent and continue to represent the sometimes uneasy coexistence of heterogeneous epistemological and ontological approaches to doing disability research in geography (Golledge, 1996, Gleeson, 1996, Imrie, 1996). Gleeson and Imrie critiqued Golledge's vision of the future direction of disability research, primarily for its reductionist and ableist conceptualizations of disability and for its exclusion of a social constructionist perspective (Imrie, 1996, Gleeson, 1996, Golledge, 1993). Golledge, a positivist behavioral geographer, challenged the criticism leveled against him because he was blind while Gleeson and Imrie were not disabled (Golledge, 1996). "From his lofty position of ableism, Imrie (whom I believe is very able) accuses me of denigrating the very group to which I belong and try to represent!" (Golledge, 1996, p. 404, emphasis original). Reflecting a broader postmodernist destabilization of authorship, these exchanges brought up challenging questions about representation and authenticity, and about how disability geographers negotiate ableisms in their research practices. Disability as a keyword in geography, then, moves apprehensively between the medical and the non-medical, the positivistic and the humanistic, and the social and material. A major challenge is finding common ground and productive possibilities between the two dominant, competing research traditions.

The Potential of Mixing Methods

Quantitative and qualitative methods are commonly seen as epistemological and ontological dualisms as they serve knowledge creation following different principles. The philosophical foundation of quantitative research is usually (post-)positivism whereas (post-)humanism characterizes the underlying philosophies for qualitative research (Baxter, 2018). Rather than reify these dualisms, we consider how mixing qualitative and quantitative methods can generate more multifaceted understandings of everyday life compared to the implementation of one type of method alone. Quantification and qualification of dimensions of disabilities are both necessary as they both provide important insights into the production of disability and circumstances under which disabilities figure in society and in individual lives. Indeed, we argue it is precisely geography's habit of thinking with and through spatiality that is key in the amalgamation of various heterogeneous relations that produce disability or facilitate its manifestation. In this section, as we summarize disability geography praxis and navigate the possibilities of mixed method approaches, we hope our ideas inspire readers to consider imaginative and productive ways to blend methods in their own research.

Quantitative methods in geography often include mapping via Geographic Information Systems (GIS) and statistics, which generate particular insights on disability. In their most rudimentary form, quantitative methods enlighten how disability can be quantified in terms of severity and frequency, which follows assertions that disability produces suffering since it deviates from what is constructed as 'normal' or 'able-bodied.' Most accessibility research falls within this body of work as it includes examinations and mappings of obstacles in the physical landscape as well as a broader set of impossibilities for disabled people navigating the taken-for-granted non-disabled world. Here, quantitative methods are used to calculate the impact of a plethora of barriers to different aspects of life. Compared to non-disabled people, disabled individuals experience further challenges, including added time, money, effort, and discomfort, as well as a lower availability of necessary care, physical and social access, and technology. Examples of relevant scholarship include the papers of: Church and Marston (2003) on measuring accessibility for disabled populations; Higgs (2005) on access to health care services; Metzel and Giordano (2007) on employment services accessibility for people with disabilities; Rosychuk et al., (2015) on mental health follow-up meetings for people with mood disorders after attending emergency departments; Gao, Foster, and Liu (2019) on spatial associations between people with disabilities' place of residence and the location of rehabilitation services; Planey (2019) on the distribution of US-based audiologists as correlated with the socio-demographic characteristics of older adults reporting difficulty hearing and structural factors; and Planey et al., (2019) on mental health help-seeking among African American youth and their families.

Qualitative methods allow answering questions on how disabled life differs from non-disabled life and challenges the taken-for-granted notions of living without disability. This methodology also attends to how different disabilities are characterized by unique sets of lived experiences (Berger and Lorenz, 2015). Qualitative methodology is thus predominantly used to understand the lived realities of disability (Chouinard, 2012, Parr, 1999, Kitchin et al., 1998), including the geographical underpinnings of what it means for people to be disabled and how this shapes other dimensions of life, for instance romantic relationships, (school) career, or in sports, how they experience medicalization and various socio-material processes in particular spaces (Butler and Bowlby 1997, Mol 2002), how and where they encounter forms of stigma, inaccessibility, and intersecting racist or sexist violence (Davidson, 2010), and how technologies, landscapes, and nonhuman animals affect disabilities (Segrott and Doel, 2004, Beljaars, 2020, Gorman, 2017, Macpherson, 2009). Such inductive approaches, informed by primarily humanist and feminist epistemologies, often employ semi or unstructured interviews that allow for participants to get across how they understand and navigate their disability. Interviews can be accompanied by participative observation (Davidson, 2017), participatory action research, and autobiographies (Davidson and Henderson, 2010).

Different research questions in disability geography ostensibly lend themselves more to either quantitative or qualitative methods. For instance, questions regarding inequalities in spatial patterns tend to be underpinned by numerical data, while studies that aim to understand how social relationships produce injustices tend to be supported by nominal data. Typically, quantitative methods are more suitable for questions related to generalizable trends while qualitative methods are more fitting for questions regarding underlying processes. However, these are simply generalizations – it does not always have to be the case that methods traditionally aligned with certain research foci remain that way. There are no hard rules for which methods must go with what research goals, and that, in fact, creative usage of methods could lead to valuable new insights and strengthen the evidence base.

In my (Wong) study, I employed quantitative and qualitative methods to analyze the circumstances that impede or facilitate the everyday mobility of people with visual impairments residing in the San Francisco Bay Area, a major metropolitan region in the U.S. (Wong, 2018a, 2018b). I used surveys and interviews (both sit-down and mobile). From the data collected, I was able to delineate the locations my participants traveled to and the routes they took to get from one location to another. Using GIS and simple statistical methods, I was able to map my participants' travels and quantify their activity spaces. Generally, I was able to understand the where, when, and how of their mobility using spatial and statistical analysis. As for qualitative methods, I used qualitative analysis and a deductive approach for coding survey and interview data. In contrast to an inductive approach where themes are generated 'in vivo' during data analysis, a deductive approach utilizes 'a priori' themes. I wanted to investigate the extent to which Hägerstrand's (1970) space-time framework aligned with the experiences of my study participants. From this approach, I was able to glean the relational processes influencing people's identities and mobilities in relation to their socio-spatial environments. For all participants, their mobility was not just a function of their physical impairment but also tightly linked to their social support networks, varying levels of societal ableism, and institutional structures and resources. A mixed method approach generated a more comprehensive and complex narrative about individual mobility as it relates to the accessibility and inaccessibility of people's social and built environments.

Holt, Bowlby, and Lea's (2019) study combined quantitative and qualitative methods to examine the relations between disability, poverty, and secondary education (pupils aged 11–16) in Southeast England. Quantitatively, they used data from the National Pupil Database to generate descriptive statistics summarizing the characteristics of pupils with Special Educational Needs and Disability (SEND). Their qualitative methods included participant observation, interviews with key educational personnel and parents, and an abductive analytic approach combining 'a priori' and 'in vivo' themes. Their quantitative results revealed that a higher proportion of students with SEND were eligible for free school meals compared to non-SEND pupils, affirming the dominant narrative that students with SEND live in poverty and come from socioeconomically disadvantaged backgrounds. The qualitative findings, on the other hand, countered and complicated this unidimensional narrative by uncovering that pupils with SEND had diverse backgrounds across the socioeconomic spectrum that differentially influenced their access to educational resources and cultural capital. Holt, Bowlby, and Lea found strikingly different types of socio-spatial educational environments for students with SEND. One type operated as spatial containers with high concentrations of poor students, narrow curricula, restricted opportunities for accessing cultural capital, and low expectations of pupils' future prospects. The other type functioned as networked hubs with varied curricula and greater access to resources, including cultural capital. Here, students were more likely to come from a higher socioeconomic group and have parents who would navigate the SEND system and advocate for their children. The results allowed the authors (2019, p. 1) to argue that "the intersecting experience of SEND, class, and capitals can (re)produce socio-economic inequalities through school spaces".

Another illustrative example of a mixed methods approach is Hamraie's (2018) critical accessibility mapping project, in which qualitative GIS is used to examine the accessibility of Vanderbilt University's campus, located in Nashville, Tennessee. In doing so, the study integrated qualitative data and methods with a spatial analytic technique that is often considered as belonging to quantitative methods. They conceived their project as a socio-spatial practice that reconceptualized geospatial data and mapping as products of a cultural process, rather than as objective representations of reality. In addition to using GIS to map data, they used R to generate statistical results and develop an interactive web application, and used Research Electronic Data Capture (REDCap) to collect data and visualize maps in real-time. Their critical accessibility mapping project sought to disrupt and go beyond depoliticized accessibility maps traditionally created to aid navigation and identify ADA (the Americans with Disabilities Act of 1990) compliance standards for narrow definitions of disability. The ADA is a U.S. civil rights laws that prohibits discrimination against people with disabilities. For their project, they posed a key question: "what counts as access, for whom, and under what conditions?" (Hamraie 2018, p. 460). Towards answering it, they used GIS and other tools to generate counter-maps that interrogated the built environment, recognized data as situated, expanded definitions of access to engage with intersectionality, and generated fresh narratives about the accessibility of university spaces.

Researcher Positionality

Regardless of the choice of method, disability geography intellectual histories have taught us that researcher positionality and research implications remain vitally important. Research design, conduct, and analysis require extensive consideration on what claims researchers can make regarding the realities of disabled people and of living with particular conditions. While this may seemingly pertain mostly to qualitative methods and interpretive modes of analysis, studies using quantitative research designs also need to carefully investigate the assumptions upon which research questions and statistical models are based, as well as how research construes disabled people (Titchkosky, 2011).

As researchers who do not identify as disabled and study people who do, we strive to be reflexive during the process of knowledge creation. At the start of my study, I (Wong) felt very much an outsider as most of my knowledge about disability came from academic readings rather than from personal experience. However, while I was physically and psychologically distant from the experiences of disabled people, I wanted to learn more, to hear directly from them, and to help share their stories. From the debates between Golledge, Gleeson, and Imrie, I knew that I needed to proceed with caution as a non-disabled researcher who had just begun to reflect on her internalized ableism. I regarded my study participants as the experts and sought to be neutral during the research process – but as my research progressed, my relationships with my participants became more friendly than distant; I developed a greater understanding of the embodied experiences of people with visual impairment during my travel interviews; and I became more aware and angered by the everyday indignities that they encountered. While still not an insider, I became less of an outsider along the outsider-to-insider continuum. I sought to be careful with the analysis of the information that people shared with me by weaving commonalities in experiences and by striking a balance between constraints as derived from visual impairment compared to those stemming from our social environments –and to do so without perpetuating ableist misconceptions. My reflexivity remains a work in progress as I endeavor (imperfectly) to be introspective about how my non-disabled identity may influence the knowledge claims that I make.

I (Beljaars) designed my study on compulsivity in Tourette syndrome and the role played by the sociomaterial surroundings of the body in inciting interactive compulsions following several principles. Focusing on the compulsive acts, rather than on the people, avoided participants having to explain these acts as they are experienced as meaningless, purposeless, and unrelated to the self or psyche. Preventing the risk of me attributing meaning to the content of compulsions where there was none, I also did not employ an interpretative analysis. The combination of interviews, observations, and mobile eye-tracking allowed for an analysis that produced spatial patterns of individual acts, which in collected form had implications for wellbeing. Also, as a researcher without Tourette syndrome, I ensured that the study did not extend the medical and clinical conceptualization of compulsivity as an inherent deficit, nor did it claim to represent life with Tourette's. Rather, I attended to disability by assembling a new formation of disabling circumstances. I mediated not having lived experience of compulsions through profiling myself with my sister who has Tourette syndrome, in accordance with her full permission to mention her and her Tourette's. It positioned me as a very well-informed outsider. Indeed, having lived with her gave me certain insights and sensitivities into the everyday realities and related sets of concerns. In effect, on several occasions, through my descriptions of her compulsivity she became a 'proxy' to help remind research participants of their own compulsions or provide some reassurance that their experiences were shared.

Given our experiences, we do not believe that studying disabilities that are not a reality for a researcher are unethical and should be avoided, because not studying a pressing matter is also unethical. However, it does mean that (even) without intending, a lack of nuanced insights can (re)produce narrow depictions of lives with disabilities, or even perpetuate stereotypes and ableist norms, and result in a curtailment of emancipatory efforts (Hansen, 2009). Therefore, rigorous self-examination is recommended of the researcher's underlying assumptions, as well as meaningful and extensive collaboration with people and communities for whom the disability under investigation is a reality, is imperative.

Future Research Directions

Along with mixing qualitative and quantitative methods, we propose that a new wave of disability geography scholarship should include deeper and more constructive engagements with intersectionality, social activism, and medicine. Therefore we join calls for a greater consideration of processes of racialization in the spatialities of disability, both in terms of more empirical work shedding light on racialized inequities and how the legacies of colonialism as emergent from critical race theory and Black intellectual histories and theories have been woven into the construction of disability (Price, 2010, Robinson, 2000, Fanon, 2008, Bell, 2011, Weinstein and Colebrook, 2017, Tolia-Kelly, 2010). See Chapter 15: Black Geographies in this book for further discussion. Also, integrating disability in Indigenous epistemes and spatial concerns would be very welcome (Sundberg 2014), in particular utilizing qualitative approaches to counter quantitative Western biomedical reformulations of indigeneity (Vawer et al., 2013, TallBear, 2013). Additionally, as Jampel and Bebbington (2018, p. 1) assert, such engagements would benefit from collaborations with development geography to improve our understanding of the "disability–poverty nexus, interventions in the name of improvement and cure, and mobility and migration" to further decolonize disability geography (also see Campbell, 2011).

We would also emphasize the importance of the intersectional overlap of disability and queerness. Indeed, recognizing that sexuality and gender identities in the context of disability also produce different vulnerabilities, systemic injustices, and exclusionary practices in a variety of spaces that could be brought to light in mixed methodologies, we encourage queer approaches to disability (McRuer, 2006, 2011, Gahman, 2017). Queering disability itself may also destabilize it as a construct and allow us to continue moving beyond the medically invoked dualism of (dis)ability and theorizations of disability through identity and representation to attend to the fluidity and situatedness of conditions (Campbell, 2009).

Mixed methodologies could analytically account for different presences, absences, and temporalities of disability, without diminishing the premise for state support. We return here to Hamraie's (2018) mixed method approach that they developed from critical access theories. 'Mapping Access' is extremely potent for its capacity to demonstrate multiple modes of inaccessibility beyond those formally recognized by the state. They explain that it "offers a new method of socio-spatial practice, with distinct benefits over compliance mapping: it recognizes marginalized experts; redefines data, crowdsourcing, and public participation; offers new stories about disability and public belonging; and materializes the principles of disability justice, an early twentieth-century movement emphasizing intersectionality and interdependence" (Hamraie, 2018, p. 456).

This feeds into the imperative to address the problems of inaccessibility and exclusivity in our research praxis, which are tied to longstanding issues of representation and ableism in geography. As Hansen and Philo (2007) remind us, to address ableism requires directly challenging notions of normalcy in a wide array of geographical research dimensions. This could involve a critical examination of how traditional and novel research methods uphold and sustain ableism in human geography, for instance in (unintentional) theorizations of the body and practice; reinventing them as enabling by challenging the taken-for-granted logical underpinnings that exclude people with (particular) disability(ies). By extension, future research in disability geography could improve by interrogating the inaccessibility of analytic techniques themselves; for instance, the inaccessibility of GIS software, such as ArcGIS, to visually impaired geographers who use screen readers. The way forward is to identify the ways in which our research praxis can be more inclusive and accessible.

Another recommendation is to build on earlier digital work that emphasizes the multitude of ways in which the spatialities of disability can be studied, such as through public or community Internet forums (e.g., Campbell and Longhurst, 2013), but also through other online networks. These digital collectives have amplified advocacy through bringing disability concerns to the forefront of more debates and challenging multiple forms of ableism and dehumanization (e.g., demonstrating physical, mental, and social inaccessibility, and responding to disability rendered as 'inspiration porn'). Social media outlets such as Twitter, Facebook, Instagram, and Tiktok offer new types of democratic performance that facilitate new types of activism, for instance through hashtags. An example here is '#geendorhout', Dutch for 'no dead wood.' This made visible the millions of people at high risk of serious illness or death from COVID-19, and was a response to the assertive views by some that high-risk groups could be sacrificed in order for people perceived as less at risk to resume the individual freedoms they had pre-pandemic. Multiple kinds of analysis that combine quantitative and qualitative methods could be used to examine these digital spaces and could provide new possibilities for more involvement of disabled people in geographical research.

With the increasingly cellular level at which the biomedical sciences study bodily, behavioral, and perceptual difference between people, and a sharp increase in neuroscientific influence in the knowledge production of social life (Davies, 2016, Jones, Pykett, and Whitehead, 2013, Pykett, 2015), it is key that the dialogues improve between disability geography and the biomedical, pharmacological, and clinical sciences. This is particularly vital at a time when states of predominantly Western countries increasingly incorporate biopolitics in governance strategies. This is in conjunction with the intensification of neoliberalist policies that result in increasing restrictions to, *and* medicalization of, difference from the able-bodied, white, male, cisgender, heterosexual archetype. We also see a rise in eugenicist ideology, which incites violence against disabled folks (Hall, 2019). These societal challenges require in-depth engagements with the scientific debates that are used as a premise upon which these ideologies grab hold. Mixed methods approaches to such endeavors could be of particular importance given the possibilities for challenging universalist claims that emerge from positivist assertions (Beljaars, 2022). Such approaches would also demonstrate the value of geographical knowledge in patient emancipatory endeavors, expansions of gentler treatment options, and adaptations of current pharmacological and behavioral therapeutic regimes.

Conclusion

To date, disability geography has utilized a broad range of methodologies and theories for studying the everyday experiences of people with disabilities in relation to their socio-spatial environments. To move disability geography forward, we echo Worth's (2008) proposal that a productive direction is to develop and expand on current theories beyond the subfield and engage more critically with the broader discipline of geography. We add to her proposal to suggest a similar vision for methods in disability geography. In particular, we contend that blending qualitative and quantitative methodologies offers exciting possibilities for further destabilizing the definition of disability, enriching our understandings of the socio-spatial practices and contexts that disabled people navigate and produce, and for envisioning and enacting liberatory futures.

References

Anderson, B. and Harrison, P. (2010) *Taking-Place: Non-Representational Theories and Geography*. Routledge.

Baxter, J. (2018) 'Qualitative Health Geography Reaches the Mainstream' in Crooks, V. A., Andrews, G. J. and Pearce, J. (eds.) *Routledge Handbook* of Health Geography. Routledge, pp. 301–310.

Beljaars, D. (2020) 'Towards Compulsive Geographies'. *Transactions of the Institute of British Geographers* 45 (2), 284–298.

Beljaars, D. (2022) *Compulsive Body Spaces*. Routledge.

Bell, C.M. (ed.) (2011) *Blackness and Disability: Critical Examinations and Cultural Interventions*. Michigan State University Press.

Berger, R. J. and Lorenz, L. S. (2015) *Disability and Qualitative Inquiry: Methods for Rethinking an Ableist World*. Routledge.

Butler, R. and Bowlby, S. (1997) 'Bodies and Spaces: An Exploration of Disabled People's Experiences of Public Space'. *Environment and Planning D: Society & Space* 15 (4), 411–433.

Campbell, F. K. (2011) 'Geodisability Knowledge Production and International Norms: A Sri Lankan Case Study'. *Third World Quarterly* 32 (8), 1455–1474.

Campbell, F. K. (2009) *Contours of Ableism: The Production of Disability and Abledness*. Palgrave MacMillan.

Campbell, R. and Longhurst, R. (2013) 'Obsessive-Compulsive Disorder (OCD): Gendered Metaphors, Blogs and Online Forums'. *New Zealand Geographer* 69 (2), 83–93.

Chouinard, V. (2012) 'Mapping Bipolar Worlds: Lived Geographies of "madness" *in Autobiographical Accounts'*. *Health & Place* 18 (2), 144–151.

Chouinard, V. (2018) 'Mapping Life on the Margins' in Crooks, V. A., Andrews, G. J. and Pearce, J. (eds.) *Routledge Handbook of Health Geography*. Routledge, pp. 172–178.

Chouinard, V., Hall, E., and Wilton, R. (eds.) (2010) *Towards Enabling Geographies: 'Disabled' Bodies and Minds in Society and Space*. Routledge.

Church, R. L. and Marston, J. R. (2003) 'Measuring Accessibility for People with a Disability'. *Geographical Analysis* 35 (1), 83–96.

Crooks, V. A., Dorn, M. L., and Wilton, R. D. (2008) 'Emerging Scholarship in the Geographies of Disability'. *Health & Place* 14 (4), 883–888.

Davidson, J. (2010) '"It Cuts Both Ways": A Relational Approach to Access and Accommodation for Autism'. *Social Science & Medicine* 70 (2), 305–312.

Davidson, J. (2017) *Phobic Geographies: The Phenomenology and Spatiality of Identity*. Routledge.

Davidson, J. and Henderson, V.L. (2010) Coming out' on the spectrum: Autism, identity and disclosure. *Social and Cultural Geography* 11 (2), 155–170, https://doi.org/10.1080/14649360903525240

Davidson, J., Bondi, L., and Smith, M. (2005) *Emotional Geographies*. Routledge.

Davies, G. (2016) 'Remapping the Brain: Towards a Spatial Epistemology of the Neurosciences'. *Area* 48 (1), 125–128.

Dear, M. J. and Wolch, J. R. (2014) *Landscapes of Despair: From Deinstitutionalization to Homelessness*. Princeton University Press.

Dorn, M. L., Keirns, C. C., and Del Casino, V. J. (2010) 'Doubting Dualisms' in Brown, T., McLafferty, S. and Moon, G. *(eds.) A Companion to Health and Medical Geography*. Blackwell Publishing Ltd, pp. 55–78.

Fanon, F. (2008 *[1952]*) *Black Skin, White Masks*. Grove Press.

Gahman, L. (2017) 'Crip Theory and Country Boys: Masculinity, Dis/Ability, and Place in Rural Southeast Kansas'. *Annals of the American Association of Geographers* 107 (3), 700–715.

Gant, R. L. and Smith, J. A. (1988) 'Journey Patterns of the Elderly and Disabled in the Cotswolds: A Spatial Analysis'. *Social Science & Medicine* 27 (2), 173–180.

Gao, F., Foster, M., and Liu, Y. (2019) 'Disability Concentration and Access to Rehabilitation Services: A Pilot Spatial Assessment Applying Geographic Information System Analysis'. *Disability and Rehabilitation* 41 (20), 2468–2476.

Gleeson, B. (1999) *Geographies of Disability*. Routledge.

Gleeson, B. J. (1996) 'A Geography for Disabled People?' *Transactions of the Institute of British Geographers* 21 (2), 387–396.

Golledge, R. G. (1993) 'Geography and the Disabled: A Survey with Special Reference to Vision Impaired and Blind Populations'. *Transactions of the Institute of British Geographers* 18 (1), 63–85.

Golledge, R. G. (1996) 'A Response to Gleeson and Imrie'. *Transactions of the Institute of British Geographers* 21 (2), 404–411.

Gorman, R. (2017) 'Therapeutic Landscapes and Non-Human Animals: The Roles and Contested Positions of Animals within Care Farming Assemblages'. *Social & Cultural Geography* 18 (3), 315–335.

Hägerstrand, T. (1970) 'What About People in Regional Science?'. *Papers of the Regional Science Association* 24, 621.

Hall, E. (2019) 'A Critical Geography of Disability Hate Crime'. *Area* 51 (2), 249–256.

Hall, E. and Wilton, R. (2017) 'Towards a Relational Geography of Disability'. *Progress in Human Geography* 41 (6), 727–744.

Hamraie, A. (2018) 'Mapping Access: Digital Humanities, Disability Justice, and Sociospatial Practice'. *American Quarterly* 70 (3), 455–482.

Hansen, N. (2009) 'Remapping the Medical Terrain on Our Terms'. *Aporia* 1 (3), 28–34.

Hansen, N. and Philo, C. (2007) 'The Normality of Doing Things Differently: Bodies, Spaces and Disability Geography'. *Tijdschrift Voor Economische En Sociale Geografie* 98 (4), 493–506.

Higgs, G. (2005) 'A Literature Review of the Use of GIS-Based Measures of Access to Health Care Services'. *Health Services & Outcomes Research Methodology* 5, 119–139.

Holt, L., Bowlby, S., and Lea, J. (2019) 'Disability, Special Educational Needs, Class, Capitals, and Segregation in Schools: A Population Geography Perspective'. *Population, Space and Place* 25 (4). e2229.

Imrie, R. (2013) 'Shared Space and the Post-Politics of Environmental Change'. *Urban Studies* 50 (16), 3446–3462.

Imrie, R. (1996) 'Ableist Geographies, Disablist Spaces: Towards a Reconstruction of Golledge's "Geography and the Disabled"'. *Transactions of the Institute of British Geographers* 21 (2), 397–403.

Jampel, C. and Bebbington, A. (2018) 'Disability Studies and Development Geography: Empirical Connections, Theoretical Resonances, and Future Directions'. *Geography Compass* 12 (12), e12414.

Jones, R., Pykett, J., and Whitehead, M. (2013) *Changing Behaviours: On the Rise of the Psychological State.* Edward Elgar Publishing.

Kitchin, R. M., Blades, M., and Golledge, R. G. (1997) 'Understanding Spatial Concepts at the Geographic Scale without the Use of Vision'. *Progress in Human Geography* 21 (2), 225–242.

Kitchin, R. M., Jacobson, R. D., Golledge, R. G., and Blades, M. (1998) 'Belfast Without Sight: Exploring Geographies of Blindness'. *Irish Geography* 31 (1), 34–46.

Lovett, A. A. and Gatrell, A. C. (1988) 'The Geography of Spina Bifida in England and Wales'. *Transactions of the Institute of British Geographers* 13 (3), 288–302.

Macpherson, H. (2009) 'The Intercorporeal Emergence of Landscape: Negotiating Sight, Blindness, and Ideas of Landscape in the British Countryside'. *Environment & Planning A* 41 (5), 1042–1054.

Marston, J. R., Golledge, R. G., and Costanzo, C. M. (1997) 'Investigating Travel Behavior of Nondriving Blind and Vision Impaired People: The Role of Public Transit'. *The Professional Geographer* 49 (2), 235–245.

Mayer, J. D. (1981) 'Geographical Clues About Multiple-Sclerosis'. *Annals of the Association of American Geographers* 71 (1), 28–39.

McRuer, R. (2011) 'Disabling Sex: Notes for a Crip Theory of Sexuality'. *GLQ: A Journal of Lesbian and Gay Studies* 17 (1), 107–117.

McRuer, R. (2006) *Crip Theory: Cultural Signs of Queerness and Disability.* NYU Press.

Metzel, D. S. and Giordano, A. (2007) 'Locations of Employment Services and People With Disabilities'. *Journal of Disability Policy Studies* 18 (2), 88–97.

Mol, A. (2002) *The Body Multiple: Ontology in Medical Practice.* Duke University Press.

Moss, P. and Dyck, I. (2003) *Women, Body, Illness: Space and Identity in the Everyday Lives of Women with Chronic Illness.* Rowman & Littlefield Publishers.

Park, D. C., Radford, J. P., and Vickers, M. H. (1998) 'Disability Studies in Human Geography'. *Progress in Human Geography* 22 (2), 208–233.

Parr, H. (1999) 'Delusional Geographies: The Experiential Worlds of People during Madness/Illness'. *Environment and Planning D: Society & Space* 17 (6), 673–690.

Philo, C. (1987) '"Fit Localities for an Asylum": The Historical Geography of the Nineteenth-Century "Mad Business" in England as Viewed through the Pages of the Asylum Journal'. *Journal of Historical Geography* 13 (4), 398–415.

Planey, A. M. (2019) 'Audiologist Availability and Supply in the United States: A Multi-Scale Spatial and Political Economic Analysis'. *Social Science & Medicine* 222, 216–224.

Planey, A. M., Smith, S. M., Moore, S., and Walker, T. D. (2019) 'Barriers and Facilitators to Mental Health Help-Seeking among African American Youth and Their Families: A Systematic Review Study'. *Children and Youth Services Review* 101, 190–200.

Price, P. L. (2010) 'At the Crossroads: Critical Race Theory and Critical Geographies of Race'. *Progress in Human Geography* 34 (2), 147–174.

Pykett, J. (2015) *Brain Culture: Shaping Policy Through Neuroscience*. Policy Press.

Robinson, C. J. (2000) *Black Marxism: The Making of the Black Radical Tradition*. University of North Carolina Press.

Rosychuk, R.J., Newton, A.S., Niu, X., and Urichuk, L. (2015) Space and time clustering of adolescents' emergency department use and post-visit physician care for mood disorders in Alberta, Canada: A population-based 9-year retrospective study. *Canadian Journal of Public Health* 106, e10–e16.

Segrott, J. and Doel, M. A. (2004) 'Disturbing Geography: Obsessive-compulsive Disorder as Spatial Practice'. *Social & Cultural Geography* 5 (4), 597–614.

Shakespeare, T. (2006) *Disability Rights and Wrongs*. Routledge.

Sontag, S. (2001 [1978]) *Illness as Metaphor and AIDS and Its Metaphors*. Macmillan.

Sundberg, J. (2014) 'Decolonizing Posthumanist Geographies'. *Cultural Geographies* 21 (1), 33–47.

TallBear, K. (2013) *Native American DNA: Tribal Belonging and the False Promise of Genetic Science*. University of Minnesota Press.

Titchkosky, T. (2011) *The Question of Access: Disability, Space, Meaning*. University of Toronto Press.

Tolia-Kelly, D. P. (2010) 'The Geographies of Cultural Geography I: Identities, Bodies and Race'. *Progress in Human Geography* 34 (3), 358–367.

Vawer, M., Kaina, P., Leonard, A., Ogata, M., Blackburn, B., Young, M., and Seto, T. B. (2013) 'Navigating the Cultural Geography of Indigenous Peoples' Attitude toward Genetic Research: The Ohana (family) Heart Project'. *International Journal of Circumpolar Health* 72.

Weinstein, J. and Colebrook, C. (2017) *Posthumous Life: Theorizing Beyond the Posthuman*. Columbia University Press.

Wilton, R. and Evans, J. (2020) 'Disability' in Kobayashi, A. (ed.) *International Encyclopedia of Human Geography*. Elsevier, pp. 357–362.

Wolch, J. (1980) 'Residential Location of the Service-Dependent Poor'. *Annals of the Association of American Geographers* 70 (3), 330–341.

Wong, A. (2020) 'Introduction' in Wong, A. (ed.) *Disability Visibility: First-Person Stories from the Twenty-First Century*. Vintage Books, pp. xxv–xxii.

Wong, S. (2018a) 'Traveling with Blindness: A Qualitative Space-Time Approach to Understanding Visual Impairment and Urban Mobility'. *Health & Place* 49, 85–92.

Wong, S. (2018b) 'The Limitations of Using Activity Space Measurements for Representing the Mobilities of Individuals with Visual Impairment: A Mixed Methods Case Study in the San Francisco Bay Area'. *Journal of Transport Geography* 66, 300–308.

Worth, N. (2008) 'The Significance of the Personal within Disability Geography'. *Area* 40 (3), 306–314.

20

METHODOLOGIES FOR ANIMAL GEOGRAPHIES

Approaches Within and Beyond the Human

Guillem Rubio-Ramon and Krithika Srinivasan

Introduction

The study of human-animal interactions is no longer new to human geography and cognate fields. From efforts of early geographers to catalogue recently discovered nonhuman animal species, to their modern study as signifiers within broader social and cultural human systems, nonhuman animals had conventionally featured in one way or another in geographic depictions of the world. However, it was not until the 1990s that a fresh understanding of geography's potentialities to study nonhuman animals not as passive objects, but as agentive beings in more-than-human worlds emerged (Whatmore, 2006, p. 604).

For the last two decades, animal geographers have looked at a wide variety of human-animal (direct and indirect) contact zones. Broadly speaking, research in animal geography seeks to capture two aspects of human-animal relationships: human perspectives and experiences, and nonhuman animal perspectives and experiences. To this end, as a field rooted in the study of the social (i.e., a subdiscipline of human geography), animal geographies have employed a range of methods conventionally used in the social sciences (e.g., surveys, interviews, texts, and ethnography) as well as, more recently, experimental methods (e.g., multispecies methods) that cross disciplinary boundaries and which aim to unsettle nature-culture epistemological divides. In what follows, we critically review the ways in which these various methods have been deployed in the study of human-animal relations in geography and beyond. While doing so, we follow recent scholarship in primarily using the term 'nonhuman animals' to refer to animals other than humans. Besides reasons of clarification, by using 'nonhuman animals' we revise the prevalent human exceptionalist use of the terms 'animal' and 'human' designating two distinct categories, as if the latter were not a subcategory of the former. However, we do occasionally use the term 'animal' to refer to nonhuman animals for reasons of brevity (e.g., animal wellbeing) or where it is part of established terms such as 'animal geographies'/'animal welfare', or within direct quotations.

DOI: 10.4324/9781003038849-23

Engaging in Conversations About Nonhuman Animals

Interviews, focus groups and quantitative tools such as surveys are some commonly used methods in the social sciences, and therefore animal geographies. At the most basic level, all of these entail having conversations with people (about nonhuman animals), though with varying emphases on scale and detail.

Qualitative methods such as semi-structured interviews and focus group discussions provide the opportunity to develop in-depth understandings of participants' perspectives, seek their views on specific documents or issues, or to probe and ask for clarifications, thereby producing rich and detailed understandings of human-animal relationships. Such methods have been extensively used to study first-hand experiences where researchers explore how humans understand, interact with, and care for companion nonhuman animals as well as more conflictual relations. For instance, Power (2008, p. 535) employs interviews and diaries recorded by new dog 'owners' to understand "the everyday practices that constitute more-than-human families" and, similarly, Fletcher and Platt (2018) and Bulsara et al., (2007) analyse data derived from interviews and focus groups respectively to explore a diversity of human-dog relationships.

In a different vein, after interviewing a wide variety of companion animal keepers and professionals involved in nonhuman animal health nutrition, Fox and Gee (2019) describe how these relationships contain aspects of human-animal surveillance and control, alongside, and sometimes, embedded within, relations of care. Likewise examining conflicting interests, geographers working at the intersection of conservation and more-than-human geographies have employed focus groups alongside interviews and document analysis to explore their participants' accounts and experiences of living along and sharing space with wildlife (Doubleday, 2018; Margulies and Karanth, 2018; Ojalammi and Blomley, 2015).

Human-animal interactions in animal-based industries have also been studied through accounts of the first-hand experiences of human stakeholders. Holloway and Morris (2012), for instance, have explored through in-depth interviews, how farmers accept or resist forms of techno-scientific knowledge about cows. In addition, focus groups have been similarly employed to analyse intracultural and cross-cultural relationships between humans and nonhuman animals through animal-based foods (Happer and Wellesley, 2019; Kupsala, 2018).

Nonhuman animals are part of more-than-human infrastructures far away from farms and slaughterhouses. Doherty's (2019) research interviewing salvagers and informal waste collectors in Kampala illuminates the shared vulnerabilities between the workers and Marabou storks. And in contrast with a generalized focus on human perceptions and life stories, researchers have asked people about their first-hand experiences with nonhuman animals to grasp and better understand nonhuman animal lived experiences and subjectivities. For instance, interviews with laboratory animal technologists and caretakers in general have proven important in understanding these animals' lived experiences and agencies (Greenhough and Roe, 2019) and how these constitute and are affected by infrastructures in which they are forceful components (Zhang, 2020).

Unlike interviews and other qualitative methods, surveys have not been commonly used in animal geographies. Surveys are good instruments to explore societal opinions and perceptions on concepts like animal welfare or to track historical changes in human-animal relationships. Surveys have traditionally focused on broad patterns in, and usually one side – the human – of human-animal interactions, revealing less about the interaction itself (Hamilton and Taylor, 2017, p. 162). For instance, Franklin's (2007) research on recent changes in human-animal relationship, draws on a survey of 2000 Australians to track the different ways in which they perceive nonhuman animals. Less common is research using surveys to explore the 'animal' side of the relationship by, for instance, asking caregivers what they think their cats do for fun (Shyan-Norwalt, 2005). In the use of such methods, careful attention to sampling is important to avoid oversimplification of findings and the reinforcement of the lack of visibility of marginalized communities and their relationships with nonhuman animals (Doubleday, 2018, p. 152; Lassiter and Wolch, 2002).

All these methods rely on people to speak about and for nonhuman animals, and while useful in understanding human perspectives, these approaches can pose challenges for understanding animal experiences, interests, and worldviews. This is because they often implicitly prioritise the human side of the human-animal relationship, given that we only listen to the answers of humans, at best narrated to us as first-hand experiences with specific nonhuman animals. One way of addressing this challenge is by incorporating contrasting (human) perspectives on the relation under study. By being explicit about methodological limitations and researcher positionality, these methods, rather than restraining the scope of the research, can help illuminate shared more-than-human worlds in which nonhuman animals not only shape human systems but are also shaped by them in both benign, positive and harmful ways.

Studying Animal Cultural Texts

Animal geographers and scholars studying nonhuman animals have frequently turned to the qualitative analysis of written documents and cultural texts to study past and present human-animal interactions. From written records such as legal texts, wills, and diaries to digital documents like websites, documentaries and TV shows, people have documented their real and fictitious interactions with nonhuman animals for centuries. Given this vast amount of potential data, some animal geographers have employed historiographic methods to understand our past and present relationships with nonhuman animals, developing what has been called historical animal geographies (Wilcox and Rutherford, 2018). The analysis of present and past cultural texts has also been used to explore multiple topics beyond the scope of historical animal geographies (e.g., Elder et al., 2008; Salter, 2018).

Like with interviews and surveys, a key problem with exploring human-animal relations through cultural texts is that researchers mostly rely on human-produced documents. Nevertheless, it is precisely the fact that most of these documents are produced by humans that makes it possible to track, reconstruct and at least partially understand, changing human relationships, perceptions, and attitudes towards nonhuman animals.

Texts produced independently of the research context such as print media have been used to explore the broad sociocultural orderings in which we place nonhuman animals. Jerolmack's (2008) analysis of newspaper articles tracks the recent construction of pigeons as a pest, while Franklin and White's (2001) article on the changing human-animal relations in Australia is informed by the content of over 1000 newspaper articles from 1949 to 1998. Scholars like Dibley and Hawkins (2019) and Probyn (2014) have questioned the ways in which our visual consumption of nonhuman animals through documentaries might have affected our perception of them. Furthermore, to understand how the lives of nonhuman animals are mediated by specific contexts, Srinivasan's (2013) work compares a diverse range of cultural texts such as laws, official statistics and digital materials to discuss the differences between dog lives in the UK and India. Scholars have also analysed data produced through diaries and essays (Power, 2008). For instance, Margulies (2019) uses photovoice and photo-essays generated by research participants to explore affective encounters between wild and domesticated animals and humans.

Widespread national newspapers, photo-essays and TV documentaries are relatively modern phenomena. But is it possible to understand our relationships with animals before their appearance? Historical animal geographers criticize conventional historical research for rendering nonhuman animals into an "absent presence" and aim to fill this void by "think[ing] the ways in which human and nonhuman agencies have shaped each other through time" (Wilcox and Rutherford, 2018, pp. 2–3). Fudge's (2018) work on the relationship between animals and the humans who farmed them in early modern England addresses these absences by analysing a dataset of more than 4000 wills. Similarly, Collard (2018) has explored the impacts of human political-economic systems and historical events such as the colonial fur trade and petro-capitalist oil spills on sea otters, illuminating not only nonhuman animals' experiences in different historical human systems of production, but also how the same systems have constrained their agencies (Blattner, Donaldson, and Wilcox, 2020).

Furthering the study of nonhuman animal agencies, the history and evolution of modern-day animal-based food systems and industries have been investigated employing a diverse set of historiographic methods and cultural texts. The work of Atkins (2016) and Nimmo (2010) on the dairy industry focuses on the materiality of this bodily liquid produced by animals for feeding their young by analysing a variety of historical texts such as scientific and agricultural journals. Also relating to the role of bovines in circumscribed sociocultural and historical contexts, Yarwood and Evans (2016) discuss the histories and geographies of the expansion of breed societies in Wales. And amid a rise of ultranationalism in India, Yamini Narayanan (2019) develops a multispecies analysis of cow protectionism through a combination of methods –including the analysis of cultural texts such as religious texts, newspapers, and historical events.

Human-animal ethnographies

Favouring "thick descriptions" (Geertz, 1973), ethnography delivers a dense and multi-faceted understanding of complex phenomena. For the last two decades, ethnographic methods have been incorporated and revamped within the sub-discipline of animal geographies. By using these methods the researcher is able to witness, track over time and place, and even directly participate in human-animal encounters (Seymour and Wolch, 2010, p. 7). Ethnographic methods such as participant and non-participant observation remain one of the cornerstones of the research within the field. By observing from a distance or by taking part in various human-animal interactions, researchers have aimed to better understand the ways in which these relationships unfold in specific contexts. Such approaches require careful attention to researcher positionality, personal experiences and a constant process of reflexivity inspired by feminist scholarship. They include narrative methods such as autoethnography and/or narrative analysis which have enabled researchers to reflect on their own biographic experiences to understand broader issues escalating beyond the anecdotal (Jones, 2013; Syse, 2013).

However, as Hamilton and Taylor (2017, p. 3) argue, ethnography has traditionally been rooted in a humanistic idea of the social, excluding nonhuman life and cemented in concepts such as people, personhood or identity, usually framed as pertaining to a human-exclusive domain. The approach of multispecies ethnography has emerged in response to this problem. To Ogden, Hall, and Tanita (2013, p. 6), multispecies ethnography can be defined "as ethnographic research and writing that is attuned to life's emergence within a shifting assemblage of agentive beings". Therefore, multispecies ethnography is less than a label encompassing a well-defined catalogue of methods than a relational strategy to challenge humanist epistemology and overcome dualist ontologies reinforcing anthropocentrism and human exceptionalism (Hovorka, 2019, p. 751). Multispecies ethnography, thus, is a framework encompassing a methodological attentiveness to the more-than-human, encompassing both " 'life's emergence' within multispecies social, political, economic and cultural specific contexts" (Gillespie, 2019, p. 71) and an unfinished understanding of 'the human' as becoming relationally in these same multispecies contexts (Ogden, Hall, and Tanita, 2013).

These characteristics, however, are not confined to multispecies ethnography alone. While multispecies ethnography has been used interchangeably with other concepts like multispecies methods or human-animal methods, this conflation can be problematic given that non-ethnographic work can also achieve multispecies or non-anthropocentric goals (e.g., Collard, 2018; Narayanan, 2018).

A Dialogue of Methods

The average animal geographer's methodological toolkit is one full of methods common across the social sciences such as participant observation, interviews, surveys, and archival research, which have been used both independently and in combination to obtain varying descriptions of the world. From within this toolkit, it has, however, been ethnography, especially participant and non-participant observation, which have recently attracted attention in terms of engagements with interdisciplinary dialogue. As Buller (2015, p. 377) points out, after identifying the limitations of representational accounts and conventional ethnography, animal geographers have attempted to expand ethnography's methodological repertoire.

Scholars have combined ethnographic methods with walking and motion studies to understand animal places and beastly places (Philo and Wilbert, 2000) along with their mobilities and rhythms, whether in harmonious or discordant relation to those of people. Holmberg's (2019) work focuses on the role of movement and rhythm in the creation of intimate bonds between humans and dogs through practices of walking, eating and sleeping. In a different vein, there is scholarship that examines the commodified movements of nonhuman animals (Bull, 2011, p. 29; Collard 2020) and the different impacts of human mobility infrastructures like those described by Fishel (2019) after her experience of running over a deer.

The use of visual methods in addition to ethnographic methods has also been embraced by scholars. This includes new ethnomethodological approaches that analyse video data on practices such as people walking dogs or fly fishing (Laurier, Maze, and Lundin, 2006) and the behaviour of canine lifelong friends (Goode, 2007). Jamie Lorimer's (2010) account of 'moving image methodologies' – be it recorded in fieldwork or as popular media works – also points to this direction, exploring elephant-human encounters by combining methods from film analysis, anthropology and the study of animal behaviour: ethology.

In the last decade, ethology (the study of animal behaviour under 'natural' conditions) has emerged as a methodological influence on the field, especially in conversation with geographic and ethnographic methods (Hodgetts and Lorimer, 2014, p. 287). With the objective of gaining direct access to nonhuman animals' emotions, cultures, perceptions and socialities, animal geographers have started to include animal behaviour as a crucial component of their analyses. For instance, this has been done through a direct dialogue between geography and ethology, resulting in work like Barua and Sinha (2019) that explores what the process of urbanization in Indian cities means to Macaques.

Methodological dialogues with ethology have been mediated through historiographic materials. Hayden Lorimer's (2012) reflections on (literally) running into seals track the historic linkages of geography and ethology through historiographic materials produced by the naturalist Frank Fraser Darling. Scholars have also engaged in conversations with non-scientists in direct contact with nonhuman animals, like fishers (Bear and Eden, 2011) and hunters (Boonman-Berson, Driessen, and Turnhout, 2019) to understand their (animal) behaviour.

In this methodological blend, sometimes left out of the picture is the subjectivity of the individual nonhuman animals. Scholars who investigate nonhuman animal experiences have often faced accusations of anthropomorphism, i.e., "the danger of projecting human(?) norms and sensibilities onto others" (van Dooren, Kirksey, and Münster, 2016, p. 8). However, as de Waal (1999) pointed out, avoiding this maligned phemonenon at all costs can become as much a naïve position as anthropomorphism (Karlsson 2012, p. 711), with an even more problematic mechanomorphism "projecting characteristics of machines onto forms of life" as a result (van Dooren, Kirksey, and Münster 2016, p. 8). Following Burghardt's (1991, p. 86) call for a critical anthropomorphism which combines "our human characteristics and abilities with various kinds of knowledge and keep the question-asking in bounds but still creative", some scholars have proposed subjecting the creative potential of anthropomorphism to well-established and shared critical evaluation standards such as careful replicable observation (Karlsson, 2012, p. 719; Burghardt, 2004). Ultimately, the fear of projecting our human values towards nonhuman animals should not become a deterrent to explore more-than-human subjectivities, including animal wellbeing. Equally, there is the need to acknowledge that mechanomorphism or approaches that deny or avoid engaging with animal subjectivity can be equally imbued with human norms, interests and values. Indeed, one of the most widely known proponents of mechanomorphism, Descartes (1991), justified his claim that animal lack sentience and subjectivity, and that that they are akin to 'machines', by arguing that to believe otherwise would make many human interactions with animals 'criminal' in character.

Overcoming the aforementioned concerns through a blend of ethnographic and animal welfare science methods, Geiger and Hovorka (2015) have studied the lived experiences and societal roles of donkeys in Botswana by combining interviews with their owners and individual welfare assessments. Patter and Hovorka's (2018) work on feral cats in southern Ontario similarly studies feral cat behaviour drawing on methods from ethology. In sum, these interdisciplinary dialogues involve experimenting with methodologies from beyond the social sciences coupled with a critical and responsible anthropomorphism (Bear, 2011) to attend "more closely to understandings of nonhumans garnered from the practice and experience of co-relationality" (Johnston, 2008, p. 646).

The Ethics and Politics of Methods

Researching nonhuman animals presents specific ethical challenges of representation and power imbalances. As Rosemary Collard (2015, p. 133) argues, research on the nature-society interface has usually presented animals either as natural resources in plans of capital accumulation or objects of conflict in projects of uneven development and distribution of power and wealth. In some cases, certain nonhuman animals are not even considered as authentic 'nature' and are disqualified as 'artefacts' not worthy of study (Arcari, Probyn-Rapsey, and Singer, 2020). Methodologically speaking, this neglect both reinforces and is shaped by particular uses of methods perhaps too centred on human individuals, communities, infrastructures and the nonhuman animals they consider important, and which frame nonhuman animals as components which do not make "any ethical demands, especially on the researcher" (Collard, 2015, p. 133).

As Patter and Blattner (2020) have discussed, this is especially the case when research does not involve direct interaction with nonhuman animals. In this situation, most ethics committees will, at best, abide by cost-benefit or non-harm principles, or at worst, render nonhuman animal life as disposable (Gillespie & Collard, 2015), leaving to the researcher any deeper ethical concerns. Even if social scientists have not typically included methods involving physical harm or control over nonhuman animals, the choice of methods should not be exempt from the ethical and political deliberation of the potential consequences of research to nonhuman animals.

Patter and Blattner (2020) have recently articulated three core ethical principles to guide scholars' methodological choices when conducting research on nonhuman animals: non-maleficence, beneficence and voluntary participation. In this section, these ethical principles serve as a prism to rethink the ethical and political implications of methods commonly used in animal geographies.

The non-maleficence principle requires that nonhuman animals not be harmed in research; therefore, researchers must anticipate and prevent, if necessary, the harms their research might cause nonhuman animals (Patter and Blattner, 2020, p. 174). Research in animal geographies can involve witnessing and even taking part in such harm. For instance, researchers have conducted participant and non-participant observation in places of violence towards nonhuman animals such as hunts (Boonman-Berson, Driessen, and Turnhout, 2019) or slaughterhouses (Pachirat, 2013) as well as in apparent spaces of care such as wildlife rehabilitation centres (Collard, 2015).

These practices raise questions about the necessity of participating in obvious or hidden violence or even watching it to analyse it. While "bearing witness and treating grief and our affective entanglements [with nonhuman animals] seriously enable us to build a transformative politics of shared understanding, care, and nonviolent social relations" (Gillespie, 2016, p. 578), the task of witnessing injustice contrasts with other forms of participant or non-participant observation in these spaces – which are perhaps not essential to understand structural violence. What is at stake in some of these practices is an ethical justification of methods cemented in the epistemological question of "what counts as knowledge" in particular research cultures (Gillespie, 2018, p. 102). There is an implicit expectation that researchers get their 'hands dirty' and participate in or witness practices of harm – or at the very least, a valorization of such 'immersive' methods. The danger is that the choice of such methods, informed by prevalent norms about what is justifiable in doing research, might end up rationalizing the practice(s) under study, preventing the fulfilment of the non-maleficence principle or, at best, addressing benefits for the nonhuman animals of concern only as abstract advances and in the long-term (Gillespie, 2018, p. 99).

This brings us to the second principle: beneficence. Beneficence entails that nonhuman animals must always – directly or indirectly – benefit from our research. According to the proponents of these principles (Patter and Blattner, 2020, p. 178), one of the areas that we must consider when thinking about beneficence is representation. Given that most ethnographic and geographic methods were originally designed by and for humans, some researchers have criticized the conventional uses of these human-centred methods because "animals themselves tend to be written out of the story" (Hamilton and Taylor, 2017, p. 2) – they tend to generate anthropocentric depictions of nonhuman animals, and, at worst, humans might use them to speak for nonhuman animals in ways that support and reinforce human interests.

Yet, meaningful and reflexive attempts to share the stories of nonhuman animals, and trying to ethically represent them, as well as their agencies and subjectivities, is perhaps better than the silencing of their experiences, cultures and socialities (Patter and Blattner, 2020, p. 178). The incorporation of new methods from other disciplines such as ethology has attempted to provide a more direct answer to the inner worlds of nonhuman animals. But at the same time, these methods have raised new ethical issues such as how to represent nonhuman animal behaviours and agencies in ways that do not reductively depict them as indistinct members of species with genetically scripted behaviours (Blattner, Donaldson, and Wilcox, 2020, p. 19).

Finally, the recent inclusion of ethology and methods from different disciplines poses additional questions in terms of voluntary participation. Patter and Blattner (2020, p. 180) advise researchers to be ready to seek mediated informed consent from nonhuman animals' human representatives as well as their embodied assent or dissent. This is relevant to current dialogues with ethology in the field. As the evaluation of this assent/dissent will rely on the researchers' interpretation of animal behaviour, it will require not only ethological assessments but informed, responsible anthropomorphism as well. Moreover, with the increasing use of methods involving technologies like GPS tracking tools (Pivot et al., 2019, p. 30), digital human-animal encounters (Turnbull, Searle, and Adams, 2020, p. 6.7) and genetic analysis (Hodgetts and Lorimer, 2014, 289), p. about direct harm to individual nonhuman animals, as well as more-than-human privacy and confidentiality, will have to be rethought outside a humanist perspective, with more concentrated scrutiny of possible negative outcomes for nonhuman participants.

Conclusions

In this chapter, we have presented a non-exhaustive review of methods employed in animal geography research, specifically focusing on the scholarship produced during the last decade. We have tracked how established methods in the social sciences have been adopted and revamped to study nonhuman animals. On some occasions, the humanist foundations of these methods have translated into an obscuring of nonhuman animals' unique perspectives, feelings and lived experiences. In response, animal geographers have turned to immersive and observational methods, as well as to the methodological toolkits of other disciplines, with the objective of listening to nonhuman animals' voices directly, instead of seeing them as characters in human-narrated stories.

Our engagements in animal geographies suggest that it is useful to understand all methods as partial, and as offering different windows into human-animal relations. For instance, conversation-oriented methods such as surveys, interviews and focus groups can illuminate perspectives and events that are not easily accessible for research, such as those that have already happened, that occur in private, or those experienced by marginalized communities. Cultural texts can be useful to understand structural and historical contexts that shape human-animal relations. Those such as diaries or photovoice have been employed for researching issues that are not easily discussed orally or that happen outside of the researcher-participant interaction. Ethnographic methods, together with methods from other disciplines, offer a way of seeing nonhuman animals directly in relation to people.

Crucial to navigating the methodological platter is understanding that there is no unmediated (by humans) way of producing knowledge on nonhuman animals, and that therefore, knowledge production in animal geographies is always political. The choice of methods shapes the knowledge that is produced: with every methodological choice, there are some voices heard, and others silenced. For instance, observational methods can overlook broader structural factors, even while conversational and textual methods can make it challenging to capture the details of nonhuman animal experiences. Methods are deeply political.

An ongoing issue for animal geographies and its methodological journey is the balance between capturing human and nonhuman animal perspectives, and the degree to which nonhuman animal subjectivities should become central in the field and, therefore, its methods. A related question is that of whether and to what degree the field should retain the anthropocentrism of the social sciences and their established methods. In this context, there is an implicit view within the field that the move from conventional methods (such as interviews) to newer, experimental ones (such as multispecies ethnography) constitutes a progression in achieving the field's more-than-human ambitions. However, reconfiguring methods so as to generate information on dimensions that are not easily accessed by conventional social science methods does not necessarily mean that older problems, including the reproduction of anthropocentrism, do not resurface in new forms. Keeping this in mind, and to conclude, we would like to consider some of the implications of the field's methodological developments, reflecting on the epistemological, ethical, and political limits that methods pose when researching nonhuman animals geographically.

First, as we have mentioned, nonhuman animals are overwhelmingly inserted in human systems, practices, and languages and, while emergent, experimental methods might illuminate new perspectives on animals' agencies, lives and experiences, they can neglect the structures in which these unfold. It is no accident, thus, that observational and immersive ethnographic methods (barring some exceptions such as Ginn (2014)) have tended to produce analyses that highlight positive human-animal relations such as living with companion animals or exceptional caring relationships within structures of violence (Srinivasan, 2016, p. 77). Even if these methods have advanced towards an attentiveness to nonhuman animals as individual agents, they can only be considered partial openings to the diversity of and within human-animal relationships. We, therefore, need a matching diversity of methods to capture the different scales (e.g., micro/meso/macro; encounter/structural; embodied/discursive) at which these relationships are constituted.

Crucially, while choosing methods, it is vital to consider the trade-offs between focusing on nonhuman animals' everyday relations with people, and understanding the social and political structures shaping, inhibiting or suppressing nonhuman animal agencies (Giraud 2019). This balance will be crucial when evaluating and proceeding with further methodological developments in the field.

Methodological choices and debates in animal geographies therefore pertain to two broad, interrelated aspects: (1) the varied strengths and limitations of different methods in relation to their capacity to offer distinctive windows to more-than-human worlds; (2) ethical and political questions about the knowledge produced with and on animals. While the first has received much attention within the field, the second has only recently started provoking meaningful thought and debate, and requires further sustained reflection.

An important implication of recognizing the ethical standing of nonhuman animals is setting limits on our research, just as we do with human participants (Patter and Blattner, 2020). This means that we must relearn to ask what can be researched and what must be left out of our projects and designs, given the varied direct and indirect consequences of our choice of methods. This might set limits to our research objectives and make certain issues more difficult to explore. But such limitations could also be considered as new horizons to look differently at the field's development of methods – they contribute to the epistemological legitimization of ways of knowing nonhuman animals that are not often included in academic cultures such as indigenous and non-Eurocentric knowledges on the more-than-human (Loivaranta 2020). The establishment of boundaries to human aspirations, even if these are for knowledge, might therefore become a pathway to cultivating alternatives to dominant research cultures – alternative paradigms which engage with nonhuman animal subjectivities and their lived experiences alongside those human systems that ultimately shape them.

References

Arcari P., Probyn-Rapsey, F. and Singer, H. (2020) Where species don't meet: Invisibilized animals, urban nature and city limits. *Environment and Planning E: Nature and Space*. DOI: 10.1177/2514848620939870

Atkins, P. (2016) *Liquid Materialities: A History of Milk, Science and the Law*. Milton Park, Abingdon, Oxon; New York: Routledge.

Barua, M. and Sinha, A. (2019) 'Animating the Urban: An Ethological and Geographical Conversation'. *Social & Cultural Geography* 20 (8): 1160–80. https://doi.org/10.1080/14649365.2017.1409908.

Bear, C. (2011) 'Being Angelica? Exploring Individual Animal Geographies'. *Area* 43 (3): 297–304.

Bear, C. and Eden, S. (2011) 'Thinking like a Fish? Engaging with Nonhuman Difference through Recreational Angling'. *Environment and Planning D: Society and Space* 29 (2): 336–52. https://doi.org/10.1068/d1810.

Blattner, C. E., Donaldson, S., and Wilcox, R. (2020) 'Animal Agency in Community'. *Politics and Animals* 6: 22.

Boonman-Berson, S., Driessen, C., and Turnhout, E. (2019) 'Managing Wild Minds: From Control by Numbers to a Multinatural Approach in Wild Boar Management in the Veluwe, the Netherlands'. *Transactions of the Institute of British Geographers* 44 (1): 2–15. https://doi.org/10.1111/tran.12269.

Bull, J. (2011) *Animal Movements-Moving Animals: Essays on Direction, Velocity and Agency in Humanimal Encounters.* Centre for Gender Research, Uppsala University.

Buller, H. (2015) 'Animal Geographies II: Methods'. *Progress in Human Geography; London* 39 (3): 374–84. https://doi.org/10/cdd3.

Bulsara, M., Wood, L., Giles-Corti, B., and Bosch, D. (2007) 'More Than a Furry Companion: The Ripple Effect of Companion Animals on Neighborhood Interactions and Sense of Community'. *Society & Animals* 15 (1): 43–56. https://doi.org/10.1163/156853007X169333.

Burghardt, G. M. (1991) 'Cognitive Ethology and Critical Anthropomorphism: A Snake with Two Heads and Hognose Snakes That Play Dead.' In *Cognitive Ethology: Essays in Honor of Donald R. Griffin,* edited by Carolyn A. Ristau. Comparative Cognition and Neuroscience Series. Lawrence Erlbaum Associates, Inc.

Burghardt, G. M. 2004. 'Ground Rules for Dealing with Anthropomorphism'. *Nature* 430 (6995): 15–25. https://doi.org/10.1038/430015b.

Collard, R-C. (2015) 'Ethics in Research beyond the Human'. In *The Routledge Handbook of Political Ecology,* edited by T. Perreault, Bridge, G., and McCarthy, J., 127–39. London: Routledge.

Collard, R-C. (2018) 'Disaster Capitalism and the Quick, Quick, Slow Unravelling of Animal Life'. *Antipode* 50 (4): 910–28. https://doi.org/10.1111/anti.12389.

Collard, R-C. (2020) *Animal Traffic: Lively Capital in the Global Exotic Pet Trade.* Durham and London: Duke University Press.

Descartes, R. (1991) *The Philosophical Writings of Descartes: Volume 3: The Correspondence.* Edited by John Cottingham, Dugald Murdoch, Robert Stoothoff, and Anthony Kenny. Vol. 3. Cambridge: Cambridge University Press. https://doi.org/10.1017/CBO9781107340824.

Dibley, B. and Hawkins, G. (2019) 'Making Animals Public: Early Wildlife Television and the Emergence of Environmental Nationalism on the ABC'. *Continuum* 33 (6): 744–58. https://doi.org/10.1080/10304312.2019.1669533.

Doherty, J. (2019) 'Filthy Flourishing:Para-Sites, Animal Infrastructure, and the Waste Frontier in Kampala'. *Current Anthropology* 60 (S20): S321–32. https://doi.org/10.1086/702868.

Dooren, T-V, Kirksey, E., and Münster, U. (2016) 'Multispecies Studies: Cultivating Arts of Attentiveness'. *Environmental Humanities* 8 (1): 1–23. https://doi.org/10.1215/22011919-3527695.

Doubleday, K. F. (2018) 'Human-Tiger (Re)Negotiations: A Case Study from Sariska Tiger Reserve, India'. *Society & Animals* 26 (2): 148–70. https://doi.org/10.1163/15685306-12341498.

Elder, G., Wolch, J., and Emel, J. (2008) '"Le Pratique Sauvage: Race, Place, and the Human–Animal Divide"'. In *The Cultural Geography Reader,* 253–62. Routledge. https://doi.org/10.4324/9780203931950-42.

Fishel, S. R. (2019) 'Of Other Movements: Nonhuman Mobility in the Anthropocene'. *Mobilities* 14 (3): 351–62. https://doi.org/10/gf7fbz.

Fletcher, T. and Platt, L. (2018) '(Just) a Walk with the Dog? Animal Geographies and Negotiating Walking Spaces'. *Social & Cultural Geography* 19 (2): 211–29. https://doi.org/10.1080/14649365.2016.1274047.

Fox, R. and Gee, N. R. (2019) 'Great Expectations: Changing Social, Spatial and Emotional Understandings of the Companion Animal–Human Relationship'. *Social & Cultural Geography* 20 (1): 43–63. https://doi.org/10.1080/14649365.2017.1347954.

Franklin, A. (2007) 'Human-Nonhuman Animal Relationships in Australia: An Overview of Results from the First National Survey and Follow-up Case Studies 2000-2004'. *Society & Animals* 15 (1): 7–27. https://doi.org/10.1163/156853007X169315.

Franklin, A. and White, R. (2001) 'Animals and Modernity: Changing Human–Animal Relations, 1949–98'. *Journal of Sociology* 37 (3): 219–38. https://doi.org/10.1177/144078301128756319.

Fudge, E. (2018) *Quick Cattle and Dying Wishes: People and Their Animals in Early Modern England*. Ithaca and London: Cornell University Press. https://muse.jhu.edu/book/61606.

Geertz, C. (1973) 'Thick Description: Toward an Interpretetive Theory of Culture'. In *The Interpretation of Cultures*, 1–30. New York: Basic Books.

Geiger, M. and Hovorka, A. J. (2015) 'Animal Performativity: Exploring the Lives of Donkeys in Botswana'. *Environment and Planning D: Society and Space* 33 (6): 1098–117. https://doi.org/10.1177/0263775815604922.

Gillespie, K. (2016) 'Witnessing Animal Others: Bearing Witness, Grief, and the Political Function of Emotion'. *Hypatia* 31 (3): 572–88. https://doi.org/10.1111/hypa.12261.

Gillespie, K. (2018) *The Cow with Ear Tag #1389*. Chicago: University of Chicago Press. http://ebook central.proquest.com/lib/ed/detail.action?docID=5355460.

Gillespie, K. (2019) 'For a Politicized Multispecies Ethnography'. *Politics and Animals* 5 (0): 1–16.

Ginn, F. (2014) 'Sticky Lives: Slugs, Detachment and More-than-Human Ethics in the Garden'. *Transactions of the Institute of British Geographers* 39 (4): 532–44. https://doi.org/10.1111/tran.12043.

Giraud, E. H. (2019) *What Comes after Entanglement?: Activism, Anthropocentrism, and an Ethics of Exclusion*. Durham: Duke University Press Books.

Goode, D. (2007) *Playing with My Dog Katie: An Ethnomethodological Study of Dog-Human Interaction*. Purdue University Press.

Greenhough, B. and Roe, E. (2019) 'Attuning to Laboratory Animals and Telling Stories: Learning Animal Geography Research Skills from Animal Technologists'. *Environment and Planning D: Society and Space* 37 (2): 367–84. https://doi.org/10.1177/0263775818807720.

Hamilton, L. and Taylor, N. (2017) *Ethnography after Humanism: Power, Politics and Method in Multi-Species Research*. London: Palgrave Macmillan UK. https://doi.org/10.1057/978-1-137-53933-5.

Happer, C. and Wellesley, L. (2019) 'Meat Consumption, Behaviour and the Media Environment: A Focus Group Analysis across Four Countries'. *Food Security* 11 (1): 123–39. https://doi.org/10.1007/s12571-018-0877-1.

Hodgetts, T. and Lorimer, J. (2014) 'Methodologies for Animals' Geographies: Cultures, Communication and Genomics'. *Cultural Geographies* 22 (2): 285–95. https://doi.org/10.1177/1474474014525114.

Holloway, L. and Morris, C. (2012) 'Contesting Genetic Knowledge-Practices in Livestock Breeding: Biopower, Biosocial Collectivities, and Heterogeneous Resistances'. *Environment and Planning D: Society and Space* 30 (1): 60–77. https://doi.org/10/drchn8.

Holmberg, T. (2019) 'Walking, Eating, Sleeping. Rhythm Analysis of Human/Dog Intimacy'. *Emotion, Space and Society* 31 (May): 26–31. https://doi.org/10.1016/j.emospa.2019.03.002.

Hovorka, A. J. (2019) 'Animal Geographies III: Species Relations of Power'. *Progress in Human Geography* 43 (4): 749–57. https://doi.org/10.1177/0309132518775837.

Jerolmack, C. (2008) 'How Pigeons Became Rats: The Cultural-Spatial Logic of Problem Animals'. *Social Problems* 55 (1): 72–94. https://doi.org/10.1525/sp.2008.55.1.72.

Johnston, C. (2008) 'Beyond the Clearing: Towards a Dwelt Animal Geography'. *Progress in Human Geography* 32 (5): 633–49. https://doi.org/10.1177/0309132508089825.

Jones, O. (2013) '"Who Milks the Cows at Maesgwyn?" The Animality of UK Rural Landscapes in Affective Registers'. *Landscape Research* 38 (4): 421–42. https://doi.org/10.1080/01426397.2013.784246.

Karlsson, F. (2012) 'Critical Anthropomorphism and Animal Ethics'. *Journal of Agricultural and Environmental Ethics* 25 (5): 707–20. https://doi.org/10.1007/s10806-011-9349-8.

Kupsala, S. (2018) 'Contesting the Meat–Animal Link and the Visibility of Animals Killed for Food: A Focus Group Study in Finland'. *Food, Culture & Society* 21 (2): 196–213. https://doi.org/10.1080/15528014.2018.1427928.

Lassiter, U. and Wolch, J. R. (2002) 'Sociocultural Aspects of Attitudes toward Marine Animals: A Focus Group Analysis'. *The California Geographer* 42 (6). http://dspace.calstate.edu/handle/10211.2/2744.

Laurier, E., Maze, R., and Lundin, J. (2006) 'Putting the Dog Back in the Park: Animal and Human Mind-in-Action'. *Mind, Culture, and Activity* 13 (1): 2–24. https://doi.org/10.1207/s15327884mca1301_2.

Loivaranta, T. (2020) 'Post-Human Lawscapes of Indigenous Community Forests in Central India'. *The Geographical Journal* 186 (3): 288–99. https://doi.org/10.1111/geoj.12342.

Lorimer, H. (2012) 'Forces of Nature, Forms of Life: Calibrating Ethology and Phenomenology'. In *Taking-Place: Non-Representational Theories and Geography*, edited by Ben Anderson and Paul Harrison, 55–78. Milton Park, Abingdon, Oxon; New York: Routledge.

Lorimer, J. (2010) 'Moving Image Methodologies for More-than-Human Geographies'. *Cultural Geographies* 17 (2): 237–58. https://doi.org/10.1177/1474474010363853.

Margulies, J.D. (2019) On coming into animal presence with photovoice. *Environment and Planning E: Nature and Space*, 2 (4): 850–873.

Margulies, J. D. and Karanth, K. K. (2018) 'The Production of Human-Wildlife Conflict: A Political Animal Geography of Encounter'. *Geoforum* 95 (October): 153–64. https://doi.org/10.1016/j.geoforum.2018.06.011.

Narayanan, Y. (2018) 'Cow Protection as "Casteised Speciesism": Sacralisation, Commercialisation and Politicisation'. *South Asia: Journal of South Asia Studies* 41 (2): 331–51. https://doi.org/10.1080/00856401.2018.1419794.

Narayanan, Y. (2019) '"Cow Is a Mother, Mothers Can Do Anything for Their Children!" Gaushalas as Landscapes of Anthropatriarchy and Hindu Patriarchy'. *Hypatia* 34 (2): 195–221. https://doi.org/10.1111/hypa.12460.

Nimmo, R. (2010) *Milk, Modernity and the Making of the Human: Purifying the Social.* New York: Routledge.

Ogden, L. A., Hall, B., and Tanita, K. (2013) 'Animals, Plants, People, and Things: A Review of Multispecies Ethnography'. *Environment and Society* 4 (1): 5–25. https://doi.org/10.3167/ares.2013.040102.

Ojalammi, S. and Blomley, N. (2015) 'Dancing with Wolves: Making Legal Territory in a More-than-Human World'. *Geoforum* 62 (June): 51–60. https://doi.org/10.1016/j.geoforum.2015.03.022.

Pachirat, T. (2013) *Every Twelve Seconds: Industrialized Slaughter and the Politics of Sight.* Yale Agrarian Studies Series edition. New Haven, Conn.: Yale University Press.

Van Patter, L. E. and Blattner, C. (2020) 'Advancing Ethical Principles for Non-Invasive, Respectful Research with Nonhuman Animal Participants'. *Society & Animals* 28 (2): 171–90. https://doi.org/10/gg29qd.

Van Patter, L. E. and Hovorka, A. J. (2018) '"Of Place" or "of People": Exploring the Animal Spaces and Beastly Places of Feral Cats in Southern Ontario'. *Social & Cultural Geography* 19 (2): 275–95. https://doi.org/10.1080/14649365.2016.1275754.

Philo, C. and Wilbert, C. (2000) *Animal Spaces, Beastly Places: New Geographies of Human-Animal Relations.* Florence, United States: Taylor & Francis Group.

Pivot, A-L., Rohbacher, A., Vimal, R., and Ferrer, L. (2019) 'Estive en partage: interactions entre ours et pastoralisme dans les Pyrénées. Rapport de projet.' *Dissonances, GEODE* , 63.

Power, E. (2008) 'Furry Families: Making a Human–Dog Family through Home'. *Social & Cultural Geography* 9 (5): 535–55. https://doi.org/10.1080/14649360802217790.

Probyn, E. (2014) 'The Cultural Politics of Fish and Humans: A More-Than-Human Habitus of Consumption'. *Cultural Politics* 10 (3): 287–99. https://doi.org/10/gg3f5t.

Salter, C. (2018) 'Our Cows and Whales'. *M/C Journal* 21 (3). http://journal.media-culture.org.au/index.php/mcjournal/article/view/1410.

Seymour, M. and Wolch, J. (2010) '"A Little Bird Told Me …": Approaching Animals Through Qualitative Methods'. In *The SAGE Handbook of Qualitative Geography*, 305–20. London: SAGE Publications Ltd. https://doi.org/10.4135/9780857021090.

Shyan-Norwalt, M. R. (2005) 'Caregiver Perceptions of What Indoor Cats Do "For Fun"'. *Journal of Applied Animal Welfare Science* 8 (3): 199–209. https://doi.org/10.1207/s15327604jaws0803_4.

Srinivasan, K. (2013) 'The Biopolitics of Animal Being and Welfare: Dog Control and Care in the UK and India.' *Transactions of the Institute of British Geographers.* 38 (1): 106–19.

Srinivasan, K. (2016) 'Towards a Political Animal Geography?' *Political Geography* 50 (January): 76–78. https://doi.org/10.1016/j.polgeo.2015.08.002.

Syse, K. (2013) 'Otters as Symbols in the British Environmental Discourse'. *Landscape Research* 38 (4): 540–52. https://doi.org/10.1080/01426397.2013.784244.

Turnbull, J., Searle, A., and Adams, W. A. (2020) 'Quarantine Encounters with Digital Animals: More-than-Human Geographies of Lockdown Life'. *Journal of Environmental Media* 1 (2): 6.1–6.10. https://doi.org/10.1386/jem_00027_1.

De Waal, F. (1999) 'Anthropomorphism and Anthropodenial: Consistency in Our Thinking About Humans and Other Animals'. *Philosophical Topics* 27 (1): 255–80. https://doi.org/philtopics1999927122.

Whatmore, S. (2006) 'Materialist Returns: Practising Cultural Geography in and for a More-than-Human World'. *Cultural Geographies* 13 (4): 600–09. https://doi.org/10.1191/1474474006cgj377oa.

Wilcox, S. and Rutherford, S. (2018) *Historical Animal Geographies*. First edition. Routledge Human-Animal Studies. Milton Park, Abingdon, Oxon; New York, NY: Routledge.

Yarwood, R. and Evans, N. (2016) 'A Lleyn Sweep for Local Sheep? Breed Societies and the Geographies of Welsh Livestock': *Environment and Planning A*, July. https://doi.org/10.1068/a37336.

Zhang, A. (2020) 'Circularity and Enclosures: Metabolizing Waste with the Black Soldier Fly'. *Cultural Anthropology* 35 (1): 74–103. https://doi.org/10.14506/ca35.1.08.

21

URBAN GEOGRAPHIES

Comparative and Relational Urbanisms

Kevin Ward

Introduction

This chapter is written from Manchester in the North West of England. This is a city with a strong and distinct industrial past that includes its role in generating and supporting an extractive relationship under colonialism (Kitchen, 1997; Peck and Ward, 2002). That it was the birth-place of the first Industrial Revolution that continues to be marshalled by many of those with a role in its promotion, while at the same time some in the city wrestle with how best to acknowledge and mark in the present its past role in a system of oppression and violence. Examples of its once central location in an emerging global economy puncture the landscape, infrastructural inheritances of a sort, rediscovered since the late 1980s as Manchester has experienced significant physical revitalization of its urban core. You probably know the thing, right? Residential and retail gentrification has been driven by wave after wave of capital invested in the city's built environment, capturing the attention of national and international media (Harris, 2015; Hale and Bounds, 2019), not least as its origins increasingly raise geo-political issues for the UK government. In addition to numerous pieces on the city's recent experiences, a documentary series on the BBC titled *Manctopia: Billion Pound Property Boom* sought to capture "Manchester's remarkable housing boom" (www.bbc.co.uk/programmes/m000lwn5). Only last year a promotion video claimed to demonstrate *How Manchester Fixed its Industrial Decline* (www.youtube.com/watch?v=fuTIDc5ug-Y&ab_channel=TheB1M). And it is not just those who believe in what Manchester claims it has achieved who have made their views clear. Most recently, the means through the local government has sold off land in the centre of Manchester to private developers has been challenged, as part of a bigger questioning of who owns the city (Gillespie and Silver, 2021).

Whether critical or supportive, each of these pieces has reaffirmed that the last couple of decades have seen Manchester continue to experience spatially and socially uneven economic development. While it is now home to a glitzy and shiny downtown, punctured by a growing number of taller buildings, the city also continues to contain within it a considerable number of deprived and marginalized communities. So it's a city, like many others, that consist of multiple and overlapping urban geographies, remaining a tale of at least two cities (Taylor et al., 1996; Peck and Ward, 2002; Symons and Lewis, 2017).

DOI: 10.4324/9781003038849-24

Those of us who reside in Manchester continue to wrestle with how best to live our lives under restricted COVID-19 lock-down conditions. The anxiety and the stress, the stop-start-stop of what you can and cannot do, where you can do it and with whom you can do it. As is the case across many cities of the world, those agencies that are charged with the governance and planning of Manchester face the potentially contradictory challenges of both addressing the immediate effects of the global pandemic while anticipating its likely longer term effects. What does seem clear is that no aspect of urban life is untouched by the consequences of COVID-19 (Acuto, 2020; Chen et al., 2020; Connolly et al., 2020; Foreign Policy, 2020; McCann, 2020). This is also an intellectual challenge to the urban elements of a number of disciplines – including geography.

Urban geography is one of the more well-established sub-disciplines in human geography. Where I am based, in the UK, its origins date back over seventy years and the 1947 publication of *City, Region, and Regionalism* by R E Dickinson and the 1951 publication of *Urban Geography* by G Taylor. In the US the origins were just over a decade later with the 1959 publication of *Readings in Urban Geography* by H M Mayer and C F Kohn (Berry and Wheeler, 2004). The journal I currently edit, *Urban Geography*, did not emerge out of the US until 1980. Other countries have different sub-disciplinary histories, troubling a single sub-disciplinary timeline. Nevertheless, and while acknowledging a degree of Anglo-centrism, many of the national histories of urban geography are organised through certain paradigms. Even if there is variation in the precise timings between the emergence of each paradigm, the conceptual and methodological overlaps between each one, and the extent to which influential figures moved across them, means it is possible to identify periods when certain approaches came to dominate. Analytically, positivist urban geography (from the 1960s), structuralist political economy (from the 1970s), Feminism (from the 1980s) and post-structuralism (from the 1990s) are often used to organise a history of urban geography. Note there are no end dates to these periods. An approach might dominate in a country, eventually taking on the mantle of "paradigm". However, even as a newer approach emerges, so others continue to be practiced. For example, in the third decade of the twenty-first century at the journal *Urban Geography* we receive submissions of work that would fit within each of these paradigms, emerging out of countries from around the world.

The brief for the chapters in this section was to "critically appraise key methodological and theoretical challenges/opportunities that are shaping contemporary research in various parts of human geography" and to "offer a spotlight on trends, tensions, and emerging questions", in my case, within urban geography. Of course, academic exercises that seek to capture and distil a wide range of work reveal as much about their authors as the sub-disciplines or sub-fields they claim to represent. There is nothing wrong with this inevitable partiality, as revealed by relatively recent exciting and insightful reviews of urban geography (Lees, 2002; Jacobs, 2012a; Derickson, 2015, 2017, 2018). And so it is here. This chapter is not exhaustive. Rather, it is based on a particular reading of urban geography, and of urban studies more generally. It is organised by two "key" methodological and theoretical challenges. The argument is that these are likely to continue to characterise work within urban geography and they are the intellectual fault-lines along which the future of the field is being made. In introducing the contributions and developments in the areas of comparative urbanism and relational urbanisms the chapter concludes with no single, overarching argument. That would be at odds with much of the material presented and discussed. For many in urban geography these are exciting, interesting and – yes – awkward and challenging times. Existing understandings of the urban are being questioned. The past "narrow" geographies of urban theory production are being slowly replaced by a wider set of locations out of which new concepts and theories are being generated. This has sparked lively debates in the field (for just a small selection, see Roy, 2009; Brenner and Schmid, 2014; Peck, 2015; Scott and Storper, 2015; Robinson and Roy, 2016; Jazeel, 2018; Scott, 2021). Emerging, slowly, are a more creative, experimental, diverse and global set of comparisons (McFarlane and Robinson, 2012; Sheppard et al., 2013; Müller and Trubina, 2020; Lancione and McFarlane, 2021). The future of the sub-field is not there waiting to be discovered but is, rather, being made through the work currently underway in universities around the world.

Comparative Urbanism

What does it mean to imagine or think "comparatively" about cities as an urban geographer in the twenty-first century? Asking and then answering this question lies behind the first "key" methodological and theoretical challenge introduced and discussed in this chapter. The last two decades have witnessed a growth in the work that has paid explicit attention to the epistemological and ontological, conceptual and geographical, underpinnings of the comparison of cities (Robinson, 2006, 2016; Ward, 2008, 2010; McFarlane, 2010; McFarlane and Robinson, 2012; Peck, 2015; Hart, 2018). Stemming from a wide range of concerns and interests, these contributions – and many more – have coalesced around a rethinking of the notion of comparative urbanism. This "upsurge in comparative urban research" (McFarlane, 2010, p. 730), in which has been "witnessed both a revival and a reorientation" (McFarlane and Robinson, 2012, p. 765) has come after a number of decades in which, according to Nijman (2007, p. 1), "comparative methodologies largely disappeared from view" in geography, perhaps because, "[e]veryone knows that comparative urbanism is difficult" (Dear, 2005, p. 247).

Difficult or not, of course there is an intellectual history comparing cities in the arts and the social sciences. For example, a series of urban anthropological studies between the 1940s and 1960s were comparative while seeking to avoid universalizing on the basis of a relatively small number of cases. Subsequent decades, however, saw urban studies "divided by the hierarchical categorisation of different kinds of cities as developed or undeveloped" (Robinson 2006, p. 41). This shift reflected wider trends in social and urban theorization as a particular form of scientific method emerged as dominant. As Pickvance (1986, p. 166) argued:

> [t]he 1970s was a period of grand theory when theoretically-driven models … which at best corresponded to realities in one country, were advanced as having general applicability … Comparative analysis is a necessary complement to such theorizing since it gives priority to testing the models over its whole scope of application.

An important aspect of this more general shift over the course of the 1970s and 1980s occurred in the rationale underpinning the selection of those discrete and bounded cases – the cities. Walton's (1973) "standardized case comparison" lay behind much of the comparative urban studies during this era. In this approach the selection of the cases was based on the findings and results of previous studies. Whole cities were chosen on the basis of what was already known about them. So cities needed to be rendered knowable before they were considered comparable. Data on cities needed to be collected and analysed, accessible and publishable in a form that created the pre-conditions for comparison. Where this revealed certain shared features or similarities then cases – the cities in question – were deemed appropriate for comparison. The comparison of eight cities – Detroit, Glasgow, Houston, Liverpool, Milan, Naples, New York and Paris – by Savitch and Kantor (2002) is a prime example of this sort of study. It contrasted the political economy of urban regimes, comparing their bargaining context, as a way of connecting local specifics to wider institutional contexts. The concern was to "[specify] how different national and international circumstances shape regimes and their policy biases" (Kantor et al., 1997, p. 349), the choice of cities reflecting a concern to examine "the Western industrial system of advanced democracies."

Of course, this strategy of case study selection tended to stifle the comparative imagination, generally limiting the focus to only cities that were understood in advance to be known, knowable and similar. While this generated a series of useful and fruitful insights, there were a number of cities that for the most part were "off the map" (Robinson, 2006). In particular, the experiences of a large and growing number of cities failed to register in the comparative studies of cities. Where cities from elsewhere were selected as cases – that is beyond those geographically located in what is known in shorthand as the global North – they were often compared against more known cities that were deemed the norm, rendering the less-known cities as 'abnormal', 'backward', or 'under-developed'. This was both intellectually and often politically debilitating and undermining. Moreover, and as Jacobs (2012b, p. 904) has so eloquently written about regarding the comparative urban studies of this era, this selection of cases:

> has been put to work in a model of sampling that, by dint of there being more than one case, the scientist is able to understand something more singular in character: a unitary cause, a common trajectory, and a shared logic. This singularity is often built around the qualities of commonality between the cases, what is referred to as the *tertium comparationis* or the "third part" of comparison. The comparative method's third part often goes under the name "pattern," and sometimes pattern can be taken as evidence for understanding connection and even causality.

While the case for the incapacity of conventional comparative methods to be able to fully articulate, and to be open to, a world of cities has been already convincingly and persuasively made (Robinson, 2006; Roy, 2009; McFarlane, 2010), another manoeuvre might also necessary. In the context of the trends in global urbanization, a more imaginative and perhaps cosmopolitan comparison of cities would not be one 'from above' as has traditionally been the way. Rather it would be one based on a significant revising of the ontological basis of comparison.

For example, thinking "comparatively" about cities through the twenty-first century might require comparisons from below and through. *From below* in the sense of being open to the diversity and the multiplicity of either specific elements or whole cities against which other cities, or parts of them, compare themselves. This involves uncovering the processes and practices in and through which certain cities or bits thereof, are brought forward as places against which other cities should benchmark themselves, sometimes quite literally (Acuto et al., 2021). And this, of course, means that some cities or bits thereof are pushed back. Implicit in this mode of comparison is a set of power geometries in which some cities are situated as places from which others should learn through comparison while others are located as those for whom comparison should involve seeking to learn, and often emulate, others (McCann, 2004; Gonzalez, 2011). Here we are to understand learning in its broadest sense, as a diverse and internally heterogeneous field with a variety of policy expertise, knowledge and understanding, fully and part-formed, complete and incomplete (McFarlane, 2011). *From through* in the sense of being open to the diversity and the multiplicity of ways – the actually practices – in and through which comparison occurs involving a variety of sites and spaces. This means uncovering the comparative architecture comprising the circulations, connections, networks and webs which shape how cities are imagined and planned, financed and governed and the diversity of urban contexts – current, past, and in the future – upon which those making urban policy draw.

Taken together, these manoeuvres involve "an effort to rethink the Euro–American legacy of urban studies" (Robinson and Roy, 2016, p. 181) and to "speak back against, and [thereby], contesting, mainstream global urbanism" (Sheppard et al., 2013, p. 893). This is about going beyond empirical addition to use comparison politically to challenge the taken-for-granted sites of urban theory generation, And, in turn, questioning what counts as urban theory and who gets to do the counting and validating.

Relational Urbanism

> Instead of taking as given pre-existing objects, events, places and identities, I start with the question of how they are formed in relation to one another and to a larger whole.
>
> *Hart 2002: 14*

Urban geography has not been immune to the "relational turn" that has challenged some of the assumptions underpinning the wider discipline of geography (Massey, 1993; Amin and Thrift, 2002; Jones, 2007; Jayne et al., 2013; Binnie, 2014). This "turn" is the second "key" methodological and theoretical challenge introduced in this chapter. This work questions the bounded and delimited conceptualization of place that seems to have once characterized much of regional and urban geography (and their wider fields, regional and urban studies). For, if it were ever enough to account for change in the nature of the urban on the basis of analysis generated solely from within cities and the countries of which they are part, then this work argues that this time has passed. Recent work from across the social sciences has instead argued for an approach to the theorization of space that is both relational and territorial (Allen et al., 1998; Allen and Cochrane, 2007; Massey, 2005; McCann and Ward, 2010). It has highlighted the increasingly open, porous and inter-connected configuration of territorial entities. Instead of topography, this work advances a topological theorization in which both the near and the far, the close and the distant, are constitutive of the urban (Jacobs, 2012a). For many contemporary cities are increasingly open, porous and inter-connected in ways that troubles once-taken-for-granted theorizations of them as given, internally coherent, and stable territorial entities.

According to Amin (2004, p. 34) "cities come with no automatic promise of territorial … integrity, since they are made through the spatiality of flow, juxtaposition, porosity and relational connectivity." Drawing on work by Massey (1993, 2005), this perspective argues that "cities exist in an era of increasing geographically extended spatial flows" (Jacobs, 2012a, p. 412). That is, cities as entities to study don't respect the formal boundaries assigned to them through governments. The relations, between institutions or people, stretch out across space. Often these relations bring places that are geographically distant closer together. As a result, this relational way of thinking about cities argues that the contemporary city needs to be conceptualized as "open, discontinuous, relational and internally diverse" (Allen et al., 1998, p. 143). This is also argued to have implications for how one goes about researching cities: you do not take for granted the city as a territory. Where do you study the city in which you are interested? Inside or outside of the city, or perhaps, do you try to trace out some of the networks of which the city is part? Or, if you are interested in migration, do you look at where people have come from as well as from where they have travelled?

It is possible to understand cities both as territories and as sites of relations with else-where. For MacLeod and Jones (2007, p. 1186) "all contemporary expressions of territory … are, to varying degrees, punctuated by and orchestrated through a myriad of trans-territorial networks and relational webs of connectivity." So, contemporary cities have boundaries of sorts, of course, but these are open to contestation, manipulation and refinement. Power is expressed through how these are made and remade. Many times these boundaries stem from government entities, in areas of policy such as education, health or policing. These boundaries also appear on maps of one sort or another. They are not natural, however. Rather, much like any boundaries, they are the outcome of various decisions over many years. And they are not fixed. Moreover, these boundaries are in part constituted through relations that cities have with other cities: cities are nodes in wider flows of ideas, finances, people and so on. So the city you may live or work in is likely involved in a number of different sorts of networks and relations with other cities, some shorter, others longer, some more important, others less so. Some are high-profile and promoted, such as those through "twinning" programmes (Jayne et al., 2013). Others are more discrete, such as capital investments by overseas governments or questionable corporations.

Implicit in this mode of comparison is then a manoeuvre that acknowledges the reconfiguring of what is understood as 'the urban'. That is holding onto cities as territorial expressions, while also understanding contemporary cities as 'multiplex', comprised of elements of relationships that stretch far beyond their boundaries (Amin and Graham, 1997, p. 418), understanding the city as "a set of spaces where diverse ranges of relational webs coalesce, interconnect and fragment." This approach to theorizing the urban holds in productive tension the inter-connectedness between relational and territorial geographies and is generative. It produces an understanding of the city that acknowledges that "all contemporary expressions of territory … are, to varying degrees, punctuated by and orchestrated through a myriad of trans-territorial networks and relational webs of connectivity" (MacLeod and Jones, 2007, p. 1186).

Conclusion

The sub-discipline of urban geography continues to experience an exciting reckoning and renewal, energized by the on-going urbanization of the planet and the various challenges facing both the more well-established and the still emerging and growing cities. The COVID-19 pandemic has only served to underscore the relational-comparative element to urban living and place-making. The focus of the chapter was on the inter-connected and overlapping fields of comparative urbanism and relational urbanism. Both are sites of lively debates across much of urban geography. They are present in a range of popular areas of the sub-discipline, including the work, for example, on the climate crisis, culture, economic development, gentrification, infrastructure, race and transport.

Robinson and Roy (2016, p. 164) argue that the sub-discipline is "undergoing a phase of rich experimentation, with a proliferation of paradigms and exploration or invention of various methodologies inspired by the diversity and shifting geographies of global urbanization." They are not alone in making this case. The last decade has witnessed a series of back and forth debates over what gets to be labelled as Theory with a capital "T" and the terms of appropriation and incorporation into the mainstream of those theories traditionally marginal. Roy (2020, p. 19) has recently argued "that despite the Southern turn in urban studies, the epistemologies and methodologies of inquiry remain more or less untouched by forms of knowledge designated as the 'other' of dominant Theory." Whether one agrees with her or not, the immediate future of urban geography seems set to be one characterized as much by disengaged as engaged pluralism (Barnes and Sheppard, 2010; van Meeteren et al., 2016; Brenner, 2018; Oswin, 2020).

References

Acuto, M. (2020) Engaging with global urban governance in the midst of a crisis, *Dialogues in Human Geography* 10 221–224.

Acuto, M., Pejic, D., and Briggs, J. (2021) Taking city rankings seriously: Engaging with benchmarking practices in global urbanism, *International Journal of Urban and Regional Research* 45 363–377.

Allen, J. and Cochrane, A. (2007) Beyond the territorial fix: Regional assemblages, politics and power, *Regional Studies* 41 1161–1175.

Allen, J., Massey, D., Cochrane, A., Charlesworth, J., Court, G., Henry, N., and Sarre, P. (1998) *Re-thinking the region: Spaces of neoliberalism*, Routledge: London.

Amin, A. (2004) Regions unbound: Towards a new politics of place, *Geografiska Annaler* 86B 33–44.

Amin, A. and Graham, S. (1997) Ordinary city, *Transactions of the Institute of British Geographers* 22 411–429.

Amin, A. and Thrift, N. (2002) *Cities: Reimaging the urban*, Polity Press: Cambridge.

Barnes, T. and Sheppard, E. (2010) "Nothing includes everything": Towards engaged pluralism in Anglophone economic geography, *Progress in Human Geography* 34 193–214.

Berry, B.J.L. and Wheeler, J.O. (2004) *Urban geography in America, 1950 to 2000: Paradigms and personalities*, Routledge: London.

Binnie, J. (2014) Relational comparison, queer urbanism and worlding cities, *Geography Compass* 8 590–599.

Brenner, N. (2018) Debating planetary urbanization: Towards an engaged pluralism, *Environment and Planning D: Society and Space* 36 570–590.

Brenner, N. and Schmid, C. (2014) The 'urban age' in question, *International Journal of and Regional Research* 38 731–755.

Chen, B., Marvin, S. and While, A. (2020) Containing COVID-19 in China: AI and the robotic restructuring of future cities, *Dialogues in Human Geography* 10 238–241.

Connolly, C., Harris, A.S. and Keil, R. (2020) On the relationships between COVID-19 and extended urbanization, *Dialogues in Human Geography* 10 213–216.

Dear, M. (2005) Comparative urbanism, *Urban Geography* 26 247–251.

Derickson, K.D. (2015) Urban geography I: Locating urban theory in the "urban age", *in Human Geography* 39 647–657.

Derickson, K.D. (2017) Urban geography II: Urban geography in the age of Ferguson, *Progress in Human Geography* 41 230–244.

Derickson, K.D. (2018) Urban geography III: Anthropocene urbanism, *Progress in Human Geography* 42 425–435.

Foreign Policy (2020) How life in our cities will change after the coronavirus pandemic. May 1. https://foreignpolicy.com/2020/05/01/future-of-cities-urban-life-after-coronavirus-pandemic/.

Gillespie, T. and Silver, J. (2021) *Who owns the city? The privatisation of public land in Manchester*, available at: www.gmhousingaction.com/wp-content/uploads/2021/05/Who-Owns-The-City-v1.9.pdf.

Gonzalez, S. (2011) Bilbo and Barcelona "in motion": How regeneration models travel and mutate in the global flows of policy tourism, *Urban Studies* 48 1397–1418.

Hale, T. and Bounds, A. (2019) The Manchester model: Universities lead urban revival, Financial Times, available at: www.ft.com/content/e5b37280-82e2-11e9-b592-5fe435b57a3b.

Harris, J. (2015) The great reinvention of Manchester: "It's far more pleasant than London", *The Guardian*, available at: www.theguardian.com/cities/2015/nov/03/the-great-reinvention-of-manchester-its-far-more-pleasant-than-london.

Hart, G. (2002) *Disabling globalization: Places of power in post-Apartheid South Africa*, Berkeley, CA: University of California Press.

Hart, G. (2018) Relational comparison revisited: Marxist postcolonial geographies in practice, *Progress in Human Geography* 42 371–394.

Jacobs, J.M. (2012a) Urban geographies I: Still thinking cities relationally, *Progress in Human Geography* 36 412–422.

Jacobs, J.M. (2012b) Commentary – comparing comparative urbanisms, *Urban Geography* 33 904–914.

Jayne, M., Hubbard, P. and Bell, D. (2013) Twin cities: Territorial and relational geographies of "worldly" Manchester, *Urban Studies* 50 239–254.

Jazeel, T. (2018) Urban theory with an outside, *Environment and Planning D: Society and Space* 36 405–419.

Jones, M.R. (2007) Phase space: Geography, relational thinking, and beyond, *Progress in Human Geography* 33 487–506.

Kantor, P., Savitch, H.V. and Vicari, H.S. (1997) The political economy of urban regimes: A comparative perspective, *Urban Affairs Review* 32 348–377.

Kitchen, T. (1997) *People, politics, policies and plans: The city planning process in contemporary Britain*, London: Sage.

Lancione, M. and McFarlane, C. (2021) *Global urbanism: Knowledge, power and the city*, Routledge: London.

Lees, L. (2002) Rematerializing geography: The "new" urban geography, *Progress in Human Geography* 26 101–112.

MacLeod, G. and Jones, M.R. (2007) Territorial, scalar, networked, connected: In what sense a "regional world"? *Regional Studies* 41 1177–1191.

Massey, D. (1993) Power-geometry and a progressive sense of place. *In Mapping the futures: Local cultures, global change,* editors Bird, J., Curtis, B., Putman, T., Robertson, G. and Tickner, L., 59–69, Routledge: London.

Massey, D. (2005) *For space*, Sage: London.

McCann, E. (2004) "Best places": Interurban competition, quality of life and popular media discourse, *Urban Studies* 41 1909–1929.

McCann, E. (2020) Spaces of publicness and the world after the coronavirus crisis, *Society and Space*, available at: www.societyandspace.org/articles/spaces-of-publicness.

McCann, E.J. and Ward, K. (2010) Relationality / territoriality: Toward a conceptualization of cities in the world, *Geoforum* 41 175–184.

McFarlane, C. (2010) The comparative city: Knowledge, learning, urbanism, *International Journal of Urban and Regional Research* 34 725–742.

McFarlane, C. (2011) *Learning the city: Translocal assemblages and urban politics*, Wiley Blackwell: Oxford.

McFarlane, C. and Robinson, J. (2012) Introduction – experiments in comparative urbanism, *Urban Geography* 33 765–773.

Müller, M. and Trubina, E. (2020) The global Easts in global urbanism: Views from beyond North and South, *Eurasian Geography and Economics* 61 627–635.

Nijman, J. (2007) Introduction – comparative urbanism, *Urban Geography* 28 1–6.

Oswin, N. (2020) An other geography, *Dialogues in Human Geography* 10 9–18.

Peck, J. (2015) Cities beyond compare? *Regional Studies* 49 160–182.

Peck, J. and Ward, K. (editors) (2002) *City of revolution: Restructuring Manchester*, Manchester: Manchester University Press.

Pickvance, C. (1986) Comparative urban analysis and assumptions about causality, *International Journal of Urban and Regional Research* 10 162–184.

Robinson, J. (2006) *Ordinary cities: Between modernity and development*, Wiley-Blackwell: Oxford.

Robinson, J. (2016) Comparative urbanism: New geographies and cultures of theorizing the urban, *International Journal of Urban and Regional Research* 40 187–199.

Robinson J. and Roy, A. (2016) Global urbanisms and the nature of urban theory, *International Journal of Urban and Regional Research* 40 164–180.

Roy, A. (2009) The 21st-century metropolis: New geographies of theory, *Regional Studies* 43 819–830.

Roy, A. (2020) "The shadow of her wings": Respectability politics and the self-narration of geography," *Dialogues in Human Geography* 10 19–22.

Savitch, H.V. and Kantor, P. (2002) *Cities in the international marketplace: The political economy of urban development in North America and Western Europe*, Princeton University Press: Princeton.

Scott A.J. (2021) The constitution of the city and the critique of urban theory, *Urban Studies* (on line first).

Scott, A.J. and Storper, M. (2015) The nature of cities: The scope and limits of urban theory, *International Journal of Urban and Regional Research* 39 1–15.

Sheppard, E., Leitner, H. and Maringanti, A. (2013) Provincializing global urbanism: A manifesto, *Urban Geography* 34 893–900.

Storper, M. and Scott, A.J. (2016) Current debates in urban theory: A critical assessment. *Urban Studies* 53 1114–1136.

Symons, J. and Lewis, C. (2017) *Realising the city: Urban ethnography in Manchester*, Manchester University Press: Manchester.

Taylor, I., Evans, K., and Fraser, P. (1996) *A tale of two cities: Local feeling and everyday life in the North of England*, Routledge: London.

van Meeteren, M., Bassens, D. and Derudder, B. (2016) Doing global urban studies: On the need for engaged pluralism, frame switching and methodological cross-fertilization, *Dialogues in Human Geography* 6 296–301.

Walton, J. (1973) Standardized case comparison: Observations on method in comparative sociology. In Armer, M. and Grimshaw, A., editors, *Comparative social research*. New York: John Wiley and Sons 173–188.

Ward, K. (2008) Editorial – toward a comparative (re) turn in urban studies: Some reflections, *Urban Geography* 29 405–410.

Ward, K. (2010) Towards a relational comparative approach to the study of cities, *Progress in Human Geography* 34 471–487.

22

ECONOMIC GEOGRAPHIES

Navigating Research and Activism

Kelly Dombroski and Gerda Roelvink

Introduction

The way we investigate the economy in many ways *produces* the economy. This is the starting point for researchers in the sub-field of economic geography known as diverse economies (Gibson-Graham and Dombroski, 2020). In this approach to economic geography, methodology is not about 'getting accurate data', but instead considers the kinds of economic realities our research is representing and thus producing. Within the diverse economies field, community economies scholars aim to both *represent* and *produce* a world where capitalism is not the sole dominant economic force shaping peoples' lives. This chapter outlines how this understanding of economic performativity is expressed in ethical and critical methodologies of economic geography, and in particular the field of diverse and community economies research.

Economic geography as a subdiscipline focuses on the ways in which places and spaces *shape* and *are shaped by* economic activities. Tracing the history of the discipline, Trevor Barnes (2009) notes that, like many subdisciplines in geography, economic geography has been influenced by trends in thinking more broadly. In the 1950s, descriptive regional geographical work was challenged by a new generation of economic geographers who aimed to create a spatial science that could map and predict relationships between economic activities and locations. These geographers used quantitative methods influenced by economics and physics, including mathematical modelling. One example is William Alonso (1964) who modelled the cost of land against the distance from a city 'centre' in *Location and Land Use: Toward a General Theory of Land Rent*.

Since the 1970s, these spatial science models in economic geography have been challenged by Marxist political economy approaches. Trevor Barnes highlights some high profile 'defections' from the modelling approach to the explanatory approach offered by Marxist influenced thinking, such as prominent economic geographer David Harvey, who went on to develop key theoretical contributions to economic geography using Marx. Here, methods were not seen as 'neutral' tools revealing the natural patterns underlying the economy, but as already influenced by the frameworks of thinking the researcher brought to the object of research, in this case, economic patterns in space and place. Critical realism was a methodological approach developed to explain economic geographies of place with reference to both the 'external necessary relations' such as economic structures and the 'internal, contingent relations', which both shape place

DOI: 10.4324/9781003038849-25

(Sayer, 1982). Andrew Sayer (1982) argued that in-depth research was required to separate out these two, and statistical spatial science was not able to do this without engaging with theory. Doreen Massey's work in the 1980s was an important step in engaging with both 'big picture' theories of political economy and rich, place-based case-study investigations (see particularly Massey, 1984).

In the 1990s, feminist economic geographers Julie Graham and Katherine Gibson (writing as JK Gibson-Graham) took aim at both the spatial science versions of economic geography *and* the Marxist-informed versions, which had previously been seen as 'opposites'. In *The End of Capitalism (as we knew it): A feminist critique of political economy*, Gibson-Graham (1996) argued that both approaches continually described and re-described the economy as a capitalist monolith, ignoring what did not fit that description. Both approaches thus worked to *produce* and reproduce the economy as ever more capitalist, especially as alternative and diverse economic practices were constantly positioned as insufficient and hopeless, even by those scholars who were anti-capitalist. For Gibson-Graham, ethical and effective methodologies of economic geography would pay more attention to the power of economic representation, in the same way that feminist and queer methodologies paid attention to the power dynamics of gender and sexual representation. Feminist theorists had challenged 'phallocentrism', pushing back against the organising of all gender thinking with cis-men as the norm or centre. Gibson-Graham drew this thinking into their analysis of economic geography, arguing that it was 'capitalocentric', organising all thinking, even Marxist thinking, around the 'norm' or centre of capitalism (1996). Gibson-Graham sought to intervene in capitalocentric understandings of the economy by representing it as an iceberg, where the capitalist economic activities examined by economic geographers (including Marxists) were only the 'tip' above the waterline. By representing all the heterogenous and invisible economic activities below the waterline, a researcher was making an informed, ethical decision to make those 'lost' activities more visible and thus more likely to continue. While this approach drew on rich place-based descriptions of economic activities similar to economic anthropology, Gibson-Graham refused to represent non-capitalist economic activities as necessarily doomed for co-option or destruction. Instead, they intentionally searched for the possibilities of different kinds of economies that were already present, refusing to fold everything into a teleological story of destruction or 'progress'. Gibson-Graham drew on the work of Eve Sedgwick, who asked:

> What if instead there were a practice of valuing the ways in which meanings and institutions can be at loose ends with each other? What if the richest junctures weren't the ones where everything means the same thing?
>
> *Sedgwick, 1994, p. 6*

This approach to research came to be known as the diverse economies approach, and scholars working in this tradition use methods that deliberately represent these 'othered' economic practices as 'at loose ends', not completely woven into a single story. This approach left room for possibly different futures from what a 'strong' capitalocentric theory of capitalism would predict for them (Gibson-Graham and Dombroski, 2020). In subsequent work, Gibson-Graham and diverse economies scholars have sought to develop visions of the economy that built on these diverse practices, in particular the ones which can contribute to 'surviving well together' as people and planet (McKinnon, Healy, and Dombroski, 2019). This is the work of community economies.

What does this research really look like in practice? In this chapter we take up this question, beginning by outlining how economic geographers in the field of diverse economies work to inventory economic diversity in an open, non-deterministic, way. This inventorying makes present a wide range of economic activities and, in doing so, increases the possibilities for economic action. We then turn to outlining some of these actions, focusing on the work of community economies scholars who are working with others to make a claim for some economic activities over others. We conclude by considering what this means for the kinds of methods economic geographers might adopt in the twenty-first century, when the changes we face are complex and the future is unknown.

Embracing and Inventorying the Mess

In his thought provoking book *After Method: Mess in Social Science Research*, John Law (2004) wrestles with the problem of 'mess' in social science. What he means by mess is the complexity and contingency of our social realities, the ways in which they are multiple and overlapping, and ultimately, very difficult to both *know* and *represent* in research. He invites us to understand methods as assemblages of materialities, spatialities and socialities that help us understand reality but also participate in *making* reality. Similarly, Boaventura de Sousa Santos (de Sousa Santos, 2014) highlights the ways in which knowledge from the Global North erases realities present in the Global South, a process which he calls 'epistemicide'. De Sousa Santos calls for ecologies of knowledge, where interacting multiple realities are acknowledged in our scholarship and fieldwork. Bringing this into the field of economic development, Arturo Escobar (2018) in *Designs for the Pluriverse* proposes a conceptualisation of reality as a pluriverse, one in which multiple realities coexist in space, and where our attempts at knowing and designing interventions must be fully embedded in this multiplicity. In economic geography, Doreen Massey's (2005) book *For Space* argues that space is a 'simultaneity of multiple trajectories'. Yet, as economic geographer Gerda Roelvink (2016) points out in her book *Building Dignified Worlds*, it seems as if sometimes in doing research we keep picking out the same constellations again and again, ignoring the vast multiplicity of the night sky for the familiar patterns that we know, such as capitalism, neoliberalism and so on.

These arguments are important to acknowledge because our understanding of what is possible in the world, our *ontology*, affects our understanding of what can be known in the world, our *epistemology*. If capitalism is seen as the only game in town, for example, then resistance to capitalism becomes the main option for political struggle and our knowledge production will be centred on showing the destructive nature of capitalism in order to motivate and aid this resistance. Our understanding of what can be known in the world will therefore also affect our *methodology*, or the way in which we structure and rationalise our research methods. But the reverse is also true: if our methodologies are too narrow and do not create space for multiplicity, what are we erasing in our research process? What other possibilities might there be for political struggle and other ways of surviving well?

The erasure of reality through the process of knowledge production is well known to Indigenous peoples around the world. In her ground-breaking book *Decolonising Methodologies* (1999), Māori scholar Linda Tuhiwai Smith describes the violence that colonial assumptions about reality have wrought over Māori in the area of research. While this book does not focus on economic geography, we can see many of the same patterns in our discipline. In the context of Aotearoa New Zealand, Māori realities were – and often still are – overwritten by the 'realities' of the coloniser, particularly in unequal research relations. What is 'possible' is described and delineated by Western science, and the diverse relationships with place and beings in place are erased. Ngāti Whakaue author Amanda Yates, for example, discusses the importance of the concept of *mauri* or intrinsic life force in Māori realities, where the distinction between 'animate' and 'inanimate' is not normalised. The *mauri* or life force of a river is important in the holistic, more-than-economic wellbeing of a community and a place (Yates, 2016). For Smith, the approach to research with Māori must be one grounded in kaupapa Māori, or Māori agendas and protocols, in collaboration with community.

For others working in economic geography or other disciplines in Aotearoa, it is about making space for multiple realities within whichever space one is working (Waitoa and Dombroski 2020). Our methodologies must make space for complexity, for 'mess', for multiple ontologies (McKinnon, Healy, and Dombroski, 2019). For those of us who have been pushed to the margins of economic geography, this awareness of multiplicity is not new. For those of us (the authors included) who have grown up in positions of ontological privilege, we must learn how to develop this capacity for multiplicity in research.

Over the last three decades, diverse economies scholars have developed methodologies that provide space for multiplicity. As Gibson-Graham and Dombroski (2020) note in their essay 'inventory as ethical intervention', in economic geography and other disciplines that study the economy, we are often encouraged to study the economy primarily through the lens of capitalism, asking 'what is capitalist about this or that' (6). In a diverse economies approach to research, we are, however, invited to 'stand back and assess whether the effects of our representations are opening up spaces for change or transformation or not' (6). For Gibson-Graham and Roelvink, this 'standing back' comes through an attentiveness to a spacious silence and slowing down, an embodied response that deliberately pays attention to what is possible, what is present, and allows the researcher to be rendered as 'newly constituted' in a 'newly constituted world' (Gibson-Graham and Roelvink, 2010, p. 322).

Formulating this into a methodology, diverse economies scholars have picked up on Eve Sedgwick's (2003) idea of 'weak' theory and used this to develop a reparative approach to economic research. Rather than confirming that which is already known (such as capitalism is the only reality), the diverse economies researcher looks for surprise, difference and that which might provide self-nourishment. Taking this approach, a researcher resists inserting that which they see into an existing framework of explanation and instead offers a loose (but rich) description of what they find.

This loose and rich description approach can help us see things which are obscured by a 'strong theory' approach. Reflecting on their different research journeys, Anmeng Liu, SM Waliuzzaman, Huong Do, Ririn Haryani and Sonam Pem (Liu et al., 2020) describe how the diverse economies framework enabled them to see and loosely describe a world in which Westernised development was not the only thing present in their home countries of China, Bangladesh, Indonesia, Vietnam and Bhutan. This process was transformative for these authors; Liu, for example, wrote an ethnographic thesis about the multiple identities and possibilities for small town youth in China's Northwest, using what she calls 'a methodology of openness' that refused to bind youth into the same old stories of backwardness (Liu, 2021). SM Waliuzzaman used GPS trackers to create rich maps of slum-dwellers' daily activities in Kallyanpur slum in Dhaka, Bangladesh, fleshing these out with in-depth interviews and showing how slum-dwellers are engaged in diverse economic and commoning activities across the city (Waliuzzaman, 2020). Ririn Haryani went on to write about the economic practice of *arisan* in Indonesia not as a dying practice on the road to modernity but as a legitimate strategy for women's economic resilience in the face of disaster (Haryani and Dombroski, 2022). As Haryani writes with Liu et al., (2020, p. 448), she used to 'perceive *arisan* as a wasting time activity' and had not previously considered even studying it: prior to putting on the 'lens' of diverse economies, it was invisible. Similarly, Huong Do used embodied participatory methods working the rice fields, which helped her come to value the traditional knowledge beliefs and practices of her farming in-laws in rural Vietnam as forms of climate change adaptation that had been overlooked in the rush to develop quantitative indicators based on 'modern' assumptions (Dombroski and Do, 2022). Finally, in Bhutan, Sonam Pem turned to oral histories and walking pilgrimage interviews to explore the diverse river care practices of Buddhist thinking, particularly as used by youth and nuns to develop forms of environmental activism grounded in tradition (Pem, 2018). What is important here is that it is not the methods as such, but the ways in which what can be seen through the methods is informed by the 'lens' of diversity that one has developed before ever going into the field. These young scholars framed these stages as 'the moment of noticing', 'the moment of revaluing', and 'the moment of reconnecting', in doing so describing their journeys from a colonised capitalocentric thinking into something more open to what is already there and what could be grown further if nurtured.

For diverse economies scholars, these moments of noticing diversity – where everything does not have to mean the same thing – are grounded in an understanding of the world which is 'more than capitalist'. Gibson-Graham developed a loose framework of diverse economic practices, based on the simple image of an iceberg mentioned earlier. In Figure 22.1, the iceberg shows what is normally 'noticed' in economic geography: market transactions, waged labour, capitalist enterprise. Then, strikingly, below the waterline lie all the other things that could be part of our economy – and our economic geographies – which usually go unrecognised.

This image has been used in participatory action research with a variety of communities around the world. One of the most vividly instructive has been the floating coconut diagrams developed by diverse economies scholars in the Pacific (Carnegie et al., 2012). Here, Pacific men and women worked in groups with researchers to document hundreds of economic activities they were involved in, paying attention to gendered divisions of labour in particular. What was clear from the resulting images was the finding that much more of the women's economic activities were 'under the waterline', but also men's economic activities were being overlooked by capitalocentric understandings of what 'counted' as economic (see Figure 22.2: floating coconut economy).

A less visually striking, but perhaps a more analytical framework, is the diverse economies table, which uses a box form with capitalist, alternative capitalist and non-capitalist sections as a prompt for thinking and seeing other forms of economic activity in a community. Gibson-Graham use it in the Philippines (Gibson-Graham, 2005b) and in the La Trobe valley of Australia (Gibson-Graham, 2006) with economically marginalised communities, Maria Bargh uses it with the tribal enterprises of her iwi (tribe) Te Arawa (Bargh, 2011), and Emma Sharp with diverse food initiatives in Auckland (2020). In Table 22.1, for example, you can see the diverse economic energy generation activities of Te Arawa in Aotearoa New Zealand. The goal is to deliberately bring to light and make more visible economic activities that might be overlooked when researchers focus only on processes of capital accumulation and neoliberalisation, a strong temptation in economic geography research.

While the table has sometimes been used in a more static way than Gibson-Graham first intended (Gibson-Graham and Dombroski, 2020), it provides an important starting point for economic geographers interested in inventorying the kinds of economic activities present in a place. Moving away from representing 'broad trends' or 'average citizen' statistics, the intervention of the table was to create space for the multiple realities already present in place, without enrolling them in a story of what 'must' come next (Gibson-Graham, 2005a). This two-step process of 'seeing diversity' and then 'multiplying possibility' (Dombroski, 2015, 2016) is a

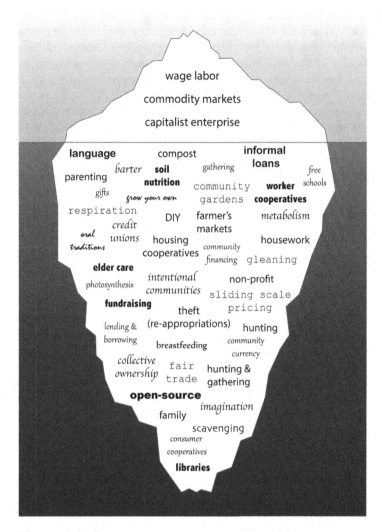

Figure 22.1 The economy iceberg.

broad brush methodology to doing research in economic geography with an eye to transformation. While capitalism is a process some can certainly see and describe, bringing other
processes to light is a way of making them seem more 'possible' as building blocks for different
kinds of economies. For scholars in diverse economies, activism builds further on this shift from
the descriptive, inventorying mode of research exemplified in the icebergs and tables into something more intentionally constructive: the possibility of taking the many things represented in
the diverse economy inventory, and working with communities to select those which foster
life, enabling communities to survive well together. This is when diverse economies moves
into community economies research, extending from inventorying to co-creating different
kinds of economic realities that are not just 'more than capitalist' but actively postcapitalist or
non-capitalist. Research methodology for this kind of economic geography thus moves from
documenting to *activating*.

Figure 22.2   Floating coconut economy.

Source: Carnegie, M., Rowland, C., Gibson, K., McKinnon, K., Crawford, J., Slatter, C. 'Floating coconut poster' in Monitoring gender and economy in Melanesian communities: Resources for NGOs, government and researchers in Melanesia, University of Western Sydney, Macquarie University and International Women's Development Agency, November, 2012.

Activating New Economies Through Performative Research

Documenting and inventorying economies through the ontological lens of diverse economies reveals a whole range of economic activities. However, these activities will not all enhance well-being, equality or sustainability. Many economic geographers have documented oppression, exploitation and homogenising forms of globalisation, but equally important is inventorying the things that *are* more life-giving, sustainable and equitable. How can we imagine and per-form realistic change if we are not documenting the experiments with doing economy differ-ently, the traditions of economic exchange that support interdependence, and the connecting patterns of solidarity economies throughout the world? If economic geographers do wish to participate in such change, they are faced with a choice in their research methodology: to stand back and document only as an observer or to 'muck in' (Wright, 2017), seeking to enhance the 'reach' of certain forms of economy. In other words, can we actively participate in the different kinds of economies we document?

In their book *Take Back the Economy* (2013), Gibson-Graham, Cameron and Healy provide a set of methodological tools for community groups and researchers to analyse and activate

Table 22.1 The diverse economy of Tūaropaki Geothermal Enterprise

Transactions	Labour	Enterprise	Property and resource ownership
Market	*Wage*	*Capitalist*	*Capitalist*
Power sold to national grid. As per joint venture, Mighty River Power provides long-term support for the sale of electricity.	Salary As per joint venture, Mighty River Power conduct maintenance of steamfield, power-plant, and transmission systems and marketing. Contracting of Strettons Chartered Accountants to manage enquiries and accounts.	Board of Directors (3 from Tūaropaki Trust and 1 representative of the joint venture). Dividends paid to shareholders.	As per joint venture, ¼ of Mokai II is owned by Mighty River Power.
Alternative Market	*Alternative Paid*	*Alternative Capitalist*	*Alternative Capitalist*
Koha (including for corporate gifts to trading partners)	In-kind. Long term pay off-Meeting with key stakeholders. Environmental – replenishing of water in geothermal area beneath power plant.	Scholarships for education and *kaumātua* grants (aim is for equitable distribution).	Right to extract granted to Tūaropaki Power Co. because of *whanau* link to Tūaropaki Trust. Land (Tūaropaki E block) owned by Tūaropaki Trust. Approx. 30 hectares for steamfield wells, pipelines, and electricity generating plant.

Non-market	Unpaid	Non-capitalist	Non-capitalist
State allocations. State appropriations = corporate tax, local government rates? *Whanaungatanga* (care for family). *Kaitiakitanga* (guardianship). Potentially the sustainability focus and activities of the company could be understood as transactions with papatūānuku (Mother Earth) – replenishing.	Presentations made at conference by Directors and Chairperson. Attendance at various *marae* meetings by Directors and Chairperson which they connect to. Advice to other Māori businesses.	Tūaropaki Trust comprised of 7 Mokai *hapū*. The Trust is an Ahu Whenua Trust, managing Māori freehold land under Te Ture Whenua Māori Act 1993. Tūaropaki Power Co. is wholly owned by Tūaropaki Trust. Māori Land Court appoints trustees to Tūaropaki Trust and holds Trust Order/Deed (provides legal framework) and requires the Trust administer "the lands for the benefit of the beneficial owners" (Tūaropaki). Tūaropaki Trust maintenance of *marae* community complex.	State – resource consents from Environment Waikato (the regional council). *Whakapapa* genealogy) – descendants of the *hapū* involved have genealogical and spiritual ties to each other and the lands involved.

Source: see Bargh 2011, p. 64.

Note: This analysis shows the variety of economic practices that are part of the wider economy of Tūaropaki, including more-than-capitalist practices.

economic change wherever they are. The book draws on a variety of research and community activist projects from around the world, constructing a vision of what a life-giving community economy might look like. Rather than provide a set of answers as to what 'good' economies might look like; however, Gibson-Graham, Cameron and Healy ask a set of questions that build on the previous work thinking about postcapitalist politics, asking, among other things, how do we survive well together? (2013, p. xiii).

Gibson-Graham, Cameron and Healy go on to answer this question in a speculative way, arguing that researchers and communities need to think up ways that we can take back the economy through collective ethical action – both in action research and community action. For these scholars and others in community economies, ethical action means:

- *Surviving* together well and equitably;
- *Distributing surplus* to enrich social and environmental health;
- *Encountering others* in ways that support their well-being as well as ours;
- *Consuming* sustainably;
- *Caring for* – maintaining, replenishing, and growing – our natural and cultural *commons*; and
- *Investing our wealth in future generations* so that they can live well (Gibson-Graham, Cameron, and Healy, 2013, p. xix).

Researching economies in this frame is thus centred on ethical action for researchers. There are two kinds of ethical action with relevance to economic geography methodologies here. Firstly, paying attention to what our research is 'making more real' and making ethical decisions about representation and research focus. Secondly, using research time to actively work with communities to co-create different kinds of economies that enable surviving well, distribution of surplus, encountering of others, sustainable consumption, care for commons, and investment in future generations. We look at both in turn.

Firstly, behind the idea that research makes some realities more real than others is the theory of performativity. Here language is seen as a force that can bring into being that which it names (Austin, [1955]1962). The force of these language acts relies on a whole host of actors and apparatus, including witnesses (Sedgwick, 2003). For research to be performative, then, we need to look at research methodologies as an assemblage stretching from methods with participants, to academic journals and networks, and on to policy makers (Law, 2004). We can thus talk about "hybrid research collectives" involving a vast arrangement of co-researchers, tools or methods, and more (Roelvink, 2016). Within these collectives, agency is not only already distributed among a range of actors. Working at increasing the agency of those involved can also become a key part of the research process, such as in the training and employment of community researchers or providing other necessities such as a medical clinic within the research project. Similarly, in his work with a community college, Leo Hwang shows diverse economies action research projects can "generate agency" in contrast to research from a distance that examines communities for what they lack, ultimately disempowering them (Hwang, 2020). Abby Templer Rodrigues used art-based methods to work directly with artists and artisans in Franklin County, Massachusetts, intentionally shifting the focus away from market-based understandings of artist economic lives into diverse economies framings of how artists and artisans can and do make a living (Templer Rodrigues, 2020). These projects are thus already responding to Judith Butler's (2010) call for those interested in economic performativity to tackle questions about how economies should work. Community economies performative research practice is guided by a political intention to build "other worlds" and the ethical commitments already listed above (Roelvink, St. Martin, and Gibson-Graham, 2015).

A key part of this research agenda is the creation of new economic subject positions. It is often very hard for people to see themselves outside of the capitalist framing, even if they are against capitalism. Those 'outside' of capitalism, as Gibson-Graham (1996, p. 6) suggest, are:

> … often understood primarily with reference to capitalism: as being fundamentally the same as…, or as being deficient or substandard imitations; as being opposite to capitalism; as being the complement of capitalism; as existing in capitalism's space or orbit.

While the diverse economies framing performs economic diversity, disrupting this capitalocentrism, seeing oneself outside of this frame requires a change in subjectivity (Gibson-Graham, 2006, p. xxxv). Here community economies researchers explore the power of affect to bring about a change in economic subjectivity. Shifts in affect have the potential to disrupt habits of thought, creating moments for new relationships and opportunities to emerge. Roelvink's (2016) research on the World Social Forum, for example, shows how stories of struggle and survival shifted participants emotionally and affectively, disrupting narratives of neoliberalism and enabling the exploration of other ways of living (in that case, enabling the commoning of recently privatised resources and the opportunity for individualised citizens to become commoners).

The work of performing more life sustaining economies is not only directed at transforming economies and opening up economic subject positions. It also transforms the researcher. When researchers deliberately seek to change the world for the better, they seek out reparative possibilities for the self. This intention draws on the work of Eve Sedgwick (2003) who argues that we can accept the horrors of the world without continuing to confirm them through our research. Instead, we can seek out possibilities for the world to be otherwise, which are also possibilities for joy and pleasure, and ultimately a form of self-nourishment for the researcher. Compassion, the "feeling of shared suffering combined with an orientation toward reparative action" (Roelvink, 2020), for others and the self lies at the heart of this "reparative stance" (Sedgwick, 2003). Through this compassion a researcher is open to "learning to be affected" (Latour, 2004; Gibson-Graham and Roelvink, 2010), a process by which one is transformed through their encounters with others in a research process centred on transforming the world. Importantly, through this process the hybrid research collective (which includes the researcher) is put into action. In their research on community gardens, for example, Jenny Cameron, Craig Manhood and Jamie Pomfrett (2011) have intentionally created opportunities for gardeners to learn to be affected by each other and other gardens, thereby increasing the possibilities for gardens as the climate changes:

> *community* gardening potentially amplifies the process of registering and responding to the world around us as more contrasts are added with gardeners working alongside others (even in gardens that use individual allotments). We contend that performative and collective research can add even more contrasts and intensify the ways in which we can learn to be affected by the world around and propelled to act.
> *Cameron, Manhood and Pomfrett, 2011, p. 502, emphasis in the original*

In the second aspect of performative research, scholars are using research time to actively work with communities to co-create different kinds of economies in action research projects. In recent times, community economies scholars have been exploring ways to constructively engage with communities, funders, state services, Indigenous tribal organisations, and other kinds of partners in action research. In Aotearoa New Zealand, scholars have worked within National Science Challenge structures to co-produce economic tools with communities. Under the hubris of the Building Better Homes Towns and Cities fund, community economies scholars are working with urban farms, charitable trusts, and government organisations to co-create and test a version of a 'Community Economy Return on Investment' tool (Healy et al., 2019), as first proposed by Gibson-Graham, Cameron, and Healy (2013) and further developed by Petrescu et al., (2021). Under this project, PhD student and urban food systems activist Bailey Peryman has begun to work with soil and society to implement new urban composting initiatives in Ōtautahi Christchurch as part of his action research. The Community Economies team is now part of the government funded Huritanga Urban Wellbeing project subtitled *Towards Socio-ecological Wellbeing-led Urban Systems in an Era of Emergency*. This action research project aims to change the way Aotearoa New Zealand plans for urban wellbeing, with the community economies researchers focusing on how return on investment is calculated for local government and community enterprise, particularly in a time of ecological emergency.

Other examples of action research in the community economies tradition might be seen as more 'direct action'. Anna Kruzynski both participates in and documents the direct action of La Pointe Libertaire (the Anarchist Point), which has sought to create, expand and protect urban commons in the Point St. Charles area of Montreal, Quebec (Kruzynski, 2020). Jarra Hicks founded an organisation that finances and builds community-owned renewable energy enterprises as part of an ongoing action research lifestyle (Hicks, 2020). Bradley Wilson and his students at West Virginia University started a fair trade coffee cooperative, connecting with farmers in Nicaragua through action research crossing the Americas (Williamson 2017). Others, such as Emma Sharp and Ann Hill, have participated in alternative food systems as a way not just only of studying the changes but investing in them (Hill, 2011; Sharp, 2020). Others, such as Maria Bargh, Matt Scobie and Suli Vunibola research with their own Indigenous communities and organisations as a way of making research fit decolonising methodologies, where community aspirations are front and centre in the research process (Bargh, 2011, 2020; Scobie, Lee, and Smyth, 2020; Vunibola, Steven and Scobie, 2022). As Cameron writes,

> If we are now living in a planetary experiment, perhaps social research could also take a more experimental approach. By this I do not mean the sort of carefully controlled experiment where we isolate and test variables to try and determine cause and effect. Rather I'm thinking of more open, even playful forms, of experimentation to try out new ways of living in the Anthropocene. Such an approach would mean setting aside the idea of research as neutral and objective activity in which there is a critical distance between the researcher and the object of study. Instead, research would entail making a stand for certain worlds and for certain ways of living on the planet, and taking responsibility for helping to make these worlds more likely and these ways of living more widespread.
>
> *Cameron, 2015, pp. 99–100*

Action research has traditionally been about collaborating with marginalised groups to produce research that directly impacts lives positively (Cameron and Gibson, 2020). In the community economies and diverse economies tradition, it has also been expanded to include action and activist research that reframes the economy to create possibility, reframes economic subjectivity to further enable possibility, and contributes to direct economic actions that change lives and economies now and into the future.

Concluding Thoughts: New Challenges for Economic Geography

Economic geography has often focused on describing and theorising the effects of capitalism. In their work transforming the capitalocentric orientations of economic geographies, JK Gibson-Graham have invited researchers to consider the performative aspects of their research methodologies. In this chapter we have discussed the methodological challenges of 'embracing the mess' and 'activating new economies' in a particular strand of economic geography. An emerging challenge is how our methodology, and use of particular methods, is created and enacted with human and more-than-human others. This includes working with those with other traditions and different ontologies as well as more-than-human participants whose agency is often not recognised. One emerging trend in community economies and decolonial forms of economic geography is the expanding engagement with ontological traditions that do not privilege the human species. We finish up this chapter looking to the future: new forms of action research that work with the more-than-human to think economy differently.

In recent times, economic geographers have begun to turn their attention to the problem of doing economies differently, for the sake of human wellbeing but also for the sake of planetary wellbeing and species survival. Community economies researchers have been at the forefront of this task. Many have drawn inspiration from the work of ecofeminist philosopher Val Plumwood (2007), who plainly states that 'we', humans living in highly industrialised societies, must radically change our mode of humanity if we are to survive the ecological crisis that has begun. Action research has thus been expanded to include research that actively seeks to transform human relationships with our environments and to live more harmoniously in place with others – both human and more-than-human others. In this vein, community economies researchers and economic geographers more generally are really just catching up with Indigenous interdisciplinary researchers who have long recognised the importance of being together in relationship with place with many others (Kimmerer, 2017). The challenge is to respectfully learn from scholars and elders in these traditions, while doing the work of transforming economies where we are and with what we have – the scholar's tools, and the activist's tools.

A few examples of where this might go seem appropriate here. Making space for Indigenous leadership and ways of doing things means that projects might not end up being framed as economic geography, as diverse economies or community economies, but might end up fully embedded in Indigenous knowledge frameworks and carried out with appropriate protocols for those communities (Waitoa and Dombroski, 2020). While projects we and others have engaged in can surely be framed as a community economies project, we can also perform different forms of academic scholarship by giving way to Indigenous ontological frameworks in our writing, project reporting and co-creation. In the end, these are best adapted to the environment in which we find ourselves agitating for cultural, social and economic change with the more-than-human.

In economic geography more generally, recognising the agency and contribution of more-than-human entities is a fraught process that requires careful scholarly and embodied work. Ethan Miller has written extensively about the work that non-humans do to support the lives of humans, and we can certainly add more generalised ecological processes to that awareness (Miller, 2019). However, framing these processes and this work as 'ecological services' has performative effects: does it then frame ecology as at the service of economies and humans? As Miller argues, we need to reframe our understanding of *economy* to fit in *ecology* rather than the other way round. There is still much to go in this process, and new methods of working with more-than-human research participants must be developed. We can work closely with ecologists and other representatives of more-than-human entities as a matter of course, drawing on multi-species ethnographic methods (Tsing, 2015) and in partnership with Indigenous activists and scholars (Bawaka Country et al., 2016; Yates, 2016). In all this, we need to remember that we too are an embodied species and draw on research methodologies that place this embodiment at the forefront.

To recap, then, embracing and inventorying the mess is a stance towards methodology that acknowledges the tangled interactions of multiple realities: multiple ontologies, epistemologies and methodologies. Economic geographers in the diverse economies tradition use multiple methods to 'see' and document the diverse economic realities already present, thus multiplying the possibilities for building different kinds of economies based on the activities already present, capitalist, non-capitalist, alternative capitalist – what Gibson-Graham and Dombroski (2020) call 'more-than-capitalist'. In hybrid research collectives they also actively seek to build some economies over others, taking a stake in the kind of economies that provide for a good life for us, other species and the planet. In this work we remain open, keeping our work as a question rather than offering an answer, and in doing so we can continue to learn from others.

References

Alonso, W. (1964) *Location and Land Use Toward a General Theory of Land Rent*. Oxford: Oxford University Press.

Austin, J. L. ([1955]1962) *How to Do Things with Words: The William James Lectures Delivered at Harvard University in 1955*. Oxford: Clarendon Press.

Bargh, M. (2011) The triumph of Maori entrepreneurs or diverse economies? *Aboriginal Policy Studies* 1 (3):53–69.

Bargh, M. (2020) Indigenous Finance: Treaty settlement finance in Aotearoa New Zealand. In *The Handbook of Diverse Economies*, eds. J. K. Gibson-Graham and K. Dombroski, 362–269. Cheltenham: Edward Elgar.

Barnes, T. (2009) Economic Geography. In *International Encyclopedia of Human Geography*, eds. R. Kitchin and N. Thrift, 315-327. Oxford: Elsevier.

Bawaka Country, S., Wright, S., Suchet-Pearson, K., Lloyd, L., Burarrwanga, R., Ganambarr, M., Ganambarr-Stubbs, B., Ganambarr, D., Maymuru, and Sweeney, J. (2016) Co-becoming Bawaka: Towards a relational understanding of place/space. *Progress in Human Geography* 40 (4):455–475.

Butler, J. (2010) Performative Agency. *Journal of Cultural Economy* 3 (2):147–161.

Cameron, J. (2015) On Experimentation. In *Manifesto for Living in the Anthropocene*, eds. K. Gibson, D. B. Rose and R. Fincher, 99–102. Earth: Punctum Books.

Cameron, J., and K. Gibson. (2020) Action research for diverse economies. In *The Handbook of Diverse Economies*, eds. J. K. Gibson-Graham and K. Dombroski, 511–519. Cheltenham: Edward Elgar.

Cameron, J., C. Manhood, and J. Pomfrett. (2011) Bodily learning for a (climate) changing world: Registering differences through performative and collective research. *Local Environment* 16 (6):493–508.

Carnegie, M., C. Rowland, K. Gibson, K. McKinnon, J. Crawford, and C. Slatter. (2012) Gender and economy in Melanesian communities: A manual of indicators and tools to track change: University of Western Sydney, Macquarie University and International Women's Development Agency.

de Sousa Santos, B. (2014) *Epistemologies of the South: Justice against epistemicide*. Boulder: Paradigm Press.

Do Thi, H., & Dombroski, K. (2022) Diverse more-than-human approaches to climate change adaptation in Thai Binh, Vietnam. *Asia Pacific Viewpoint*, 63 (1): 25–39.

Dombroski, K. (2015) Multiplying Possibilities: A postdevelopment approach to hygiene and sanitation in Northwest China. *Asia Pacific Viewpoint* 56 (3):321–334.

Dombroski, K. (2016) Seeing diversity, multiplying possibility: My journey from post-feminism to postdevelopment with JK Gibson-Graham. In *The Palgrave Handbook of Gender and Development*, ed. W. Harcourt, 312–328. Basingstoke: Palgrave Macmillan.

Escobar, A. (2018) *Designs for the pluriverse: Radical interdependence, autonomy, and the making of worlds*. Durham and London: Duke University Press.

Gibson-Graham, J. (2005a) Building community economies: Women and the politics of place. In *Women and the Politics of Place*, eds. W. Harcourt and A. Escobar, 130–157. Bloomfield: Kumarian Press.

Gibson-Graham, J. K. (1996) *The End of Capitalism (As We Knew It)*. Minneapolis: University of Minnesota Press.

Gibson-Graham, J. K. (2005b) Surplus possibilities: Postdevelopment and community economies. *Singapore Journal of Tropical Geography* 26 (1):4–26.

Gibson-Graham, J. K (2006). *A Postcapitalist Politics*. Minneapolis and London: University of Minnesota Press.

Gibson-Graham, J. K., J. Cameron, and S. Healy. (2013) *Take Back the Economy: An ethical guide for transforming our communities*. Minneapolis: University of Minnesota Press.

Gibson-Graham, J. K., and K. Dombroski. (2020). Introduction to The Handbook on Diverse Economies: Inventory as ethical intervention. In *The Handbook of Diverse Economies*, eds. J. K. Gibson-Graham and K. Dombroski, 1–25. Cheltenham: Edward Elgar.

Gibson-Graham, J. K., and G. Roelvink. (2010) An economic ethics for the Anthropocene. *Antipode* 41:320–346.

Haryani, A. R., and K. Dombroski. (2022). Arisan: Producing economies of care in Yogyakarta, Indonesia. In *Community economies in the global south: Case studies of rotating savings, credit associations and economic cooperation*, eds. C. S. Hossein and P. Christabell, 167–186. Oxford: Oxford University Press.

Healy, S., K. Dombroski, D. Conradson, G. Diprose, J. McNeill, and A. Watkins (2019) More than monitoring: Developing impact measures for transformative social enterprise. Paper read at Implementing the Sustainable Development Goals: What role for social and solidarity economy?, at Geneva.

Hicks, J. (2020) Community finance: Community-owned renewable energy enterprises. In *The Handbook of Diverse Economies*, eds. J. K. Gibson-Graham and K. Dombroski, 370–377. Cheltenham: Edward Elgar.

Hill, A. (2011) A helping hand and many green thumbs: Local government, citizens and the growth of a community-based food economy. *Local Environment* 16 (6):539–553.

Hwang, L. (2020) Focusing on assets: Action research for an inclusive and diverse workplace. In *The Handbook of Diverse Economies*, eds. J. K. Gibson-Graham and K. Dombroski, 520–526. Cheltenham: Edward Elgar.

Kimmerer, R. (2017) *Braiding Sweetgrass*. Minneapolis: Milkweed Press.

Kruzynski, A. (2020) Commoning property in the city: The ongoing work of making and remaking. In *The Handbook of Diverse Economies*, eds. J. K. Gibson-Graham and K. Dombroski, 283–291. Cheltenham: Edward Elgar.

Latour, B. (2004) How to talk about the body? The normative dimension of science studies. *Body and Society* 10:205–229.

Law, J. (2004) *After Method: Mess in Social Science Research*. London and New York: Routledge.

Liu, A. (2021) *An Ordinary China: Reading 'Small-Town Youth' for Difference in a Northwestern County Town, School Of Earth And Environment*, University of Canterbury, Christchurch, NZ.

Liu, A., S. M. Waliuzzaman, H. Do, R. Haryani, and S. Pem. (2020) Journeys of post-development subjectivity transformation – a shared narrative of scholars from the majority world. In *The Handbook of Diverse Economies*, eds. J. K. Gibson-Graham and K. Dombroski, 444–451. Cheltenham: Edward Elgar.

Massey, D. (1984) *Spatial Divisions of Labour: Social Structures and the Geography of Production*. London: Macmillan.

Massey, D. (2005) *For Space*. Los Angeles, London, New Delhi, Singapore, Washington DC: Sage.

McKinnon, K., S. Healy, and K. Dombroski (2019) Surviving well together: Postdevelopment, maternity care, and the politics of ontological pluralism. In *Postdevelopment in Practice: Alternatives, Economies, Ontologies*, eds. E. Klein and C. Eduardo Morreo, 190–202. London and New York: Routledge.

Miller, E. (2019) *Reimagining Livelihoods: Life Beyond Economy, Society, and Environment*: University of Minnesota Press.

Pem, S. (2018) *Negotiating Gross National Happiness as Community Economy: A Case Study of the Thimphu River, Waterways Centre for Freshwater Management*, University of Canterbury, Christchurch, NZ.

Petrescu, D., C. Petcou, M. Safri, and K. Gibson. (2021) Calculating the value of the commons: Generating resilient urban futures. *Environmental Policy and Governance* 31 (3):159–174.

Plumwood, V. (2007) *A review of Deborah Bird Rose's Reports from a Wild Country: Ethics for Decolonisation*.

Roelvink, G. (2016) *Building Dignified Worlds: Geographies of Collective Action*. Minneapolis: University of Minnesota Press.

Roelvink, G. (2020) Framing essay: Diverse economies methodology. In J. K. Gibson-Graham and Kelly Dombrosk, eds. *The Handbook of Diverse Economies*. Cheltenham: Edward Elgar, Chapter 50, 453–466.

Roelvink, G., K. St. Martin, and J. K. Gibson-Graham eds. (2015) *Making Other Worlds Possible: Performing Diverse Economies*. Minneapolis: University of Minnesota Press.

Sayer, A. (1982) Explanation in economic geography: Abstraction versus generalization. *Progress in Human Geography* 6 (1):68–88.

Scobie, M., B. Lee, and S. Smyth. (2020) Grounded accountability and Indigenous self-determination. *Critical Perspectives on Accounting*:102198.

Sedgwick, E. (2003) *Touching, Feeling: Affect, Pedagogy, Performativity*. Durham, N.C.: Duke University Press.

Sedgwick, E. K. (1994) *Tendencies*. London: Routledge.

Sharp, E. L. (2020) Care-fully enacting diverse foodworlds in Auckland, Aotearoa New Zealand. *Gender, Place & Culture* 27 (8):1214–1218.

Templer Rodrigues, A. (2020) Techniques for shifting economic subjectivity: Promoting an assets-based stance with artists and artisans. In *The Handbook of Diverse Economies*, eds. J. K. Gibson-Graham and K. Dombroski, 419–427. Cheltenham: Edward Elgar.

Tsing, A. L. (2015) *The Mushroom at the End of the World: On the Possibility of Life in Capitalist Ruins*: Princeton University Press.

Tuhiwai Smith, L. (1999) *Decolonizing Methodologies: Research and Indigenous Peoples*. London & New York, Dunedin: Zed Books & University of Otago Press.

Vunibola, S., Steven, H., & Scobie, M. (2022) Indigenous enterprise on customary lands: Diverse economies of surplus. *Asia Pacific Viewpoint* 63 (1), 40–52.

Waitoa, J., and K. Dombroski. (2020) Working with Indigenous methodologies: Kaupapa Māori meets diverse economies. In *The Handbook of Diverse Economies*, 502–510. Cheltenham: Edward Elgar Publishing.

Waliuzzaman, S. M. (2020) *A Commons Perspective on Urban Informal Settlements: A Study of Kallyanpur Slum in Dhaka, Bangladesh, School of Earth and Environment*, University of Canterbury, Christchurch, NZ.

Williamson, T. (2017) Firsthand coffee: Bringing together mountain communities. Barista Magazine. www.baristamagazine.com/firsthand-coffee/. Last accessed 15 June 2021.

Wright, S. (2017) Critique as delight, theory as praxis, mucking in. *Geographical Research* 55 (3):338–343.

Yates, A. (2016) Mauri-Ora: Architecture, indigeneity, and immanence ethics. *Architectural Theory Review* 21 (2):261–275.

23

GEOGRAPHIES OF EDUCATION
Data, Scale/Mobilities, and Pedagogies

Yi'En Cheng and Menusha De Silva

Introduction

Geographies of education have grown from a nascent strand of research within human geography into a vibrant subfield that addresses a wide range of topics connected to social, economic, cultural, and political geographies. This is reflected in how geographies of education continue to be defined in a broad and inclusive manner:

> Geographies of education and learning consider the importance of spatiality in the production, consumption and implications of formal education systems from pre-school to tertiary education and of informal learning environments in homes, neighbourhoods, community organisations and workspaces. Between them, these geographies foreground the wider political, economic, social and cultural processes shaping and being reshaped through formal and informal spaces of education across the globe, and the ways they are experienced, embraced and contested by educators and diverse subjects of education, including children, young people, parents and workers (subject positions that clearly are not mutually exclusive).
>
> *Holloway and Jons, 2012, p. 482*

A diverse body of literature has arisen exploring how education is tied to the role of political-economic restructuring (Thiem, 2009), experiences of children, youth, and families (Holloway et al., 2011), mobilities, and immobilities (Waters, 2017), informal and alternative enactments of education (Mills and Kraftl, 2016), emotions and affects (Kenway and Youdell, 2011), citizenship and politics (Cheng and Holton, 2019), and international study spaces (Madge, Raghuram, and Noxolo, 2015). These works have enriched both the sub-fields in which these writings are located as well as the wider discipline of human geography. Yet, relatively less attention has been paid to a review of the methodological approaches that inform these geographical writings.

This chapter addresses such a gap by providing a critical appraisal of the key methodologies used in generating this exciting body of scholarship on geographies of education, with a focus on data, scale/mobilities, and pedagogy. A critical discussion of methodologies necessitates a closer inspection of the nature of the information that is to be collected for research as well as the empirical and analytic sites that comprise the object of study. In other words, within the

DOI: 10.4324/9781003038849-26

extant geographical research of education, how do researchers understand "data" and their sources and which methods have been deployed due to the particular knowledge shaping what counts as data? How do particular imaginations of "scale" shape the ways in which research considerations around theory and methodology are addressed? Additionally, we consider how might thinking about pedagogy serve as a useful lens to improve the type of data generated for geographical analyses of education?

Data: Embodied, Co-Constructed, and Digital

In this section, we focus on the ways researchers understand data alongside the methods which are deployed to gather such information. The discussion begins with an appreciation of how qualitative research on education and learning using both conventional and more innovative forms of the interview method have helped to assert the significance of embodied experiences. This is followed by an analysis of the use of participatory techniques to ascertaining knowledge formation as a co-constructed process but at the same time raised new questions about 'voice' and 'agency' of young people. We end the discussion with a note on missed opportunities for geographies of education to tap into quantitative and digital methodologies.

Contemporary geographical writings about education and learning have mobilized a range of qualitative methodologies that treat data as embodied and emergent. Data is therefore understood as produced from specific subject positionings such as class, gender, sexuality, race, and more as well as through interlocking sets of spatial relationships, as opposed to a set of facts that exist outside of value systems and worldviews. This understanding of data allows researchers to take seriously the perspectives of different subjects and actors while appreciating how individual experiences are informed by and connected through wider social structures. Within this approach, subjective experiences serve as a lens to answer research questions about formations of identity and citizenship, social reproduction and inequality, and geopolitical relationships that are intricately connected to the scale of the everyday. More specifically, the predominant use of interview and ethnographic methods have improved understandings of how young people – and their families and other actors in the field – experience education across formal settings such as state schools and universities to informal and alternative sites of learning. For example, Gergan and Smith (2020) demonstrate how young Ladakhi who migrate to cities for university education learn to engage with novel political ideas on campuses and in city spaces and bring with them critical ways of thinking about gender and mobility back home to Ladakh villages. The educational experiences of these Ladakhi youths intertwine different learning genres and spaces, contributing to the blurring of the 'formal' and the 'informal' as well as the 'mainstream' and the 'alternative' (Mills and Kraftl, 2016).

To capture even more subtle aspects of embodied experiences and spatialities, some researchers have modified conventional modes of qualitative interviewing and ethnographic fieldwork with emerging participatory and digital methods. For example, social media analysis of Instagram images has been used alongside interviews to tease out the slippages between what has been promoted in Singapore schools through the model of Social and Emotional Learning and the everyday emotional messiness of student behaviors (Ang and Ho, 2019). The images not only reveal students' representation of their emotions but also produce affective forces that subvert hegemonic aspirations in the cultivation of a 'proper' student. This raises an important parallel to Brown's (2011) conceptualization of the emotional geographies of aspirations as well as the diverse ways emotions are transformed into spatial discourses, practices and embodied experiences (Kenway and Youdell, 2011). Extending the analytic lens on emotions to examine the mobile worlds of education, Holton and Finn (2017) incorporate mobile methods, such as go-along and walking interviews (Holton, 2015) to capture the shifting senses of belonging as students move across cities and educational sites, both nationally and transnationally. Such approaches reveal the emergent and existing spatial connections across different actors and sites that connect young people's embodied experiences to institutional framings in the family, school, and campuses, and interpreted within national and transnational contexts.

Geographical research in education has also played an important role in demonstrating how knowledges are co-constructed and produced from research interactions which are bound in power relations involving young people and adults. Influenced by ideas from the New Social Studies of Childhood (NSSC) paradigm, a child-centered approach has become an important lens for researchers working with children in educational settings to gain access to their everyday worlds (see Chapter 24 for further discussion of Children's Geographies). The use of participatory methods has generated critical insights about young people's knowledges leading to a reconsideration of adultist notions of what counts as valid and legitimate in the realm of politics (Hakli and Kallio, 2013; Skelton, 2010) as well as an arena of creative theorizations (Horton and Kraftl, 2006). In a study of children's politics, Elwood and Mitchell (2012) involved young students in an after-school activity program to produce spatial maps that express the ways the children exert autonomy and challenge adult claims over spaces and boundaries imposed upon them. Also using children's creative representations of their educational worlds, Cranston (2020: 9) stresses the innovative prospect of cartoon illustrations produced by international school students, where "cartoon characters and landscapes allows us to explore their interpretation of what the global looks and feels like at the scale of their everyday lives". As such, the analysis afforded by cartoons is taken to be representative of socio-spatial relations beyond the local environments to include the more abstract but embodied notions of the global and the transnational.

> Yet, scholars have also challenged the tendency to over-emphasize children's voices and actions. This is because such a focus may not capture the manifold ways in which powerful adult-related processes and actions continue to define city, national, and transnational discourses that are folded into young people's everyday lives.
>
> *Ansell, 2009*

For example, Cheng (2014) has demonstrated how a segment of youths in Singapore who turn to private higher education as a way to secure middle-class aspirations and self-worth are produced in and through state-endorsed ideas about heteronormative adulthood and life transition. This is a reminder to be careful about conceptualizations of young people's agency as autonomous and liberal, as they are historically privileged constructs of youth creativity and individual capacity for resistance (Durham, 2008; see also Holloway, Holt and Mills, 2019). So, while participatory methodologies allow for a deeper and wider understanding of the range of young people's agentic voices and practices, these experiences need to be critically examined for interpellations that mirror ideas from the adult worlds. Additionally, researchers need to be reflexive about their relationships with young people as research subjects, even as they claim to use child-centered and participatory approaches, and actively work towards minimizing various forms of adultism, didacticism, and judgmentalism (see Cope, 2009). Instead of straightforward claims about giving voices to young people and allowing their voices to be heard through our research endeavors, there needs to be reflexivity in how (adult) researchers and other non-youth actors and ideas are informing such voices.

In comparison to advances in qualitative methodologies within recent studies of geographies of education, there has been less attention to quantitative approaches and their opportunities and limits in researching the spatial dimensions of education and learning. Geographers have used quantitative approaches, which are centrally concerned with measurable (and therefore numerical) data, to statistically analyze spatial variations and consequences in the operation, provision, and quality of education systems and resources (Bradford, 1991; Johnston et al., 2006; Webber and Butler, 2007). While these studies have shown rigorous quantitative evidence of the ways education are tied to social production and inequality, the insights need to be balanced by critical qualitative analyses that examine the processes that lead to such unequal spatial outcomes and further augmented by social and cultural theories (Holloway et al., 2011). A recent significant study of higher education (im)mobilities in the United Kingdom which has skillfully combined large-scale mapping of young people's spatial movements across multiple locations with in-depth qualitative interviewing and mapping provides an exemplary research example (Gamsu and Donnelly, 2020). Putting this amalgam of information into social network analysis and data visualization, they presented arguments about how "hierarchies of universities are embedded in a distinct archipelago of elite geographies connecting particular institutions and locales concentrated in England" (Gamsu and Donnelly, 2020, p. 15). What is particularly striking is the way the study engages deeply with theories of power, space, and education that cross quantitative and qualitative divides.

Geographical analyses of education and learning informed by serious engagements with – and reflections about – quantitative methods have an uncharted potential to extend debates about the relationships between education, digital technologies, and data. This is especially relevant in light of the COVID-19 pandemic which has prompted an accelerated ushering in of virtual platforms and online technologies into higher education spaces as a predominant solution to the threat of contagion by proximity and movement (Sidhu et al., 2021). For example, geographical accounts that examine how data can be generated, mobilized, and used as a mode of education governance in the contexts of schools have begun to emerge (Finn, 2016; Gulson and Sellar, 2018). While the main focus of these studies is on how data constitutes a novel form of architecture of governance in education, they also pave the ground to harness information from and to critique data infrastructures through diverse experimentations of methodological approaches from computational social sciences to digital humanities. What counter-topographies to educational inequalities, for instance, might be built from insights generated from the volumes of data that were previously not available? Such forays will bring geographies of education into conversation with an even wider range of human geographical inquiry such as big data and internet cultures (Graham and Shelton, 2013; Miller and Goodchild, 2015) as well as social scientific research into digitalization and platform spatialization of education (Williamson, Eynon, and Potter, 2020; Jarke and Breiter, 2019). Nevertheless, all these opportunities to further stretch extant definitions of what comprises data and to enhance analyses informed by qualitative data need to be matched with an even more rigorous discourse in human geography – and geographies of education alongside allied subdisciplines – on the methodological challenges and risks of handling and interpreting digital information (see Kitchin, 2013).

Scale: Educational Sites and Mobilities

The notion of scale has been a vital and contested concept in human (and physical) geographical research cutting across all topics of interest. Within the geographies of education, considerations about scale have featured around the need to conceptually grasp education as an activity with interconnections to different levels of institutions and processes. In their editorial piece on the embodied geographies of education, Cook and Hemmingway (2011, p. 4) argue that "[p]laces cannot be bounded but are inexplicably linked to wider scales, with particular interactions and articulations of social relations, through a mixing of local and larger-scale processes". Brock (2016) offers a focused critique of educational research which he argues has been fixated on a singular scale of the national and explicates how attention to the spatial concepts of space, place, and scale can reveal educational activities as constituted across – "local, regional, national, international and global levels". These considerations reveal how, on the one hand, spatial scales play a critical role in the research imagination in terms of defining the object of analysis both theoretically and empirically and, on the other hand, raise complex questions about what scale means, how to map it onto space and time, and how multiple scales are constituted in relation to each other (Marston, Jones and Woodward, 2005; Leitner and Miller, 2006).

The contested dynamic of scale and its implication on geographical research imagination has been reflected in a recent debate within geographies of education about the inward and outward-looking approaches to critical analyses of education. Thiem (2009: 168) writes that a "new geography of education should also be much broader empirically than it has been to date" and that "multiscalar and multi-institutional definitions of education need to be adopted more widely if geographers are to realize the full significance of contemporary educational formations". Holloway et al., (2011, p. 594) provide a counter-argument that such an assessment excludes a rich vein of work emerging from geographies of children, youth, and families, and that a more capacious research approach is needed to link "an inward-looking focus on educational spaces with an outward-looking approach that assesses their importance to other time/spaces". In our review of the diverse geographical writings on education, the scholarship presents another set of scalar bifurcation around, firstly, the institutional geographies of schools and university campuses and, secondly, the geographies of transnational education and student mobility. While these two strands of research may not preclude each other, there remains insufficient engagement across the two sets of literature in terms of theoretical and methodological considerations.

Research on the geographies of schooling and university campuses has emphasized the institution as the primary analytic scale in which everyday social interactions and wider processes fold into each other. For example, Hopkins (2011, p. 159) advances critical geographies of the university campus which draws attention to "both the microgeographies of campus as well as the connections between global and national issues and how these influence and shape the ways in which the university campus is encountered". Placing a similar emphasis on the role of the institution, Nguyen, Cohen, and Huff (2017) call for critical geographies of education centered on schools as a site of analysis and for geographical theory generation. Departing from such views, Henry (2020, p. 184) argues that the widely accepted notion of formal schooling as a universal social good needs to be re-examined, and that an examination of the diverse meanings of education "from socialist, Afrocentric, and other critical perspectives would undoubtedly yield productive challenges to the hegemony of mass, compulsory schooling". A particular challenge posed to researchers by such an argument is the question of what other spatial (and temporal) scales might be imagined and come to matter when the research moves beyond mainstream understandings of education and learning, and whether the dominant and conventional scale of the modernist nation should take precedence.

The scholarship on transnational education and student mobilities provides one way of rescaling the analytic lens beyond the nation-state. Many of these writings draw upon migration and mobilities thinking to make sense of education in cross-border contexts, ranging from studies of internationally mobile actors such as students (Sidhu, Ho, and Yeoh, 2020; Prazeres, 2013) and academics/teachers (Leung, 2017; Koh and Sin, 2020) to the circulation of scientific and cultural knowledges via various periods, channels, and artifacts (Jöns, Heffernan and Meusburger, 2017). Also, geographical research that helps to establish these multiple relationships between education and mobilities cast a spotlight on the limits of methodological nationalism (Wimmer and Glick-Schiller, 2002). Instead of a focus on nationalism, research that is framed through methodological *trans*-nationalism offers a 'both/and' viewpoint towards identities, attachments, and memberships instead of the 'either/or' perspective of methodological nationalism (Faist et al., 2013, p. 142). This approach highlights how mobile actors occupy a space that links them to not only the socio-political and economic structures but also situates them with their peers in both their home and host countries. Research in this area has contributed towards an expanded notion of education and its attendant relationships to knowledge production and identity formation vis-à-vis a more mobile and circulatory notion of the analytic scale.

An approach that has gained significant attention by scholars seeking to address methodological nationalism is the multi-sited ethnography. This approach may either involve the researcher following the research subject and object across multiple locations or employing a multi-sited research team. However, conducting multi-sited research poses practical difficulties including increased challenges to secure funding and institutional support, especially for research in the humanities and social sciences, and new travel restrictions placed upon researchers in a tumultuous time of the pandemic and escalating geopolitical tensions. In the face of these difficulties, new techniques that tap into cost-effective and digital means, for example, the use of 'integrative and iterative methods' such as diary- interviews (Collins and Huang, 2012), may offer rich insights into transnational lives and spatial connections of students and other educational activities that would otherwise be inaccessible.

Beyond multi-sited ethnography, researchers have tackled the challenge of doing multi-scalar research by drawing upon methodologies that focus on how individuals are connected within various networks formed through the flow of ideas, people, and things. Based on the example of a specialist agri-food and land-based rural university, Robinson (2018, p. 280) shows how learning spaces involving affective encounters with animals, plants, and buildings are embedded in an assemblage where universities have to continuously expand in "student numbers and infrastructure without losing the soul of the campus and its unique blend of quaint rurality and state-of-the-art technology in building design and equipment". Integrating time and/or temporality into scalar analyses provides another fresh way of thinking about spatial connections. Mills' (2016) work on the historical geographies of voluntarism engages with archival research and the life course approach to study the case of the Woodcraft Folk in the UK. The study reveals that the youth organization positioned itself as provider of training that is deemed essential for a specific life stage, and how the intergenerational relationships among children and parents, folk 'elders', and adult volunteers connect both formal and informal spaces. Other scholars adopt retrospective and prospective reflections on life paths, such as De Silva, Woods and Kong (2020) whose study on an underground Christian international school in China conveys how alternative education leads to alternative visions of life and aspirations that are global and less rooted in local social norms. These examples signal the need for more methodological experimentation to analyse the complex and multiscalar geographies of education.

Pedagogy: Synergies and Reciprocities

The focus on education as an object of inquiry within geographical research on education plays a critical role in informing educators of the various discourses and inequalities that shape their teaching and the circumstances of their students. In the light of recent calls to enjoin "geographies of education and education of geography" (West et al., 2020), we recast the geographies of education through the lens of pedagogy – teaching and learning – to consider how methodologies adopted in these two areas of inquiry may be considered reciprocal. As succinctly conveyed by Chang and Kidman (2019, p. 1), geography education describes "the phenomenon that is concerned with the subject content that is taught, the way it is taught and how we will know if students have learned what has been taught". Methodologies adopted in pedagogical research focus on two interdependent stages of the teaching process, firstly on the geographical traditions and theories that inform the educator's practice and how curricula are designed, and secondly pedagogical approaches that frame the research design to ascertain various factors that shape students' learning.

Methods that are employed in geographical research on education have helped to inform research on geographical pedagogy. For instance, Stokes et al., (2012) assess the pedagogical value of remote/virtual study of field sites by using surveys to quantify the students' learning alongside semi-structured group interviews to consider the role of tutors and the nature of the assessment. Visual methodologies have also been incorporated into the teaching of geography in higher education. Wee et al., (2013) argue visual techniques in teaching can flesh out students' diverse notions of sense of place and prior experiences that shape their spatial thinking. Baylina Ferré and Rodó de Zárate (2016, p. 608) propose 'relief maps' as a visual tool to help students understand "the relation between power structures, places, and the lived experience", and also create an opportunity for students to recognize privilege and how the complex intersection of structures perpetuate inequality. Other data sources that are frequently used in pedagogical research, such as student assessments, feedback, and curricula materials, also have the potential to inform research on student experiences. For example, Ramdas (2019) examines qualitative comments in 11 years of students' feedback on their self-guided fieldwork linked to a first-year module to reveal feelings of uncertainty and adventure in students' transition from high school to university.

The practice of enhancing students' geography education requires an attentiveness to the particular spatial contexts and the subject positionings that shape their learning experiences. Therefore, researching pedagogical practices requires a degree of reference to scholarship on the educational milieu that students are situated in and, thus by extension, the methodologies that frame geographies of education (Smith, 2002). We focus on two pedagogical approaches to highlight the synergies between methodologies adopted in the geographies of education and geographical education.

Firstly, transition pedagogy focuses on students' (potential) challenges with adjusting to new learning environments and considers how they link to the various milestones of a person's life course such as the shift from school to university. A focus on the 'distant' relationships that shape the curriculum and teacher-student interactions undergird this body of work. While early studies on transition pedagogy adopted a more top-down approach, including longitudinal studies that seek to understand causes for college dropout (Tinto, 1987 in Ramdas, 2019), recent methodological approaches are aimed at centering students' voices, recognizing transition not as single events but as continuities, and understanding how everyday activities and external circumstances such as being a first or second-generation graduate shape these transitions (Barnes et al., 2011). Indeed, the gradual methodological evolution of transition pedagogy highlights how geographical research on the temporalities of students' social-cultural transitions and their differentiated access to supportive learning environments have implications on pedagogical considerations (Holdsworth, 2009).

Secondly, educators of geography also have increasingly examined the emotionality of learning and teaching. These methodologies informed by feminist pedagogy assert that education is a "fluid process whereby the student is empowered to act for social change" (Dowler, 2002, p. 68). Through reflexive accounts on the teaching experiences, and analysis of teaching material, course outlines, and students' assessments, educators highlight the value of discomfort as a pedagogy that challenges students to disrupt stereotypes, and/or broach emotionally sensitive topics (see Simon, 2009 on gender; Maddrell and Wigley, 2019 on death). Also, learning environments are shaped by different classroom contexts and the positionalities of, and power relations amongst, the persons who occupy the space. Burke et al., (2017) draw upon inputs from two instructors and eight students to discuss how a protest that occurred in another country becomes a 'generative space' for the authors to reflect on their emotional responses to dissent and solidarity, and the spatial relations that connect them to the event, which in turn offers a more nuanced understanding of geographies of intimacy. In doing so, this pedagogically informed research provides a rare example of how pedagogical tools can help produce disciplinary-based research. Pertinently, there are also ongoing debates on the epistemic tensions of moving away from the settler-colonial perspective within university geography, and teaching diverse and decolonized geography has generated a renewed interest on how students understand their place within colonial-capitalist power geometries (Daigle and Sundberg, 2017).

Conclusions

A review of the diverse scholarship on the geographies of education reveals a multitudinous range of methodological approaches. However, there are at least two salient commitments displayed by researchers that are reflected within and across these writings. Firstly, qualitative methodologies have been prioritized and used in excavating and highlighting the voices of children and young people in their respective educational contexts. This emphasis is strongly intertwined with parallel developments in the sub-field of geographies of children and young people, where there has been a sustained engagement with critical social theories around childhood and youth as well as an attentiveness to feminist research underpinnings that stress the significance of bringing to the fore subjects and voices from the margins (Holloway, Holt and Mills, 2019; Kraftl, 2013). Secondly, researchers have also paid heed to spatial connectivities through the geographical concepts of scales, sites, and mobilities, as well as linking the analyses of these spatial connections to broader issues around structural inequalities and socio-cultural change. Arguably, a hallmark of contemporary geographical research, such an approach has informed critical analyses of education in ways that permit a more capacious appreciation of the systemic configurations, power geometries, and cultural productions in a range of formal and informal learning spaces.

Throughout this chapter, we have provided a critical appraisal of selected geographical writings on education with respect to methodological underpinnings and concerns. The focus on data, scale/mobilities, and pedagogy has allowed us to introduce some unsettled but productive tensions concerning the ways in which research practices and the theories that inform them are implicated in knowledge production – and consumption in the case of knowledge transfers to students. We suggest accordingly three considerations for future geographical research on education. Firstly, there is a need to expand the capacity for researching across not just children and young people's voices but also older youths and adults, as the world begins to adopt lifelong learning and where adult education is becoming increasingly common. This is not to lose sight of the ethical commitment to underlining children and young people's voices, but a call to pull in and mobilize more techniques – and technologies – to help us understand and reimagine geographies of education. Secondly, the empirical and analytical scale needs to move beyond its singular focus on schools and campuses and incorporate more flexible and mobile affordances that allow tracing of and jumping across scales to follow flows, interconnections, and movements related to education. Last, greater cross-pollination between methods in "geographies of education" and "education of geography" will help broaden and open up experimental spaces for imagining more ways of doing research as well as confronting the ethical dilemmas that come along with new methodologies.

•

References

Ang, C. and Ho, E. L. E. (2019) Feeling schools, affective nation: The emotional geographies of education in Singapore, slippages as tactical manoeuvres. *Emotion, Space and Society*, 32: 100589.

Ansell, N. (2009) Childhood and the politics of scale: Descaling children's geographies? *Progress in Human Geography*, 33(2): 190–209.

Barnes, L., Buckley, A., Hopkins P. and Tate, S. (2011) 'The transition to and through university for non-traditional local students: Some observations for teachers', *Teaching Geography*, Summer: 70–71.

Baylina Ferré, M. and Rodó de Zárate, M. (2016) New visual methods for teaching intersectionality from a spatial perspective in a geography and gender course. *Journal of Geography in Higher Education*, 40(4): 608–620.

Bradford, M. (1991) School-performance indicators, the local residential environment, and parental choice. *Environment and Planning A*, 23(3): 319–332.

Brock, C. (2016) *Geography of Education: Scale, Space and Location in the Study of Education*. London: Bloomsbury.

Brown, G. (2011) Emotional geographies of young people's aspirations for adult life. *Children's Geographies*, 9(1): 7–22.

Burke, S., Carr, A., Casson, H., Coddington, K., Colls, R., Jollans, A., ... and Urquhart, H. (2017) Generative spaces: Intimacy, activism and teaching feminist geographies. *Gender, Place and Culture*, 24(5): 661–673.

Cheng, Y. (2014) Biopolitical geographies of student life: Private higher education and citizenship life-making in Singapore. *Annals of the Association of American Geographers*, 105(5), 1078–1093.

Cheng, Y. and Holton, M. (2019) Geographies of citizenship in higher education: An introduction. *Area*, 51(4): 613–617.

Collins, F. L. and Huang, S. (2012) Introduction to special section on migration methodologies: Emerging subjects, registers and spatialities of migration in Asia. *Area*, 44(3): 270–273.

Cook, V. A. and Hemming, P. J. (2011) Editorial Education spaces: Embodied dimensions and dynamics. *Social and Cultural Geography*, 12(1): 1–8.

Cope, M. (2009) Challenging adult perspectives on children's geographies through participatory research methods: Insights from a service-learning course. *Journal of Geography in Higher Education*, 33(1): 33–50.

Cranston, S. (2020) Figures of the global: Mobility journeys of international school pupils. *Population, Space and Place*, 26(6): e2305.

Daigle, M. and Sundberg, J. (2017) From where we stand: Unsettling geographical knowledges in the classroom. *Transactions of the Institute of British Geographers*, 42(3): 338–341.

De Silva, M., Woods, O., and Kong, L. (2020) Alternative education spaces and pathways: Insights from an international Christian school in China. *Area*. DOI: 10.1111/area.12634

Dowler, L. (2002) The uncomfortable classroom: Incorporating feminist pedagogy and political practice into world regional geography. *Journal of Geography*, 101(2): 68–72.

Durham, D. (2008) Apathy and agency: The romance of agency and youth in Botswana. In: Cole, J. and Durham, D. (eds) *Figuring the Future: Globalization and the Temporalities of Children and Youth*. Santa Fe: SAR Press, 151–178.

Elwood, S. and K. Mitchell. (2012) Mapping children's politics: Spatial stories, dialogic relations and political formation. *Geografiska Annaler Series B*, 94(1): 1–15.

Faist, T., Fauser, M. and Reisenauer, E. (2013) *Transnational Migration*. Cambridge, UK: Polity Press.

Finn, M. (2016) Atmospheres of progress in a data-based school. *Cultural Geographies*, 23(1): 29–49.

Gamsu, S. and Donnelly, M. (2020) Social network analysis methods and the geography of education: regional divides and elite circuits in the school to university transition in the UK, Tijdschrift voor Economische en Sociale Geografie. doi.org/10.1111/tesg.12413

Gergan, M. and Smith, S. (2020) 'The path you will choose won't be a fairy tale': Urban prefiguration and mountain Nostalgia in India's Northwest Himalayas. *Space and Polity*, 24(1): 77–92.

Graham, M. and Shelton, T. (2013) Geography and the future of big data, big data and the future of geography, *Dialogues in Human Geography*, 3(3): 255–261.

Gulson, K. and Sellar, S. (2018) Emerging data infrastructures and the new topologies of education policy. *Environment and Planning D: Society and Space*, 37(2): 350–366.

Hakli, J. and Kallio, K. P. (2013) Subject, action and polis: Theorizing political agency. *Progress in Human Geography*, 38(2): 181–200.

Henry, J. (2020) Beyond the school, beyond North America: New maps for the critical geographies of education. *Geoforum*, 110: 183–185.

Holdsworth, C. (2009b) 'Going away to uni: Mobility, modernity, and independence of English higher education students', *Environment and Planning A*, 41(8): 1849–1864.

Holloway, S. L., Holt, L., and Mills, S. (2019) Questions of agency: Capacity, subjectivity, spatiality and temporality. *Progress in Human Geography*, 43(3): 458–477.

Holloway, S. L. & Jons, H. (2012) Geographies of education and learning. *Transactions of the Institute of British Geographers*, 37(4): 482–488.

Holloway, S. L., Brown, G., and Pimlott-Wilson, H. (2011) Editorial introduction: Geographies of education and aspiration. *Children's Geographies*, 9(1): 1–5.

Holton, M. (2015) Adapting relationships with place: Investigating the evolving place attachment and "sense of place" of UK higher education students during a period of intense transition. *Geoforum*, 59: 21–29.

Holton, M. and Finn, K. (2017) Being-in-motion: The everyday (gendered and classed) embodied mobilities for UK university students who commute. *Mobilities*, 13(3): 426–440.

Hopkins, P. (2011) Towards critical geographies of the university campus: Understanding the contested experiences of Muslim students. *Transactions of the Institute of British Geographers*, 36(1): 157–169.

Horton, J. and Kraftl, P. (2006) What else? some more ways of thinking and doing 'Children's Geographies.' *Children's Geographies*, 4(1): 69–95.

Jarke, J. and Breiter, A. (2019) Editorial: The datafication of education. *Learning, Media and Technology*, 44(1): 1–6.

Johnston, R. J., Burgess, S., Wilson, D., and Harris, R. (2006) School and residential ethnic segregation: An analysis of variations across England's local education au-thorities. *Regional Studies*, 40(9): 973–990.

Jöns, H., Heffernan, M., and Meusburger, P. (2017) *Mobilities of Knowledge: An Introduction.* Singapore: Springer.

Kenway, J. and Youdell, D. (2011) The emotional geographies of education: Beginning a conversation. *Emotion, Space and Society*, 4(3): 131–136.

Kitchin, R. (2013) Big data and human geography: Opportunities, challenges and risks. *Dialogues in Human Geography*, 3(3): 262–267.

Kraftl, P. (2013) Beyond 'voice', beyond 'agency', beyond 'politics'? Hybrid childhoods and some critical reflections on children's emotional geographies. *Emotion, Space and Society*, 9: 13–23.

Kraftl, P. (2013) Towards geographies of 'alternative' education: A case study of UK home schooling families. *Transactions of the Institute of British Geographers*, 38(3): 436–450.

Koh, S. Y. and Sin, I. L. (2020) Academic and teacher expatriates: Mobilities, positionalities, and subjectivities. *Geography Compass*, 14(5): 12487.

Leitner, H. and Miller, B. (2006) Scale and the limitations of ontological debate: A commentary on Marston, Jones and Woodward. *Transactions of the Institute of British Geographers*, 32(1): 116–125.

Leung, M. W. H. (2017) Social Mobility via academic mobility: Reconfigurations in class and gender identities among Asian scholars in the global north. *Journal of Ethnic and Migration Studies*, 43(16): 2704–2719.

Maddrell, A. and Wigley, E. (2019) Teaching challenging material: Emotional geographies and geographies of death. In H. Walkington, J. Hill and S. Dyer (eds.), *Handbook for Learning and Teaching in Geography*, London: Edward Elgar, pp. 241–255.

Madge, C., Raghuram, P., and Noxolo, P. (2015) Conceptualizing international education: From international student to international study. *Progress in Human Geography*, 39(6): 681–701. doi:10.1177/0309132514526442

Marston, S. A., Jones, J. P. and Woodward, K. (2005) Human geography without scale. *Transactions of the Institute of British Geographers*, 30: 416–432.

Miller, H. and Goodchild, M. (2015) Data-driven geography. *GeoJournal*, 80(4): 449–461.

Mills, S. (2016) Geographies of Education, Volunteering and the Lifecourse: The Woodcraft Folk in Britain (1925–75). *Cultural Geographies*, 23(1): 103–119. www.jstor.org/stable/26168706

Mills, S. and Kraftl, P. (2016) Cultural geographies of education. *Cultural Geographies*, 23(1): 19–27.

Nguyen, N., Cohen, D., and Huff, A. (2017) Catching the bus: A call for critical geographies of education. *Geography Compass*, 11: e12323.

Prazeres, L. (2013) International and intra-national student mobility: Trends, motivations and identity. *Geography Compass*, 7(11): 804–820.

Ramdas, K. (2019) Fieldwork as transition pedagogy for non-specialist students in geography: Promoting collaborative learning amidst uncertainty. In H. Walkington, J. Hill and S. Dyer (eds.), *Handbook for Learning and Teaching in Geography*, London: Edward Elgar, pp. 87–98.

Robinson, P. A. (2018) Learning spaces in the countryside: University students and the Harper assemblage. *Area*, 50(2): 274–282.

Sidhu, R., Cheng, Y., Collins, F., Ho, K. C. and Yeoh, B. S. A. (2021) International student mobilities in a contagion: (Im)mobilising higher education? *Geographical Research*. Doi: 10.1111/1745–5871.12471

Sidhu, R. K., Ho, K. C., and Yeoh, B. S. A. (2020) *Student Mobilities and International Education in Asia Emotional Geographies of Knowledge Spaces*. Singapore: Palgrave.

Simon, S. (2009) "If you raised a boy in a pink room...?" Thoughts on teaching geography and gender. *Journal of Geography*, 108(1): 14–20.

Skelton, T. (2010) Taking young people as political actors seriously: Opening the borders of political geography. *Area*, 42(2): 145–151.

Smith, L. (2002) The "cultural turn" in the classroom: Two examples of pedagogy and the politics of representation. *Journal of Geography*, 101(6): 240–249.

Stokes, A., Collins, T., Maskall, J., Lea, J., Lunt, P., & Davies, S. (2012) Enabling remote access to fieldwork: Gaining insight into the pedagogic effectiveness of 'direct' and 'remote' field activities. *Journal of Geography in Higher Education*, 36(2), 197–222.

Thiem, C. H. (2009) Thinking through education: The geographies of contemporary educational restructuring, *Progress in Human Geography*, 33(2): 154–173.

Waters, J. L. (2017) Education unbound? Enlivening debates with a mobilities perspective on learning. *Progress in Human Geography*, 41(3): 279–298. doi:10.1177/0309132516637908

Webber, R. and Butler, T. (2007) Classifying pupils by where they live: How well does this predict variationsin their GCSE results? *Urban Studies*, 44: 1229–1253.

Wee, B., DePierre, A., Anthamatten, P., and Barbour, J. (2013) Visual methodology as a pedagogical research tool in geography education. *Journal of Geography in Higher Education*, 37(2): 164–173.

West, H., Hill, J., Finn, M., Healey, R. L., Marvell, A., and Tebbett, N. (2020) GeogEd: A new research group founded on the reciprocal relationship between geography education and the geographies of education. *Area*. doi:10.1111/area.12661

Williamson, B., Eynon, R., and Potter, J. (2020) Pandemic politics, pedagogies and practices: Digital technologies and distance education during the coronavirus emergency. *Learning, Media and Technology*, 45(2): 107–114.

Wimmer, A. and Glick-Schiller, N. (2002) Methodological nationalism and beyond: Nation-State building, migration and the social sciences. *Global Networks*, 2(4), 301–334.

24

CHILDREN'S GEOGRAPHIES
Playing with Participatory Methods

Nicole Yantzi and Janet Loebach

Introduction

In the inaugural issue of *Children's Geographies*, Matthews stated that in research and decision-making children were and, in many cases, still are, "little more than silent (or silenced) bystanders and touchline spectators to projects that have attempted to disentangle their worlds through the observations of adults" (Matthews, 2003, p. 4). The value and use of participatory methods to understand children's geographies are highlighted in this chapter through an examination of participatory approaches for examining children's uses, preferences, experiences and opinions concerning outdoor play environments. The subtitle *playing with participatory methods* is not intended to trivialize methods that involve children in the research process, but instead to emphasize the importance of including children as active participants in research exploring their lived experiences. This is not a chapter on how to use participatory methods with children (for this type of information, see Derr et al., 2018), but rather to examine the value of participative methods, and to examine the strengths and challenges of participative work with children. The comic in Figure 24.1 captures the main tenets of this chapter concerning who has valid knowledge about children's play and play geographies, and what are the appropriate ways to collect knowledge. In trying to understand children's play the adults in the comic solely rely on adult knowledge and have not considered involving children in the process. In the comic the adults' frustration only dissipates after their eureka moment – when they consider collecting the experiences and opinions of contemporary children and including them in the process.

The first part of the chapter examines key intellectual shifts within the sub-discipline of human geography, the social sciences, and broader society which have fostered growth in the use of participatory methods with children. Part two examines several barriers to children's participation overall and especially for specific sub-groups of children. The third part provides two examples of participatory methods to show how the shifts and opportunities detailed in part one enriched two research projects with children.

Why Focus on Participatory Methods and Play?

Play, which happens across and within diverse environments, is essential for children's health, development, and well-being. Children that have access to high quality play environments,

DOI: 10.4324/9781003038849-27

Figure 24.1   The Eureka moment.

Note: This comic was co-created by Nicole Yantzi and her 16 year old nephew Nathan Yantzi who have enjoyed playing in many different types of environments.

offering rich and diverse play experiences, are more likely to have well-developed skills in memory, language, and behaviour regulation which assist with school adjustment and academic learning (Bodrova & Leong, 2005; Pramling Samuelsson & Johansson, 2006) .The need to value and prioritize children's play environments should be a concern throughout the world including in areas of conflict and crisis such as after human and environmental disasters, and in highly marginalized and displaced communities (International Play Association, 2017). The United Nations Convention on the Rights of the Child (UNCRC) (1999) recognizes the right to play for all children, which is further emphasized by UN General Comment 17 article 3 "Play and recreation are essential to the health and well-being of children and promotes the development of creativity, imagination, self-confidence, self-efficacy, as well as physical, social, cognitive and emotional strength and skills" (United Nations, 2013, p. 4).

Play is researched in many different disciplines (e.g., psychology, sociology, education), but in children's geographies the focus has been on the physical environments that children play in (e.g., Murnaghan, 2019), how these environments impact children's behaviours and development (e.g., Woolley, 2018), and how children can themselves influence their play environments (e.g., Rasmussen, 2004). The way in which children's geographies scholars examine these three areas has evolved to include more roles and power for children as research participants. As alluded to in the introductory comic, to understand children's play geographies, it is essential for researchers and planners to involve children in the research process. According to Beazley et al., (2009, p. 370) there are four articles in the UNCRC that together mean that children "have the right to be properly researched". For example, Article 12 of the UNCRC states that children have a right to express their views on issues affecting their lives, and to have these views listened to by decision-makers and considered in policy and practice (UN General Assembly, 1989); however, as Bessel (2017) rightly points out the UNCRC does not specifically discuss research. Children's geographies have evolved to embrace and integrate these notions within research processes and outputs more explicitly.

Part 1: Shifts Inside and Outside Human Geography

In the past children's experiences were largely excluded in qualitative and quantitative human geography research which instead used adult proxies to collect data on children's behaviours and experiences. Children were not asked, rather, their parents/guardians, teachers, educational assistants etc. were questioned on the children's behalf. However, a critical turn in human geography, beginning in the 1970s, would significantly affect approaches to the study of children's geographies, and the role of children themselves. Critiques of the predominantly quantitative and positivist focus of human geography facilitated the development of social geography which attempted to highlight power differences and inequalities in society. This attention to societal differences led human geographers to begin questioning the lack of diversity of human experiences that was being represented in research (Norton, 2009). This critique emphasized the need to "expose the hegemonic values which underpin these differential positionings and to raise consciousness that within western societies many aspects of life are the outcome of white, ableist, adult, male, middle-class decision-making" (Matthews & Limb, 1999, p. 62; Sibley, 1995).

The observational research conducted by Bunge (1973), Hart (1979), Moore (1986), and Katz (1986) laid important early foundations in geographic research on children's play environments. Then in the 1990s there was a significant surge of work examining children's experiences of the spaces where they live, learn and play (see McKendrick, 2000 for an extended bibliography). Yet, Freeman's (2020) reflection on twenty-five years of work in children's geographies highlights many outstanding issues including under-researched children, spaces and experiences; the continuing need to critique how methods can strengthen and marginalize participation; and using research to draw attention to children's precarious lives due to environmental, social, economic and cultural marginalization.

Within the sociology of childhood, the new social studies of childhood movement (NSSC) was a catalyst for the paradigm change in how children are viewed and included more broadly within research processes in the social sciences (James & Prout, 1997; James et al., 2014). Some of the key tenets of the NSCC included: (1) children's experiences are diverse and intersect with other social positionings such as socioeconomic status, ethnicity, gender and ability; (2) children's experiences and perceptions are different than adults and valuable and worthy of study in and of themselves; and (3) that "[t]he overarching reach of institutional processes to define and separate children as a group apart emphasizes the hegemonic control that concepts of 'childhood' – what is thought right and proper for children – exercise over children's experiences at any point in time" (James & James, 2004, p. 21). According to James (2009, p. 34) the shift to seeing children as social actors, with unique experiences and opinions from adults, represents one of "the most important theoretical developments in the recent history of childhood studies". The key premise of considering a child as an individual social agent reinforces the bi-directional relationship between children and their environments; they are not only impacted by their physical settings, but they can also shape and impact these environments in return. For example, recent research demonstrates how youth assume societal discourses surrounding identity and belonging, yet at the same time they challenge, re-imagine, and change what these discourses mean in their everyday lives (Vanderbeck & Dunkley, 2004). There is ample evidence that demonstrates children influence how their social spaces and everyday environments are used (James & James, 2004; Valentine, 2011). Children can develop solutions, show creativity, and come up with thoughtful suggestions and reflections on their daily experiences (James et al., 2014).

Increasingly, viewing children as social agents who have different experiences and opinions concerning their environments than adults, is argued to require that research be conducted in a more participatory way. As Hopkins and Bell state "[a] researcher's philosophical understandings about the roles of children and young people in society and space is connected to their understanding of research ethics and the way that they believe children should be treated in research" (2008, p. 3).

Part 2: Critical Perspectives

For an in-depth overview of ethical and methodological strengths and challenges of using participative approaches with children see the special issue in Children's Geographies, 2007, 5(3). Here, we will focus on critical dimensions that impact the ability of children to participate in the geographic research process. We will examine the agency literature and what it means for research with children, and how unequal power relationships between children and adults, different social positionings, and children's limited involvement in all phases of research can have profound impacts for conducting 'truly participatory' research with children. Participatory research is less about the choice of method and more focused on "the social relations involved in the data production and analysis, particularly with respect to where the locus of control and power lies" (Ansell et al., 2012, p. 169).

While we recognize the contention in the literature with defining agency, and particularly how this applies to children, this chapter follows current geographies of children and NSCC praxis in recognizing that children are active agents that shape and influence their surroundings (James and James, 2004; Markström and Halldén, 2009; Valentine, 2011). However, we acknowledge that children may not have equal opportunities for agency in all their daily environments; children may be limited in their agency given practices, social and cultural values, and assumptions in some spaces and contexts. This can have consequences for research processes and praxis emphasizing the importance of examining both the role of adult gatekeepers (Clark & Richards, 2017) and how demographic and social characteristics such as gender, ethnicity and family income can facilitate or hinder children's involvement (Hoggett, 2001; Markström & Halldén, 2009). It is important to recognize that there are significant barriers related to power relationships and social positionings that manifest inequities in participation in research amongst children. Researchers must negotiate and address the adult-centric policies and rules embedded in specific spaces and contexts, "as research projects cannot erase the context of adult power that children face daily in their homes, schools and communities" (Abebe, 2009, p. 458). For example, research conducted in schools can be impacted by the many adult gatekeepers who must give their permission for children to participate, including relevant boards, principals, teachers, and finally the child's parents or guardians. As adults choose which children will be involved, some children's perspectives remain invisible (Horgan, 2017).

Scholars conducting field work with children in poorer, developing countries, and underserved communities face additional challenges in understanding how the historically and geographically contingent social constructions of participation, childhood, and adult-child relationships impact research (Abebe, 2009; Twum-Danso, 2009). The socialization of children in many cultural communities to be submissive, respect their elders, and not bring shame to their parents can conflict with the UNCRC's values of children's rights and participation. As Twum-Danso (2009) discovered this can make it very difficult to address adult-child power differences in their research in Ghana. Although the researcher preferred for children to speak freely, the practice of having to stand up or raise their hands to speak was so ingrained that it made child participants extremely uncomfortable not to follow this practice. The children also struggled when asked their opinion as this was something they had never experienced (Twum-Danso, 2009).

Intersecting identity characteristics can hinder the participation of specific groups of children in research which in turn impacts which stories and experiences are represented. Therefore, it is crucial to focus on "the process producing those voices" which involves "power dynamics as well as … broader cultural, societal, ideological and institutional influences and how children's voices are heard and represented" (Horgan, 2017, p. 250). The statement of a young woman of colour in Cahill's (2007, p. 300) research that "The most important thing for me to be able to do this work was it not feeling like school" emphasizes the importance of how researchers position themselves and discuss the research project. In selecting New Zealand children for the 'Kids in the City' project, the team carefully selected schools based on neighbourhood walkability scores, socio-economic status, and ethnic make-up of the school population including indigenous children (Oliver et al., 2011). Researchers play an essential role in determining what stories and whose voices will be represented. Dodman (2004) encourages researchers to move from a sole focus on the problematic aspects of youth in developing countries such as teen pregnancy and street children and instead ask them to evaluate their daily spaces. Research conducted with 11- to 15-year-old participants in Kingston, Jamaica demonstrates the importance of examining the complexities of their perspectives about their home, school and city and contextualizing this in terms of socio-economic status (Dodman, 2004).

A central aim of participatory research with children and youth is to provide mechanisms, which include not only tools but social structures and relations by which we can better hear and understand the lived experiences of children – including calls for these understandings to be co-produced with young people themselves (Horgan, 2017). Researchers must ensure that this participatory work does not take on a tokenistic or performative nature, masking what are still agendas and narratives constructed by adults (Horgan, 2017; Clark & Richards, 2017).

Several scholars including Hart (1992) and Lansdown (2005) have questioned how children are involved in the research process and emphasize the importance of critically thinking about the word participation. Hart's (1992) original ladder of participation depicts different approaches to involving children, with the higher rungs depicting participation that is more meaningful for the children as they are more genuinely engaged in or contributing to research practices and decision-making. The lower rungs, which represent manipulation, decoration, and tokenization, are examples of 'participation' which only superficially involves children, or worse, undermines young people's rights and agency (Hart, 1992). Similarly, Lansdown's (2005) model terms the degrees of children's participation as consultation, participatory processes, and self-initiation. At the highest degree, children are not only involved in the decision-making processes but also "define the work and are not merely responding to an adult agenda" (Eckhoff, 2019, p. 7). Figure 24.2 demonstrates that meaningful roles for young people are based on increasing the number of opportunities and the number of children that can be involved while at the same time increasing the influence and role of young people in the organization or community. Applying this to participatory research means increasing the opportunities for diverse young people to be involved and expanding their power within research and decision-making processes.

Some scholars in geography and other disciplines facilitating participatory research with young people are more consciously trying to deepen the involvement of children within and across the research process. 'Deep participation' reflects an approach which seeks to involve child and youth participants in as many stages of the research process as possible, beginning with defining the scope and key research questions, through to identifying methods, conducting data collection, analysis, and interpretation of the research, and ending with the dissemination of findings and research outputs (Ansell et al., 2012; Horgan, 2017). This approach means that youth participants themselves are genuinely collaborating to drive the research aims and processes (Kesby, 2007). Cahill (2007) provides a powerful example of what can happen when researchers involve youth participants at the very beginning. In introducing the project to participants, it was presented in an open and vague way focused on the experiences of young

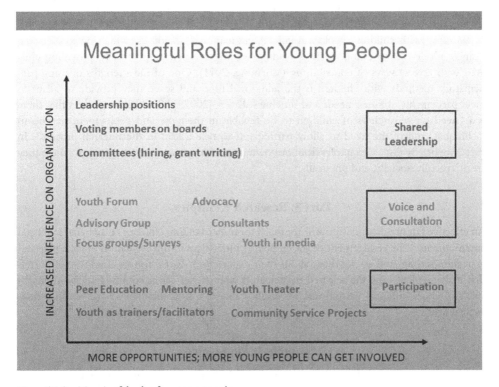

Figure 24.2 Meaningful roles for young people.
Source: Adapted from the Youth Commission, Hampton, Virginia.

women in the city. Discussion with participants about what was important to them, culminated in a richer research project titled "Makes me mad: Stereotypes of young womyn of colour", which examined how they were shaped by, resisted, and ultimately tried to change the stereotypes that affected their experiences in the city.

Child participants are rarely given the opportunity to determine the products of the research, how they are produced and where and how they are disseminated. This situation, however, may be changing as more social science researchers are conducting work within action research frameworks which seek to not only better understand children's lived experiences, but to help affect positive change within children's social and environmental realms. In Cahill's (2007) study the participants, viewed as co-researchers, developed a website (www.fed-up-honeys. org), sticker campaign and a report (Rios-Moore et al., 2004).

There are several potential benefits to involving youth in all research phases including that: (1) they bring ideas and perspectives to the table which adults cannot provide, including an understanding of contemporary youth experience and culture; (2) children are typically less mired in regulatory or political frameworks or adherence to past practices, and may be more likely to think creatively and out-of-the-box (Derr et al., 2018); and (3) young people are often eager to play a more substantive role in improving their communities and local play and recreation environments. Derr et al., (2018), drawing on more than a decade of participatory work with youth in Colorado, United States, remind us that "young people themselves want to be seen as valued contributors, and to be included in urban decision-making and public places within their communities" (p. 8). When youth are involved in planning, building, and maintaining play and recreation places in their community, they are more likely to experience those environments as welcoming, which can translate to increased use and care of these local settings (Loebach et al., 2020; Melcher, 2020).

Working with children involves using a combination of methods, which do not rely on a single mode of expression, such as reading/writing or answering verbally, providing children with diverse ways of contributing (Warming, 2011). One of the strengths of using participatory methods with children is the additional time and space they provide to adapt to meet participants' abilities, needs and timelines. Kesby (2007, p. 200) acknowledges that there is a "need for researchers of children to be flexible in their use and deployment of various techniques, and of the need to allow participants to take a lead in the research process". In Part 3 we briefly discuss examples from our own research that meet the participants 'where they are at' spatially, socially, and physically.

Part 3: Research Examples

An effective but highly underutilized approach for understanding children's play behaviours and environments is the engagement of young people themselves in the evaluation of local play and recreation settings and asking them about their experiences. Part 3 features two research briefs that make evident how the selected participatory approaches can accommodate the strengths and challenges different participants experience in expressing their views and sharing their experiences.

Example 1: Photovoice

There are numerous studies which highlight the benefits of summer camp experiences for youth with disabilities including increasing feelings of hopefulness, security, and self-efficacy, but it is rare for such studies to include first-hand accounts from the youth campers. In

Ontario, Canada, children with disabilities can attend a sleepaway camp, either one for children with any kind of disability or else one for children with specific disabilities (herein referred to as disability-specific sleepaway camp). In studies to date of the impact of these summer camp programs, research participants are primarily parents/guardians, support workers and camp counsellors, and therefore their views and experiences of camp as respite tend to dominate. As part of her master's thesis research, Jenna Simpson (supervised by the first author) demonstrated the value of Photovoice in capturing the voices of camp participants themselves, and how adolescents (14–17 years) with neurodevelopmental conditions at a disability-specific sleepaway camp in Ontario, Canada make meaning of their summer camp experiences (Simpson, 2019).

Camp Neuro, a pseudonym, accepts children and adolescents with neurodiversities, whose severe behavioural challenges mean that they are unable to attend other camps and/or have been removed from other camps. Simpson's (2019) research was the first to consider the views of Camp Neuro held by the youth campers themselves and, in so doing, adapted Photovoice to meet the needs of the neurodiverse participants.

Photovoice "can be altered to fit diverse partnerships, community contexts, participant characteristics and research or intervention interests" (Catalani & Minkler, 2010, p. 448). This flexibility means that Photovoice is a promising approach for capturing the experiences of individuals with severe intellectual and cognitive challenges who may struggle to express themselves verbally (Stafford, 2017). Photovoice typically engaged participants to take and then discuss their own photographs in small or large group interviews to explore their lived realities. This approach was not optimal for the Camp Neuro participants as each participant had unique behavioural challenges which could make group talk difficult (Humphrey & Lewis, 2008; Kaehne & O'Connell, 2010), and the participating teens had varying communication abilities (Simpson, 2019). At Camp Neuro the teens are encouraged to set up their own daily schedule; the study methodology allowed the teen campers to determine when they wanted to take photographs of the camp, as well as how and where they wanted to take photographs. Participants were guided to take photos that would help to express their answer to the question "What does Camp Neuro mean to you?" (Simpson, 2019). The teens were also given the option to participate in individual interviews to discuss their photos, which allowed the interview format to be specifically tailored to their needs and preferences (Simpson, 2019). The flexibility inherent in Photovoice provided the ability to work with participants' strengths and help them feel included in and share power within the research process. As Stafford (2017, p. 611) states "when researching with children with diverse impairments what is important is that they are offered flexibility and choice in how they are researched". In terms of their participation, the campers had the freedom to choose which parts of the research they wanted to participate in, determine their own schedule for data collection, and help to decide how findings were to be shared.

Often researchers employing a Photovoice protocol will use the participants' photographs as discussion prompts but do not actually analyse and present the content of the photographs themselves. Simpson's (2019) original plan for the Camp Neuro study stipulated that those participating in the interview phase were to select five of their photographs to discuss. Simpson (2019) discovered that the participants felt proud of all their photographs and wanted to talk about each of them; the study plan was therefore amended to let participants talk about all their photographs if desired. To capture campers' experiences who had difficulty vocalizing the photographs were treated as data (see Simpson, 2019 for more details).

Photovoice provided the opportunity to work with youth with neurodiversities who can struggle with abstract thinking and expressing their emotions (American Psychological

Figure 24.3 Purple's photograph.

Association, 2013). During unstructured interviews, the photographs provided a tool to help participants move from a concrete thing – a picture of something at camp – to more abstract thoughts and even life lessons; participants that were interviewed expressed moving and profound sentiments about Camp Neuro and life outside camp (Simpson, 2019). Two of the participants learned that they have intrinsic value. One camper, when asked about a photograph she took (Figure 24.3: Purple's photograph), expressed "camp taught me that... even if it's flawed, it's still beautiful. And before camp, I had trouble realizing that but at camp it kinda opened me up to new possibilities" (Simpson, 2019, p. 96).

Often youth with neurodiversities are not included in research about their life experiences as their capacity for communication and understanding is questioned (Stafford, 2017). This Photovoice project revealed that adolescents with severe neurodevelopmental conditions can still share powerful experiences and learnings from their play and recreational environments when flexible participatory tools and diverse modes of expression are integrated (Simpson, 2019). The results were used to co-create an exhibit at a major fundraiser for Camp Neuro attended by peers, caregivers, funders, and decision-makers. By helping to select the photographs and interview excerpts to display at the exhibition, campers also participated in the dissemination and translation of the research findings. Funders and decision-makers who attended indicated that they left with an enhanced awareness of the value of Camp Neuro for the campers themselves (Simpson, 2019).

Example 2: Youth-Performed Play Space Assessments

While children can spend a significant amount of time in community play spaces, including local parks and public schoolyards, they are rarely involved in the assessment of these places or decision-making processes designed to improve them (Loebach et al., 2020; Corkery & Bishop, 2020, Flanders Cushing, 2016). This second example highlights how participatory methods can be used to meaningfully engage groups of children in research to assess and inform the development of their community play spaces.

The goal of the research was to understand student and community needs for the renovation of their school play yard, and to develop design priorities via consensus decision-making informed by the students' insights from their own research and evaluation. In London, Canada, university researchers, with ongoing support from city planners, worked with a group of Grade 8 students (aged 12–13 years) to carry out a series of assessments of their school play yard over three months. Diverse, child-driven methods to evaluate, for example, how and when the space was being used by students and the broader community, existing amenities which were considered valuable, and to highlight gaps in play provision or environmental trouble spots, such as areas where trash or water collects (see Loebach, 2020).

One key objective of the adult facilitators was to minimize the inherent power imbalances that typically exist between adults and children in research processes (Schafer & Yarwood, 2008; Todd, 2012), particularly when the engagement takes place in an institutional setting such as a school (Spyrou, 2011), as well as within peer groups (Kellett, 2010). Several strategies were utilized, including working to build genuine relationships with the children (Warming, 2011) by engaging with them each week over multiple months. Over time, this allowed the facilitators to become 'familiar figures' with whom the young people were comfortable and did not interact with in the same way that they might be expected to with teachers or other adult authorities (Mayall, 2008; Horgan, 2017). This helped to position the students as legitimate co-researchers alongside adult facilitators who consciously worked to avoid taking on an authoritative role (Kellett, 2010).

The students were divided into small 'design firms', each of which were responsible for assessing the existing play space and the needs of its current users, including themselves, but also other students, educational and maintenance staff, and community users, and putting forth design recommendations which reflected their research findings. In this way the power was largely shifted from the adult facilitators to the student groups (Ansell et al., 2012), and students acted as co-researchers as well as key informants themselves (Jones, 2004; Spyrou, 2011). Setting up the youth in peer groups, with no adult present for the discussion and decision-making components, allowed for youth-driven ideas and insights to emerge. Each student in a 'firm' also took on differing roles, diffusing the responsibilities to help minimize potential power dynamics within the peer groups themselves (Kellett, 2010).

While the adult facilitators presented multiple vehicles through which students could assess their current play environment, including environmental audits, behaviour mapping, and interviews with other students and staff, as co-researchers, the students were involved in decisions about how these methods would be operationalized. For example, identifying as a group that it was necessary to learn more about the wishes and needs of other child users, the facilitators recommended that they consider interviews or focus groups. The students decided to conduct focus groups and worked together to develop the questions which would be asked of other students; each firm then conducted a 30-minute focus group with one other grade then reported back, allowing the large group to identify priorities and wishes for varying aged students at the school.

Similarly, to better understand the condition of and opportunities provided by the existing space, the students chose to conduct environmental audits of their play yard. Facilitators provided the firms with an initial set of prompting questions to consider but student groups were then encouraged to develop additional research questions that might be relevant to pose and explore through their in-situ assessment of the space. Each firm then prepared maps of the yard which reflected both their unique questions and their findings (see example maps in Figures 24.4 and 24.5). To examine how students at the school currently use the space, the students chose to carry out behaviour mapping or 'activity mapping' in the play yard; facilitators

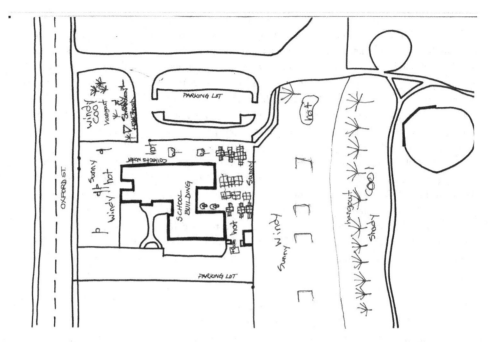

Figure 24.4 Environmental audit map produced by student co-researchers.

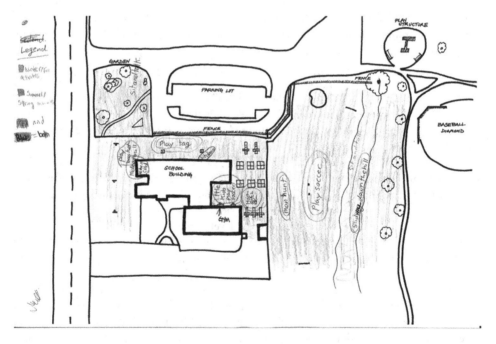

Figure 24.5 Environmental audit map produced by student co-researchers.

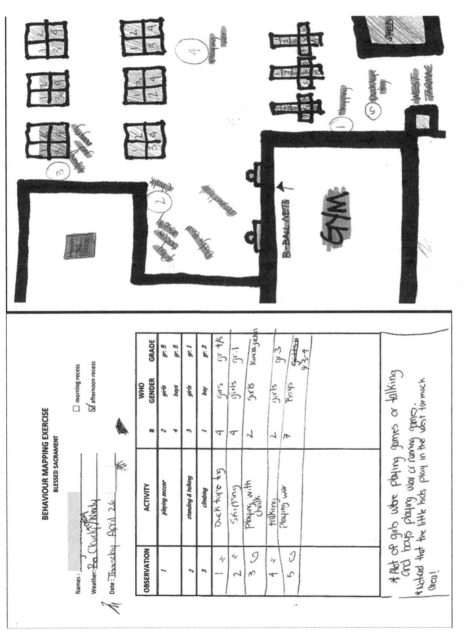

Figure 24.6 Activity map of play yard activity produced by student co-researchers.

helped them to develop a simple protocol for observing and documenting common activities taking place in the yard (Figure 24.6 shows an Activity Map and Data Sheet produced by one of the groups). Students also conducted 'intercept interviews' with other students while they were playing to learn more about what they did in the yard, and what they liked or disliked about the space. Afterwards, in a large group debrief, each group shared their insights from these activities and facilitators helped the students to organize their collective observations and insights to inform their next steps.

Environmental audits, behavior mapping and focus groups conducted and adapted by the students were part of a larger, iterative framework whereby the youth participants produced their own knowledge of and narratives about the play space; they used their individual and collective findings from their research to inform their own design plan and recommendations for the space. Each firm presented their findings and plans to the larger group; the priorities and key insights of each group were recorded on a master list. All of the ideas and priorities were then discussed and confirmed as a group via consensus decision-making strategies, which resulted in a final set of recommendations for the play space renovation.

These activities were also supplemented by other components of the project framework which worked to deepen youth participation. One of these strategies was the integration of multiple, interactive and highly visual methods to provide youth participants with many ways to evaluate the play space and capture the diversity of experiences and wishes within the group. The use of mixed, visual, and activity-focused methods, along with opportunities to engage as individuals as well as in small and large groups, provided multiple mechanisms and modalities for the youth to safely express their individual insights as well as to contribute to collective analyses (Hill, 2006; Elden, 2012, Quiroz et al., 2014).

Another strategy for facilitating deeper participation is through capacity building; that is, providing participating children with relevant training and knowledge building opportunities. Multiple workshops were conducted with student participants to provide them with foundational knowledge and tools that would allow them to conduct their research activities in a more independent and informed manner. For example, the project started out with multiple sessions aimed to introduce them to the city officials as well as the social, environmental and economic policies normally driving decision-making around community spaces, and steps in the design process. Later in the process, when students decided to conduct focus groups with other students, they were given tips and guidelines for conducting effective focus group discussions and analyzing the results. In this way, the adult researchers looked to provide the youth with the knowledge, language and tools necessary for them to genuinely serve as co-researchers and analysts, rather than simply carrying out research activities that were actually adult-framed. Gaining the tools and language for participation and demonstrating to city planners their capacity for substantial and creative contributions to environmental assessment and planning, may also pave the way for the youths' continued interest and engagement in research and civic activities.

Conclusions

The chapter attests to the importance of ensuring that children are given opportunities to share their experiences and ideas about their play and recreational environments. With participatory methods it is not enough that the method supports the research question or hypothesis, researchers must also ensure that the approach to data collection, analysis and sharing of results also supports the participants' abilities which are facilitated and constrained by the temporal and geographic context in which they live. The flexibility and creativity inherent to participatory methods make them well suited for capturing children's complex and diverse play experiences and behaviours. Children value the opportunity to share "their own words and images [to] convey the authenticity of their ideas and the specificity of their visions" (Derr et al., 2018, p. 244). We encourage all researchers to seek those eureka moments in which children's engagement is meaningful, valuable and they are given opportunities to impact their play and recreational geographies. Processes and tools for soliciting the voices and lived experiences of children can be challenging and resource intensive; operationalizing them in a way that also allows for and respects differing voices and experiences to be heard, is even more so. However, despite the challenges of facilitating children's deeper participation in research and decision-making, this chapter highlights multiple examples which effectively and genuinely engage children in evaluating their own play behaviours, preferences, and geographies.

References

Abebe, T. (2009) Multiple methods, complex dilemmas: Negotiating socio-ethical spaces inparticipatory research with disadvantaged children. *Children's Geographies*, 7(4), 451–465. https://doi.org/10.1080/14733280903234519.

American Psychiatric Association. (2013) *Diagnostic and statistical manual of mental disorders* (5th ed.). American Psychiatric Association Pub.

Ansell N., Robson, E., Hajdu, F., & Van Blerk, L. (2012) Learning from young people about their lives: Using participatory methods to research the impacts of AIDS in southern Africa. *Children's Geographies*, 10(2), 169–186.

Beazley, H., Bessell, S., Ennew, J., & Waterson, R. (2009) The right to be properly researched: Research with children in a messy, real world. *Children's Geographies*, 7(4), 365–378. https://doi.org/10.1080/14733280903234428

Bessell S. (2017) Rights-based research with children: Principles and practice. In T. Skelton, R. Evans, & L. Holt (Eds.), *Methodological Approaches. Geographies of Children and Young People*, vol 2. Singapore: Springer. https://doi.org/10.1007/978-981-287-020-9_17.

Bitou, A. & Waller, T. (2017) Participatory research with very young children. In T. Waller, E. Ärlemalm-Hagsér, E. Sandseter, L. Lee-Hammond, K. Lekies, & S. Wyver, *The SAGE Handbook of Outdoor Play and Learning* (pp. 378–394). London: Sage. https://doi.org/10.4135/9781526402028.n25.

Bodrova, E. & Leong, D. J. (2005) The importance of play: Why children need to play. *Early Childhood Today*, 20(1), 6–7.

Bunge, W. (1973) The geography of human survival. *Annals of the Association of American Geographers*, 63(3), 275–295. https://doi.org/10.1111/j.1467-8306.1973.tb00925.x

Cahill, C. (2007) Doing research *with* young people: Participatory research and the rituals of collective work. *Children's Geographies*, 5(3), 297–312. https://doi.org/10.1080/14733280701445895

Catalani, C. & Minkler, M. (2010) Photovoice: A review of the literature in health and public health. *Health Education & Behavior*, 37(3), 424–451. https://doi.org/10.1177/1090198109342084

Clark, J. and Richards, S. (2017) The cherished conceits of research with children: Does seeking the agentic voice of the child through participatory methods deliver what it promises?, *Researching Children and Youth: Methodological Issues, Strategies, and Innovations. Sociological Studies of Children and Youth*, 22. Emerald Publishing Limited, Bingley, pp. 127–147. https://doi.org/10.1108/S1537-46612018000 0022007

Corkery, L. & Bishop, K. (2020) The power of process: Shaping our public domains with young people. In J. Loebach, P. Owens, S. Little, & A. Cox. (Eds.). *Handbook for designing public spaces for young people: Processes, practices and policies for youth inclusion*. New York, NY: Routledge.

Derr, V., Chawla, L., & Mintzer, M. (2018) *Placemaking with children and youth: Participatory practices for planning sustainable communities* (First edition). New York: New Village Press.

Dodman, D. R. (2004) Feelings of belonging? Young people's views of their surroundings in Kingston, Jamaica. *Children's Geographies*, 2(2), 185–198. https://doi.org/10.1080/14733280410001720502

Dotterweich, J. ACT for Youth Center of Excellence, Bronfenbrenner Center for Translational Research, Cornell University. (2015) From report "Positive Youth Development 101".

Eckhoff, A. (2019) Participation takes many forms: Exploring the frameworks surrounding children's engagement in participatory research. In A. Eckhoff (Ed.) *Participatory Research with Young Children. Educating the Young Child (Advances in Theory and Research, Implications for Practice)*, vol. 17. Cham: Springer. https://doi.org/10.1007/978-3-030-19365-2_1

Elden, S. (2012) Inviting the messy: Drawing methods and 'children's voices'. *Childhood*, 20(1), 6–81.

Freeman, C. (2020) Twenty-five years of children's geographies: A planner's perspective. *Children's Geographies*, 18(1), 110–121. https://doi.org/10.1080/14733285.2019.1598547

Flanders Cushing, D. (2016) Youth master plans as potential roadmaps to creating child- and youth-friendly cities. *Planning Practice & Research*, 31(2), 154–173.

Hart, R. (1979) *Children's Experience of Place*. New York: Irvington.

Hart, R. (1992) *Children's participation: From tokenism to citizenship*, Innocenti Essay no. 4. International Child Development Centre. www.unicef-irc.org/publications/100-childrens-participation-from-tokenism-to-citizenship.html

Hill, M. (2006) Children's voices on ways of having a voice: Children's and young people's perspectives on methods of consultation. *Childhood*, 13, 69–89.

Hoggett P. (2001) Agency, rationality and social policy. *Journal of Social Policy*, 30, 37–56.

Hopkins, P. E. & Bell, N. (2008) Interdisciplinary perspectives: Ethical issues and child research. *Children's Geographies*, 6(1), 1–6.

Horgan, D. (2017) Child participatory methods: Attempts to go 'deeper'. *Childhood*, 24(2), 245–259.

Humphrey, N. & Lewis, S. (2008) 'Make me normal': The views and experiences of pupils on the autistic spectrum in mainstream secondary schools. *Autism*, 12(1), 23–46. https://doi.org/10.1177/1362361307085267

International Play Association. (2017) Access to play in crisis. *Play Rights Magazine*, 1(17). https://ipaworld.org/playrights-magazine/

James, A. & Prout, A. (1997) A new paradigm for the sociology of childhood? Provenance, promise and problems. In *Constructing and reconstructing childhood* (2nd ed., pp. 7–33). London: Routledge-Falmer.

James, Allison. (2009) Agency. In J. Qvortrup, W. Corsaro, & M. Honig (Eds.), *Handbook of Childhood Studies*. New York: Palgrave.

James, Allison, & James, A. L. (2004) *Constructing children, childhood and the child*. In *Constructing Childhood: Theory, Policy and Social Practice* (pp. 10–28). New York: Palgrave Macmillan.

James, Allison, Jenks, C., & Prout, A. (2014) *Theorizing Childhood*. Polity Press.

Jones, A. (2004). Involving children and young people as researchers. In: S. Fraser (Ed.) *Doing Research with Children and Young People* (pp. 113–130). London: SAGE; Open University Press.

Kaehne, A. & O'Connell, C. (2010) Focus groups with people with learning disabilities. *Journal of Intellectual Disabilities*, 14(2), 133–145. https://doi.org/10.1177/1744629510381939

Katz, C. (1986) Children and the environment: Work, play and learning in Rural Sudan. *Children's Environments Quarterly*, 3(4), 43–51.

Kesby, M. (2007) Methodological insights on and from children's geographies. *Children's Geographies*, 5(3), 193–205. https://doi.org/10.1080/14733280701445739

Kellett, M. (2010) *Rethinking Children and Research: Attitudes in Contemporary Society*. London; New York: Continuum International Publishing Group.

Lansdown, G. (2005) Can you hear me? The right of young children to participate in decision affecting them (Working Papers in early childhood development, no. 36). The Hague, Netherlands: Bernard van Leer Foundation.

Loebach, J., Owens, P., Little, S. & Cox, A. (Eds.). (2020) *Handbook for Designing Public Spaces for Young People: Processes, Practices and Policies for Youth Inclusion*. New York, NY: Routledge.

Markström, A.-M. & Halldén, G. (2009) Children's strategies for agency in preschool. *Children & Society*, 23(2), 112–122. https://doi.org/10.1111/j.1099-0860.2008.00161.x.

Matthews, H. (2003). Coming of age for children's geographies. *Children's Geographies*, 1(1), 3–5.

Matthews, H., & Limb, M. (1999) Defining an agenda for the geography of children: Review and prospect. *Progress in Human Geography*, 23(1), 61–90.

Mayall, B. (2008) Conversations with children: Working with generational issues. In: P. Christensen and A. James (Eds.), *Research with Children* (pp. 120–135). London: Falmer Press.

McKendrick, J. H. (2000) The geography of children: An annotated bibliography. *Childhood*, 7(3), 359–387.

Melcher, K. (2020) Youth-built projects: Involving youth in the construction of public places. In J. Loebach, A. Cox, S. Little & P. Eubanks Owens (Eds.) *Designing Public Spaces for Young People: Processes,*

Practices and Policies for Youth Inclusion. New York, NY: Routledge. www.taylorfrancis.com/chapters/edit/10.4324/9780429505614-9/youth-built-projects-katherine-melche

Moore, R. (1986) *Childhood's Domain: Place and Play in Child Development*. London: Croom Helm.

Murnaghan, A. M. F. (2019) Play and playgrounds in children's geographies. In T. Skelton (Ed.), *Establishing geographies of children and young people* (Vol. 1, pp. 407–425). Springer, Singapore.

Norton, W. (2009) *Human Geography* (7th ed). OUP Canada.

Oliver, M., Witten, K., Kearns, R. A., Mavoa, S., Badland, H. M., Carroll, P., Drumheller, C., Tavae, N., Asiasiga, L., Jelley, S., Kaiwai, H., Opit, S., Lin, E.-Y. J., Sweetsur, P., Barnes, H. M., Mason, N., & Ergler, C. (2011) Kids in the city study: Research design and methodology. *BMC Public Health*, 11, 587. https://doi.org/10.1186/1471-2458-11-587.

Pramling Samuelsson, I. & Johansson, E. (2006) Play and learning – Inseparable dimensions in preschool practice. *Early Child Development and Care*, 176(1), 47–65. https://doi.org/10.1080/0300443042000302654.

Quiroz, P.A., Milam-Brooks, K. and Adams-Romena, D. (2014) School as solution to the problem of urban place: Student migration, perceptions of safety, and children's concept of Community. *Childhood*, 21(2), 207–225.

Rasmussen, K. (2004) Places for children- Children's places. *Childhood*, 11(2), 155–173.

Rios-Moore, I., Allen, S., Arenas, E., Contreas, J., Jiang, N., Threatts, T., & Cahill, C. (2004) *Makes me mad: Stereotypes of young urban womyn of color*. Center for Human Environments, City University of New York. http://englishamped.weebly.com/uploads/1/0/4/0/10404073/makesmemadreport.pdf.

Schäfer, N. and Yarwood, R. (2008) Involving young people as researchers: Uncovering multiple power relations among youths. *Children's Geographies*, 6(2), 121–135.

Sibley, D. (1995) *Geographies of Exclusion*. New York: Routledge.

Simpson, J. (2019) "A place where kids like me can be ourselves": Exploring the meaning of summer camp with adolescents with neurobiological disorders. [Unpublished master's thesis]. Laurentian University.

Spyrou, S. (2011) The limits of children's voices: From authenticity to critical, reflexive representation. *Childhood*, 18(2), 151–165.

Stafford, L. (2017) 'What about my voice': Emancipating the voices of children with disabilities through participant-centred methods. *Children's Geographies*, 15(5), 600–613. https://doi.org/10.1080/14733285.2017.1295134

Todd, L. (2012) Critical dialogue, critical methodology: Bridging the research gap to young people's participation in evaluating children's services. *Children's Geographies*, 10(2), 187–200.

Twum-Danso, A. (2009) Situating participatory methodologies in context: The impact of culture on adult–child interactions in research and other projects. *Children's Geographies*, 7(4), 379–389. https://doi.org/10.1080/14733280903234436

UN General Assembly. (1989) *United Nations Convention on the Rights of the Child*. www.ohchr.org/en/professionalinterest/pages/crc.aspx

United Nations. (2013) UN Committee on the Rights of the Child (CRC), *General comment No. 17 (2013) on the right of the child to rest, leisure, play, recreational activities, cultural life and the arts* (art. 31). UN Committee on the Rights of the Child (CRC). www2.ohchr.org/english/bodies/crc/docs/GC/CRC-... · DOC file.

Valentine, K. (2011) Accounting for agency. *Children & Society*, 25(5), 347–358. https://doi.org/10.1111/j.1099-0860.2009.00279.x

Vanderbeck, R. M. and Dunkley, C. M. (2004) Introduction: Geographies of exclusion, inclusion and belonging in young lives. *Children's Geographies*, 2(2), 177–183.

Warming, H. (2011) Getting under their skins? Accessing young children's perspectives through ethnographic fieldwork. *Childhood*, 18(1), 39–53.

Woolley, H. (2018) How to grow a playspace: Development and design. *Children, Youth and Environments*, 28(2), 191–193. https://doi.org/10.7721/chilyoutenvi.28.2.0191

25

ANARCHIST RESEARCH WITHIN AND WITHOUT THE ACADEMY

Everyday Geographies and the Methods of Emancipation

Richard J. White and Simon Springer

Introduction

Anarchism has always been anti-ideological; anarchists have always insisted on the priority of life and action to theory and systems.

Wieck, 1972, p. 10

If anarchism is a spirit, it is the spirit of revolt... The challenges of our time require us to rebel against the disabling faith in the idea that oppression, hierarchy and captivity are somehow the national consequences of human evolution. Our revolt is our emancipation.

White et al., 2016, p. 1

In popular conversation, any reference to anarchy or anarchism will provoke stereotyped images of malevolent individuals consumed with misanthropic disgust and terroristic intent into being. Typically, these *anarchists* bring with them violent intentions, either concealed about their person (the bomb-sewn jacket of 'the Professor' in Conrad's Secret Agent perhaps), or as its visible messenger: their nihilistic hands bearing Molotov cocktails, ready to set aflame the night sky. Propaganda by deed, perhaps (see Fleming, 1980)? It is far less likely, that any mention of anarchy, anarchism or an anarchist will immediately spark connections with (i) established body/ies of highly respected academic research, and (ii) geographical research. What image, we wonder, might be unleashed if a researcher walked down the high street, clipboard in hand, and invited passers-by to describe to them what an *anarchist geographer* looks like?

However, as we hope to demonstrate in this chapter, there are many rich and exciting intersections that exist between anarchism, geography and scholarly research, and have been so for many years. As Springer (2013, p. 46) notes, geographers began engaging with anarchism

DOI: 10.4324/9781003038849-28

at least as far back as the nineteenth century, with key figures like Élisée Reclus and Peter Kropotkin rejecting the discipline's preoccupation with imperialism by developing an emancipatory vision for geography. The research of both Kropotkin (1842–1921) and Reclus (1875–1894) still continues to exert a significant influence across many contemporary aspects of critical geographical knowledge and understanding, not least those which are concerned with addressing the current ecological, economic and political crises and challenges we face today (see Ferretti, 2019; Pelletier, 2009; MacLaughlan, 2016). That notwithstanding, it is also important when thinking about anarchist geographies that we extend our reading to fully embrace other key contributions that anarchist geographers have made, and continue to make. However foundational their work is within geography and anarchism, we are not proffering a *Kropotkinist* Geography or a *Reclusist* Geography, in the way that some other radical geographies might have rarified a single individual! There is no time like the present, and the present moment represents an ideal opportunity to bring our research questions and ideas into a full and productive conversation with anarchism. Anarchist geographies are very much in the ascendency: the twenty-first century has certainly witnessed a resurgence of anarchist thought and practice, seen within the discipline and critical social sciences, and through expressions of activists and activism across the world (see Springer et al., 2016; de Souza et al., 2016, White et al., 2016).

Of all the radical traditions, given its *a priori* refusal to privilege one form of domination (e.g., class, gender, race) over another, anarchists have sought to engage with *all* types of injustices, wherever and in whatever form they manifest themselves. As Wigger (2016, p. 133) notes:

> Anarchists seek social change beyond the point of capitalist production by extending social struggles to sexual, ecological, racial, patriarchal and gendered forms of inequality and oppression.

Unsurprisingly therefore, we can clearly see how these struggles have animated anarchist geographies, with this body of research consistently raising consciousness and awareness to the reality that social (in)justice is always spatial (see White and Springer, 2018). To this end, research in anarchist geographies have made numerous critical interventions across geographical subdisciplines, including political geographies (see A Collective of Anarchist Geographer, 2017; Ince and de la Torre, 2016); neoliberalism (Springer, 2016b); diverse economics (White and Williams, 2017); political ecology (Brock, 2020); queer space making (Rouhani, 2012); radical pedagogies (Ferretti, 2018; Stenglein and Mader, 2016); more than human communities, particularly critical animal geographies (Reclus, 1901; White and Springer); intersectional forms of activism and organization (Springer, 2014; Ferretti, 2016; Veron, 2016); and researching activism (Ince and White, 2020). Certainly, as Sidaway et al., (2017, p. 281) noted: "the dynamic currents that animate anarchist geographical praxis today, as geographers (re)map the possibilities of what anarchist perspectives might yet contribute to understandings of geography, and in turn, what geography might yet contribute to how we understand, appreciate, and practice anarchism, are flourishing".

These anarchist lines of flight will undoubtedly continue to push geography further in ever radical and critical directions. For example, given the ongoing and devastating global impacts of COVID-19, we certainly envisage some of the most urgent areas of attention for anarchist geographers will be those that coalesce around post-capitalist, post-statist and critical posthuman/vegan geographies.

With the overall aim of trying to raise consciousness and awareness about anarchist geography in the context of research, the main body of this chapter is divided into two key sections. While always taking care not to obscure the highly inter-disciplinary, pluralistic and dynamic essence of both geography and anarchism, the first section focuses on 'anarchism, anarchist geographies and research'. The second section addresses the more comprehensive inter-locking question of, "What are anarchist methodologies in human geography, and how should anarchist research be disseminated?" The question of dissemination research often weighs heavily on the shoulders of many radical and critical researchers, and invariably brings into play a myriad of dilemmas, some more easily negotiated than others. Greenaway (2011, p. xvii) captures the essence of this when observing that:

> Another problem for those who work as academics is how to do research and writing in a way that reaches out to a variety of audiences, and bridges the perceived gap between theory and activism. This is not just a question of the accessibility of ideas and language, but of where to publish or speak, when only certain publications and venues are academically acceptable.

Bringing these dilemmas under the focus of the chapter, we ask what is the point of anarchist research if it is exclusionary, only accessible to those who can converse in English, heard by those who are privileged enough to afford to pay (or have their institutions pay) exorbitant conference fees, and/or read by those able to spend £100+ on a Handbook produced by a private publishing company for profit? Tackling these issues also raises important questions of ownership and copyright. Here we also reflect on our own seeming hypocrisy and complicit nature in writing this piece for this volume. The chapter, however, concludes on an optimistic note. As anarchist geographies continue to gather momentum in our time of crisis and crises, they carry with them the hope of a better world. It is our hope that many readers will seriously reflect on how they may connect, or more fully align, their research with these anarchist lines of flight in mind.

Anarchism, Anarchist Geographies and Research

> ...an anarchist approach to geography embraces partial, fragmented, and overlapping worlds, wherein empowerment and emancipation become possible as shifting islands of reflexivity between theory and practice.
>
> *Springer, 2016, p. 1*

Given just how poorly understood anarchism is in popular circles, it is important to provide some sense of what anarchism means, and what is understood by anarchist geographies in the context of this chapter. In doing so this then creates a foundational space from which it is possible to critically question what anarchist geographies and anarchist methodologies are all about. We swim with caution in these definitional waters, mindful of the contesting currents of thought that swirl strongly beneath them! One of the salient qualities of anarchism has been its ability to resist being pinned down, labelled and neatly categorized. Indeed, it contains an essence which escapes – is truly beyond – language, though not understanding. Certainly, we have no desire to curtail this free spirit by imposing definitional boundaries – still less in policing (!) them.

One constructive way of approaching the question, 'what is anarchism?' would be to identify some of its core and adjacent concepts (see Franks et al., 2018). These would certainly include: an anti-hierarchical stance, and a commitment to prefigurative praxis, direct action, mutual aid, and horizontalism (see Text Box 25.1).

Text Box 25.1: Working Definitions of Several Core and Adjacent Anarchist Concepts

- *Antihierarchical stance.* For Amster (2018, p. 15) "an anti-hierarchical perspective is evident in anarchist theory and action alike. Indeed, it might be said that a robust notion of anti-hierarchy is the sine qua non of anarchism, the core concept that differentiates it at root from other ideologies".
- *Prefigurative praxis.* "The praxis of prefiguration defines the appropriate sphere of action as here and not whilst looking to the future and directing our agency towards reassessing the contexts and structures that shape and constrain individual lives". (Honeywell, 2011, p. 67)
- *Mutual aid* has enjoyed a long association with anarchism, and anarchist geography in particular – not least through Kropotkin's (1902) classic work "Mutual Aid". For Kropotkin, mutual aid embodies the highest ethical principle. "In the practice of mutual aid, which we can retrace to the earliest beginnings of evolution, we thus find the positive and undoubted origin of our ethical conceptions; and we can affirm that in the ethical progress of [humankind], mutual support-not mutual struggles-has had the leading part" (1902/1915, pp. 222–223).
- *Horizontalism.* "While this slippery term has meant slightly different things for different people, it generally connotes a form of "leaderless", autonomous, directly democratic movement building whose adherents consider it to be non-ideological". (Bray, 2018, p. 101)

Reading anarchism in this way is also helpful in thinking about the areas of research and enquiry that anarchist research(ers) are likely to see as being particularly relevant and important. Certainly, the commitment to praxis rejects appeals to 'High Theory', and instead places greater emphasis on engaging with social relationships 'at the human scale'. It is no surprise then that significant branches of anarchist research have focused on everyday acts that "make the human community possible" (Ward, 1996, p. 8), such as informal coping strategies, altruism, mutual aid and reciprocity in households and within local communities (e.g., White and Williams, 2017; Springer, 2020).

A commitment to some or all of these concepts, practices and principles can also be detected in other radical approaches and 'isms'. For example, anarcha-feminism, "emerged as a 'school of thought in the late nineteenth-century" (Kowal, 2019, p. 265). This conjunction of anarchism and feminism can be seen, and "understood in multiple ways: Anarchist feminism might be anarchists sympathetic to feminism or feminists for whom anarchism is a necessary corollary of their politics" (Kinna, 2017, p. 254). Thus, rather than calculating exactly how and where these overlaps may be, we'd encourage a more intuitive approach: as Uri Gordon (2008, p. 3) argued, "You know it when you see it". There is certainly great potential for further fertile intersections between anarchism and feminist research, non-representational research and participatory action research to emerge (see Chapters 31, 30, and 28, respectively, in this volume, respectively).

Back in the 1970s, Elaine Leeder, in her paper 'Feminism as an Anarchist Process' offered some important critical reflections in this context.

> I have come to realize that the interaction in an all womens' groups [sic] has a unique flavor and style and that this is particularly true of feminist groups. ... It could in fact be called Anarchist because the values of leaderlessness, lack of hierarchy, non-competition and spontaneity have historically been associated with the term Anarchism. They are also Feminist values. ... It is clear to me from my experience with women in varying groups that the time has come for Feminists to make clear and articulate the Anarchism in our Feminism.
>
> *Leeder, n.d., pp. 1, 3*

It is really encouraging to think that a more appropriately *contemporary* understanding and narration of anarchism might yet encourage more people – more researchers – to dare to speak its name, and explicitly recognize the influence that anarchist praxis has had/is having on their own attitude, experiences, and research. Indeed, there are parallels here to be found with Smith's (2015, p. 1) exploration of the "productive links between the stance of anarchism and recent work in non-representational theorising in the social sciences". His paper argued for a "revalorization of anarchism in the social sciences" (ibid), stressing that "the time may be ripe for anarchism to be viewed with more relevance, particularly in a post-representational social scientific milieu" (Smith, 2015, p. 3).

A further important consideration comes in the form of the self-aggrandizing scholar, that potentially finds refuge in the midst of card-carrying radical geographers (anarchist or otherwise). We can clearly identify researchers who consider themselves – and are viewed by others as – anarchist geographers. However, as anarchist geography continues to gather in momentum and visibility, as with all radical approaches there is the dilemma of how this radical intent can be maintained while being ushered ever more tightly into the web of mainstream academic circles. What are the perils of anarchist geographical research becoming 'edgy' or 'trendy', and thereby attracting a particular type of academic creature who seeks to gain from this (through publishing or research grants), but possessing no deep or real commitment to anarchism generally? On the flipside, we have been accused of turning "anarchist geography" into a brand simply by doing what all scholars do, namely publishing our work. Responding to these thoughts ushers in the question of "why" anarchist research is being undertaken: it is always worth critically evaluating the main motivations that underpin a particular research agenda, and the importance that of this for methodology and methods.

A commitment to anarchism and anarchist praxis is not an optional extra, something that just comes into play in an explicitly academic or research environment, perhaps because anarchism is seen to be 'trendy'. Rather, anarchism is something that should influence every fibre of our thoughts and being, and be present in all the spaces we encounter, whether the university, the home, the wider community, and so on. It should certainly inform our everyday relations with others. An ethics of care is a key characteristic of anarchist approaches insofar as they aspire to seek authentic and genuine ways of empathizing, sympathizing, and demonstrating compassion for others.

This attitude and desire work to dissolve any formal relations that differentiate one (the researcher) from the other (the researched), and places both on a much more equal and 'human' footing. In contesting *authority*, we might also expose the hubris that comes with the notion of *authorship*, as though ideas are ever bolts from the blue as opposed relationally constructed fractals. Drawing on the figure of Élisée Reclus once again, these social qualities are beautifully captured by Woodcock (1988, p. 13):

> One of the reasons why Reclus flourished so happily in a movement that had become radically decentralised into small groups depending on personal affinities was the power he had of empathising with individuals. The gentleness of manner, the lack of pride or pretention, that he combined with an almost ferocious integrity... allowed him to mingle without affectation among workers whose manners and whose education were very different from his own.

Within the context of "academic" research (indeed life generally!) while it is particularly easy to *say* that we empathize, care etc. with those whom we interact with, there is certainly no shame in acknowledging how difficult it may be to put into practice. Context is obviously important here, not least our positionality and entangled experiences, being particularly mindful of the broader social relationships and communities we are embedded within. We must always be conscious that, even if we try and do 'the right' thing, a range of unintended consequences may present themselves. And that is okay insofar as it acknowledges that we are never 'perfect' in any situation. Perhaps the best we can hope for is captured in this insightful summary by the Brazilian anarchist/left-libertarian geographer, Marcelo Lopez de Souza (2019, p. 21, emphasis added):

> At the end of the day, the 'moral of the story' could be summarised as follows: there are subtle and unconscious ways to patronise and subalternise those whom we want to understand and help but whose culture and history (and/or social class) are different from ours, and it is precisely these subtle and unconscious ways -very much related to the influence of problematic 'biographical atmospheres', and sometimes also to the insufficient immunisation provided by some 'ideological soils' - that generate problems of coherence among radical scholars. *In order to avoid incoherence, we do not need to 'romanticise' those whom we show our solidarity and offer our support after all, they are all fallible human beings, too. Moreover, true dialogue presupposes horizontality, and true horizontality - something not easy to be achieved - presupposes the right to disagree but at the same time the obligation to do so without arrogance.*

In short, a healthy scepticism is important, and particularly justifiable when thinking about anarchism and research for the following reasons: "One simply cannot rely on self-identification. Just because someone claims to be an anarchist does not make it so. By the same token, just because someone never identified [themselves] an anarchist does not mean that [their] ideas cannot be qualified as anarchist" (Graham, 2015, p. 3). The same critique – as we have drawn attention to earlier – holds fast for the considerable research that takes place which can clearly be read as illustrative of 'anarchist geographical praxis' yet has not been defined or framed as either.

Foregrounding Methodology: Methods, Ethics and Dissemination Strategies

We can secure a meaningful foothold in thinking through anarchist approaches toward methodology and methods by drawing on the provocative and captivating figure of Paul Feyerabend. Feyerabend is particularly well known through his foundational publication "Against Method"

(1993), where he advances a conceptualizing of anarchistic theory of knowledge (epistemology) as being 'against scientific method'. To understand this position, it is worth reflecting on this passage he wrote in his later work 'The Tyranny of Science' (2011). Feyerabend invites us to:

> ... Imagine a wood, or a field, with its delicately balanced ecology including the humans who have to live off the products of both. Is there a way of understanding such a system? Of finding out about its robustness and its limits? Of discovering what will be tolerated and what leads to irreversible change? Yes, there is. Whoever has lived in the region for generations has learned its peculiarities and its life rhythms and has stored this knowledge in eyes, ears, in the sense of smell, in feelings, in the mind, in the stories that are old to the community. In short, whoever has stored the knowledge not just in her/his mind but in her/his whole being, possesses information that is not contained in the results of a scientific appraisal. Only a little of this information can be written down or otherwise articulated – it shows itself in how things look and how they feel and it cannot be transferred to a person lacking the appropriate experience. However the knowledge is there – and it should be used.
>
> *2011, p. 48*

Thus, an anarchist approach to what is 'the best' or most appropriate methodology and methods to harness, should, at the very least, begin with another question: In what ways can we hope to enable our research to connect with, recognize, value and celebrate the individual essence of things – and relationships – in ways that the positivist scientific gaze and 'objective' evaluation cannot? Here we must think creatively and be brave enough to cast off the shackles of past wisdom if the situation at hand calls for it. Ley (2019, p. 4) appeals to (our) use of 'the anarchist imagination', namely an imaginary which "allows the social scientist and historian to undergo...' 'moments of madness [sic]' in her research". Feyerabend (2011, p. 130) stressed the importance of the imaginary – and maintaining original thought when it serves to attract pessimism and derision from others: "Those who think that new things can be found only by wandering along a precisely defined path are wrong... 'anything goes' means only 'don't restrict your imagination' because a very silly idea can lead to a very solid result. Also, don't restrict your imagination by logic".

Given this reading of methodology and methods through an anarchist lens, it is unsurprising to find a keen depth of research focused on the 'human' scale, seeking to dive deeply into the richness and messiness of life in the here and now. Wigger (2016, p. 139) for example, acknowledges and extends this focus, noting that:

> anarchism through its methods of transformative praxis, frequently gives ontological primacy to micro-level relations and everyday life. Decentralised, bottom-up grassroots struggles that aim at changing micro-relations in everyday life are considered the crux for changing macro-structures. The transformation of social structures hence evolves cumulatively through enlarging social spaces with alternative organisational forms, preferably decentralised organisational structures based on affinity groups that work together on an ad hoc basis, or more engrained and enduring voluntary associations.

Focusing on the question of how anarchists might look to develop their methods for their research, they would do so in a way that takes a critical look at any 'good practice' that emerges from 'ready-made', pre-determined, 'This is how you must harness (a method)' literature. An inherent prescription for 'doing methods' comes with the considerable risk of closing down a fuller realization of experimentation and innovation. When we are bound by systematic rules it becomes difficult to think and act outside the box. Resisting this, an anarchist approach would desire to harness methods in ways that are intrinsically exploratory, processual, always unfolding, and open to serendipity and experimentation. Read the research guidebooks by all means, but do so in a way that does not curtail you in unleashing your own methodological imaginary! (Orenstein and Luken, 1978).

An anarchist researcher will do everything possible to avoid framing others as 'subjects' of their enquiry. For example – and aligned with a methodological position long held within feminism (e.g., England, 2006) – whenever engaging other human participants, it is vitally important that all hierarchical relationships (researcher/researched) are resisted and headed off at every stage (DeLeon, 2019; Munn-Giddings, 2006). This commitment, yet again, emphasizes the need to be 'in the moment' when engaging with others. However well you prepare, you will never be able to anticipate everything that you might experience – in terms of relationships to others – beforehand. Therefore, it is vital that good practices of reflexivity are followed here (see Berger, 2015; Lumsden, 2019). For example, how do you act if the respondent assumes that they are supposed to behave passively, and reactively in your presence? It is quite plausible to assume this if they have designated you as 'the expert intellectual' in the relationship, based on their own (poor/limited) experiences of academics, rooted in unequal teacher-pupil relationships. The irony of course is that it is the researcher who is asking the researched to share their knowledge. The responsibility and challenge comes in the form of asking 'how can' this research encounter be used not just to gain knowledge and insight about a particular issue, but also as a way of establishing trust and rapport. If successfully done so, then the potential to forge new and important solidarities in this way – in harnessing anarchism as research process – comes into play. If research is understood as a moment of reciprocal learning, community building, and an expression of shared respect – an idea shared by many critical scholars – it becomes decidedly anarchist inasmuch as such thinking moves us towards recognizing research as an actual practice of mutual aid.

Disseminating Anarchist Research Within and Beyond the Academy

... we need to sidestep the polarisation of 'activism' and 'academia', theory and practice. History, theory, reading and writing can all be forms of resistance and activism. A more constructive response is to find ways of bringing together different perspectives, analyses, ways of doing: not answers, but questions; not a single, smooth, impenetrable surface, but rough edges which can spark off one another, provide new points of access. Standard methods of propagating ideas – meetings, conference, books and articles – can be subverted in form and content to become spaces where past, present and future are reimagined and new ways of thinking become possible.

Greenaway, 2011, p. xvii

An instinctive answer, when responding to the question of where anarchist research should take place, would be 'outside the academy'. Indeed, a great deal of what constitutes the diverse landscapes of anarchist research takes place in meaningful ways outside of the academy, and certainly by researchers and authors without academic appointments. This rootedness of knowledge and understanding outside of institutions is not at all surprising when one considers that "Anarchy sought to make manifest tribes without rules, architecture without architects, and *education without schools*" (Levy, 2019, p. 9). This response also applies when our conceptualization of anarchist research is extended to embrace the everyday experimentation and agency of communities, and explore different ways of collectively being in the world. Reinforcing this point, Springer (2016, p. 64) argues that,

> Committing radical geography to an anarchist agenda would necessitate a negation of the false dichotomy the disciplines maintain between the academy as a space of knowledge production, on the one hand, and wider society as the domain of social struggle, on the other...

That notwithstanding, any response that uncritically asserts that anarchist research should – or does – take place 'beyond the academy', is problematic. Such a narrative, for example, risks fetishizing a particularly skewed reading of the academic and academia. In this reading, the university is typically envisaged as enjoying an ivory-tower like existence; a place where knowledge is pursued for the sake of itself, divorced and desensitized from the crises and struggles that are present in the world.

Rejecting this framing of the university is important. Instead, a more nuanced reading is encouraged, one that recognizes that many of the struggles that social justice activists rail against – brutal authoritarianism and managerial bureaucrats in the thralls of neoliberalism, to give two example – are struggles that are very much alive and relevant 'on the university campus'. The realities of being an academic researcher (at any level) in increasingly neoliberalized spaces of higher education (see Nocella and Juergensmeyer, 2017; Parker, 2014) come at an enormous cost for many. The constant struggle to secure tenure, help students/learners publish in 'the right' journals, continually chase funding to satisfy externally-driven targets – in a climate of ever diminishing resources and autonomy – is the brutal and unforgiving reality that increasingly casts a shadow across 'the 21st century university'. Such a dystopic scenario often leads to an ultimatum: either pursue the commodified aspirations of neoliberal academia or get out. For some the decision is a stark one: to make a difference in the world means turning their back on the discipline they love. It is sobering to note that these dilemmas have a long history, and Kropotkin himself found himself at such a crossroads; being forced to choose between following academic geography or 'the people':

> Reared in Russia and steeped in the populist mystique of the "going to the people" movement, Kropotkin found himself morally bound to devote his time and energy to the cause of the masses, to the detriment of his scientific interests, which he felt it would be a selfish indulgence to pursue. So he set out on a course that was bound to destroy his geographical career, and though in later years he made use of his training to write important books, like Mutual Aid, which gave a scientific support to anarchist arguments, he never again returned professionally to geography, even though he was treated with great respect by English geographers for his brilliant work, many years ago, on the geography of East Asia.
>
> *Woodcock, 1988, p. 2*

Careful consideration of how anarchist research ought to be disseminated should come at the very beginning and certainly not bolted on as an afterthought. In a perfect world *everybody* would be able to – should they want to – access anarchist research: either through reading the author's work, or listening to them speak to it. In such a world, any financial contribution would be voluntary, and certainly not a prerequisite; non-English translations of the research would be in place; the text would be available to download from wherever the individual found themselves in the world. There would be no copyright, and open-access would be subject to fair use policy. It should be noted that many anarchist and radical publishers do actively try to operate in ways that aspire to meet these goals wherever possible.

However, for those who are in academia, seeking to publish their research along these 'open access' lines can be intensely problematic for them. For this open, inclusive, community-orientated emphasis on disseminating knowledge flies in the face of the increasingly privatized and commodified worlds of academia and academic publishing houses. The dominance of these neoliberal metrics of assessment of both research and writing is so pervasive in contemporary academia there are precious few scholars, if any, who escape being put under considerable and continuous pressure to publish their research only in 'the highest ranking' journals for example. Failure to do so carriers with it enormous risks both for the academic's immediate future (particularly if they are in precarious employment, seeking tenure), and their longer-term pro-spective in terms of making a case for internal promotion, or being an attractive applicant for other research posts.

All is not lost, however, and there are several tactics and strategies that the anarchist researcher potentially has open to them. These might include: buying (their) books and donating them to the groups/ communities directly; providing PDF versions of final versions of their work to those who might reach out to them; uploading files to a range of 'academic' websites where individuals, though they might have to formally register can access the documents without paying. Though not without their faults, this would include Academia.edu or ResearchGate. Other possible options include having their work published on The Anarchist Library https:// theanarchistlibrary.org/special/index) or the Humanities Commons https://hcommons.org/

So why, you might justifiably ask, did we agree to author a chapter for a prohibitively expensive book (certainly in its Hardback form), one which is only likely to be purchased by university libraries? Are we guilty not only of being hypocritical – and complicit in writing this piece? We justify our contribution here in two ways. First, we see this chapter as embedded as part of a much wider richer and more complex tapestry of social geography, crucially a geography which is still to fully recognize the potential that anarchist praxis can offer. In that sense, it might serve as a small, but important intervention, which can be taken forward by others: particularly, we hope, student researchers. Second, it is important to note that the invitation to contribute came in the shape of a wonderfully crafted email from one of the editors. In this way – on a social level – the decision to say "yes" was markedly influenced by this initial invitation. Hopefully by engaging with the editors, and fellow authors of this book, there will be several meaningful and interesting collaborative experiences and opportunities that can now envisaged and looked forward to in future.

Conclusion: Futures

It would appear that we are entering not only the Anthropocene, a geological period during which human activity is the dominant influence on climate and environment, but the Necrocene, the Age of Death, in which Empire works inevitably against the Earth.

Marshall, 2019, p. x

We finish this chapter by calling for more *anarchist* geographies to be brought into being: geographies that will be folded into an energized emancipatory praxis that can actively work to address the crises that threaten to end the world as we know it. The success of this vision will certainly be dependent on ongoing support and contributing to the intersectional anarchist struggles including "anti-capitalism, radical environmentalism, queer liberation, anti-militarism, prison abolition, information freedom and free speech, freedom of movement, anti-racist, anti-fascist and labour struggles" (Wigger, 2016, p. 133). In this context, and linked to an ongoing desire to further decolonize the geography curriculum (see Lopes de Souza, 2019; Ferretti, 2020), and the Eurocentric canon of classical anarchism, we hope to see productive engagements with Black anarchism – and contributions from – Black anarchists in these critical dialogues and debates come to the fore (see Black Rose Anarchist Federation (2016). Here it is important not to overlook the several important interventions that have been intent on decolonizing anarchism: each of which deserve greater recognition, and further attention from critical geographers (see Galian, 2020). This certainly includes Ramnaths' alternative perspectives on anticolonial movements in India (2011); Galvan-Alvarez et al., focus on decolonizing the state (2020); and broader attempts "...to develop an anarchism that can both fight white supremacy and articulate a positive vision of cultural diversity and cultural exchange" (Black Ink, 2018, n.p.).

In this context, we also see the task of decolonizing euro-centric anarchist though and practice, and foregrounding Indigenous movements in defence of the commons (Clark, 2019) as being central to developing a wider inter-species justice agenda. We would certainly welcome anarchist geographies embracing more fully the struggles for inter-species justice; struggles that fold in the more than-human worlds that we are dependent upon. Indeed, research that contributes to anarchism's posthuman futures (Cudworth and Hobden, 2018; White, 2015) might prove to be some of the most vital and urgent contributions of them all. Some examples to think about here include zoonotic diseases (Cudworth et al., 2021), capitalism and farmed animals (White, 2017), insect population collapse (Gunderman and White, 2021), appeals for total liberation (Anonymous; Springer, 2021). Such ambition will also present a new cluster of challenging theoretical, methodological and ethical research dilemmas that are only just beginning to be through in the critical social sciences. How can we gain consent of other-than-human beings? How can we create meaningful spaces to represent their voices and experiences, desires and fears? As importantly – and indeed a question that accompanies all research focused on social and spatial justice – it is never just enough to disseminate knowledge, but to do so in a way that gets people to really *care*, and then act.

Perhaps the COVID-19 crisis will prove a catalyst to new and powerful expressions of inter-species care and relationality? Certainly, it provides with absolute clarity that we must not return to the old norms that have wrought misery, havoc, and destruction on the world and all life within it. Research in this context must be potentially everywhere, seeking to transform and overturn all sources of oppression, domination, and exploitation where they emerge. What is certain is that we need new imaginaries, new questions, new methods, new ways of seeing and being in the world. And if we are not prepared to embrace this radical direction now, then when? As the editorial collective for the journal *ephemera* (2020), writing in response to the current Covid-19 pandemic, argued:

> Crisis is not only something new and transitory, it also crystallizes and reinforces existing inequalities. But crises may also hold a potential to create other political imaginaries and new organizational realities. We hope that the decision to slow down the journal enables the ephemera community to make time and space to collect itself, to listen, to read, to reflect in a way that makes us part of a transformation rather than being simply controlled by it. Crises raise important questions: what have we been doing and what are we doing now as academics, as citizens, as friends, as parents, and as members of different communities? This crisis may be a moment to pause, to (re)

think what kind of life we want, what kind of work we want, and what kind of care we can give and receive.

Let us be emboldened by this invitation to begin anew, and do so by taking strength and inspiration from others where it is found. If we can consciously, and creatively orientate ourselves and our research and our methodologies in ways that demand the impossible, we can act as a catalyst through which new forms of emancipatory praxis can to be brought forward into the world. That would have something to do with anarchist geography, wouldn't it?

References

A Collective of Anarchist Geographers (2017) Beyond electoralism: Reflections on anarchy, populism, and the crisis of electoral politics. *ACME: An International Journal for Critical Geographies,* 16(4), 607–642. Retrieved from https://acme-journal.org/index.php/acme/article/view/1571

Anonymous (2019) *Total Liberation*. Active Distribution.

Amster, R. (2018) Anti-Hierarchy. In Franks, B., Jun, N., and Williams, L. (2018). *Anarchism: A Conceptual Approach*. Routledge: London (pp. 15–28).

Berger, R. (2015). Now I see it, now I don't: Researcher's position and reflexivity in qualitative research. Qualitative Research, 15(2), 219–234. https://doi.org/10.1177/1468794112468475

Black Ink, (2018) *Ashanti Alston's Black Anarchism*. Available at https://blackink.info/2018/03/08/ashanti-alstons-black-anarchism/ [last accessed 01.07.2021]).

Rose, B. Anarchist Federation (2016) Black Anarchism.. A Reader by Black Rose Anarchism Federation.

Bray, M. (2018) Horizontalism. In B. Franks, N. Jun and L. Williams (Eds). *Anarchism: A Conceptual Approach*. Routledge: London (pp 101–114).

Brock, A., (2020) 'Frack off': Towards an anarchist political ecology critique of corporate and state responses to anti-fracking resistance in the UK. *Political Geography*, 82, 1–15.

Clark, J. (2019) *Between Earth and the Empire: From the necrocence to the beloved community*. PM Press: Edinburgh.

Clare, N. White, G. and White, R.J. (2017) "Striking Out! Challenging academic repression in the neo-liberal university through alternative forms of resistance: Some lessons from the United Kingdom. In A.J. Nocella and E. Juergensmeyer (ed) (2017) *Fighting academic repression and neoliberal education: Resistance, reclaiming, organizing, and Black Lives Matter in education* Peter Lang Publishing: New York (pp. 15–32).

Cudworth, E. and Hobden, S. (2018) Anarchism's Posthuman Future. *Anarchist Studies* 26.1. pp 79–104.

Cudworth, E., Boisseau, W. and White, R.J. (2021) "Guest editorial", *International Journal of Sociology and Social Policy*, 41(3/4), 265–281. https://doi.org/10.1108/IJSSP-04-2021-514.

de Souza, M. L., White, R. J., and Springer, S. (eds) (2016) Subverting the meaning of "theory". *Theories of Resistance: Anarchism, Geography and the Spirit of Revolt*. London: Rowman & Littlefield.

de Souza, M. L (2019) Decolonising postcolonial thinking. *ACME: An International Journal for Critical Geographies*, 18(1), 1–24. Retrieved from www.acme-journal.org/index.php/acme/article/view/1647

DeLeon, A. P. (2019) Anarchy and qualitative methods. In *Oxford Research Encyclopedia of Education*. https://doi.org/10.1093/acrefore/9780190264093.013.345

ephemera collective (2020) ephemera's response to the coronavirus crisis. *ephemera*. Retrieved from www.ephemerajournal.org/contribution/ephemera%E2%80%99s-response-coronavirus-crisis

England, K. (2006) Producing feminist geographies: Theory, methodologies and research strategies. In S. Aitken, & G. Valentine (Eds.), *Approaches to Human Geography*. SAGE: London. www.doi.org/10.4135/9781446215432.n27 (pp. 286–297)

Ferretti, F. (2016) Organisation and formal activism: Insights from the anarchist tradition. *International Journal of Sociology and Social Policy*, 36(11/12), 2016, 726–740.

Ferretti, F. (2018) Teaching anarchist geographies: Elisée Reclus in Brussels and "The Art of Not Being Governed", *Annals of the American Association of Geographers*, 108:1, 162–178, DOI: 10.1080/24694452.2017.1339587

Ferretti, F. (2020) History and philosophy of geography I: Decolonising the discipline, diversifying archives and historicising radicalism. *Progress in Human Geography*, 44(6), 1161–1171. doi:10.1177/0309132519893442

Ferrell, J. (1998) Against the law: Anarchist criminology. *Social Anarchism*, 25, 5–23.

Ferretti, F. (2019) *Anarchy and Geography: Reclus and Kropotkin in the UK*. Routledge: London.

Feyerabend, P. (1993) *Against Method*. London and New York: Verso.

Feyerabend, P. (2011) *The Tyranny of Science*. Cambridge: Polity Press.

Fleming, M. (1980) Propaganda by the deed: Terrorism and anarchist theory in late nineteenth-century Europe. *Studies in Conflict & Terrorism*, 4(1–4), 1–23.

Franks, B., Jun, N., and Williams, L. (2018) *Anarchism: A Conceptual Approach*. Routledge: London.

Galián, L. (2020) Decolonizing Anarchism. In: *Colonialism, Transnationalism, and Anarchism in the South of the Mediterranean. Middle East Today*. Palgrave Macmillan: Cham. https://doi.org/10.1007/978-3-030-45449-4_2

Galvan-Alvarez, E. Birk Laursen, O. and Ridda, M. (2020) Decolonising the state: Subversion, mimicry and criminality, *Postcolonial Studies*, 23(2), 161–169, DOI:10.1080/13688790.2020.1752356

Graham, R. (2015) *We do not fear anarchy, we invoke it: The First International and the origins of the anarchist movement*. AK Press: Oakland and Edinburgh.

Greenaway, J. (2011) Preface: Sexual anarchy, anarchophobia and dangerous desires. In J. Heckert and R. Cleminson (eds). *Ethics, Relationships and Power*. Routledge: New York.

Gordon, U. (2008) *Anarchy Alive! Anti-Authoritarian Politics from Practice to Theory*. Pluto: London.

Gunderman, H. and White, R. (2021) "Critical posthumanism for all: A call to reject insect speciesism", *International Journal of Sociology and Social Policy*, 41(3/4), 489–505. https://doi.org/10.1108/IJSSP-09-2019-0196

Honeywell, C. (2011) *Anarchism*. Polity Press: Cambridge.

Ince, A. and de la Torre, G.B. (2016) For post-statist geographies. *Political Geography*, 55, 10–19.

Ince, A. and White R.J. (2020) Activist Geographies. In H. Wilson and J. Darling (eds) *Research Ethics for Human Geography*, SAGE: London.

Kinna, R., (2017) Anarchism and feminism. In N. Jun (Ed.) *Brill's Companion to Anarchism and Philosophy*. Brill: Leiden (pp. 253–284).

Kowal D.M. (2019) Anarcha-Feminism. In: Levy C., and Adams M. (eds) *The Palgrave Handbook of Anarchism*. Palgrave Macmillan: Cham (pp. 265–279).

Kropotkin, P. (1902/1915) *Mutual Aid: A factor of evolution*. William Heinemann: London.

Leeder, E. (n.d./ 1977) Feminism as an anarchist process: The practice of anarcha-feminism. The anarchist library: Anti-copyright. Retrieved from https://theanarchistlibrary.org/library/elaine-leeder-feminism-as-an-anarchistprocess

Levy, C (2019) Introduction. In C. Levy and S. Newman (eds) *The Anarchist Imagination: Anarchism Encounters in the Humanities and the Social Sciences*. Routledge: London.

Lopes de Souza, M. (2019) Decolonising postcolonial thinking. *ACME: An International Journal for Critical Geographies*, 18(1), 1–24. Retrieved from https://acme-journal.org/index.php/acme/article/view/1647

Lumsden, K. (2019) *Reflexivity: Theory, Method, and Practice*. Routledge: London.

MacLaughlin, J. (2016) *Kropotkin and the Anarchist Intellectual Tradition*. Pluto Press: London.

Marshall, P. (2019) 'Foreword'. In J.P. Clark. *'Between Earth and Empire: From the Necrocene to the beloved Community'*. PM Press: Edinburgh.

Munn-Giddings, C. (2006). Links between kropotkin'theory of 'mutual aid'and the values and practices of action research. *Educational Action Research*, 9(1), 149–158.

Nocella, A.J. and Juergensmeyer, E. (eds) (2017) *Fighting Academic Repression and Neoliberal Education: Resistance, Reclaiming, Organizing, and Black Lives Matter in Education*. Peter Lang Publishing: New York.

Orenstein, D. and Luken, P. (1978) Anarchistic methodology: Methodological anti-authoritarianism as a resolution to paradigmatic disputes in the social sciences. *Sociological Focus*, 11(1), 53–68.

Parker, M. (2014) University, Ltd: Changing a business school. *Organization*, 21(2), pp. 281–292.

Parker, M., Cheney, G., Fournier, V., & Land, C. (2014) Advanced capitalism: Its promise and failings. In *The Routledge Companion to Alternative Organization* (pp. 27–41). Routledge: London.

Pelletier, P. (2009) *Élisée Reclus, géographie et anarchie*. Les Editions du Monde Libertaire: Paris.

Ramnath, M. (2011) *Decolonizing Anarchism: An Antiauthoritarian history of India's liberation struggle*. AK Press: Edinburgh.

Reclus, E. (1901 [1996]) On Vegetarianism. Available at https://theanarchistlibrary.org/library/elisee-reclus-on-vegetarianism

Rouhani, F. (2012). Anarchism, geography, and queer space-making: Building bridges over chasms we create. *ACME: An International Journal for Critical Geographies*, 11(3), 373–392. Retrieved from https://acme-journal.org/index.php/acme/article/view/938

Sidaway, J.D., White, R.J., Barrera de la Torre, G., Ferretti, F., Crane, N.J., Loong, S., Knopp, L., Mott, C., Rouhani, F., Smith, J.M. and Springer, S. (2017) The anarchist roots of geography: Toward spatial emancipation. *The AAG (American Association of Geographers) Review of Books*, 5(4), 281–296.

Smith, T.S.J. (2015) Anarchism and non-representational theory in the social sciences. *E-International Relations* .www.e-ir.info/2015/08/07/anarchism-and-non-representational-theory-in-the-social-sciences/ pp.1–6.

Springer, S. (2012) Anarchism! What geography still ought to be. *Antipode*, 44*(5)*, 1605–1624.

Springer, S., (2013) Anarchism and geography: A brief genealogy of anarchist geographies. *Geography Compass*, 7(1), 46–60.

Springer, S. (2014) Human geography without hierarchy. *Progress in Human Geography*, 38(3), 402–419.

Springer, S. (2016a) *The Anarchist Roots of Geography*. Minnesota Press: Minneapolis.

Springer, S. (2016b) Fuck Neoliberalism. *ACME: An International Journal for Critical Geographies*, 15(2), 285–292. Retrieved from https://acme-journal.org/index.php/acme/article/view/1342

Springer, S. (2020) Caring geographies: The COVID-19 interregnum and a return to mutual aid. *Dialogues in Human Geography*, 10(2), 112–115.

Springer, S., White, R. J., and de Souza, M. L. (eds) (2016) *Anarchism, Geography, and the Spirit of Revolt. The Radicalization of Pedagogy (Vol.1)*. Rowman & Littlefield: Lanham.

Springer, S. (2021) Total liberation ecology: Integral anarchism, anthroparchy, and the violence of indifference. In S. Springer, J. Mateer, and M. Locret-Collet (eds). *Undoing Human Supremacy Anarchist Political Ecology in the Face of Anthroparchy*. PM Press: Oakland (pp. 235–253).

Stenglein F. and Madar, S. (2016) Cycling diaries: Moving towards an anarchist field trip pedagogy. In S. Springer, R.J. White and M.L. De Souza (eds) *Anarchism, Geography, and the Spirit of Revolt. The Radicalization of Pedagogy (Vol.1)*. Rowman & Littlefield: Lanham (pp. 223–246).

Veron, O (2016) (Extra) ordinary activism: Veganism and the shaping of hemeratopias. *International Journal of Sociology and Social Policy*, 36(11/12), 756–773.

Ward, C. (1996) *Anarchy in Action*. Freedom Press: London.

Wieck, D.T. (1972) 'Preface'. In G. Baldelli, *Social Anarchism*. Penguin: London.

Wigger, A. (2016) Anarchism as emancipatory theory and praxis: Implications for critical Marxist research. *Capital & Class*, 40(1), 129–145. https://doi.org/10.1177/0309816816628247

White, R.J. (2015) Following in the footsteps of Élisée Reclus: Disturbing places of inter-species violence that are hidden in plain sight. In A. Nocella, R.J. White, and E. Cudworth (ed) *Anarchism and Animal Liberation: Critical Animal Studies, intersectionality and total liberation*. McFarland Press: Jefferson.

White, R.J. (2017) Rising to the challenge of capitalism and the commodification of animals: Post-capitalism, anarchist economies and vegan praxis. In D. Nibert (eds) *Animal Oppression and Capitalism*. Praeger: Conneticut.

White, R.J., and Cudworth, E. (2014) Taking it to the streets: Challenging systems of domination from below. In A. Nocella, J. Sorenson, K. Socha. A. Matsuika (eds) *Critical Animal Studies Reader: An Introduction to an Intersectional Social Justice Approach to Animal Liberation*. Peter Lang Publishing Group: New York (pp. 202–219).

White, R.J. and Springer, S. (2018) Making space for anarchist geographies in critical animal studies. In J. Sorenson and A. Matuoka (eds). *Critical Animal Studies: Towards Trans-species Social Justice*. Rowman and Littlefield International: London.

White, R.J., Springer, S., and de Souza, M.L (2016) Introduction. Performing anarchism, practicing freedom, pursuing revolt. In R.J. White, S. Springer and M.L. de Souza (eds) *The Practice of Freedom: Anarchism, Geography, and the Spirit of Revolt*. Roman & Littlefield: New York (pp. 1–22).

White, R.J. and Williams C.C. (2017) Crisis, capitalism and the anarcho-geographies of community self-help. In A. Ince and S.M. Hall (eds) *Sharing Economies in Times of Crisis: Practices, Politics and Possibilities*. Routledge: London.

Woodcock, G. (1988) '*Élisée Reclus: An introduction*'. In M. Fleming. *The Geography of Freedom*. Black Rose Books: Quebec (pp. 11–18).

PART III

Cross-Cutting Issues in Human Geography Methodologies

Stephanie E. Coen

Part III zooms out from the sub-disciplines of human geography to explore some of the cross-cutting methodological challenges and debates linked to the implicit and explicit issues that human geographers contend with in doing research. Bringing together key points of methodological concern that shape both the design and practice of geographical inquiry, the chapters in this section prompt us to think about the values and goals that underpin our work. While many of these issues are applicable to social sciences research more widely, a geographical perspective brings attention to the contextualized nature of methodologies and methodological choices. The chapters here each grapple with dimensions of the 'situatedness' of the research process and what this means for methodology in terms of ethics, representation, participation, agency, power, gatekeepers, positionalities, tokenism, audience, and impact. At the heart of these issues, is the question of community: Who are we engaging in our research? Who is our research 'for'? What are our relationships with communities? The chapters in this section offer insights into the ways that geographical methodologies can centre this fundamentally human—or rather humane—question.

One of the defining methodological features of geography as a discipline is the tradition of fieldwork. Chapter 26 by Hopkins and Finlay and Chapter 27 by Kasanga, Bisung, and Luginaah interrogate the complexities of urban and rural fieldwork, respectively. Without wishing to reproduce an urban-rural binary, these chapters in tandem highlight particular methodological considerations for deploying geographical research in the fluid terrains of urban and rural places. Hopkins and Finlay draw our attention to the ethical and methodological complexities of engaging with local institutions (e.g., schools, government offices, religious centers) when researching sensitive and politicized issues in a minority world Anglo-urban setting. Reflecting on 'the rural' as a constantly evolving socio-spatial construct for geographic research, Kasanga et al., critically consider the porous boundaries of insider/outsider identities, navigating local norms, and building trust within the context of shifting and socio-political relations. Illustrated through first-hand examples of lived experience and fieldwork in Kenya, Malawi, and Ghana, the authors put forward key considerations, grounded in practical examples, for navigating everyday challenges of fieldwork in multifaceted rural areas.

Participatory methodologies have become an increasingly common feature of human geography research, but many thorny questions remain about how to realize meaningful

DOI: 10.4324/9781003038849-29

participation. In Chapter 28, Castleden and Sylvestre prompt us to reflect on what we mean by developing equitable research relationships in participatory frameworks that aim for justice and positive change. They take us into the balancing act between "tea-drinking" diplomacy and "shit-kicking" that occupying activist-scholar positions in the academy can entail. This chapter reminds us that geographical research is situated in layers of sometimes competing contexts, between the neoliberal university and the social change it seeks to affect that ultimately shape our approaches in, at times, necessarily strategic ways.

Along with participation, representation is a perennial methodological question for human geographers. In Chapter 29, Thompson invites us to critically interrogate the concept of 'lived experience' in human geography and consider what it means to represent experiences as 'lived.' Thompson illustrates how a geographical sensibility is necessary for contending with the multiplicity of (re)presenting lived experience in different material, social, and cultural circumstances. Taking another angle on representation, Gorman and Andrews (Chapter 30) tackle the tricky methodological question of 'going beyond representation' in geographical research. They offer a comprehensive treatment of non-representational theory, digestible for the less initiated, and put forward several methodological provocations for ways to empirically grapple with non-human agency and the in-the-moment 'energy' of phenomena.

The practice of geographical research is invariably situated within the context of our lives as academics and individuals in the wider world. In Chapter 31, Loyd—together with a co-author collective of graduate students and tenured faculty from US and Canadian universities—asks us to think about what it means methodologically to undertake geographical research in a moment of crisis through the prism of the COVID-19 pandemic. Indeed, this book was written and edited during the pandemic—a context that shaped our collective capacity and pace in realizing this project. Using the concept of feminist slow scholarship, Loyd and team identify feminist care ethics, attention to temporality, and embodied emotional labor as dimensions of how feminist slow scholarship offers a way of doing "care-full" research to change the university and world around us. As geographers, we are especially well-positioned to consider the contextual impacts of our ways of working and doing research—but making decisions that support our wellbeing, and the wellbeing of our collaborators and participants, can be challenged by the neoliberal academy that sometimes persists as if our lives and the world outside of the academy do not apply.

This section concludes with the overarching methodological questions of audience and impact. What change do we want our research to affect, and for whom? In thinking through moving research beyond the academy, DeLoyde and Mabee (Chapter 32) present tenets for translating geographical research into effective policy development. They highlight actionable points of intervention in traditional research methodologies to engage decisionmakers and policy makers in the knowledge production and dissemination cycle. This chapter offers practical ways that human geographers seeking to influence policy can improve the utility and relevancy of their work for decision-makers and, in particular, draws attention to compatibility issues across scales of research and the scales at which policymakers operate. As the concluding section to this volume, Part III leaves us to consider how our research practices are embedded in wider networks of relationships—with our communities, our collaborators, our institutions, decision-makers, and ourselves. These chapters point to the need for dialogue, collaboration, and co-production across academia and other sectors.

26

POLITICS, INSTITUTIONS AND PLACE

Researching Sensitive Subjects in Urban Contexts

Peter Hopkins and Robin Finlay

Introduction

Urban research often involves understanding and negotiating complex political relationships and institutions, together with the social, political, cultural and economic relations that shape them. In this chapter, we explore some of the ethical and methodological challenges involved in conducting research in urban contexts. We pay attention specifically to the relationships among politics, institutions and place. We focus on the interconnected scales of local, national and global politics in urban contexts and the political figures who are involved. We next explore the contested role of urban institutions as places in which to do research. The political complexity of urban areas means that, while they are a deeply interesting and engaging context in which to do research, many methodological conundrums are likely to present themselves.

When conducting research in urban contexts, it is crucial to demonstrate a sensitivity to everyday political complexities. Failing to be attentive to the issues can lead to gatekeepers restricting access to specific communities, to research foci being challenged and misrepresented, and worse still, to researchers engaging in ethically questionable practices that threaten the wellbeing of both their participants and themselves. In this chapter, we draw upon our own experiences of urban fieldwork with marginalised groups, discussing sensitive topics in order to set out some of the interwoven challenges involved in doing research in cities.

Politics, Politicians and Power

A crucial set of ethical and methodological considerations regarding researching urban contexts concerns the need for in-depth knowledge and understanding of complex political issues that relate to various scales and specific politicians or political parties. The sensitive or politicised nature of the research topic often adds a layer of ethical and methodological complexity to any study; this is heightened when working with marginalised groups, as a result of their gender, age, sexuality, race, ethnicity, religion, nationality, immigration status or political persuasion, or a combination of these identities. We often see tensions play out in elections and referenda,

DOI: 10.4324/9781003038849-30

such as in the build up to polling, then in its aftermath. Consider the ways in which, for example, race and migration were brought into debates in the United Kingdom about the Brexit vote (Burrell et al., 2019) when the UK voted in 2016 to leave the European Union, or in the Scottish Independence Referendum of 2014, when some ethnic minority groups were forced to question their sense of belonging in urban Scotland (Botterill et al., 2016). Many of the tensions associated with these elections live on afterwards and can create divisive and politically charged contexts in which to do research.

An example of these complexities is a project where we explored young people's narratives of politics and place with a specific focus on Sunderland, a post-industrial city in North East England often referred to as 'left behind' and where 61 per cent voted to leave the European Union. Some 52 young people aged 14 to 25 participated in the study, together with a group of six adults and six local stakeholders who either provided services in the local area or were engaged in local politics (Finlay et al., 2020). We found that far-right political activity was a key feature of the urban landscape; more than 13 far-right demonstrations were reported over the 13-month period from 2016. The groups targeting the city included the English Defence League (EDL), Britain First, and the Democratic Football Lads Alliance (DFLA). As such, there was a volatility, sensitivity and instability to social relations in this largely white city, with young people regularly discussing racism and Islamophobia as serious concerns in their everyday lives (Finlay et al., 2020).

In a study about Islamophobia and anti-Muslim hatred in North East England (Hopkins and Clayton, 2020), almost three-quarters of the survey respondents indicated that they felt that Islamophobia was 'getting worse'. As the report noted about this situation:

> To account for the worsening Islamophobia, there was mention of the rise of far-Right politicians (both mainstream and fringe figures), as well as Brexit (the North-East voted to leave, with 58% of the vote: only Newcastle voted to remain) and the role of news and social media in emboldening and normalising acts of racism, including attacks on individuals and mosques. Reference was made to international and national politics, the visibility of far-Right groups and the weak reaction of some politicians in local areas.
>
> *Hopkins and Clayton, 2020, p. 6*

We can see in this example that layers of political issues are shaping people's experiences of racism and Islamophobia. Therefore, when doing urban research, the researcher's awareness of and a sensitivity to such issues is vital.

Often, as in the case just described, there are politicians who play a leading and divisive role in generating tension in the cities, regions or nations that they have been elected to serve. In his first few days in office, former US President Donald Trump introduced several executive actions that further marginalised and stigmatised minority populations. These included an indefinite ban on Syrian refugees, a 'Muslim ban' that aimed to prevent travel from Muslim-majority countries and cuts in funding to 'sanctuary cities', those that did not agree to detain and deport immigrants (Gökariksel and Smith, 2017). As a result:

> By 6 March 2017, he had signed 34 executive orders, presidential memoranda, or proclamations that restrict the rights of women, immigrants, Muslims, and Native Americans while relaxing regulations on manufacturing companies, increasing support for law enforcement and the military, and moving towards dismantling the 'administrative state'.
>
> *Gökariksel and Smith, 2017, p. 630*

There may also be locally elected representatives, councillors, senators and city mayors who are divisive and introduce problematic policies and practices that marginalise already excluded groups. For example, in his role as New York Mayor, Rudy Giuliani introduced an aggressive form of 'broken windows' policing. Although crime rates may have fallen as a result, Cahill et al., (2019) exposes how this approach to law enforcement, coupled with global urban restructuring, served to criminalise, marginalise and dehumanise young people of colour in the city. In another case, we can see exclusion at play in Cape Town's changes to urban governance, leading to the marginalisation of street youth (van Blerk, 2013). The tendency to alter policies to force street youth into less visible parts of the city made these people more open to abuse, restricted their access to city services and risked assumptions that the 'problem' of street youth had been addressed.

Doing research about sensitive topics in cities involves careful interpersonal negotiations, as in these examples, because researchers often explore the issues of social groups to which they do not necessarily belong. Debates on reflexivity and positionality have problematised the assumption of a simplistic outsider/insider binary (Mohammad, 2001), recognising that researchers occupy a position of 'inbetweenness' (Nast, 1994) and are frequently neither a complete insider nor a complete outsider. Nonetheless, in the field, people may make assumptions about the political persuasion or positionality of the researcher or team (and vice versa). These may require diplomatic conversations with gatekeepers and potential participants to establish trust in the researcher's intentions. During a focus group forming part of a study with young Pakistani Muslim men in urban Scotland, for example, the first author was aggressively quizzed by a potential participant over whether he was affiliated to the far-right British National Party (Hopkins, 2007a); he reassured them about the anti-racist approach being taken and his disdain for far-right politics and politicians.

In relation to positionality, some topics may encourage participants to demonstrate a denial about their involvement in urban social and political movements or issues that might be stigmatising or potentially controversial. For example, it is widely accepted that in the West of Scotland there is a problem with inter-Christian bigotry and prejudice. This tends to take the form of tension between Catholics and Protestants, often associated with the rivalry between the Rangers and Celtic football clubs in Glasgow and thus associated with white, working-class masculinities. In contrast, Lindores and Emejulu (2017) find that care work is political and that women are active agents in perpetuating sectarianism through family and community relationships, including on the issues of marriage and the upbringing of children. Despite the presence of sectarianism, Goodall et al., (2015) find that there is a discursive deficit: some participants are not used to talking about sectarianism and others claim to be unsure what the term means. In a review by Morrow (2017) of the work by the Independent Advisory Group on Tackling Sectarianism in Scotland, the key finding is the challenge of denial, blame and defensiveness that often arises when sectarianism comes up in discussion.

Sectarianism, bigotry and prejudice within cities can often be traced back to higher level policies. Decisions relating to immigration policy and practice can shape the ethnic and cultural diversity of cities, thereby influencing urban social and political relations, service provision and inter-community relations. An example can be found in the UK dispersal programme for asylum seekers and refugees, where new arrivals are dispersed to cities across the United Kingdom that have the capacity to accommodate new arrivals (Phillips, 2006). The challenge is that some refugees find themselves in large, diverse, multicultural cities with services that are relevant to them, experienced service providers who are knowledgeable about their issues and perhaps even refugee community groups of interest to them. By contrast, others find themselves where racism and hostility are prevalent, services and expertise are lacking, and it takes longer to find out about relevant organisations and opportunities. Some cities have a combination of such attributes, with variations across and within specific neighbourhoods.

Some cities and urban neighbourhoods develop a reputation as 'cities of sanctuary' for refugees and asylum seekers or as offering refugees high-quality services that are held up as examples of good practice. In others, not afforded this status, white nationalism and racism flourish. Indeed, sometimes these contexts are co-located and the claims of welcome, inclusion and integration are contested. This presents diverse ethical and methodological challenges for urban researchers. In our study in Sunderland mentioned above, the presence of far-right political groups and marches contrasted with that of anti-racist groups and collectives such as Sunderland Unites, which operates against all forms of discrimination under the motto 'Love Sunderland, hate racism'.

Adding to the complexity of urban political geographies is the continued unfolding of the COVID-19 pandemic. In cities around the world, we are seeing a reconfiguration and often worsening of social and economic inequalities and a growing vulnerability of marginalised and minority populations (Ho and Maddrell, 2021). These trends are causing a range of political struggles and political activism around the rights and welfare of urban populations. For instance, our work with asylum seekers and refugees in Newcastle and Glasgow during the pandemic has revealed complex urban socio-political struggles regarding the welfare and governance of asylum-seeker populations (Finlay et al., 2020). In particular, the way in which the pandemic has increased the vulnerability and exclusion of displaced people is creating a volatility and sensitivity in the urban politics of asylum and migration. In these moments of significant crisis, negotiating the urban political landscape in research projects may become fraught and complex. At the same time, the evolving situation provides a fast-changing landscape where there are fruitful opportunities to conduct valuable and critical research. In a sense, the way in which inequalities are extended and hardened by the pandemic lends ethical urban research on COVID-19's exclusionary impacts on specific urban neighbourhoods and communities an increased urgency and importance.

In this section, we have given examples to illustrate how research in urban contexts encounters various localised political issues and political claims, and to show how these intersect with broader national structures and global political affairs. This raises a range of ethical and methodological considerations about how to conduct urban research. To forge ethical and productive collaborations with local institutions and local interlocutors, it is crucial to have an understanding and sensitivity to a range of urban political matters.

Negotiating the Contested Spaces of Urban Institutions

We now focus on the effect of urban institutions on undertaking sensitive qualitative research in urban contexts. Philo and Parr (2000, p. 514) distinguish between the geographies *of* institutions (the spread of types of institutions) and geographies *in* institutions (the internal social and spatial relations within institutions). An awareness of both is often vital in conducting high-quality urban research; however, we contend that appreciating the relationality between the various urban organisations is also important, as there are often complex political, territorial, financial and competitive politics at play.

Diverse institutional contexts have been the focus of geographical research. Examples are research about schools (Collins and Coleman, 2008), universities (Hopkins, 2011) and youth work settings (Mills, 2015), also women's refuges (Bowstead, 2019), call centres (Blumen, 2019) and religious institutions (Ehrkamp and Nagel, 2017). Many types of institutions are located in cities, particularly those recognised as more formal or state-sponsored, for instance schools, colleges, universities, town halls and local government offices. In addition, there are third-sector, voluntary and community groups. For many people, these are increasingly critical institutions, given the neoliberal withdrawal or privatisation of many services previously provided by the state (van Lanen, 2017). These changes often lead to widening urban social and spatial inequalities, and local action groups, religious organisations and community associations have to step in with the intention of providing support that is no longer available from the state.

Whether working directly with the voluntary and community sector, schools, colleges and universities or with religious organisations such as churches, mosques, gurdwaras and synagogues, institutions have an important role in the urban centre in which they are based. Many play a key part in people's negotiations of the city (or have previously done so); even if they have not been involved with a specific institution, they undoubtedly have feelings, views and opinions about each institution, its work and its service (or not) to the community. It is important to note that many of these institutions are not only located within and implicated by the political issues mentioned in the previous section but may provide challenging yet fruitful settings in which to conduct urban research.

A key methodological consideration here is the relationships with gatekeepers; that is, those who hold the power and control in institutions and with whom you need to negotiate to gain access to specific groups or people in order to do your research. In some cases, there are layers or tiers of gatekeepers with whom you might need to liaise, so this can be a time-consuming and complex process. Indeed, many gatekeepers will decline to participate and will refuse access to colleagues, clients or service users. There can be complex politics between city institutions. Some may compete with each other for funding, others may have recently merged due to concerns about resources, and others still may have changed direction under new management and moved into an area or issue that they previously did not serve. Moreover, some may refuse to work with each other or, for ethical or political reasons, with a researcher whom they associate with an organisation for which they do not work. Alongside these issues, some may have a specific motivation for undertaking the work that they do. They may have a statutory or legal responsibility to local, provincial or state authority to provide certain services (e.g., education or housing for specific groups). Others have a feminist motivation to provide spaces of safety and security for women in dangerous situations (e.g., women's refuges), while other organisations still are driven by religion or belief to provide assistance to the less fortunate, such as those experiencing homelessness or who are short of food.

Sanghera and Thapar-Bjökert (2008) observe that the relationship between the researcher and the gatekeeper 'is important, as the gatekeeper is often the first point of contact in the field research process, yet it is a relationship that is fraught with inconsistencies and instabilities'. They point to the increased significance of such relationships when the research takes place in a contested or controversial political context. Furthermore, the establishment of such relationships is 'not easily quantifiable; rather it is much more about perceptions, fears and distrust of those "inside" and those perceived as "outsiders"' (p. 548). Many gatekeepers have a strong desire to protect the communities that they work with, especially when these already occupy a marginalised place in the city. This is likely to be heightened in relation to political events and decisions such as those associated with austerity cutbacks or the COVID-19 pandemic. Moreover, many gatekeepers have pressures on their time and availability that may not be obvious to researchers; even when access is granted and a gatekeeper appears cooperative with the aims of the research, further negotiations may be needed with other layers or permissions, or with senior managers or heads of division.

Building effective relationships with gatekeepers requires sensitivity to these issues. As previously discussed, this is often informed by, and interwoven with, the role of positionalities in the field (see England, 1994; Fisher, 2015; Mohammad, 2001). 'Positionality frames social and professional relationships in the research field and also governs the "tone" of the research' (Sanghera and Thapar-Bjökert, 2008). Indeed, the issue of positionality represents an ethical issue for urban researchers (Hopkins, 2007b). While focusing too much on the self poses a risk of thinking reflexively about positionality (Kobayashi, 2003), considerations of positionality are important as they alert us to the ways in which a researcher's social identities may have an impact vis-à-vis the researched, and the related ideas of power, privilege and social positioning. Positionality is therefore also about politics and place.

The need to build relationships with gatekeepers and considerations of positionalities in the field mean that we have to think of a means of working with urban institutions that brings mutual benefit to all involved. This often relies on the researcher engaging in additional work, as it may be that an additional layer of approval, additional advice or support to the organisation is required. It may be that an alternative output is more appropriate to the remit of the organisation that is being worked with. Some gatekeepers may be sensitised to what Alexander (2000: 228) refers to as 'the more usual parachute-in-interview-and-escape route to empirical research', hence in return for providing access to potential participants they will want to know what you are offering.

An example that brings these debates to the fore is our project with ethnic and religious minority young people in urban Scotland (Hopkins et al., 2017). Much of this project took place in secondary or high schools, where access must be negotiated with the local authority office before specific schools are approached directly. Upon receiving approval from Glasgow City Council and then from the headteachers of the schools, we worked in several schools to conduct focus groups and interviews with pupils. In one, the senior member of staff wanted to know more about the research that we were doing, why we were doing it and what types of relationships we wanted to develop with the pupils. The local social and political context was crucial, as this school is a non-denominational state school that is highly diverse in terms of pupil ethnicity and migration history. It was important that we demonstrated our knowledge and understanding of the diversity of the school and also our approach to anti-racism. This gatekeeper arranged for a small group of students to be consulted on whether the research should be allowed to take place. It was agreed that we would ultimately provide the school with the final report and discuss our findings with the teaching staff at a teacher training day.

Upon receiving approval from the students and from the headteacher, this senior member of staff spent a considerable amount of time arranging several focus groups and over 20 interviews with young people from diverse ethnic and religious backgrounds, securing parental consent for the younger pupils. At a meeting following the completion of the final interview, the gatekeeper with whom we had now developed a close working relationship said, 'Well, what are you going to do for us now? I've spent the last few weeks running around arranging focus groups and interviews for you!'

Given our commitment to ethical research that is mutually beneficial to all, we discussed this in detail to explore the possibilities and consider what would be of most interest to the school. It was agreed that we would work with this gatekeeper and one of her classes to co-develop a protocol for how pupils would like to be treated by researchers. This would include a series of sessions in class, a visit to the university for all involved and the design of the final protocol. In addition, the students were able to launch the protocol with Glasgow City Council and with the Children and Young People's Commissioner for Scotland, and with their teacher and the researcher they co-authored a short publication (see Hopkins, Sinclair and Shawlands Academy Student Research Committee, 2017).

The example above illustrates how changing urban landscape is such that many services previously provided by the state are now either not available, or only on a restricted basis. In some cases, voluntary, community and religious organisations now provide some of these services (e.g., van Lanen, 2017). Many urban institutions have had to restructure and work with significantly reduced resources and staffing. This often creates a complicated ethical landscape that researchers must negotiate when seeking to collaborate. Given the frequent lack of resources, care needs to be taken about the types of requests that researchers make. An institution that is already financially stretched and understaffed may struggle to accommodate certain requests, which may raise questions about the value of the research to the communities that they work with. Thus, it is important for researchers to be well informed about the political dynamics at play in cities and institutions and to use this knowledge to draw up their collaboration strategies.

In this context, we contend that careful consideration needs to be given to the outcomes of the research and how these may benefit the aims and objectives of the institution being worked with, not just leading to the research team completing its publications or to the doctoral researcher finalising their thesis. Many organisations and institutions are unlikely to find academic conferences – often expensive to attend – and complex journal articles – which you normally have to pay for, unless at a university that subscribes – of much use to their ongoing practice or to providing services in the city. It is more ethical in such contexts to provide multiple formats of the outputs to ensure that these are appropriate to the diverse audiences to whom your work is of relevance, as in the case of the school discussed above. Perhaps the institution that you are working with would like you to write a short report, help with a funding bid, brief its board of trustees or management committee or write about the research findings for its newsletter. Perhaps it would be interested in exploring additional opportunities for its clients or service users, or it might even want to develop an exhibition, a set of online resources or a staged performance on the issues raised in the research.

Given how busy they often are, it may be that these institutions are happy just to have helped with your research; however, it may be that they want more. It is our mandate, as ethical urban researchers, to explore these possibilities and, where possible, to offer our assistance. It is not about proving the impact of the research or demonstrating its social benefits, even although these may be the most beneficial outcomes for you, as the researcher; the point is to generate mutually beneficial urban research.

Conclusions

In this chapter, we have unpacked some of the political complexities of urban research, paying attention specifically to the influence of politics and politicians and the role and function of urban institutions. We have argued that skill in the use of specific qualitative methods is only part of the story. In addition, ethical urban researchers require an in-depth knowledge and understanding of relevant politics and politicians, especially when working on sensitive topics or in politically challenging contexts. Furthermore, urban institutions often play a vital role in granting access to knowledge and an understanding of the local issues to other organisations or potential participants. Here, ethical urban researchers need the skills to work in mutually beneficial ways with gatekeepers and to demonstrate sensitivity to the roles that the various positionalities might play in these relationships.

References

Alexander, C. (2000) *The Asian Gang: Ethnicity, identity, masculinity*. Oxford: Berg.

Blumen, O. (2019) On the service frontline: Israeli Arab-Palestinian men in a call-center. *Gender, Place and Culture* 26(2), 203–226.

Botterill, K., Hopkins, P., Sanghera, G., and Arshad, R. (2016) Securing disunion: Young people's nationalism, identities and (in)securities in the campaign for an independent Scotland. *Political Geography* 55, 124–133.

Bowstead, J. (2019) Spaces of safety and more-than-safety in women's refuges in England. *Gender, Place and Culture* 26(1), 75–90.

Burrell, K., Hopkins, P., Isakjee, A., Lorne, C., Nagel, C., Finlay, R., Nayak, Anoop, B., Matthew C., Pande, R., Richardson, M., Botterill, K. and Rogaly, B. (2019) Brexit, race and migration. *Environment and Planning C: Politics and Space* 37(1), 3–40.

Cahill, C., Stoudt, B.G., Torre, M.E., Darian, S., Matles, A., Belmonte, K., Djokovic, S., Lopez, J., and Pimentel, A. (2019) 'They were looking at us like we were bad people': Growing up policed in the gentrifying, still disinvested city. *ACME: An International Journal for Critical Geographies* 18(5), 1128–1149.

Collins, D. and Coleman, T. (2008) Social geographies of education: Looking within, and beyond, school boundaries. *Geography Compass* 2, 281–99.

Ehrkamp, P. and Nagel, C. (2017) Policing the borders of church and societal membership: Immigration and faith-based communities in the US South. *Territory, Politics, Governance* 5(3), 318–331.

England, K. (1994) Getting personal: Reflexivity, positionality, and feminist research. *Professional Geographer* 46(1), 80–89.

Finlay, R., Nayak, A., Benwell, M., Hopkins, P., Pande, R. and Richardson, M. (2020) *Growing up in Sunderland: Young people, politics and place*. Newcastle upon Tyne: Newcastle University.

Fisher, K. (2015) Positionality, subjectivity and race in transnational and transcultural research. *Gender, Place and Culture* 22(4), 456–473.

Gökariksel, B. and Smith, S. (2017) Intersectional feminism beyond U.S. flag hijab and pussy hats in Trump's America. *Gender, Place and Culture* 24(5), 628–644.

Goodall, K., Hopkins, P., McKerrell, S,, Markey, J., Millar, Stephen, R.J., and Richardson, M. (2015) *Community Experiences of Sectarianism*. Edinburgh: Scottish Government Social Research.

Ho, E.L. and Maddrell, A. (2021) Intolerable intersectional burdens: A COVID-19 research agenda for social and cultural geographies. *Social and Cultural Geography* 22(1), 1–10.

Hopkins, P. (2007a) Global events, national politics, local lives: Young Muslim men in Scotland. *Environment and Planning A* 39(5) 1119–1133.

Hopkins, P. (2007b) Positionalities and knowledge: Negotiating ethics in practice. *ACME: An International Journal for Critical Geographies* 6(3), 386–394. www.acme-journal.org/index.php/acme/article/view/787

Hopkins, P. (2011) Towards critical geographies of the university campus: Understanding the contested experiences of Muslim students. *Transactions of the Institute of British Geographers* 36(1), 157–169.

Hopkins, P., Botterill, K., Sanghera, G., and Arshad, R. (2017) Encountering misrecognition: Being mistaken for being Muslim. *Annals of the American Association of Geographers* 107(4), 934–948.

Hopkins, P and Clayton, J. (2020) *Islamophobia and anti-Muslim hatred in North East England*. London: Tell MAMA.

Hopkins, P., Sinclair, C., and Shawlands Academy Student Research Committee. (2017) Research, relevance and respect: Co-creating a guide about involving young people in social research. *Research for All* 1(1), 121–127.

Kobayashi, A. (2003) GPC ten years on: Is self-reflexivity enough? *Gender, Place and Culture* 10(4), 345–349.

Lindores, S. and Emejulu, A. (2017) Women as sectarian agents: Looking beyond the football cliché in Scotland. *European Journal of Women's Studies* 26(1), 39–53.

Mills, S. (2015) Geographies of youth work, volunteering and employment: The Jewish Lads' Brigade and Club in post-war Manchester. *Transactions of the Institute of British Geographers* 40(4), 523–535.

Mohammad, R. (2001) 'Insiders' and/or 'outsiders': Positionality, theory and praxis. In Melanie Limb and Claire Dwyer (eds), *Qualitative Methodologies for Geographers*, pp. 101–114. London: Arnold.

Morrow, D. (2017) *Review of the Implementation of the Recommendations of the Advisory Group on Tackling Sectarianism in Scotland*. Edinburgh: Scottish Government.

Nast, H. (1994) Women in the field: Critical feminist methodologies and theoretical perspectives. *Professional Geographer* 46(1), 54–66.

Phillips, D. (2006) Moving towards integration: The housing of asylum seekers and refugees in Britain. *Housing Studies* 21(4), 539–553.

Philo, C. and Parr, H. (2000) Institutional geographies: Introductory remarks. *Geoforum* 31, 513–521.

Sanghera, G., and Thapar-Bjökert, S. (2008) Methodological dilemmas: Gatekeepers and positionality in Bradford. *Ethnic and Racial Studies* 31(3), 543–562.

van Blerk, L. (2013) New street geographies: The impact of urban governance on the mobilities of Cape Town's street youth. *Urban Studies* 50(3), 556–573.

van Lanen, S. (2017) Living austerity urbanism: Space–time expansion and deepening socio-spatial inequalities for disadvantaged urban youth in Ireland. *Urban Geography* 38(10), 1603–1613.

27

NAVIGATING RURALITIES IN HUMAN GEOGRAPHY RESEARCH

Reflections from Fieldwork in Complex Rural Settings

Moses Kansanga, Elijah Bisung and Isaac Luginaah

Introduction

Rural spaces have undergone significant changes over the past few decades, with prospects for further radical alteration in the coming decades. Geopolitical and environmental changes such as climate change, resource contestations and biodiversity degradation, and population decline will continue to reconfigure rural spaces, demanding greater attention from development planners and researchers. In Africa for example, yields of staple crops such as maize and wheat have decreased in recent years, resulting in food insecurity gaps (IPCC, 2019). Because of weak coping mechanisms, rural subsistence farmers are heavily impacted by reduced crop yields. For example, rural farmers in Ethiopia report poor crop yield due to climate change associated events such as severe and unpredictable rainfall patterns and rising temperatures (Mekonnen et al., 2020; Gebrehiwot, 2013).

In addition to climate change associated impacts on food security, changing population dynamics in rural areas have brought forth new challenges for researchers and policy makers. Urban migration has become a key escape route for rural dwellers who are faced with declining economic opportunities and agricultural shocks. Intensifying rural to urban mobility has led to changes in the social fabric of many rural areas. For example, rural–urban migration has negative effects on food production in rural Ghana, in part due to reduced availability of an active labour force (Luginaah et al., 2009; Rademacher-Schulz et al., 2014). With declining populations, the ability of rural areas to foster economic development and attract key services is significantly diminished. Rural environments in sub-Saharan Africa have become attractive spots for scale environmental development, such as carbon forest projects, biofuel plantations, and other large scale agricultural activities. Because of the livelihood impacts of these developments, new environmental contestations are emerging in many rural spaces.

Finding solutions to these social and environmental changes requires geographers to engage with locally grounded research to showcase the depth and breadth of the problems, understand

DOI: 10.4324/9781003038849-31

the lived and performed experiences of rural residents, and interrogate the motives and actions of macro level actors. At the same time, theoretical innovations in conceptualizing rurality and understanding rural livelihoods over the past few decades, as well as methodological advances in geographical research, have renewed interest in ways researchers navigate rural environments and contested spaces during fieldwork (Appiah, 2020; Adu-Ampong and Adams, 2019; Nyantakyi-Frimpong, 2021). In this chapter, we reflect on some of the theoretical, methodological, and practical tensions and complexities in conducting research in specific rural places. First, we address the implications of current definitions and scope of "rural", both in terms of its spatial focus and social configuration, for geographical research and fieldwork. Second, we offer some methodological reflections on our fieldwork and community engagement in rural spaces and conclude by re-echoing the broader implications for contemporary rural geography research.

The reflections in this chapter are based on fieldwork conducted in rural Kenya, Malawi and Ghana. The work in Kenya was conducted in Usoma, which can be described as a rural community with community distance (approx. 5 km) from Kisumu, the third largest city in Kenya. Households in the community number less than 500 and incomes are very low, with many residents depending on Lake Victoria for livelihood opportunities such as fishing and sand harvesting. Fieldwork in this community involved the use of household surveys, photovoice with women leaders, and focus group discussions to understand water challenges, water-health linkages, and community efforts toward addressing water challenges. Although close to a major city and nested on the shore of the second largest freshwater lake in the world, many of the residents relied on Lake Victoria for drinking needs. We also integrate methodological insights from our research with smallholder farmers in northern Malawi. Malawi is one of the poorest countries in sub-Saharan Africa, with about 80 per cent of the population relying on agriculture. That notwithstanding, food insecurity is a major concern, especially in rural areas. Our work in Malawi centred on using agroecology to improve agricultural productivity and food security. Based on a five-year intervention, we highlight experiences from the implementation of a variety of field methods including quantitative and qualitative methods in rural Malawi. We also draw insights from work with smallholder farmers in northern Ghana. Northern Ghana is a relatively impoverished area compared to southern Ghana. In this context, smallholder farming is the main livelihood. From 2015 to early 2018, we implemented a research project that explored the livelihood and environmental implications of agricultural intensification in this context using a combination of research methodologies, including interviews and focus group discussions.

Theoretical and Methodological Considerations

Since the cultural turn in the social sciences, rural research has witnessed a renaissance as evidenced by the increase in rural scholarship in geography and other social science disciplines most notably sociology, anthropology, and environmental studies. As observed by Woods (2010), as the countryside, its landscapes and its people continue to receive increased research focus, there is a constantly evolving thought process on what constitutes the rural and how ruralities—the materialities and meanings of the countryside and the identities people build around it—are produced and reproduced (Cloke, 1997). Associated with this conceptual puzzle and constantly evolving social, cultural, economic, and political fabric of rural spaces are unique practical and methodological challenges for rural geographers who must constantly and consciously navigate these complexities both as '(re)producers' and 'performers' of rurality. Thus, contemporary challenges of rural geography research are largely rooted in the ways the 'rural' is constituted and partly in how rural spaces are rapidly changing both materially and conceptually (Cloke, 2006; Woods, 2010; Mason-Renton et al., 2016; Kansanga et al., 2020).

Despite being one of the widely researched geographical categories of our time, the rural remains one of the most contested spaces whose conceptualization has assumed diverse imaginations. As observed by Cloke (2006), although the rural is often thought of as a context-specific construct, it continues to come across on the one hand as an imaginative space with diverse cultural meanings, and on the other hand as a spatial container framed as an alternative to the city including being viewed as a space to relocate, farm or visit for tourism. The notion of rurality as a performed entity that involves and requires the constant performance and reproduction of socioeconomic, cultural, and political process of everyday life is also noteworthy. Indeed, this performative dimension draws attention to the fact that rural geographers and the ways of knowing they employ are integral parts of the performance of rurality. This fluid representation of the rural and the associated uncertainties has led to the re-emergence of research questions about the conceptualization of rurality in rural geography (Mason-Renton et al., 2016). However, as many have argued, any lucid understanding of the rural area requires a careful historical reflection.

In the *Handbook of Rural Studies,* Cloke (2006, p. 21) provides a rich summary of the three key theoretical framings that embodied rural geography from the 1970s to the 1990s, moving

> from a functional perspective that sought to fix rural space through the identification of its distinctive functional characteristics; to a political-economic perspective that attempted to position the rural as the product of broader social, economic and political processes; to a perspective in which rurality is understood as socially constructed, such that 'the importance of the 'rural' lies in the fascinating world of social, cultural and moral values that have become associated with rurality, rural spaces and rural life.

Although the dominance of the last perspective has meant a deterritorialization of the rural and its visualization as a 'spatial container', concerns about the neglect of the unique materiality of rural conditions and how these conditions shape the everyday life of those dwelling there have been advanced. This limitation, in part, produced opportunities for the rematerialization of the rural. Of the many attempts, including the efforts to define the rural statistically as powered by the technological innovations in Geographic Information Systems, the conceptualization of the rural as a hybrid and networked space has evolved to be promising. Work in this area, including the novel work of Murdoch (2000, 2003), has emphasized the rural as an outcome of human and non-human co-construction, and a networked space that underscores the tangled nature of rural and urban identities and processes. As stressed by Cloke (2006), this construction of the rural as hybrid or networked has had two crucial implications for rural geography. Aside from providing an opportunity for salvaging the material and social dimensions of the rural, this hybridity or networked perspective also highlights the blurring of the spatial boundaries of the rural and the need for geographers to explore opportunities for interdisciplinary research to unpack the constantly evolving 'rural'.

These developments have implications for geographers whose methods must adapt to cater for everyday complexities in human geography research. This implies the need for spatially and conceptually 'malleable' methodological approaches capable of 'stretching' to cover rural-urban connections in a globalized world including complex rural-urban material flows and movement of people. It also implies the need for methodological depth in rural geography research that creates opportunities for the increasing important 'more than human' dimensions of ruralities. While doing all these, rural methodologies must also be sensitive to the trap of the assumption of sameness of structure and function of the rural (Woods, 2010). Rural geographers must therefore constantly engage the rural with recognition of potential differences in rural identities and meanings. The unique socio-cultural context and practices of rural communities can be both enabling and constraining in research. For example, while hospitality is a widely recognized value of rural societies across the world, certain cultural practices may make some basic questions difficult to ask. For instance, among rural cultures where the counting of household members is prohibited, questions about household size would be difficult to ask. More pressing for rural geographic research is the fact that these sociocultural practices and norms are never static, but constantly evolving in ways that can even generate new difficulties for researchers. The constantly evolving social fabric of rural societies also implies that researchers must position themselves to be able to constantly adapt research approaches to changing socio-cultural practices. The kind of questions researchers can ask must respect local context and reflect the socio-cultural realities of the rural.

Increasingly, rural geographers are also challenged to find new ways of forging trust with rural communities in the research process. This is particularly necessary given that the rural has become an arena of constant entry and (re)entry, with rural communities having to constantly relate with researchers who they may perceive to be asking the same questions. Aside from the potential for this constant visitation to generate respondent apathy in some cases, it also has the potential to compromise the quality of rural research outcomes. The latter is particularly important given that rural populations tend to have expectations that rural research will yield 'immediate' benefits to improve poverty. The quality of rural research may also be compromised due to bias in responses that may be motivated by the desire to receive external support. Rural geographers must therefore work cautiously to avoid this research-development trap.

The blurring rural-urban divide and increased spatial mobility in a globalized world also means that rural research, especially longitudinal studies, and other designs that require follow-up or longer fieldwork, are faced with the challenge of keeping track of research participants. Thus, the question as to how rural research designs can be made more flexible and responsive to the everyday realities of follow-up is crucial. The recognition that rurality is a performed relational process, and rural geographers are themselves intricately involved in the process not just as mere recorders of rural reality, but as producers and (re)producers of rurality, makes questions of positionality crucial (Cloke, 2006). Rural researchers must constantly reflect upon and identify positional conflicts in the conduct and reporting of rural research. Overall, while the changing structure of rural areas allows geographers to raise new research questions and create opportunities for interdisciplinary research, it also means the methodologies of rural geography must evolve to address these new questions and the associated challenges.

Fieldwork: Practicalities and Complexities

Building Trust with Respondents and Gatekeepers

Trust is an essential element in establishing rapport during fieldwork. As highlighted earlier, rural spaces are constantly evolving with new socio-political and cultural processes that require rural geographers to strategically navigate their positionality in order to forge trust with differently positioned individuals in the rural. Gambetta (2000, p. 218) defines trust as

> a particular level of the subjective probability with which an agent assesses that another agent or group of agents will perform a particular action, both before [they] can

monitor such action (or independently of [their] capacity ever to be able to monitor it) and in a context in which it affects [their] own action.

For example, trust can facilitate participants sharing sensitive information with the belief that the researcher will assure their anonymity. In ethnography research where the sample size is sometimes very small, trust can involve building long-term research-participants relationships. When this happens, the researcher-participant boundary is blurred, culminating in on-going negotiations to achieve desired outcomes. For example, in photovoice research in rural Kenya, women participants side-stepped instructions that required them not to take photos that identify members of the community. Some took images of their own toddlers practicing open defecation, ostensibly to illustrate sanitation challenges. When we asked one of them why she did not follow the instructions and took compromising photos of her toddler, she used her trust in the research team to justify her actions, arguing that she knew the researchers will use the photos for a "good purpose". Her sense of trust led her to believe that the research team would not exploit her photos and contributions for other purposes. Such photos were excluded from the analysis because of the ethical and moral justification to not publish them, even though we received consent from the parent.

Many rural communities in sub-Saharan Africa have dense social networks that are used to foster and facilitate everyday social transactions. When researchers enter a village to begin fieldwork, they mostly rely on these social networks to recruit participants without acknowledging and compensating people for using the community's existing stock of trust and social capital. In most rural areas, snowball sampling is a metaphor for riding on the warmth and connections of our initial contact person. As argued by Glasius et al., (2018), people meet you, the researcher, "not necessarily because they are very interested in your research, but more because they trust you or the friends who introduced you to them". The initial contact who introduces a researcher assumes a certain level of responsibility, indirectly vouching for the researcher and assuring the new respondent of their safety and the meaningfulness of the research topic. The researcher in such circumstances needs to act in ways that do not ruin the initial contact's trustworthiness and social relationships. In rural settings with small populations, there is also a danger of recruiting from a singly social circle of trusting individuals with strong bonds, which can lead to lack of divergent perspectives in the data.

In addition, building trust with gatekeepers is important for conducting fieldwork in rural areas. In these settings, gatekeepers can represent a wide array of people and institutions, including individuals or community groups that can withhold or facilitate access to respondents or other community members; individuals with political authority to take decisions on behalf of the community; traditional leaders and elders who are recognized as the "mouthpiece" of the community; faith organization leaders; and persons in charge of formal institutions such as school principals. Aside from access to respondents, gatekeepers represent people who provide direct and indirect access to other resources and logistics, such as transportation and documents (Campbell et al., 2006). Building trust with gatekeepers might require passing through higher channels and prolonged engagements through meetings, informal and formal visits, paying of traditional fees (in the case of traditional leaders), or leveraging the trust of other researchers or academic supervisors. In rural settings where traditional authorities and elders are key decision makers, recognizing and respecting their authority is key. It is equally important to recognize that approval given by a higher-level gatekeeper might not necessarily translate to access or consent by other gatekeepers at a lower level. In Ghana and Kenya for example, we needed to seek approval from respective family elders before we could get access to interview their household members, even though the village chiefs had already given their approval and support.

Indeed, higher-level keepers' decisions can sometimes run counter to the wishes and aspirations of community members, creating delicate and complex political situations for researchers to manoeuvre.

Positionality and Subjectivities

Several geographers over the past two decades have addressed the role of researcher subjectivities in the collection, analysis, and interpretation of data. Mullings (1999, p. 337) describes positionality as a researcher's "perspective shaped by his/her unique mix of race, class, gender, nationality, sexuality and other identifiers, as well as location in time and space". After reviewing the insider/outsider binary debates in the 1980s and 1990s, she argued that the 'insider/outsider' position binary

> is less than real because it seeks to freeze positionalities in place, and assumes that being an "insider" or "outsider" is a fixed attribute. The "insider/outsider" binary is, in reality, a boundary that is not only highly unstable but also one that ignores the dynamism of positionalities in time and through space.
>
> *Mullings, 1999, p. 340*

For Black researchers, who are born in and/or reside in or outside continental sub-Saharan Africa, it is almost impossible to be viewed as or present themselves as outsiders or insiders when conducting research in rural places there. The insider/outsider space becomes a fluid and interactive space that can change at any time. In our research in Malawi, it was not uncommon for a village elder to introduce us as "brothers from Africa who are researchers based in north America". Such an introduction immediately dismantles the insider/outsider binary because being a "brother" makes one an insider to a certain degree, and being a researcher in North America denotes detachment from the community.

Being an "insider" leads to shared positionality between the researcher and respondent. This happens when the researcher intentionally or unintentionally temporarily assumes the shared space or when respondents ascribe such a space to the researcher. In such situations, respondents, based on their assumptions of the mutual positionalities they share with the researcher, might decide to share certain information they might otherwise withhold from "outsider" researchers. In a similar vein, some respondents might not disclose certain critical information, with the assumption that the researcher understands the local situation. In Kenya for example, women participants during interviews often said "you know how our leaders are" without describing the (in)actions of their leaders to the researcher. They assumed that the interviewer, who was from sub-Saharan Africa, though not from Kenya, understood local development planning and political processes and the environment in Kenya. In such situations, being perceived as an outsider without knowledge of Kenyan politics might have allowed respondents to elaborate or explain their points without any prompting from the interviewer.

Related to positionality is power that comes with the positions researchers inhabit. In rural areas where education levels and incomes are low, the researcher might be rightly or wrongly perceived as an expert, an ally who can help address their development challenges, or a potential benefactor to the community. These perceptions give undue power and privilege to the researcher. Navigating researcher privileges and power, particularly during ethnographic fieldwork, demands a high sense of humility and responsibility. Otherwise, the researcher would become a "teacher" instead of a "listener". In our research, it was very common to find respondents/interviewees willing to change their work schedule (e.g., delay in going to farm) to fit our daily calendar, believing that we were very busy people with "more important" tasks to accomplish. Knowing that their schedule was (if not more) important as ours, we negotiated for a time/location that did not disrupt their work schedule.

Implementing Longitudinal Research Design and Methods

The blurring of rural-urban boundaries partly due to the increased mobility of rural populations, presents some challenges which rural geographers must reflect upon before deploying certain research designs and methods. In this subsection, we will reflect upon our experiences conducting longitudinal research in Malawi and Kenya. Mobility is an increasingly important process in rural communities. Rural residents use migration as a safety net against seasonal food insecurity especially. It is common for active household members to migrate to cities during the dry season to seek casual work. In such situations where migration is intense, longitudinal research designs that target individuals can be difficult to implement.

We implemented a five-year longitudinal study on agroecology in northern and central Malawi. This study required follow-up after the baseline study. Rural Malawi is, however, characterized by intense rural to urban migration. International migration, especially to South Africa is also common among Malawians. In this context, a longitudinal study can suffer from attrition depending on the target unit of analysis. High attrition rate in follow-up studies can render longitudinal data unusable. To navigate this potential barrier in our research, we pitched the unit of analysis as the household. This design meant that any adult member in the household could respond to the follow-up survey on behalf of the household. Although there were instances where the whole household migrated, the strategy helped reduce attrition during follow-up. In most cases where whole households migrated in search of better soil conditions, they were mostly within a traceable distance. Getting in touch with households who migrated was facilitated by another design strategy we incorporated. Our intervention was community driven involving a farmer-to-farmer knowledge sharing approach. Thus, participants were actively involved in project activities and closely connected to one another. This facilitated follow-up in situations where some households had relocated in search for better farming conditions. While navigating research design issues can be complex in rural research, a careful reflection on potential challenges may enable rural geographers to adapt their study designs in ways that can minimize some of these methodological challenges.

Navigating Socio-Cultural Norms

Rural communities are an embodiment of culture as expressed in the norms, beliefs and practices that structure rural life. While customary practices and norms may vary across geographies, they generally work to structure rural life, including how rural communities relate to outsiders. Specific to research, these norms and practices can be both enabling and constraining. In relation to the former, rural communities are typically closely knit, with a rich sense of community and hospitality which are vital resources for rural geographers, particularly outsiders. From experience conducting research in rural Malawi, Kenya and Ghana, rural hospitality is crucial for research. For instance, in the absence of important commercial services like hotels and restaurants, rural research thrives on the hospitality of rural communities who host and share with researchers. This hospitality is particularly crucial for ethnographers who must live with study participants for extended periods. Rural communities also receive from researchers in diverse ways including materially through participation in research related activities which helps ensure some level of reciprocity.

In some instances, the cultural norms and practices in rural areas can, however, be constraining in the research process. Rural norms and beliefs may not permit certain questions. For example, during research with smallholder farming households in northern Ghana, one of our basic research questions was about household size. We asked respondents directly how many people were in their households. While this question appeared simple and straightforward, the first few participants hesitated. During our third interview with a retired agricultural extension officer, we asked the same question and he explained that there is prevailing cultural belief that counting all household members could attract death to the household. This cultural belief surrounding the counting of household members introduced an element of complexity that required different ways of approaching the objective of exploring household size. In this situation we revised our research question. Instead of asking directly for the number of persons in the household, we broke down the question and integrated it into different areas of the research. For instance, when talking about education, we would ask how many children in the household were educated and how many were not. This approach was relatively appropriate since it did not require counting of all household members. In the end we could still deduce the size of each household without inconveniencing participants. While these beliefs vary across rural communities, the example above points to the potential for differences between local belief systems and the ways of knowing employed in rural geographic research. The example further highlights the opportunity for researchers to explore culturally non-invasive ways of navigating these barriers.

In terms of restrictions to particular rural spaces, in conducting agricultural research in northern Ghana, we discovered that farmers had designated sacred areas on their plots where they positioned their deities to protect the farm from theft and spiritual attacks. This is a common practice in rural communities across Africa where sacred groves are maintained as abodes of deities and the ancestors. It is believed that these deities can harm any trespassers with illnesses and calamities. Geographic research requiring the acquisition of spatial data and firsthand observation necessitates particular sensitivities for researchers to account for in the data collection process. While this may appear challenging, there may sometimes be opportunities for rural researchers to access these spaces if the necessary cultural rites are performed to seek the permission of the ancestors and gods. Researchers who may want to access some culturally restricted spaces may have to communicate with traditional authorities to perform the necessary rites before access is granted.

Another key area of sociocultural complexity in rural research is the navigation of gender norms. In most rural societies in Africa for instance, cultural norms position men as the authority figures within the household. This cultural structuring of gender relations can shape research in several ways. For instance, at the basic level of participant selection where researchers may be interested in the views of women, there is the tendency for husbands to automatically assume the position as respondents. In cases where researchers gain access to women, gendered power dynamics may still shape their responses to certain questions, especially those related to the power dynamics and control of household assets. Although a foundational way of doing successful gender research is to involve men by meeting with them to explain the objectives to them even if they are not participants, it is also crucial for rural researchers to forge convenient spaces where women respondents can feel comfortable enough to express their views. From experience conducting research with rural women in Malawi and Ghana, having a woman research assistant, or holding women-only focus groups, has been helpful in winning the trust of both husbands and wives. Men tend to feel comfortable having their wives talk to fellow women while women may be more open when discussing household dynamics with other women.

It is therefore crucial for rural researchers to ask preliminary questions about local cultural practices and customs before commencing research. Rural geographers must also be sensitive to the potential drawbacks of prevailing gender dynamics and integrate strategies for enhancing cooperation during fieldwork. This way, rural geographers can make their questions congruent with local beliefs and practices to avoid conflict in the research process.

Conclusion

In this reflective piece, we drew upon unique experiences conducting fieldwork in rural Kenya, Malawi, and Ghana to highlight the complexities of rural research and how rural geographers may navigate some of the challenges. While this chapter is not meant to be a blueprint for rural geography research, we position it as an opportunity to further stimulate ongoing discussions on the complexities of rural research. As outlined earlier, rural spaces are increasingly changing with implications for the conduct of geographic research. While this transformation comes with the opportunity for rural geographers to ask new questions, it also presents some challenges. For instance, increased mobility of rural dwellers implies that geographers interested in longitudinal research must adapt their methods to address the loss of participants to follow-up. Rural research does not happen in a cultural vacuum. It is deeply embedded in, and shaped by, sociocultural norms and practices which can be both enabling and constraining. While it is crucial for rural geographers to constantly explore the opportunities cultural norms offer in the research process, the design and execution of research must proceed in a manner that does not conflict with local norms. The kind of questions rural researchers pose, and the research procedures, must be congruent with local traditions and beliefs. This is not to suggest that rural culture is cast in stone. As discussed earlier, there are opportunities for rural researchers to navigate certain cultural constraints in the research process with the consent and trust of traditional custodians. Likewise, rural researchers must proceed on the realization that their research shapes rurality. As observed by Cloke (2006), rural geographers are themselves intricately involved in the performance of rurality not just as mere recorders of information but producers and reproducers of rurality. Thus, aside from the need for rural geographers to constantly reflect on their positionality in the research process, they must also proceed with the assumption that what they publish and their interpretation of rural phenomena in general contributes to the reproduction of rurality. The future of rural research will also be closely shaped by technology, further opening new possibilities for new virtual geographies in relation to the rural. Overall, rural spaces will continue to go through transformation and rural geographers must adapt their tools to accommodate the constraints and opportunities that come with rural change.

References

Adu-Ampong, E. A., and Adams, E. A. (2020) "But You Are Also Ghanaian, You Should Know": Negotiating the Insider–Outsider Research Positionality in the Fieldwork Encounter. Qualitative Inquiry, 26(6), 583–592.

Appiah, R. (2020) Community-based Participatory Research in Rural African Contexts: Ethico-Cultural Considerations and Lessons from Ghana. Public Health Rev 41, 27.

Campbell, L. M., Gray, N. J., Meletis, A. Z., Abbott, J. G., and Silver. J. Silver. (2006) Gatekeepers and Keymasters: Dynamic Relationships of Access in Geographical Fieldwork. Geographical Review, 96(1), 97–121, DOI: 10.1111/j.1931-0846.2006.tb00389.x

Cloke, P. (1997) Country backwater to virtual village? Rural studies and 'the cultural turn'. Elsevier.

Cloke, P. (2006) Conceptualizing Rurality. Handbook of Rural Studies:18–28.

Gambetta, D. (2000) 'Can We Trust Trust?', in Gambetta, Diego (ed.) Trust: Making and Breaking Cooperative Relations. University of Oxford, chapter 13, pp. 213–237. www.sociology.ox.ac.uk/papers/gambetta213-237.pdf

Gebrehiwot, T. and van der Veen, A. (2013) Farm level adaptation to climate change: The case of farmer's in the Ethiopian highlands. Environ Manage, 52(1):29–44, DOI: 10.1007/s00267-013-0039-3. Epub 2013 Jun 1. PMID: 23728486.

Glasius, M. et al., (2018) Building and Maintaining Relations in the Field. In: Research, Ethics and Risk in the Authoritarian Field. Palgrave Macmillan, Cham. https://doi.org/10.1007/978-3-319-68966-1_4

IPCC (2019) Climate Change and Land: An IPCC special report on climate change, desertification, land degradation, sustainable landsecurity. www.ipcc.ch/srccl/chapter/chapter-5/5-1-framing-and-context/

Kansanga, M. M., Ahmed, A., Kuusaana, E. D., Oteng-Ababio, M., and Luginaah, I. (2020) Of Waste Facility Siting and Relational Geographies of Place: Peri-urban landfills, community resistance and the politics of land control in Ghana. Land Use Policy, 96, 104674.

Luginaah I., Weis T., Galaa S., Nkrumah M.K., Benzer-Kerr R., Bagah D. (2009) Environment, Migration and Food Security in the Upper West Region of Ghana. In: Luginaah I.N., Yanful E.K. (eds) Environment and Health in Sub-Saharan Africa: Managing an Emerging Crisis. Springer. https://doi.org/10.1007/978-1-4020-9382-1_2

Mason-Renton, S. A., Luginaah, I., and Baxter, J. (2016) The Community Divide is More Detrimental than the Plant Itself: Confrontational stigma and community responses to rural facility siting. Journal of Rural and Community Development, 11(2), 22–24.

Mekonnen, A, Tessema, A, Ganewo, Z, Haile, A. (2021) Climate Change Impacts On Household Food Security And Adaptation Strategies In Southern Ethiopia. Food Energy Secur, 10, e266.

Mullings, B. (1999) Insider or Outsider, Both or Neither: Some Dilemmas of Interviewing in a Cross-Cultural Setting. Geoforum, 30(4), 337–350.

Murdoch, J. (2000) Networks—A New Paradigm of Rural Development? Journal of Rural Studies 16(4), 407–419.

Murdoch, J. (2003) Co-constructing the Countryside: Hybrid networks and the extensive self. In Paul Cloke (ed.) *Country Visions*, 263–282. Harlow: Pearson Education Limited.

Nyantakyi-Frimpong, H. (2021) How Identity Enriches and Complicates the Research Process: Reflections from Political Ecology Fieldwork. The Professional Geographer, 73(1), 38-47, DOI: 10.1080/00330124.2020.1823863

Rademacher-Schulz, C., Schraven, B. and Mahama, S. E (2014) Time Matters: shifting seasonal migration in Northern Ghana in response to rainfall variability and food insecurity. Climate and Development, 6(1), 46-52, DOI: 10.1080/17565529.2013.830955

Woods, M. (2010) Performing Rurality and Practising Rural Geography. Progress in Human Geography, 34(6), 835–846.

28

PARTICIPATORY GEOGRAPHIES

From Community-Engaged to Community-Led Research

Heather Castleden and Paul Sylvestre

Introduction

Knowledge is produced everywhere; it is not the purview of the academy alone. Yet the academy has laid claim to knowledge creation and its ownership, as well as control over its interpretation and the power to declare legitimacy (or not) of new knowledge since the Age of Enlightenment. Geography – the field and those who have populated it – has embraced this Eurocentric view throughout that epoch and continued unfettered for centuries. But in the 1970s, with the cultural turn in human geography, we started to see changes in the margins of the discipline arising with the critical, feminist, and post-structuralist influences of the time. Hints of other knowledge being 'out there' were surfacing in the ivory tower and certain types of geographers were paying attention. Geographical approaches to collecting data, and even the questions geographers were asking, and the analyses of the data we collected – be they human or physical – were not always keeping pace with evolving 'best practices'. Geographers' abilities to claim omniscience and neutrality in our interpretations of all things spatial, were not superlative, let alone accurate. For an emerging group of geographers, some working individually and some in collectives, it was time to engage *with* the people and places we were studying and not simply see them as subjects and objects under the microscope. Participatory geographies began to appear as a small part of our disciplinary landscape over 40 years ago.

The overarching goal of participatory geographies can be defined as equitably engaging with people and places in ways that reflect their priorities in an effort to create some form of positive change (Castleden, Garvin, and Huu-ay-aht First Nation, 2008). Such a goal is not easy or tidy, nor is it without issues of power and representation of voice, agency, and diversity. Along with 'community', which has long been a contested concept, so too are the notions of what makes research 'participatory', 'action oriented', 'community-based' or 'community-engaged'. These terms require some unpacking to reveal the rhetoric and reality. In this chapter we define participatory research and provide a genealogy of participatory research in and beyond the academy, before turning to its specific rise in geography. We then move into a description of research 'on community', to research 'in community', and landing on research 'with and by

DOI: 10.4324/9781003038849-32

community'. We outline the rationale for, the current debates of, and the ethical issues arising from processes associated with undertaking participatory geographical research by delving into the works of a number of human geographers working at the forefront of participatory geographies. Finally, we touch on a number of the benefits and challenges of doing this type of work by offering modest suggestions for those who are considering this methodology in their geographical inquiry based on our own experiences as geographers in attempting to do participatory research 'in a good way' (Ball and Janyst, 2008). Participatory geography is not for the faint of heart or the rigid of mind. As activist-scholars currently in the academy, we know there are times for shit-kicking (i.e., refusing to succumb to the demands of the neoliberal academy) and tea-drinking (i.e., diplomacy in navigating the colonial, patriarchal, ableist, and heteronormative tensions, and doing the work while holding our noses). This chapter explores the spectrum, the trends, the tensions, and emerging questions for participatory geographers.

What is 'Participatory' Research?

Participatory approaches differ from others in that they attempt to equitably involve people, communities, and/or organizations in research, draw on and draw in their knowledge and experience, share decision-making responsibilities about questions being asked, methods being used, and analyses being undertaken, while building community capacity for independent research (Minkler and Wallerstein, 2003). Essentially, an academic who practices participatory research should be aiming to work themselves out of a job! (But do they? A critique we explore later in the chapter.) Participatory researchers employ a broad range of data collection methods and analyses that involve some form of researcher-researched reflection, dialogue, and action (Kirby and McKenna, 1989). These methods may be conventional (e.g., interviews, focus groups, surveys) or emergent (e.g., arts-based, community archives, collaborative film). Academic researchers also aim to develop theoretical contributions from their collaborative projects, which emerge only by working closely with research partner-participants to identify the most effective ways to answer particular research questions (Minkler and Wallerstein, 2003). While the goals of conventional investigator-driven social research, including human geography, focus on relationships between and/or differences in phenomena or to focus on social structures and/or individual experiences, an overarching set of goals prevails in participatory research: to develop mutual trust and respect, equalize power differences, and co-own the knowledge produced in an effort to bring about social justice and change (Reason and Bradbury, 2001).

The goals of participatory research – balancing power, fostering trust, and collective ownership – are intrinsically linked to one another, therefore necessitating brief conceptual definitions to contextualize their usage here. 'Balancing research power' commonly refers to researchers and their research partner-participants sharing control of the research process and outcomes (Fals-Borda and Rhaman, 1991). Yet *how* to balance power is seldom adequately addressed in academic literature despite calls for participatory researchers to engage with the tensions of power differentials (Wallerstein and Duran, 2006) and to recognize that internal community politics can reinforce and obscure power relationships (Malone et al., 2006). For example, gender, age, and social position can affect how individuals act or do not act both in the community and in the research process (Malone et al., 2006). Power differentials may also influence participants' responses during interviews (Baxter and Eyles, 1997). Castleden and colleagues (2008) bring attention to two other overarching tensions about power that are worth noting. First, there is often an unequal relationship between the researcher and the partner-participants, typically (although not always) tipped in favour of the researcher. Second, while participatory research is collaborative, its applications in the academy have largely been undertaken by white, cis-gender scholars in partnership with the 'other' (e.g., Indigenous peoples, people with disabilities, people living in poverty, etc.).

Recognizing and acknowledging individual and structural power dynamics while working towards establishing a power balance can positively influence the development of trust between parties. 'Fostering trust' is thus an overlapping and related process that begins with sharing power and giving up power (Castleden et al., 2017). Specifically, trust is established when researchers work in an open, honest, and transparent manner (Minkler, 2004). Trust is also built when researchers become involved in the community's activities, listen to and address community partners' needs, and reciprocate in some way. For example, there is a growing trend towards working with communities (however defined) to building research capacity through training and employing local people in research (Castleden et al., 2012b). All of these actions are additional ways to build trust resulting in community ownership over the research process (National Aboriginal Health Organization, 2005; Kauper-Brown and Seifer, 2006). Through shared control of the decision-making process all parties involved own the research process. But what does 'owning' the research process actually mean? This question is one of many emerging tensions in the field of participatory geographies, which we will explore throughout the chapter.

Where Does Participatory Research Come From?

The roots of academic insights into participatory research, as noted above, lie in the social and political movements of the twentieth century. Before the cultural turn of the 1970s though, there was Kurt Lewin (1946), a German-American who first introduced the concept of engaging with participants in research undertakings in the 1940s as a way of confronting issues of social justice and challenging researcher 'objectivity' (Fals-Borda and Rahman, 1991). Others, particularly in the Global South, took up the call to revolutionize knowledge-to-action in and beyond the academy. "Names like Freire, Fals Borda, Rahman, Kassam, Mustafa, Mbillinyi, Vio Grossi, Cadena and [Tandon] were the persons who were referenced. Today, much of that history has been forgotten and many people believe that community-based participatory research is something that was 'discovered' in the global North" (Hall and Tandon, 2017, p. 371; see also Tandon, 2002). For example, the anti-colonial critique implicit in Paulo Freire's (1970) notion of radical pedagogy emphasized the absolute necessity of community-based identification of problems and solutions for advancing liberatory agendas. Rooted in postcolonial ethics, Freire's radical pedagogy took aim at the racial colonial violence inherent in western modes of knowledge production and expertise, arguing that "any situation in which some individuals prevent others from engaging in the process of inquiry is one of violence. The means used are not important; to alienate human beings from their own decision-making is to change them into objects" (1970, p. 85).

Around the same time that Freire was mobilizing in Brazil, participatory research was emerging in the Tanzanian context. Budd Hall, a research fellow at the University of Dar es Salaam at the time, reflected on the experimentation and creativity taking place there that "attempted to put the less powerful at the center of the knowledge creation process; to move people and their daily lived experiences of struggle and survival from the margins of epistemology to the centre" (1992, p. 16). Hall (1981) described participatory research as having three core activities: research, education, and action. Interest in power and democracy, and considerations of gender, race, ethnicity, sexual identity, physical and mental abilities are all central to participatory research (Hall, 1992). The 1970s also saw the establishment of the International Participatory

Research Network with its coordinating office in New Delhi, India (established with intention outside of the Global North to decenter the usual suspects), and Hall (1992) notes that participatory research began in community settings long before it began to take hold in the academy.

Undeniably, there was a seismic shift in the academic landscape of Indigenous-focused research in 1999 when Linda Tuhiwai Smith, a Maori scholar, published her pivotal book, *Decolonizing Methodologies: Research and Indigenous Peoples*. Finally, non-Indigenous academics could no longer ignore the harm their research had caused over the decades.

> It galls us that Western researchers and intellectuals can assume to know all that it is possible to know of us, on the basis of their brief encounters with some of us. It appals us that the West can desire, extract and claim ownership of our ways of knowing, our imagery, the things we create and produce, and then simultaneously reject the people who created and developed those ideas"
>
> *Linda Tuhiwai Smith, 1999, p. 1*

Smith's writing details how imperial thinking was, is, and – more than 20 years since her first edition – still remains embedded in all disciplines and the establishment's domination of truth. The field of Geography is an accomplice here, with its hand in the 'discovery' of 'new worlds'; the peoples and places were only 'new' to them as part of Europe's imperial expansion. Smith recognized that Indigenous peoples have long felt researched to death (and indeed research has played a role in the attempted genocide of many Indigenous peoples) and her book was a call to action that would be about bringing Indigenous peoples back to life. Others, including Mohawk scholar Marlene Brant Castellano (2004) and Cree scholar Willie Ermine (2007) evoked similar calls for engaging in ethical research with and, more importantly, by Indigenous peoples themselves.

Early Critiques of Participation

Through the late 1970s and 1980s transnational development agencies, such as the World Bank and the International Monetary Fund, vigorously adopted participatory approaches to development, transforming community-based development from marginalized radical practice to new development orthodoxy. The proliferation of participatory development was a response to the failure of large-scale, top-down international development projects of that period to achieve their stated aims. The shift toward 'community' participation was meant to rationalize development by eliminating the need for centralized bureaucratic management, to empower local and Indigenous communities by mobilizing grassroots knowledge to identify and grapple with local issues, and to overcome local resistance to outside interference (Bhatnagar and Williams, 1992; Cooke and Kothari, 2001; Nadasdy, 2005). Handing the stick to 'communities' coincided with a rapid expansion of international development non-governmental organizations as a replacement for government-run development agencies.

The institutionalization of participatory approaches prompted a series of withering critiques by anthropologists and critical development scholars (e.g., Cooke, 2003; Cooke and Kothari, 2001; Kapoor, 2005; Stirrat and Henkel, 1997). Rather than empowering communities by dismantling power asymmetries, commentators observed that participatory development merely reorganized power, concealing imperialist and neo-colonial relations of dominance under a participatory veneer (Cooke, 2003). Given the overt effort to address power imbalances among participants and its perceived responsiveness to local conditions, its ethical superiority over conventional development practices took on an aura of development 'common sense'. However, though self-styled facilitators spent considerable time working to balance power in partici-patory spaces, they spent far less time reflecting on how participation as development ortho-doxy worked to entrench and shore-up broader relations of power (Cooke and Kothari, 2001; Kothari, 2005). Far from facilitating a neutral process for co-producing knowledge, critics argued that participatory arenas functioned as a form of governmentality where latent, yet dominant western development discourses structured the terrain over which identifying and addressing problems took place.

Others have problematized taken-for-granted western assumptions that permeated institutionalized participatory schemes. For instance, homogenizing and idealized notions of community tended to gloss over very real power relations and tensions running through com-munities, silencing already marginalized voices in the facilitator's quest to find consensus among community (Kapoor, 2005; Stirrat and Henkel, 1997). Maguire (1996), in taking a feminist lens to the question of community, showed how simplistic figurations about what constituted com-munity elided gendered relations of power, such that the category community too often meant 'male community'. Critics also question the modernist assumptions at the core of the singular and self-reflexive Freirean participatory subject who seizes their individual empowerment and historical agency through a process of conscientization facilitated by the participatory project (Cooke and Kothari, 2001). In taking aim at the role of the participatory practitioner, both Kapoor (2005) and Henkel and Stirrat (2001) suggest that traits, like humility, patience, and respect, which are so valued in participatory research, are little more than a ruse. They contend that such practices are akin to the self-effacement of a quasi-missionary that, far from decen-tring the researcher, serves as a tactic of covert self-aggrandizement that consolidates researchers' power in participatory spaces. Indeed, Michelle Fine, one of the founding members of the Public Science Project that arose from the Participatory Action Research Collective at the City University of New York, touches on how participatory approaches recuperate colonial logics:

> Early in the century, 'twas noble to write of the other for purposes of creating what was considered knowledge. Perhaps it still is. But now much qualitative research is under-taken for what may be an even more terrifying aim – to 'help' them. In both contexts the effect may be othering: muted voices; 'structure' imported to local 'chaos'; others represented as extracted from their scenes of exploitation, social relationships, and meaningful communities.
>
> *Michelle Fine, 1994, p. 79*

Thus, for many, participatory approaches represented a new techno-managerialist tool that, far from dismantling relations of imperial and colonial dominance, served as their continued conditions of possibility and therefore needed to be resisted (e.g., Cooke and Kothari, 2001; Coombes, 2012; Nadasdy, 2005).

Participatory Research Comes to Geography

A small cadre of geographers developed participatory research programs during the 1990s (see for instance, Cornwall and Jewkes, 1995). Indeed, three journals had special collections on research and action in geography during that decade: *The Canadian Geographer* (1993), the *Professional Geographer* (1994), and *Antipode* (1995; see Farrow et al., 1995). At the close of the millennium, Kitchin and Hubbard (1999), who also edited a special issue of *Area* on research, action, and 'critical' geography, note that their collection's contributors focused on issues of reflexivity, positionality, power dynamics, empowerment, and praxis. In their editorial, they point out that while recognition of power imbalances in geographical research was on the rise, there were very few who were explicitly trying to support real change.

Entering the twenty-first century, participatory geographies were greatly popularized as a research approach through the mid to late 2000s. Cognizant of critiques of power and ethnocentrism leveled by development scholars (e.g., Mohan, 1999), geographers sought to develop a more spatial understanding of participation and empowerment while disentangling the socially transformative potential of participatory approaches from their tendency to reproduce dominant social formations (see, e.g., Castleden et al., 2012; Chatterton et al., 2010; Jupp, 2007; Kesby, 2007; Kindon et al., 2007; Klodowsky, 2007; mrskinpaisby, 2008; Pain, 2004; Pain and Kindon, 2007). The British-based Participatory Geographies Working Group (PyGyWG), for example, was founded in 2005 (achieving Research Group status in 2009) for the purpose of pushing forward conversations about how geographic research might be reoriented toward benefitting more than just the discipline, but also (and primarily) the groups, communities, and social struggles outside the academy, which many in the discipline had made their object of inquiry. Geography was further awakened to participatory Indigenous geographies in 2007 with Native Hawaiian geographer, Renee Pualani Louis, when she published her influential piece "Can you hear us now? Voices from the margin: using Indigenous methodologies in geographic research" in *Geographic Research*. In it, she challenges western paradigms and argues that Indigenous methodologies in Geography provide "a mechanism for Indigenous peoples to participate in and direct these research agendas ensures that their communal needs are met, and that geographers then learn how to build ethical research relationships with them" (2007:130).

Also in 2007, Rachel Pain and Sara Kindon guest edited a special issue on Participatory Geographies in *Environment and Planning A: Economy and Space*. They curated a collection of papers that showcased the work of other geographers engaging in participatory research and stimulated debate in the field. Their issue brought attention to the overlooked geographical nature of participation by noting that it is "inherently spatial… Surprisingly, however, the specific relations between participants, place, and space have received little attention" (Pain and Kindon, 2007, p. 2807). By this time, Pain and Kindon noted that participatory research was gaining in popularity because of a growing recognition that there were limits to what conventional research could do to effect positive social change, a questioning of the relevance of investigator-driven research, and the acknowledgement that people and communities who experience oppression and marginalization are best positioned to develop interventions and counter-strategies to address injustice. As Caitlin Cahill (2007) argued, the project of participatory geographies was not about revealing subjugated knowledges and subaltern silences. Rather, Cahill contended that because of its commitment to relationality, participatory praxis had the potential to generate new research and organizing practices, new knowledges and ways of knowing, and new subjectivities necessary for radical social transformation (see also, Kesby, 2005).

A key contribution of geography to participatory praxis then, as noted above, has been debate about how the spatialities of participation fundamentally conditions the structure and success of participatory research. Cornwall's (2004) early distinction between "Invited" and "Popular" spaces attempted to categorize participatory geographies based on who was creating the space and to what end. In juxtaposing invited spaces, those produced by outside, resource-bearing agents who invited community members in, against popular spaces, those organically organized by community members, Cornwall established a spectrum from empowerment to autonomy in which the latter holds more promise for radical social change, and the former tends to reproduce bureaucratic forms of domestication. Kesby (2007), however, questioned whether we can read directly from participatory spaces to politics so easily. Drawing on several of the critiques outlined above, he was quick to point out that group dynamics can make popular spaces that perpetuate oppressive and power laden relationships and that, properly conceived, invited spaces can offer radical and liberatory potential. Rather than being prescriptive about what kinds of participatory spaces result in transformational social change, Kesby (2006) suggested participatory geographers address how participatory spaces are always already imbricated in a broader power geometry, which means participation is always context dependent and contingent on the particularities of local and regional settings (see also, Blackstock et al., 2015). In addressing these methodological and epistemological challenges, participatory researchers have increasingly mobilized Mary Louis Pratt's (1991) concept of "contact zones" to foreground participatory spaces as locations where differently positioned actors encounter and collaboratively analyze the uneven power relations that differentially structure their lives (Askins and Pain, 2011; Bettencourt, 2020; Ritterbusch, 2019; Torre et al., 2012).

Aside from important theoretical interventions, geographers have also extended participatory approaches through innovative methods and pedagogies. The use of community mapping (Parker, 2006), participatory action mapping (Boll-Boss and Hankins, 2018), participatory GIS (Dunn, 2007; Elwood, 2006), and photovoice (Castleden et al., 2008) are only some examples of collaborative techniques mobilized by participatory geographers to destabilize dominant depoliticized abstract spatial representations, which entrench uneven power relations by occluding the complex webs of socio-spatial relations that constitute place. Pedagogically, participatory geographers have worked hard to destabilize the spatialities of the classroom by integrating their research programs into activist-oriented service learning (Pain et al., 2013; Ritterbusche, 2019) and immersive field schools (Castleden et al., 2014) to generate useful work for community partners and engage students intellectually, affectively, and socially in ways that work against the passivity of the university lecture hall and promotes student social justice activism. To be sure, as Ritterbusche (2019) attests, this rarely goes as planned and the time constraints of academic semesters, the risks of student voyeurism, and institutional aversion to the discomfort of student clientele can tend to foreclose the sort of meaningful relationship building central to participatory work. Furthermore, in our own experience with immersive field schools we have seen hard won transformative momentum fizzle as many students leave the field and plug back into the routine of their daily lives.

Despite the challenges of enacting participatory research and the perennial criticisms leveled at participatory approaches (which we describe below), engagement with participatory praxis has become truly vast across the discipline and beyond. Because of the radical potential inherent in its critical political and ethical commitments, participatory geographies have proliferated across places and communities where systemic violence and organized abandonment manufacture uneven landscapes of vulnerability, risk, and harm. For example, geographers have employed participatory research within carceral geographies in the context of mass incarceration to center the knowledges and experiences of victims of the carceral state, and in turn, have developed radical interventions that support incarcerated people and their families while working toward prison abolition (Gilmore, 2007; Fine and Torres, 2006; Martin et al., 2009; Ybarra, 2021). Though not always framed as such, Ruth Wilson Gilmore's (2007) foundational book *Golden Gulag* was a deeply participatory piece of geographic scholarship. Working alongside the anti-prison activist group 'Mothers Reclaiming Our Children', *Golden Gulag* showed us how the carceral state's systematic assault on communities of colour was not merely an outcome of state failure but was also an active process of state building. This knowledge, born out of struggle, solidarity, and revolutionary praxis, helped form a grassroots abolitionist movement against domestic militarism (see also Megan Ybarra's (2021) important participatory work on the siting and expansion of US immigration detention centers). Both Gilmore's and

Ybarra's work are indicative of politically engaged scholarship whose commitments to radical grassroots organizing enact socio-spatial possibilities against white supremacist and racial capitalist geographies. New scholarship is also emerging in critical participatory planning strategies (see Ashtari and de Lange, 2019; Rodriguez, 2021).

Indigenous and decolonial geographers have also taken participatory approaches as both ethical commitment and political conviction. For instance, *The Canadian Geographer/Le Géographe canadien,* which has been at the forefront of scholarship concerning Indigenous and decolonizing geographies, produced a 2012 issue on "Community-Based Participatory Geographical Research with Indigenous Peoples in Canada" edited by Castleden, Mulrennan, and Godlewska (2012a), showcasing a range of participatory geographical projects. That said, even nearly a decade ago, Castleden and colleagues (2012a) were questioning whether 'progress' in participatory approaches, ethics, and interventions are being made. In an empirical study in the same issue, Castleden and colleagues (2012b) explore how Canadian university-based researchers understand participatory geographies in theory and operationalize such approaches in practice; they revealed the risk of rhetoric as participatory research moves from the margins to the mainstream. De Leeuw and colleagues (2012) offer a critique of such research, arguing that no methodology can guarantee ethical or decolonized research. Instead, they propose that friendships forged between individuals outside the environment of research may do more to allow the kind of dialogue and exchange that can guide researchers in their moral and ethical decision-making. Though far from a comprehensive accounting of how critical geographers have taken up participatory research, our intention here is to illustrate the complexity and growing importance of participatory geographies to critical geographic scholarship.

Trends, Tensions, Emerging Questions: Where Next for Participatory Geographies?

While participatory geographers may be attuned to navigating complex power relationships within the space of a given project, they far less often consider how the resources and empowerment strategies developed within the project space are taken up in the social context they are ostensibly meant to transform. As Franks (2015), Fox (2013), and others point out, participatory researchers need to be honest and develop a shared understanding with participants about the limits of transformational change possible in community-based projects. Indeed, Kesby's (2007) concern that we must find ways to effectively "distanciate" collectively generated resources for community empowerment beyond the scale of the project remains a perennial challenge. Given the transformative aims of participatory geographies, there is a need to consider outcomes at a multiplicity of scales (Blackstock et al., 2015). Stiegman and Castleden (2015), Browne and colleagues (2017), as well as Sylvestre and colleagues (2018), extend Kesby's insights to argue that academic researchers, whose participatory aims are constrained by the disciplining power of academic spaces (e.g., tenure and promotion, publish or perish, university financial regulation), tend to carry the demands of academic spaces with them into participatory spaces, reconfiguring participatory research in important and often unacknowledged ways. Beyond merely shared research design, they highlight the importance of finding innovative ways to distribute control of academic and financial resources more equitably. Institutional uptake for participatory approaches is further evidenced by the growth of community geography programs at US academic institutions and the development of community geography as a subfield in the discipline (Robinson et al., 2018; Shannon et al., 2020). The place-based focus of such programs and institutional backing that recognizes community resources and grassroots knowledge mobilization as legitimate academic work are promising developments.

Equally exciting are the ways in which technology is reshaping the nature of participation and how participatory projects can mobilize their research to advance radical social change. Open platforms for creating and sharing data, such as OpenStreetMap (Solis et al., 2020), grassroots mapping and citizen science made possible by organization like Public Laboratory for Open Technology and Science (Breen et al., 2015), or Karrabing Collective's use of augmented reality to create digital place-based archives (Povinelli, 2011) are all examples of how technology is helping to enhance modes of participatory data collection, analysis, and knowledge mobilization in ways that early practitioners could never have imagined. In their work with LGBTQ folks, Laury Oaks and colleagues (2019) have shown the utility of social media in reaching and working with geographically dispersed communities who interact primarily online. Whereas Ashtari and de Lange (2019) use a digital interactive multiplayer game to engage stakeholders with each other's desires, values, and expectations in urban planning.

Exciting as such innovations may be, Maori scholar Brad Coombes (2012; 2017) cautions us about the extent to which participation creates radical change, asserting that participatory geographers overstate their progress. For Coombes, "[t]here is little hope that sharing in knowledge production can advance Indigenous causes; rather the keys to academic offices and publishing houses must eventually be transferred to those who know their worlds" (2012, p. 290) While in theory participatory geographers intend to work themselves out of a job, the reality is that this has never happened. De Leeuw and Hunt (2018) note that the majority of participatory geographers are white settlers who idealize a form of activist research but nevertheless often fail to foreground Indigenous voices or mobilize knowledge in a manner that is accessible to participants. While the latter comment is surprising given the demonstrated lengths that participatory geographers go to find creative ways to mobilize knowledge through, for instance, theatre (Franks, 2015), arts-based methods (Castleden et al., 2008), documentaries (Kindon, 2003), interactive co-designed learning interventions (Spry, 2020), questions about representation and settler colonial power are pressing (also see de Leeuw Chapter 8, this volume). Paralleling critiques of participatory development leveled over two decades ago, the concerns remain that for the most part participatory geographies merely reorganize, rather than challenge dominant power formations, and that in doing so they work to defer Indigenous self-determination and resurgence (Coombes, 2012; Coombes et al., 2014). Participatory approaches are clearly imperfect, but they nevertheless do yield positive albeit differentiated impacts for all involved.

Concluding Comments

The tension between participatory research as legitimating prevailing hierarchical power relations and participatory research mobilized to support radical social change, which has been debated for decades, will never cease to be an issue. As Rachel Pain and colleagues (2012, p. 120) remind us, "participation as deployed by the state itself is not the same as that enacted with radical grassroots alternatives." As struggles for liberation encounter intersecting challenges (i.e., the climate crisis, resurgent white nationalism and fascism, the extension of the carceral and security state), participatory researchers need to build and strengthen alliances with radical social movement organizing. They can collaborate in direct action while shifting the financial and technical resources of the university to the movement organizers, groups, and communities who are already leading grassroots collective responses. While we believe that ultimately "the revolution will not be funded" (INCITE! 2017), participatory geographies are well positioned to contribute meaningfully to some of the most pressing struggles of our time.

References

Ashtari, D., and de Lange, M. (2019) Playful civic skills: A transdisciplinary approach to analyse participatory civic games. *Cities, 89*, 70–79.

Askins, K. and Pain, R. (2011) 'Contact zones: Participation, materiality and the messiness of interaction'. *Environment and Planning D: Society and Space, 29*(5): 803–821.

Ball, J., and Janyst, P. (2008) Enacting research ethics in partnerships with Indigenous communities in Canada: "Do it in a good way". *Journal of Empirical Research on Human Research Ethics, 3*(2), 33–51.

Baxter, J., and Eyles, J. (1997) Evaluating qualitative research in social geography: Establishing 'rigour' in interview analysis. *Transactions of the Institute of British Geographers, 22*(4), 505–525.

Bettencourt, G. M. (2020) Embracing problems, processes, and contact zones: Using youth participatory action research to challenge adultism. *Action Research, 18*(2), 153–170. https://doi.org/10.1177/14767 50318789475

Bhatnagar, B., and Williams, A. C. (1992) *Participatory development and the World Bank: Potential directions for change.* Washington, DC: The World Bank.

Blackstock, K., Dinnie, L., Dilley, R., Marshall, K., Dunglinson, J., Trench, H., ... Griffin, A. (2015) Participatory research to influence participatory governance: Managing relationships with planners. *Area, 47*, 254–260. doi:10.1111/area.12129

Boll-Bosse, A. J. and Hankins, K. B. (2018) "These maps talk for us:" Participatory action mapping as civic engagement practice. *The Professional Geographer, 70*(2), 319–326.

Breen, J., Dosemagen, S., Warren, J., & Lippincott, M. (2015) Mapping Grassroots: Geodata and the structure of community-led open environmental science. *ACME: An International Journal for Critical Geographies, 14*(3), 849–873.

Browne, K., Banerjea, N., McGlynn, N., Bakshi, L., Banerjee, R., and Biswas, R. (2017) Towards transnational feminist queer methodologies. *Gender, Place and Culture, 24*(10), 1376–1397.

Cahill, C. (2007) The personal is political: Developing new subjectivities through participatory action research. *Gender, Place and Culture, 14*(3), 267–292.

Castellano, M. (2004) Ethics of Aboriginal research. *Journal of Aboriginal Health, 1*(1), 98e114.

Castleden, H., Garvin, T., and Huu-ay-aht First Nation. (2008) Modifying Photovoice for community-based participatory Indigenous research. *Social Science and Medicine, 66*(6), 1393–1405.

Castleden, H., Mulrennan, M., and Godlewska, A. (2012a) Community-based participatory research involving Indigenous peoples in Canadian geography: Progress? An editorial introduction. *The Canadian Geographer/Le Géographe canadien, 56*(2), 1390–1392.

Castleden, H., Sloan Morgan, V., and Lamb, C. (2012b) "I spent the first year drinking tea": Exploring Canadian university researchers' perspectives on community-based participatory research involving Indigenous peoples. *The Canadian Geographer/Le Géographe canadien, 56*(2), 160–179.

Castleden, H., Daley, K. Sloan Morgan, V., and Sylvestre, P. (2013) Settlers unsettled: Using field schools and digital stories to transform geographies of ignorance about Indigenous peoples in Canada. *Journal of Geography in Higher Education, 37*(4), 487–499.

Castleden, H., Martin, D., Cunsolo, A., Harper, S., Hart, C., Sylvestre, P., ... and Lauridsen, K. (2017) Implementing Indigenous and Western knowledge systems (part 2): "You have to take a backseat" and abandon the arrogance of expertise. *International Indigenous Policy Journal, 8*(4). https://ojs.lib.uwo.ca/index.php/iipj/article/view/7534/6178

Chatterton, P., Hodkinson, S., and Pickerill, J. (2010) Beyond scholar activism: Making strategic interventions inside and outside the Neoliberal University. *Acme: An international e-journal for critical geographies, 9*(2), 245–274.

Cooke, B. (2003) A new continuity with colonial administration: Participation in development management. *Third World Quarterly, 24*(1), 47–61.

Cooke, B. and Kothari, U. (eds.). (2001) *Participation: The new tyranny?* Zed books.

Coombes, B. (2012) Collaboration: Inter-subjectivity or radical pedagogy? *The Canadian Geographer/Le Géographe canadien, 2*(56), 290–291.

Coombes, B. (2017) Kaupapa Māori Research as Participatory Enquiry: Where's the Action? In Hoskins, T.K. and Jones, A. *Critical Conversations in Kaupapa Maori.* Wellington: Huai Publishers.

Coombes, B., Johnson, J. T., and Howitt, R. (2014) Indigenous geographies III: Methodological innovation and the unsettling of participatory research. *Progress in Human Geography, 38*(6), 845–854.

Cornwall, A. (2004) Issues of power and difference in participation in development. In S. Hickey and G. Mohan (eds.). *Participation: From tyranny to transformation: Exploring new approaches to participation in development,* 75–90. London: Zed Books.

Cornwall, A. and Jewkes, R. (1995) What is participatory research? *Social Science and Medicine, 41,* 1667–1676. doi:10.1016/0277-9536(95)00127-S

de Leeuw, S., Cameron, E. S., and Greenwood, M. L. (2012) Participatory and community-based research, Indigenous geographies, and the spaces of friendship: A critical engagement. *The Canadian Geographer/ Le Géographe canadien, 56*(2), 180–194.

de Leeuw, S. and Hunt, S. (2018) Unsettling decolonizing geographies. *Geography Compass, 12*(7), e12376.

Dunn, C. E. (2007) Participatory GIS – a people's GIS?. *Progress in Human Geography, 31*(5), 616–637.

Elwood, S. (2006) Critical issues in participatory GIS: Deconstructions, reconstructions, and new research directions. *Transactions in GIS, 10*(5), 693–708.

Ermine, W. (2007) The ethical space of engagement. *Indigenous Law Journal, 6*(1), 193–203.

Fals-Borda, O. and Rahman, M. A. (Eds) 1991. *Action and Knowledge: Breaking the Monopoly with Participatory Action Research.* London: Apex Press.

Farrow, H., Moss, P., and Shaw, B. (1995) Symposium on feminist participatory research. *Antipode, 27*(1), 71–74.

Fine, M. (1994) Working the hyphens: Re-inventing self and other in qualitative research. In Norman K. Denzin, Yvonna S. Lincoln (Eds.), *Handbook of Qualitative Research,* 70–82. Thousand Oaks: Sage.

Fine, M. and Torre, M. E. (2006) Intimate details: Participatory action research in prison. *Action Research, 4*(3), 253–269. https://doi.org/10.1177/1476750306066801

Fox, R. (2013) Resisting participation: Critiquing participatory research methodologies with young people. *Journal of Youth Studies, 16*(8), 986–999.

Franks, A. (2015) Kinder cuts and passionate modesties: The complex ecology of the invitation in participatory research. *Area, 47*(3), 237–245.

Freire, P. (1970) *Pedagogy of the oppressed.* New York: Herder and Herder.

Gilmore, R. W. (2007) *Golden gulag: Prisons, surplus, crisis, and opposition in globalizing California.* Oakland: University of California Press.

Hall, Budd L. (1981). .Participatory Research, Popular Knowledge and Power: A Personal Reflection. *Convergence, 14*(3), 6–19.

Hall, B. L. (1992) From margins to center? The development and purpose of participatory research. *The American Sociologist,* 23(4), 15–28.

Hall, B. L. and Tandon, R. (2017) Participatory research: Where have we been, where are we going? – A dialogue. *Research for All, 1*(2), 365–374.

Henkel, H. and Stirrat, R. (2001) "Participation as spiritual duty; empowerment as secular subjection". In Participation: The New Tyranny?, Edited by: Cooke, B and Kothari, U. 168–184. London: Zed.

INCITE! (2017) *The revolution will not be funded: Beyond the non-profit industrial complex.* Durham: Duke University Press.

Jupp, E. (2007) Participation, local knowledge and empowerment: Researching public space with young people. *Environment and Planning A: Economy and Space, 39,* 2832–2845. doi:10.1068/a38204

Kapoor, I. (2005) Participatory development, complicity and desire. *Third World Quarterly, 26,* 1203–1220. doi:10.1080/01436590500336849

Kauper-Brown, J. and Seifer, S. (2006) *Developing and sustaining community-based participatory research partnerships: A skillbuilding curriculum.* University of Washington: The Examining Community-Institutional Partnerships for Prevention Research Group.

Kesby, M. (2005) Retheorizing empowerment-through-participation as a performance in space: Beyond tyranny to transformation. *Signs: Journal of Women in Culture and Society, 30*(4), 2037–2065.

Kesby, M. (2007) Spatialising participatory approaches: The contribution of geography to a mature debate. *Environment and Planning A: Economy and Space, 39,* 2813–2831. doi:10.1068/a38326

Kindon, S., Pain, R. and Kesby, M., Eds. (2007) *Participatory Action Research Approaches and Methods: Connecting People, Participation and Place.* London: Routledge.

Kindon, S. (2003) Participatory video in geographic research: A feminist practice of looking? *Area, 35*(2), 142–153.

Kirby, S. and McKenna, K. (1989) *Experience, research, social change: Methods from the margins.* Toronto: Garamond Press.

Kitchin, R. M. and Hubbard, P. J. (1999) Research, action and 'critical' geographies. *Area, 31*(3), 195–198.

Kothari, U. (2005) Authority and Expertise: The Professionalisation of International Development and the Ordering of Dissent. Antipode, 37, 425–446. https://doi.org/10.1111/j.0066-4812.2005.00505.x

Lewin, K. (1946) Action research and minority problems. *Journal of Social Issues*, 2, 34e46.

Maguire, P. (1996) Considering more feminist participatory research: What's congruency got to do with it? *Qualitative Inquiry, 2,* 106–118.

Malone, R., Yerger, V., McGruder, C. and Froelicher, E. (2006) "It's like Tuskegee in reverse": A case study of ethical tensions in institutional review board review of community-based participatory research. *American Journal of Public Health, 96*(11), 1914–1919.

Martin, R. E., Murphy, K., Hanson, D., Hemingway, C., Ramsden, V., Buxton, J., ... & Hislop, T. G. (2009) The development of participatory health research among incarcerated women in a Canadian prison. *International Journal of Prisoner Health, 5*(2), 95–107.

Minkler, M. (2004) Ethical challenges for the 'outside' researcher in community-based participatory research. *Health Education and Behavior, 31*(6), 684–697.

Minkler, M. and Wallerstein, N. (2003) *Community-based participatory research for health.* San Francisco: John Wiley.

Mohan, G. (1999) Not so distant, not so strange: The personal and the political in participatory research. *Ethics, Place and Environment, 2*(1), 41–54.

mrskinpaisby (2008) 'Taking stock of participatory geographies: Envisioning the communiversity'. *ransactions of the Institute of British Geographers, 33*(3): 292–299.

Nadasdy, P. (2005) The anti-politics of TEK: The institutionalization of co-management discourse and practice. *Anthropologica, 47,* 215–232. doi:10.2307/25606237

National Aboriginal Health Organization. (2005) *Ownership, control, access, and possession (OCAP) or self-determination applied to research: A critical analysis of contemporary First Nations research and some options for First Nations communities.* Ottawa: First Nations Centre.

Oaks, L., Israel, T., Conover, K. J., and Avellar, T. R. (2019) Community-based participatory research with invisible, geographically-dispersed communities: Partnering with lesbian, gay, bisexual, transgender and queer communities on the California central coast. *Journal for Social Action in Counselling and Psychology, 11*(1), 15–32.

Pain, R. (2004) Social Geography, Participatory Research. *Progress in Human Geography, 28*(5) 652–663.

Pain, R. and Francis, P. (2003) Reflections on participatory research. *Area, 35*(1), 46–54.

Pain, R., Finn, M., Bouveng, R. and Ngobe, G. (2013) Productive tensions – engaging geography students in participatory action research with communities. *Journal of Geography in Higher Education, 37*(1), 28–43.

Pain, R., Kesby, M., and Askins, K. (2012) The politics of social justice in neoliberal times: A reply to Slater [with Reply]. *Area, 44*(1), 120–123.

Pain, R. and Kindon, S. (2007) Guest editorial. *Environment and Planning A: Economy and Space, 39*, 2807–2812. doi:10.1068/a39347

Parker, B. (2006) Constructing community through maps? Power and praxis in community mapping. The Professional Geographer, 58(4), 470–484.

Povinelli, E. A. (2011) Routes/Worlds. *E-Flux Journal* (27), 1–12.

Pratt, M. L. (1991) Arts of the contact zone. *Profession*, 33–40.

Pualani Louis, R. (2007) Can you hear us now? Voices from the margin: using Indigenous methodologies in geographic research. Geographic Research, 45(2), 130–139.

Reason, P. and Bradbury, H. (2001) *Handbook of action research.* Thousand Oaks, CA: Sage Publications.

Reynolds, K., Block, D. R., Hammelman, C., Jones, B. D., Gilbert, J. L. and Herrera, H. (2020) Envisioning radical food geographies: Shared learning and praxis through the food justice scholar-activist/activist-scholar community of practice. *Human Geography, 13*(3), 277–292.

Ritterbusch, A. E. (2019) Empathy at knifepoint: The dangers of research and lite pedagogies for social justice movements. *Antipode, 51*(4), 1296–1317. https://doi.org/10.1111/anti.12530 https://doi.org/10.1111/anti.12530

Robinson, J. A. and Hawthorne, T. L. (2018) *Making Space for Community-Engaged Scholarship in Geography.* The Professional Geographer, 70(2), 277–283.

Rodriguez, A. D. (2021). *Diverging space for deviants: The politics of Atlanta's public housing.* Athens: University of Georgia Press.

Shannon, J., Eaves, L. E., Hankins, K. B., Jung, J., Robinson, J., Bosse, A. J., ... and Fischer, H. (2020) Community geography: Toward a disciplinary framework. *Progress in Human Geography*, https://doi.org/10.1177/0309132520961468

Smith, L. (1999) Decolonizing methodologies: Research and Indigenous peoples. London: Zed Books.

Solís, P., Rajagopalan, S., Villa, L., Mohiuddin, M. B., Boateng, E., Wavamunno Nakacwa, S., & Peña Valencia, M. F. (2020) Digital humanitarians for the Sustainable Development Goals: YouthMappers as a hybrid movement. *Journal of Geography in Higher Education*, 46(1), 80–100.

Spry, L. (2020) Unsettling settlers' colonial privilege through performance: Movement, sound, participation, play, laughter. [Doctoral dissertation, Queen's University]. Qspace.

Stiegman, M. and Castleden, H. (2015) Leashes and lies: Navigating the colonial tensions of institutional ethics of research involving Indigenous peoples in Canada. *International Indigenous Policy Journal*, 6(3). https://doi.org/10.18584/iipj.2015.6.3.2

Stirrat, R. L. and Henkel, H. (1997) The development gift: The problem of reciprocity in the NGO world. *The Annals of the American Academy*, 554, 66–80.

Sylvestre, P., Castleden, H., Martin, D., and McNally, M. (2018) "Thank you very much... you can leave our community now.": Geographies of responsibility, relational ethics, acts of refusal, and the conflicting requirements of academic localities in Indigenous research. *ACME: An International Journal for Critical Geographies*, 17(3), 750–779.

Tandon, R. (2002) *Participatory research: Revisiting the roots*. New Delhi: Mosaic Books.

Torre, M. E., Fine, M., Stoudt, B. G., and Fox, M. (2012) Critical participatory action research as public science. In H. Cooper, M. Camic, L. D.L., A. T. Panter, D. Rindskopf, and K. J. Sher Eds.), *APA Handbook of Research Methods in Psychology, Vol. 2.*, 171–184. Washington DC: American Psychological Association.

Wallerstein, N. and Duran, B. (2006) Using community-based participatory research to address health disparities. Health Promotion Practice, 7(3), 312 e323.

Wynne-Jones, S., North, P., and Routledge, P. (2015) Practising participatory geographies: Potentials, problems and politics. *Area*, 47(3), 218–221.

Ybarra, M. (2021) Site fight! Toward the abolition of immigrant detention on Tacoma's tar pits (and everywhere else). *Antipode,* 53(1), 36–55.

29

THE METHODOLOGICAL IMPLICATIONS OF INTEGRATING LIVED EXPERIENCE IN GEOGRAPHIC RESEARCH ON INEQUALITIES

Claire Thompson

Lived Experience and Geography

How people perceive and form attachments to places are bound up with different experiences. Social geographers in the mid-twentieth century drew on a range of social theories to examine the relationship between experience and place (Del Casino Jr, 2017). The concept of lived experience is taken from phenomenology. It refers to a representation of the experiences of a given person and the knowledge they gain from them (Hiles, 2008). The term has become increasingly popular in recent years across disciplines and outside the academy, particularly in the third sector (Hoerger, 2016). Calls to 'champion the voices' of lived experience and 'support and value' lived experience are now commonplace in the strategies and mission statements of charities and campaign groups, especially those working around mental health, trauma, and racism. For researchers, lived experience allows them to investigate the subjective, experiential and emotional dimensions of inequality (Longhurst & Hargreaves, 2019).

While lived experience is a useful concept, it is also a somewhat problematic one (Paley, 2014). Is not all experience 'lived'? What other kinds of experience are there? Attaching the label of 'lived experience' to an account can provide an assumed, rather than justified, authenticity that does not consider the socio-historical processes or subjectivities that underpin that account (McIntosh & Wright, 2019). Experiences are situated and attempting to define what is discrete and valuable about employing the concept of 'lived' experience is not straightforward. A geographical lens offers a way of acknowledging situatedness and materiality. Lived experience can be understood as the intrinsically corporeal link between practice and consciousness (Merleau-Ponty, 1962) and, as such, social geography and the concept of embodiment have located it in the mid-point between mind and body: the intersubjective space of

DOI: 10.4324/9781003038849-33

perception (Simonsen, 2007). This presents challenges in terms of research methods. How does the researcher access this intersubjective space? And how does the researcher interpret what is found there, given that the purpose is to determine and interpret the meanings of experiences, rather than uncritically treat them as straightforward 'facts' (Lindseth & Norberg, 2004; Paley, 2014)? These challenges are amplified when lived experiences of inequality are the subject of research.

The Challenges of Researching the Lived Experience of Inequality

In the last forty years, inequality has increased both within and between nations (Dorling, 2015). Economic inequality, in particular, undermines personal and social wellbeing and impairs the safe functioning of societies (Rodriguez-Bailon et al., 2017). Inequality can be understood in terms of geographically specific experiences and perceptions of hierarchies of differences (Ķešāne, 2019). It is a multidimensional and emotionally laden phenomenon (Holmes & McKenzie, 2018), engendering a variety of feelings including stigma (Pearce, 2012), pride (Markus, 2008), shame, loneliness (Hodgetts, Radley, Chamberlain, & Hodgetts, 2007), and belonging (Dempsey, 1990) that are intertwined with attachments to place and belonging. Inequalities are rooted in socio-historical structures that also serve to shape the perceptions of those experiencing them (Panahi & Abedini, 2020). What people say in response to questions about their experiences (of inequality) is derived not only from individual meaning-making, but from social norms and collective understandings (Paley, 2014). Put simply, exploring 'lived experience' is not a straightforward matter of asking someone with that experience the 'correct' questions and, thereby, being granted unmediated access to a readily understandable, static and authentic 'truth' about the phenomena being studied.

For example, researchers frequently identify a gap between extant inequalities and lay perceptions of them (Garcia, Sanchez, Sanchez-Youngman, Vargas, & Ybarra, 2015; Irwin, 2018). In other words, people often underestimate the extent of inequalities – even the ones they are experiencing – and see themselves as being somewhere 'near the middle' rather than as marginalised or on a low-income (Irwin, 2018). Explanations for this discrepancy range from the psychological – it *functions as* a defence mechanism or coping strategy for individuals facing adversity (Davydov, Stewart, Ritchie, & Chaudieu, 2010), to the structural and ideological – it is a *function of* false consciousness and the reproduction of systemic inequalities (Jost & Banaji, 1994; Newman, Johnston, & Lown, 2015).

A further complication is that interpretations of experiences of inequality are dynamic. People can oscillate, depending on context and audience, between positive assertions of their personal agency and – at other times - articulating a critical awareness of, and vulnerability to, the socio-structural processes that drive the inequalities they face (Savage, Bagnall, & Longhurst, 2001). That is not to say that people talking about their experiences of inequality are being inconsistent or even dishonest, rather that accounts of lived experience have material consequences and different framings are appropriate for different situations.

The necessity of this dynamism can be seen when exploring how people interact with the social housing system. Affordable, good quality and secure accommodation in the form of social housing is thought to be an effective mechanism to reduce inequalities. There is a shortage of social housing in England and access to it is often dependent on narrow and rigid criteria at the local level around conditionality and 'need', which can be viewed as excluding some vulnerable groups (D. Robinson, 2013). When recounting their housing problems to researchers, as part of a larger study on Olympic Regeneration in London (Smith et al., 2012), the participants

I spoke to recounted their housing needs almost as if presenting-their-case: giving a strong sense of having had to 'perform' various versions of this narrative for state and housing officials in the past (Thompson, Lewis, Greenhalgh, et al., 2017). They told us about their years 'on the list' waiting for social housing and how their health was being affected by the substandard housing they were currently living in. We were struck by how bleak these accounts were and the lack of agency participants described.

However, when talking about other topics and experiences they typically presented a different interpretation of housing problems: they down-played the impact on their wellbeing and talked about how well they coped; they shared happy memories about their homes and neighbourhoods; they explained their resilience and resourcefulness. Both types of account revealed the lived experience of social and housing inequalities, but context and audience required different framings of them. In a more social setting, lengthy and sometimes graphic descriptions of health problems and highly personal experiences of poverty might not be appropriate and could be seen as embarrassing or even humiliating. By contrast, when interacting with social housing officials and – by association – when recounting these interactions to researchers – focusing on the positives might very likely have a detrimental effect on their chances of securing social housing.

Demonstrably, interpretations of the lived experience of housing inequalities are dynamic. They are mediated by place, in a relational sense: by the social and physical context in which they are experienced and recounted (S. Cummins, Curtis, Diez-Roux, & Macintyre, 2007). Having to navigate the social housing system and tailor narratives to increase the chances of accessing scarce resources is a complex and context specific experience of marginalisation. It requires a degree of performativity and compartmentalisation that can be difficult to understand or relate to from the outside. In order to more fully explore these experiences, researchers need to adopt approaches that are sensitive to context and place.

Geographical Approaches to Researching the Lived Experience of Inequality

Exploring the situatedness and placed-based nature of lived experiences of inequality requires a geographically informed approach. For geographers, place can be thought of not only in terms of location in physical space and position relative to other places, but also as a specific context containing distinctive idiosyncratic attributes that is imbued with social significance and values (Curtis, 2004). A relational approach to place proposes to un-fix taken for granted understandings of place and identity (Castro & Lindbladh, 2004; Cattell, 2001; S. Cummins et al., 2007; Frohlich, Corin, & Potvin, 2001). It treats place and context as relative, dynamic and fluid (S. Cummins et al., 2007). In this model, the voices of individuals and groups are accounted for and their descriptions of place are incorporated. A focus on place and context helps to avoid essentialising or simplifying lived experiences; presenting them as static. Exploring how experience is mediated by place addresses the complexities of marginalisation and inequality, providing a critical lens.

Geographers employ a range of methods to operationalise place-based approaches. Qualitative longitudinal research (QLR), visual methods, diaries, ethnography, and go-along interviews have all been used to investigate the dynamic and instersubjective spaces of lived experience (Neale, Henwood, & Holland, 2012; Thompson, Lewis, & Taylor, 2017; Thompson & Reynolds, 2019; Thompson, Smith, & Cummins, 2018). Sometimes this is done simply by spending time in different contexts with participants (ethnography). It can also include travelling with participants (mobile methods) and asking participants to recount their experiences at different time points (QLR). More recent approaches involve asking participants to generate images (visual methods) or written accounts (diaries) in different settings and/or at different time points. These approaches acknowledge and even go some way to overcoming the fact that we can only ever access partial lived experience that is socially constructed and situated. The sections below explore some of these methods in more detail.

Go Alongs

In the above example about housing, we used go-along interviews alongside other methods to explore the mediating influence of place on housing narratives (Thompson, Lewis, Greenhalgh, et al., 2017). A go-along interview is an ethnographic mixture of observation and interview, concentrated around a particular place, journey or activity. The interviewee takes the researcher on a journey or visits a place. This is a participant-led method that is intended to explore participants' interactions with place (Carpiano, 2009). When researching lived experiences of inequality, this approach helps the researcher to access the dynamic nature of accounts because the physical and social context serve as prompt – the interviewee takes you to the place or thing or activity that they are telling you about. Some of the participants we interviewed walked around their housing estates and/or neighbourhoods whilst being interviewed and letting their surroundings serve as talking point and prompt.

Mobile methods, like go-alongs, have the potential to explore embodied movement and its non-representational aspects, like sensory, emotional and affective experiences (Spinney, 2014) (see also Gorman and Andrews, this volume). As such, they can access lived experience: the mid-point between mind and body (Simonsen, 2007). Lived experience is more than perception and opinion. Theorising the body and non-representation aspects is crucial to how we understand lived experience (Moss & Dyck, 2003). The physicality and embodied nature of mobile methods highlights the non-representational aspects of lived experience by accompanying the participant as they experience and interpret them in situ. The smells, sounds, feelings, and affective states that arise from living on a particular housing estate or working in particular industry are brought into focus when the research is conducted while actively doing and moving in those settings. Go-alongs seek to explore reflexive aspects of lived experience in situ (Kusenbach, 2003).

Go-alongs are typically used in conjunction with other methods (like semi-structured interviews) and increasingly used to achieve methodological triangulation. They explore the complexity of phenomena under study by complementing several theoretical perspectives and various methods in order to generate a deeper understanding of what is studied (Carpiano, 2009; Flick, Hirseland, & Hans, 2019). Flick and colleagues (2019) used them in this way, alongside interviews, to investigate the lived experiences of integration and belonging among low-income immigrants to Germany. Data from go-alongs highlighted participants' disrupted senses of belonging and the ways in which job centres were contested spaces: dynamically functioning as places of reinforced discrimination, as inadequate and inept institutions, and somewhere that essential assistance might be accessed (albeit sporadically) (Flick et al., 2019). These conflicting interpretations and the contradictions and disruptions that mobile methods uncover, allows the researcher to delve into the nuance of lived experiences of inequality. Lived experience of phenomena, including inequality, are dynamic and vary by the different contexts and places in which they occur. Equally, places themselves are dynamic. Go-alongs allow the researcher to interpret how this dynamism unfolds in space and time.

Ethnography

Ethnography has a methodological connection with lived experience (McIntosh & Wright, 2019). Ethnographic inquiry aims to achieve intimate familiarity with the subject matter by being sensitive to the interpretive and interactive features of (human) group life (Prus, 1996). It allows the researcher to examine and understand everyday routines and practices. Lived experiences of inequality generate a range of negative emotions, such as shame and stigma that are more than residual effects of marginalisation. They are part of the practices that feeds back into unequal relationships and help reproduce inequity (Loveday, 2016). In which case, these emotions and affective states can be understood as 'affective practices' (Wetherell, 2014). Loveday (2016) used ethnographic methods (narrative-style interviews) to research the lived experience of social class and gender inequalities in higher education. The narrative-style interviews revealed that, over time, the affective practice of shame becomes performative: it feeds back into social norms legitimising inequities, and it becomes embodied and internalised (Loveday, 2016). Stigma and shame are inherent potentialities in all systems of stratification (Goffman, 1963). Where there is difference, typically, there is inequality.

The reproduction of inequalities results in identity dilemmas. Fixed identity categories, like class, ethnicity, gender, and sexuality, are both the basis for oppression and the basis for collective power (Dunn & Creek, 2015; Gamson, 1995). Our identities both fix us within hierarchies and, at the same time, offer us a source of belonging and support. Ethnographic research into the lived experience of inequalities can uncover how marginalised groups negotiate identity dilemmas. Loveday's work described not only how shame impacted upon the actions and self-perceptions of participants, but also how they made sense of and negotiated their classed and gendered identities (Loveday, 2016).

Ethnographic approaches can illuminate how lived experiences of inequality intersect with the broader structural drivers of those inequalities; they provide context. For example, we applied Wacquant's (2007) notion of 'advanced marginality' as a theoretical framing of (food) poverty in a recent study of food banking (Thompson et al., 2018). This entailed situating lived experiences of food poverty within the mechanisms of contemporary urban poverty – looking at how marginalised communities become physically, geographically and economically disconnected from wider society (I. Cummins, 2016).

We took an ethnographic approach, using a combination of observations and interviews (Thompson et al., 2018). This allowed us to focus on wider context and the access the social worlds of participants. Rather than simply asking them about food banking and food poverty, we sought to understand how these aspects were experienced in the context of their everyday practices and routines. In fact, when interviewing, participants rarely mentioned food straight away. Instead, they spoke about the events, often traumatic ones, that had first led them to using a food bank. Sometimes participants did not even talk about food in their first interviews. Uncertainty, hardship, and lack of agency around food were part of broader precarious experiences of inequality. To paraphrase the observation of a particularly reflective study participant, food poverty is about poverty, not food. Lived experiences of food poverty and food banking are both embedded in extreme precarity and marginalisation and also serve to perpetuate them (Thompson et al., 2018). Lived experience is at the intersection of intersubjective experiences and specific socio-historical and temporal locations (McIntosh & Wright, 2019).

Again, a geographic place-based approach was able to address the complexities of lived experience. Food poverty is a term and label that is externally applied to the people experiencing it by various policy, statutory, political, and academic sectors. In the years I have spent researching food poverty, none of the people I have spoken to who have experienced 'food poverty' have ever described it in those terms. Instead, they described the deprivation and income crises they had endured that led to them not being able to afford food. In these accounts, a lack of food was framed as a distressing symptom of a cluster of poverty-related problems over which they often had little or no control (including housing, welfare benefits, casual employment, caring responsibilities). Experiences specific to food were mostly talked about in relation to accessing food banks or when we were physically in food banks together. In other words, it was dependent on context. That is not to say that 'food poverty' is not a valid and useful concept. As inequality in general, and levels of food poverty in particular, continue to rise, it is vital that we understand the lived experience of them. However, a place and space sensitive approach is required to avoid over simplifying the issue.

Visual Methods

In recent years, there has been increasing interest, especially in geography, around using visual methods as a way of getting closer to lived experience (Oldrup & Carstensen, 2012). Visual methods are used to understand and interpret images and/or visual representations including photographs, video footage, drawings and paintings. While they are well established in anthropology and sociology, they are still relatively new to inequalities research (Glaw, Inder, Kable, & Hazelton, 2017). Video, in particular, has been used to elicit the embodied, multisensory and relational nature of people's place experiences that are the focus of much recent geography research (Kaley, Hatton, & Milligan, 2019). Like go-alongs, visual methodologies are typically used alongside other methods and as part of a broader ethnographic approach. Morrow (2000) used visual methods alongside freely-written accounts in a school-based project exploring young people's subjective experiences of their neighbourhoods. Young people were asked to take photographs of and then draw maps of their local neighbourhoods. The research revealed how young people were excluded from the social life of their local communities by virtue of their age: they were treated with suspicion in retail spaces and did not have access to well-maintained and resourced spaces in which they could socialise (Morrow, 2000). Visual methods are a popular way of engaging young people in research. Research within the geographies of children and young people makes ample use of participatory visual methods to explore different places and spaces from children's perspectives (Yarwood & Tyrrell, 2012).

In the study of Olympic regeneration mentioned in the go-alongs section above (Smith et al., 2012), we also used participatory visual methods to engage children and young people. To explore neighbourhood perceptions and lived experiences of relatively deprived London boroughs, we organised a series of school video focus group workshops. We split the young people into small groups and asked them to organise and film interviews with each other about topics arising from the focus group discussion (Thompson, Lewis, & Taylor, 2017). In this way, we were able to facilitate and explore the children's co-interpretation of their lived experience. The data highlighted that the young people experienced the redeveloped parts of their neighbourhood – that they described as 'not for us' – as an artificial façade that hid the area's dirt and deprivation from visitors (Thompson et al., 2015).

At the time of data collection, some of the research team – including myself – lived in neighbouring London boroughs and had lived or worked in the area at some time. In retrospect, we probably tacitly assumed that we had a reasonable grasp of issues in the area. There had been lots of coverage of the Games in the national and local press and 'securitization' had been flagged as a concern: specifically, that temporary security measures for the event were oppressive and unfairly aimed and marginalised groups. This is something we wanted to explore. To our surprise, the young people in the focus group workshops accepted and even welcomed the increased security measures. Increased police presence and CCTV was described as having a protective effect and made them feel safer in a neighbourhood that they considered to be unsafe (Thompson et al., 2015). The 'distance' between researcher and participant is fluid and can be both under and over-estimated, which can result in judgements and assumptions that undermine the research. In this instance, using participant-led visual methods challenged our assumptions about what the lived experience of young people in the borough might entail.

Implications for Research and Practice

Severe inequalities increase the prevalence of most of the health and social problems that tend to occur more frequently among those on a low income (Wilkinson & Pickett, 2017). Obesity, teenage birth rates, mental illness, homicide, low social capital, hostility, and racism are more prevalent in more unequal societies (Wilkinson & Pickett, 2007). While we know and agree that inequality is harmful, there is little understanding of the processes that make it so (Wilkinson & Pickett, 2017). A focus on the lived experience of inequality presents a useful way of investigating those processes as they unfold in time and space. Geography has much to contribute here. Geographers have long been interested in poverty and inequality (Dorling, 2015). The geographical framing of lived experience as an embodied space of perception (Simonsen, 2007) and the range of theoretically-informed geographical methods described in this chapter offer a means of exploring the dynamic nature of lived experience. As the studies discussed here illustrate, lived experience of inequality is not simple or straightforward. Caution has been expressed around how the emergent use of 'lived experience' tends to employ it as a free-floating notion, untethered from the theoretical and methodological contexts in which it originated (McIntosh & Wright, 2019).

In more practical terms, the growing practice of prioritising lived-experience input when designing and conducting research, resulting in the privileging of co-production, is laudable. It is, however, also somewhat at odds with the way higher education and research institutions function. Research institutions create moments of dissonance for precariously placed researchers and students (Barrett & Bosse, 2021), who will likely be in a short-term and/or temporary post with many other competing projects and responsibilities. There is limited time and resources to establish and maintain the community links and relationships necessary to facilitate lived-experience input and co-production in a meaningful and authentic way. Barret and Bosse (2021) explain that this tension can lead to heightened feelings of risk that may induce researchers to compromise their own ethics or values to ensure both the ongoing continuation of the partnership and the desired goals of the community partners.

Conclusion

As social, economic and health inequalities widen, so too does the urgency with which we must seek to understand how these inequalities are perpetuated and experienced. The purpose of researching lived experiences of inequality is to, ultimately, inform responses to them and effect change. However, we must retain critical engagement in the face of challenges and avoid exploitative practices, tokenism, and the creation of pressurised research environments which do not allow the proper time and scope to develop the partnerships and alliances necessary to successful integrate lived experience into research.

Geographical methods, including go-alongs, ethnography and visual methods, support this endeavour by helping to access the dynamic nature of lived experiences of inequality and, at the same time, situating them within every day routines and broader social contexts. Geography offers a range of theories and methods to investigate experience and inequities, and community geography (Shannon et al., 2020) provides a possible framework for theorising and applying them. Geographical approaches to inequalities, inequities, and lived experience can, and do, inform public health and social policy.

Along with the UK, the US is among the richest and most unequal nations in the world (Dorling, 2010). Danny Dorling has challenged British and American geographers to 'wake up' to the context of extreme income inequality in which they are situated and question how it might skew writing and research (Dorling, 2015). This is a challenge that all human geographers need to take-up. The rise of community geography programmes in the US can inform responses to this challenge. Community geography is a pragmatic approach enacted by local coalitions of academics, policy-makers, residents, and sometimes activists (Shannon et al., 2020). These coalitions focus on work which enables under-resourced communities to better address community social and development challenges (J. A. Robinson, Block, & Rees, 2017). This approach places engagement and experience at the core of social inquiry (Shannon et al., 2020), meaning that research agendas – not just research findings – can be informed by lived experiences of inequalities. The longer-term and more stable ties to partners and communities outside of the academy that this approach entails can also help counteract potentially exploitative practices of engagement and ethical compromises that prevailing systems of research can inadvertently produce (Barrett & Bosse, 2021). Geographical methods are helping us to understand the complexities of lived experiences of inequality. Perhaps geographical practices more generally, like that of community geography, can help us embed these methods and build the communities of practice needed to support their use and achieve change.

References

Barrett, E., and Bosse, A. J. (2021) Community geography for precarious researchers: Examining the intricacies of mutually beneficial and co-produced knowledge. *GeoJournal*. https://doi.org/10.1007/s10708-020-10358-2

Carpiano, R. M. (2009) Come take a walk with me: The "Go-Along" interview as a novel method for studying the implications of place for health and well-being. *Health and Place, 15*(1), 263–272. doi:10.1016/j.healthplace.2008.05.003

Castro, P. B., and Lindbladh, E. (2004) Place, discourse and vulnerability - a qualitative study of young adults living in a Swedish urban poverty zone. *Health and Place, 10*(3), 259–272. doi:10.1016/j.healthplace.2003.11.001

Cattell, V. (2001) Poor people, poor places, and poor health: The mediating role of social networks and social capital. *Social Science and Medicine, 52*(10), 1501–1516.

Cummins, I. (2016) Wacquant, urban marginality, territorial stigmatization and social work. *Aotearoa New Zealand Social Work, 28*(2), 75–83.

Cummins, S., Curtis, S., Diez-Roux, A. V., and Macintyre, S. (2007) Understanding and representing 'place' in health research: A relational approach. *Social Science and Medicine, 65*(9), 1825–1838. doi:10.1016/j.socscimed.2007.05.036

Curtis, S. (2004) *Health and Inequality: Geographical perspectives*. London: Sage.

Davydov, D. M., Stewart, R., Ritchie, K., and Chaudieu, I. (2010) Resilience and mental health. *Clinical Psychology Review, 30*(5), 479–495.

Del Casino Jr, V. J. (2017) Social Geography. In N. Castree, M. F. Goodchild, A. Kobayashi, W. Liu, and R. A. Marston (Eds.), *International Encyclopedia of Geography, 15 Volume Set: People, the Earth Environment and Technology*. London: John Wiley & Sons.

Dempsey, K. (1990) *Smalltown: A study of social inequality, cohesion and belonging*. Melbourne: Oxford University Press.

Dorling, D. (2010) *Injustice*. Bristol: Policy Press.

Dorling, D. (2015) Income inequality in the UK: Comparisons with five large Western European countries and the USA. *Applied Geography, 61,* 24–34.

Dunn, J. L. and Creek, S. J. (2015) Identity dilemmas: Toward a more situated understanding. *Symbolic Interaction, 38*(2), 261–284.

Flick, U., Hirseland, A., and Hans, B. (2019) Walking and talking integration: Triangulation of data from interviews and go-alongs for exploring immigrant welfare recipients' sense(s) of belonging. *Qualitative Inquiry, 25*(8), 799–810.

Frohlich, K. L., Corin, E., and Potvin, L. (2001) A theoretical proposal for the relationship between context and disease. *Sociology of Health and Illness, 23*(6), 776–797.

Gamson, J. (1995) Must identity movements self-destruct? A queer dilemma. *Social Problems, 42*(3), 390–407.

Garcia, J. A., Sanchez, G. R., Sanchez-Youngman, S., Vargas, E. D., and Ybarra, V. D. (2015) Race as lived experience: The impact of multi-dimensional measures of race/ethnicity on the self-reported health status of Latinos. *Du Bois Review: Social Science Research on Race, 12*(2), 349–373.

Glaw, X., Inder, K., Kable, A., and Hazelton, M. (2017) Visual methodologies in qualitative research: Autophotography and photo elicitation applied to mental health research. *International Journal of Qualitative Methods, 16*(1). https://doi.org/10.1177/1609406917748215

Goffman, E. (1963) *Stigma: Notes on the Management of Spoiled Identity.* New York: Simon and Schuster.

Hiles, D.R. (2008) Heuristic Inquiry. In L. M. Green (Ed.), *The SAGE Encyclopedia of Qualitative Research Methods* (pp. 390–393). Thousand Oaks: Sage Publications.

Hodgetts, D., Radley, A., Chamberlain, K., and Hodgetts, A. (2007) Health Inequalities and Homelessness: Considering Material, Spatial and Relational Dimensions. *Journal of Health Psychology, 12*(5), 709–725.

Hoerger, J. (2016) Lived Experience vs. Experience. Retrieved from https://medium.com/@jacobhoerger/lived-experience-vs-experience-2e467b6c2229

Holmes, M. and McKenzie, J. (2018) Relational happiness through recognition and redistribution: Emotion and inequality. *European Journal of Social Theory, 22*(4), 439–457.

Irwin, S. (2018) Lay Perceptions of inequality and social structure. *Sociology, 52*(2), 211–227.

Jost, J. T., and Banaji, M. R. (1994) The role of stereotyping in system-justification and the production of false consciousness. *British Journal of Social Psychology, 33,* 1–27.

Kaley, A., Hatton, C., and Milligan, C. (2019) Health geography and the 'performative' turn: Making space for the audio-visual in ethnographic health research. *Health & Place, 60.* https://doi.org/10.1016/j.healthplace.2019.102210

Ķešāne, I. (2019) The lived experience of inequality and migration: Emotions and meaning making among Latvian emigrants. *Emotion, Space and Society, 33,* 100597. doi:https://doi.org/10.1016/j.emospa.2019.100597

Kusenbach, M. (2003) Street Phenomenology: The go-along as ethnographic research tool. *Ethnography, 4*(2), 455–485.

Lindseth, A., and Norberg, A. (2004) A phenomenological hermeneutical method for researching lived experience. *Scandinavian Journal of Caring Sciences, 18*(2), 145–153.

Longhurst, N., and Hargreaves, T. (2019) Emotions and fuel poverty: The lived experience of social housing tenants in the United Kingdom. *Energy Research & Social Science, 56,* 101207. doi:https://doi.org/10.1016/j.erss.2019.05.017

Loveday, V. (2016) Embodying deficiency through 'affective practice': Shame, relationality, and the lived experience of social class and gender in higher education. *Sociology, 50*(6), 1140–1155.

Markus, H. R. (2008) Pride, prejudice, and ambivalence: Toward a unified theory of race and ethnicity. *American Psychologist, 63*(8), 651–670.

McIntosh, I., and Wright, S. (2019) Exploring what the notion of 'lived experience' offers for social policy analysis. *Journal of Social Policy, 48*(3), 449–467.

Merleau-Ponty, M. (1962) *Phenomenology of Perception.* London: Routledge and Kegan Paul.

Morrow, V. M. (2000) 'Dirty looks' and 'trampy places' in young people's accounts of community and neighbourhood: Implications for health inequalities. *Critical Public Health, 10*(2), 141–152.

Moss, P., and Dyck, I. (2003) Embodying social geography. In K. Anderson, M. Domosh, S. Pile, and N. Thrift (Eds.), *Handbook of Cultural Geography.* London: Sage. pp. 58–73.

Neale, B., Henwood, K., and Holland, J. (2012) Researching lives through time: An introduction to the Timescapes approach. *Qualitative Research, 12*(1), 4–15.

Newman, B. J., Johnston, C. D., and Lown, P. L. (2015) False consciousness or class awareness? local income inequality, personal economic position, and belief in American meritocracy. *American Journal of Political Science, 59*, 326–340.

Oldrup, H. H., and Carstensen, T. A. (2012) Producing geographical knowledge through visual methods. *Geografiska Annaler: Series B, Human Geography, 94*(3), 223–237.

Paley, J. (2014) Heidegger, lived experience and method. *Journal of Advanced Nursing, 70*(7), 1520–1531.

Panahi, R., and Abedini, S. (2020) The perception of gender inequality and lived experience of women: The case of Iranian women in Maragheh. *Research on Humanities and Social Sciences, 10*(1), 19–36.

Pearce, J. (2012) The 'blemish of place': Stigma, geography and health inequalities. A commentary on Tabuchi, Fukuhara & Iso. *Social Science and Medicine, 75*(11), 1921–1924.

Prus, R.C. (1996) *Symbolic interaction and ethnographic research: Intersubjectivity and the study of human lived experience*. Albany: State University of New York Press.

Robinson, D. (2013) Social housing in England: Testing the logics of reform. *Urban Studies, 50*(8), 1489–1504.

Robinson, J. A., Block, D., and Rees, A. (2017) Community geography: Addressing barriers in public participation GIS. *The Cartographic Journal, 54*(1), 5–13.

Rodriguez-Bailon, R., Bratanova, B., Willis, G. B., Lopez-Rodriguez, L., Sturrock, A., and Loughnan, S. (2017) Social class and ideologies of inequality: How they uphold unequal societies. *Journal of Social Issues, 73*(1), 99–116.

Savage, M., Bagnall, G., and Longhurst, B. (2001) Ordinary, ambivalent and defensive: Class identities in the Northwest of England. *Sociology, 35*(4), 875–892.

Shannon, J., Hankins, K. B., Shelton, T., Bosse, A. J., Scott, D., Block, D., ... Nicolas, A. (2020) Community geography: Toward a disciplinary framework. *Progress in Human Geography*, 45(5), 1147–1168.

Simonsen, K. (2007) Practice, spatiality and embodied emotions: An outline of a geography of practice. *Human Affairs, 17*, 168–181.

Smith, N. R., Clark, C., Fahy, A. E., Tharmaratnam, V., Lewis, D. J., Thompson, C., ... Cummins, S. (2012) The Olympic Regeneration in East London (ORiEL) study: Protocol for a prospective controlled quasi-experiment to evaluate the impact of urban regeneration on young people and their families. *Bmj Open, 2*(4), 11. doi:10.1136/bmjopen-2012-001840

Spinney, J. (2014) Close encounters? Mobile methods, (post)phenomenology and affect. *Cultural Geographies, 22*(2), 231–246.

Thompson, C., Lewis, D.J., Greenhalgh, T., Smith, N. R., Fahy, A. E., and Cummins, S. (2017) "I don't know how I'm still standing" a Bakhtinian analysis of social housing and health narratives in East London. *Social Science and Medicine, 177*, 27–34.

Thompson, C., Lewis, D., and Taylor, S. J. C. (2017) The use of video recording in longitudinal focus group research. In R. Barbour & D. Morgan (Eds.), *A New Era of Focus Groups*. Basingstoke: Palgrave Macmillan, pp. 207–225.

Thompson, C., Lewis, D. J., Greenhalgh, T., Smith, N. R., Fahy, A. E., and Cummins, S. (2015) "Everyone was looking at you smiling": East London residents' experiences of the 2012 Olympics and its legacy on the social determinants of health. *Health and Place, 36*, 18–24.

Thompson, C., and Reynolds, J. (2019) Reflections on the go-along: How "disruptions" can illuminate the relationships of health, place and practice. *The Geographical Journal, 185*(2), 156–167.

Thompson, C., Smith, D., and Cummins, S. (2018) Understanding the health and wellbeing challenges of the food banking system: A qualitative study of food bank users, providers and referrers in London. *Social Science and Medicine, 211*, 95–101.

Wetherell, M. (2014) Feeling rules, atmospheres and affective practice: Some reflections on the analysis of emotional episodes. In C. Maxwell and P. Aggleton (Eds.), *Privilege, Agency and Affect* (pp. 221–239). Basingstoke: Palgrave Macmillan.

Wilkinson, P., and Pickett, K. E. (2007) The problems of relative deprivation: Why some societies do better than others. *Social Science and Medicine, 65*(9), 1965–1978.

Wilkinson, P., and Pickett, K. E. (2017) The enemy between us: The psychological and social costs of inequality. *European Journal of Social Psychology, 47*, 11–24.

Yarwood, R., and Tyrrell, N. (2012) Why children's geographies? *Geography, 97*(3), 123–128.

30

WHAT ROLE FOR MORE-THAN-REPRESENTATIONAL, MORE-THAN-HUMAN INQUIRY?

Richard Gorman and Gavin Andrews

...one can say that 'geo-graphy', in its etymological connection to 'earth-writing', holds the idea of 'representation-as-text' at its practical and theoretical root.

Dubow, 2011, p. 646

We want to work on presenting the world, not on representing it, or explaining it.

Dewsbury et al., 2002, p. 438

From Representation to More-Than-Representation

Questions and concerns about what representations do are fundamental to geography, entangled with both the political and ethical promises of the discipline (Anderson, 2019). The process of representation – the act of describing and depicting landscapes, natures, spaces, places, bodies, and a myriad of other entities and the relationships we share with them – is central to geographical inquiry (Gilmartin, 2017; Wylie, 2007). Cultural geographers in particular have demonstrated how representations can act to (re)produce relations of power that reinforce structures steeped in inequality (Anderson, 2019). The act of producing representations themselves can be profoundly problematic and, arguably, inherently limiting. Alongside questions of who and what is represented or excluded, attention must also be paid to what is implicitly inhibited by representational modes of geography (Dubow, 2011), lest representation takes precedence over lived experience and materiality (Thrift, 1996).

The perceived overvaluing (or, over-extending) of the representational–referential dimensions of life prompted the development of what became known as 'non-representational theory' (NRT) (Anderson, 2011), a specific posthumanist tradition largely associated with the work of Nigel Thrift (Thrift, 1996, 2007) and colleagues in the UK. Rather than positing an external world waiting to be represented, or a text to be deconstructed, non-representational approaches attempt to move the focus of inquiry away from discursive meanings, towards the physicality and experience of the active ongoing world (Andrews, 2018; Dewsbury et al., 2002). Thus, inquiry becomes about attending to what happens in the present moment, right now in a fluid world that is in a perpetual state of becoming (Thrift, 2007). This involves a type

DOI: 10.4324/9781003038849-34

of wonderment; being open to the energy and movement of the present, but at the same time being mindful of the complexity all of the different forces (human and non-human) that form it (Andrews, 2018). Moreover it involves denying the common academic impulse to contain or reduce the messiness of life to a set of socially-constructed explanations and categorisations (Boyd and Edwardes, 2019). We might think of all this as moving beyond a form of 'post-game' interrogation and analysis that endlessly deconstructs and explains what happened in the field after it happened, to instead provide a lively and timely 'play-by-play' running commentary, that describes events in the field as they unfold with animation and energy that it true to them (Andrews, 2017; McCormack, 2013).

Taking this explanation further, for Anderson (2011), non-representational approaches are 'theories of practice' that focus on what humans and non-humans do. The argument here is that representationalism tends to leave a sizeable portion of the world unrepresented, particularly the many subtle, unspoken and often unintentional micro performances and practices involved in the reproduction of everyday life (Andrews et al., 2014; Dewsbury et al., 2002). In short, representations filter out the immediacy, intimacy and detail of life. Thus, instead, a more-than-representational approach rejects understandings of the world that are purely predicated on analysing human-made meanings (Müller, 2015). This is not to say that we cannot represent sensuous, corporeal, lived experience, but rather to recognise that the moment we do, we immediately lose something (Carolan, 2008). Representation inevitably involves processes of transformation and differentiation (Doel, 2010). Indeed, as Dubow (2011, p. 646 emphasis original) has explained, it is 'the *re* of representation' and the substitutive value that this implies that produces challenges for producing accounts of worlds as immediate and on-flowing.

Importantly, non-representational theory is not a singular theory. We might instead understand it as a series of suggestive statements (Anderson, 2019) drawing from a number of theoretical traditions. Non-representational theory has become something of an umbrella term for a body of (quite diverse) work that shares a connection through seeking to better engage with and evoke more-than-textual, multisensual, and more-than-human worlds in becoming (Lorimer, 2005).

Perhaps paradoxically given its titling, neither is non-representational theory anti-representation (Wylie, 2007). These approaches do not involve a rejection or disinterest in representations, but rather a consideration of representations for what they do, and their role in shaping ongoing everyday life (Simpson, 2017). As Carolan (2008, p. 412) argues, 'representations tell only part of the story, yet they still have a story to tell'. Indeed, the point is not to stop talking about events (McCormack, 2013). Instead, such a style of thinking and practicing geographic research is about the avoidance of reducing the world to representations alone (Simpson, 2017). Thus, within a non-representational geography the act (or, event) of representing is in itself an active part of the world and life's continuous unfolding, rather than taking place outside of, or even determinative of, worlds of embodied practice and performance (Wylie, 2007).

Lorimer (2005: 84) has (in an admittedly oft-repeated quotation) instead suggested the idea of 'more-than-representational theory', which he summarises as a focus on:

> How life takes shape and gains expression in shared experiences, everyday routines, fleeting encounters, embodied movements, precognitive triggers, practical skills, affective intensities, enduring urges, unexceptional interactions and sensuous dispositions. Attention to these kinds of expression, it is contended, offers an escape from the established academic habit of striving to uncover meanings and values that apparently await our discovery, interpretation, judgement and ultimate representation.

> In short, so much ordinary action gives no advance notice of what it will become. Yet,
> it still makes critical differences to our experiences of space and place.

The growth – and coalescing – of a specific *more-than*-representational theory can also be seen as a response to some of the critical challenges that have been levelled at non-representational theory. A new formulation, building on the 'tactical suggestions' (Dewsbury et al., 2002) of non-representational theory, whilst an attempt at reiterating (or perhaps, rehabilitating) the fact that going beyond representation, does not mean to dispense with it (Andrews, 2018). Particularly, there have been critiques regarding non-representational theory's tendency towards a universalist reduction of things, and its (arguably) limited ability to deal with the questions of social difference and political power that have material consequences on people's lives (Bondi, 2005; Colls, 2012). Markers such as gender, age, and ethnicity 'do not sit easily with a theoretical and empirical approach that is interested in how 'a' subject comes into being through processes of becoming and emergence' (Colls, 2012, p. 441). Here, Colls provides a possible solution, in the form of advocating researchers develop a 'nomadic consciousness' between representation and non-representation, 'a position that holds the gender blindness of poststructuralist thought in tension whilst also considering the potential of such work for rethinking 'the subject'" (Colls, 2012, p. 441). That is, remaining critical of the problems that inattention to difference creates, yet simultaneously open to the potential it offers (Andrews et al., 2014). Indeed, Andrews (2018) suggests that through such a process, a more-than-representational approach can aid in exposing why certain bodies become marginalised, and the non-representational processes involved.

A more-than-representational approach foregrounds everyday sensuous embodied practices and performances. It asks geographers to think about – and account for – how non-representational processes (the immediate, physical, and sensory), those which are often less-than-fully consciously experienced, overflow and interplay with what is more consciously accessible and more representational facets (social, political, and economic factors for instance) to together shape how life unfolds (Andrews, 2018). Thus, a more-than-representational approach acknowledges that forms of knowledge come about through our active embodiment in the world, through our 'doings' (Carolan, 2008).

These more-than-representational approaches also align with, inform, and are informed by, wider paradigm shifts and ontological turns taking place within geographic thinking and qualitative research more broadly (Andrews, 2018) (see also Kearns on human geography's 'turns', Chapter 5 this volume). Particularly, the 'post-qualitative' movement that presents a significant challenge to traditional notions of representationalism, subjectivity, and relationality in applied empirical research (MacLure, 2013). Post-qualitative research can increasingly be found throughout the social sciences and humanities disciplines, often mapping onto research arising from a posthumanist theoretical tradition with its decentring of the conscious human subject. Much of the discussion of the post-qualitative is about its challenge to representationalism – which we have described above. Though additionally (and with some similarities to more-than-representational approaches) post-qualitative approaches challenge the notion of the 'expert' knowing human researcher who, instead, like subjects and objects, is merely part of a research assemblage. This in turn, involves challenging the notion that subjects' knowledge should always be the point of departure, and researchers' knowledge the end point to instead recognise that the subject and subjectivity both come from an ecology of physical processes and influences. As such, a post-qualitative approach deprivileges face to face conversation as the 'gold-standard' research encounter and contests the constant reliance on language in qualitative research, instead encouraging bespoke, event-specific methodological design. That is, methods that are not 'off-the-shelf', but thoughtfully and creatively tailored to engage in the events and encounters that the researcher seeks to witness and become part of in their ongoingness. This is, in part, a means of encouraging the researcher to resist coming into the field with too many hard preconceived socially and academically constructed ideas about the people, events and relations to be encountered. This resistance to highly structured, formalised, and/or predetermined procedures also flows through to processes of analysis, with a hesitance to be caught up in habitual 'coding' of data and the creation of static representations or translations. Instead, more performative methodologies are centred that aim to use research to actively change, build, and 'boost' the world in desired directions, introducing new realities and affects into life, both in the field and through forms of knowledge translation (Andrews and Drass, 2016). By performative methodologies we refer to methodologies that generate different knowledge in different forms, methods that depend upon the *doing* of research in order for insights and understandings to emerge (Douglas and Carless, 2013).

Moving away from privileging certain types of understanding also opens up opportunities to engage with the more-than-human nature of worlds. This can particularly be seen in work that actively displaces humans as being central to producing meaning and instead highlights nonhuman agency and subjectivity, acknowledging the lively contributions of plants and nonhuman animals in social settings (Whatmore, 2002; see also Chapter 20, this volume). Indeed, like much research emerging under the banner of 'posthumanism', more-than-representational approaches can be seen to seek to trouble the very centrality of the unitary human subject (Ginn, 2017). As Dowling et al., (2017, p. 824) argue, 'this shift in recognizing and acknowledging multiple more-than-human agencies challenges researchers to do geography differently – to perform, to engage, to embody, to image and imagine, to witness, to sense, to analyse – across, through, with and as, more-than-humans'.

More-than-representational research asks that we recognise that the world is made through performative practices, always in the making, affective, more-than-human, and requires experimental and novel modes of 'presenting and presencing research' (Müller, 2015, p. 410). It is this requirement that will become the central concern of this chapter, as we turn to consider the methodological implications and challenges of a more-than-representational geography.

The Methodological Implications and Challenges of a More-Than-Representational Geography

As readers of this text may be beginning to realise, this 'style of engagement with the world' (Anderson, 2011) raises substantial questions for methodological practice and empirical research. As Carolan (2008, p. 412) asks, 'how can one hope to study and report on that which is said to be more-than-representational?'. The challenge, as Thorpe and Rinehart (2010) explain, is to evoke, rather than explain. To draw on Vannini (2015), a more-than-representational account should highlight the vitality of the inanimate and the more-than-human, it should make us feel something, and give us a sense of the ephemeral, the fleeting, and the not-quite-graspable.

However, studying often less-than-fully cognitively acted and experienced phenomenon will always have its challenges and even impossibilities. Embodied experiences and practices are neither easily accessible to reflection, nor particularly amenable to textual representation (O'Connell, 2013). A conventional methodological approach of going out into the world, reporting back, and then analysing events is inadequate to the task of engaging with the affective and processual ways in which spacetimes unfold (McCormack, 2013). Indeed, a 'sense' of events-in-process are lost when the emphasis is on capturing post-event meaning (McCormack, 2013). There are challenges to acknowledge human and non-human agency in equal or appropriate measures, complexities in retaining fidelity and authenticity to events, and struggles in accounting for researchers' and subjects' judgement filters when conveying events that often preceded them (Andrews, 2018; Vannini, 2015).

Research methods possess the potential to not only record existing practices but also to hone them and invent new ones in line with the politics and ethics of a new playful, experimental paradigm. More-than-representational research thus becomes a project of realising that our research is in every way a performative production that can create change as well as ideas (Dewsbury, 2010), whilst accepting that, ultimately, the world frequently exceeds our capacities to encounter and represent it (Roe and Greenhough, 2014).

Whilst this may require developing new approaches to conducting research, it does not mean that established methods in human geography have to be abandoned. Rather, in what follows, this chapter explores how existing methods might be tweaked, extended, or supplemented to engage with the imperatives and promises of a more-than-representational approach. Indeed, as Latham (2003, p. 2000) suggests, 'pushed in the right direction there is no reason why these methods cannot be made to dance a little'.

Practicing More-Than-Representational, More-Than-Human Research

There is no singular method for more-than-representational research (Cadman, 2009). Instead, efforts to go beyond representation have seen researchers creatively develop hybrid methods that attempt to engage with more-than-representational and more-than-human processes and phenomenon. Often this is in radically experimental ways as researchers shrug off restrictive categorical conventions and methodological conservativisms, and embrace creativity and playfulness to produce open-ended accounts. Similarly, neither is a more-than-representational approach characterised by the choice, or the rejection, of a particular method, as has often been imagined by those not familiar with it, or who extend assumptions from its more forceful theoretical statements (Vannini, 2015).

In what follows, we present a variety of methodological approaches which reflect, encourage, and enable an engagement with more-than-representational, more-than-human research: sensory and proactive ethnographies, multi-species considerations, arts-based approaches, new approaches to interviews and conversation, utilising innovative technologies, and finally, new ways of writing. Importantly, rather than segmented or exclusive categories these themes overlap each other. Indeed, the more-than-representational geographers' methodological toolkit is as overflowing and ever-expanding as the lively worlds they seek to explore and evoke.

Sensory and Proactive Ethnographies

A more-than-representational approach involves an acknowledgement that, fundamentally, certain knowledge can only be gained processually through lived embodied experiences, often less-than-fully consciously (Andrews, 2018). In other words, worlds are sensed, not just reasoned after viewing or during narrative (Greenhough, 2010). Apprehending life's immediacy and on-going nature requires finding and learning ways to witness how life takes place as an unceasing flow of heterogeneous embodied practices (Anderson, 2011) in ways that involve researchers becoming 'infected by the effort, investment, and craze of the particular practice of experience being investigated' (Dewsbury, 2010, p. 326). As Thrift (2000) suggests, this requires a paradigm shift from *participant observation* to *observant participation*.

Geographically, focusing on the senses can create a route to move beyond feelings *about* place, to instead develop 'feelings *of* place; how places are registered in the moment' (Andrews,

2011; Gorman, 2017a) – a 'sense of place' that arises in its most basic corporeal form. A means of attending to the very vitalism and generative potential of places and how they produce certain affective states and capacities. Of course, there are limits to how much sensory happenings can be translated directly into language and words, and there is a need for reflexivity as to how much is simplified, lost, or deadened in attempts to convey these embodied experiences – a matter we will return to shortly.

Practically this might involve seeing, hearing, touching, tasting, smelling, and moving – often in doings *with* others; a serious empirical involvement with practices, embodiment, and materiality (Dewsbury, 2010). Thus, rather than just an important topic to be theorised and engaged with, the body itself becomes 'something *through* which research' is done (Crang, 2005, p. 232 emphasis original). Sensorily and materially experiencing co-presence offers a means of attending to the idiosyncratic and in-the-moment experiences of encounters – the immaterial, the invisible, and the taken for granted (Pink, 2009). Researchers become familiar with aspects of life through embodied experiences, learning through their own sensorial immersion as a means of apprehending and comprehending the unfolding of more-than-representational life. For example, Longhurst et al., (2008), in a project exploring 'culturally embodied difference' describe sharing food and drink with research participants, and how this helped them to become 'attuned to others' reactions, such as a crinkling of the nose, a screwing up of the face and turning away from the plate' (212–213). Longhurst et al., describe how using our bodies as an integral part of the research process 'enables geographers to begin to talk from an embodied place, rather than from a place on high' (213). Similarly, through incorporating taste and tasting into her research, Hocknell (2016) was able to engage with the gaps between participants initial representations and explanations of their food preferences and their experiences of the tastefulness of that same product, to create a more-than-representational account of food-knowledges.

More-Than-Human Considerations

A more-than-representational geography 'transcends the human, focusing on relations amid inanimate objects, living, non-human matter, place, ephemeral phenomena, events, technologies, and much more' (Vannini, 2015, pp. 5–6). However, whilst there exists much scholarship that eloquently articulates the need to decentre the human from geography's dominant focus, there is much less work that articulates how this might be realised in coherent and compelling ways in practice (Pacini-Ketchabaw et al., 2016).

Addressing this disjuncture Tsing (2010) argues that multispecies studies require mobilising the talents and knowledge of those close to, and passionate about, non-humans. This might involve observing the entangled and asymmetrical lively connections and collective affects taking place between species, and how different bodies are interconnected, mutually dependent, and mutually response-able within common worlds (Pacini-Ketchabaw et al., 2016). As Buller (2014, p. 378) argues, 'we may not share language with non-humans but we do share embodied life and movement and [...] ways of inhabiting the world'. Animals are active social agents; they engage in relationships and have the capacity for ample non-linguistic intersubjective exchanges with humans. The challenge – as ever – is representing this in meaningful and evocative ways that avoids anthropomorphic accounts of human-animal relations and nonhuman presence and agency.

In response many geographers have drawn on what Kirksey and Helmreich (2010) term as 'multispecies ethnography'. This being a mode of research that develops a means of bringing creatures previously on the margins more vividly into the foreground, creating a more symmetrical treatment of different species within ethnographic observation and writing. Multispecies ethnography aims to provide a way of acknowledging that interactions between species are not purely mediated through a level of human involvement, evoking entanglement and co-production. However, the 're-wilding' of methodological practice must avoid the attraction of just attending to those animals which are familiar, in-view, and easy to evocate. Similarly, whilst we may find routes to present lively and vital encounters with animals, there is the whole gamut of the mundane more-than-human to consider (Gorman, 2017b). Pitt (2015) for example, explores how researchers might pay closer attention to what plants do.

Much methodological praxis that attempts to engage with the more-than-human involves a variety of styles of 'learning to be affected'. A process that requires cultivating more-than-cognitive modes of attention, attuning to 'the multifarious ways that human and nonhuman bodies are moved, disconcerted and enlivened through their common world encounters' (Taylor and Pacini-Ketchabaw, 2015, p. 514). For Taylor and Pacini-Ketchabaw (2015, p. 514), in practice this involved paying attention not only to what their research participants were saying and doing, nor even to just how participants bodies were moved, affected or enlivened by the animals they encountered, but 'also paying attention to the movements and actions of the worms, ants, water, rain boots, fingers, sticks, rocks, mud, pebbles and dust. We push ourselves to learn to be affected by and think with all of the actors'.

Arts-based Approaches

More-than-representational geography has endeavoured to work with the arts as a means to connect knowledge to everyday life, practice, and communities through political affects. Arts-based approaches seek to provide an outgrowth of knowledge in the unfolding moment through movement, expression and creativity. Creative and material acts can point us towards 'modes of bodily experience, emotional work and multi-sensuous being that might otherwise slip past' (Hawkins, 2011, p. 472), producing knowledges that are more carnal or visceral by actively intervening in the world (Boyd and Edwardes, 2019).

This has involved an engagement in a plethora of other creative practices, collaboration with creative artists and non-academic practitioners (Hawkins, 2011), and attempts to use media that is less 'fixed' in nature than print (Barnes, 2019). This might involve experimenting with sound (Gallagher and Prior, 2014), music (Andrews and Drass, 2016), film (Richmond, 2016), photography (Latham and McCormack, 2009), theatre (Raynor, 2019), dance (Pine and Kuhlke, 2013), animation (Gorman et al., 2022), craft and other 'making' practices (Boyd and Edwardes, 2019), alongside experiments with drawing (Coen, 2016), creative writing (Lorimer, 2019), and poetic registers (Coen et al., 2018). These – and other turns to arts-based approaches – can be understood as a means of attending to Wylie's (2010, p. 213) calls to practice 'cultural geography as performance'. As Sheller (2015, p. 134) notes, geographers are 'no longer bringing data back from "the field" for analysis and presentation, but are staying there and creating other kinds of work'.

A more-than-representational engagement with arts-based approaches involves recognising art as 'productive rather than representative of culture' (Hawkins, 2011, p. 473) and avoiding temptations to recast discussions – and utilisations – of creative practices and performances into a familiar textual mould (Müller, 2015). Emblematic of this is Balayannis' (2019) striking photo-essay on the visual politics of hazardous sites which aims to depart from this epistemological habit and approach photographs through a more-than-representational register to explore the practices in which images are made, circulated, and encountered. Balayannis draws on the power of photography to create a feeling and evocatively 'unsettle naturalised ways of seeing contamination' (2019, p. 573).

New Approaches to Interviews and Conversation

More-than-representational approaches are less concerned with what people say, and more with what they do; action, rather than reason (Andrews, 2018). This creates a significant challenge for methods that are heavily invested in language practices (interviews, fieldnotes, focus groups) (MacLure, 2013). A focus on the verbal can erase the more-than-human elements and neglect the material context of conversations – features that shape and influence what is said.

Müller suggests that rather than aiming for a density of words, inquiring into the more-than-representational might involve examining the absence or loss of words: 'the more-than-representational begins where research subjects fumble for words and established codes of communication break down' (Müller, 2015, p. 417). Thus, it is not about abandoning interviews, nor even their transcription, but reframing how these are analysed – and even what might be recounted and recollected from the interview itself. As MacLure notes, transcripts rarely record 'what eyebrows, hands, shoulders or crossed legs are doing, and if they do attend to such features, the aim is usually to point to what they "mean"' (MacLure, 2013, p. 664).

Thus, there is the possibility to re-imagine interviewing in a more 'lively' manner, one which opens generative possibilities for evoking the more-than-representational. In this way, using interviews as a means of showing intense, affective, emotional, and embodied relationships between more-than-human elements (Dowling et al., 2017). Rather than focusing solely on what interview utterances mean, how and whether they are true, generalisable, or filled with power, interviews might embrace what bodies do, and how they are entangled amidst acts of speaking and voicing (MacLure, 2013). This involves reconsidering what the 'data' generated by the interview is – not so much the transcript, but rather the interview event itself. Walking (or 'go-along') interviews are a well-established method within geography, yet there is potential to push these further, where the focus becomes on the matter of 'walking with', and less on the 'interview'; speaking to participants as they are actively 'doing' (Carpiano, 2009) – whatever the activity. For example, Pitt (2015) describes the opportunities to learn about gardens, gardening, and human-plant relationships through moving and working with gardeners and plants. Britton and Foley (2020) describe using a combination of 'swim-along' and even 'surf-along' interviews to help contextualise participant's experiences as they arose in place. Similarly, Denton and Aranda (2020) describe the use of 'swim-along' interviews as a means of accessing the embodied, emplaced and temporal experience of sea swimming. In these examples, and other more-than-representational approaches to interviews, the emphasis is on the performative aspects of interviews themselves, what they produce in-the-moment, not their later transcription.

New Technologies

New technologies are increasingly part of the contemporary research assemblage for geographers. Barratt (2012, p. 48) explains how 'in all spheres of life, our bodies and the places they go are technologised; some of these new technologically mediated engagements are complex and subtle, often entered into with little consideration with regard to experiential consequences'. Attempting to explore and evoke the movement and energy associated with more-than-representational life has seen a number of geographers engage and experiment with a range of technologies - such as headcams, street cams, GPS, pedometers and various bio-sensors - to obtain different, complementary registers of the body and place. Such methods offer routes to account for sensorial experiences in new ways, and to greater level of detail than before (Spinney, 2011). The crux of what makes these approaches useful for a more-than-representational approach is their ability to complement, facilitate and augment more conventional discursive methods (Bell et al., 2015; Spinney, 2015).

Barratt (2012) argues that engaging with the more-than-representational involves finding means to help respondents articulate their experiences in new ways – and that new technologies are one such route. For example, Beljaars (2020) has explored how geographers might utilise mobile eye-tracking methods to engage with the visual and auditory appearance of bodily situations in real-time. Beljaars argues that eye-tracking offers a means for more-than-representational accounts of sensuous dispositions, a real-time attunement to the unfolding of the world, and a means of conducting research *with* the body (Beljaars, 2020). Similarly, Osborne and Jones (2017) describe how 'biosensing' somatic measures such as heart rate or electrodermal activity might offer geographers a means to explore research participants' somatic responses to environments at an embodied level, one which allows researchers to move beyond the subjectivity of participants' self-reporting.

Geographers have also engaged with moving image methodologies (with a variety of technological complexity), both as a means of evoking affect, drawing attention to the more-than-human, and engaging with forms of knowledge, skill, and practice that can easily be omitted through textual and oratory accounts (Lorimer, 2010). Smith et al., (2020) explain how recording video can support a carefully cultivated practice of 'noticing', allowing researchers to re-watch encounters and move to attend to (some aspects of) the more-than-representational. The use of video, argue Bear et al., (2017), provides an innovative way of decentring the human, exploring relations-in-becoming; and moving beyond an over-reliance on textual representations. Indeed, as a means of attending to the more-than-human, video-based methods offer some means of presenting animals as animals (Philo, 1995) moving – at least some way – beyond only human representations of animals.

New Ways of Writing

More-than-representational styles of geography encourage an academic style which seeks to describe and present, rather than diagnose and represent (Cadman, 2009). This is a substantive shift in how geographers do their 'earth-writing'. As Latham (2003) argues, there is ultimately little point in cultivating a theoretical and methodological approach that moves to account for embodiment and the taking-place of socio-spatial practice, if we merely gesture towards it conceptually in our writing, and return to presenting data based on talk. One attraction though of this new 'earth-writing' is that there is a freedom gained by no longer having to fully explain or thoroughly theorise the social subject (Thrift, 2007) – though this is not an exercise in, or excuse for, trivialising or simplifying the immediacies of geographic life (Andrews, 2018).

The aim, as Vannini (2015, p. 12) so eloquently explains, is to realign one's writing so that:

> You cease to be so preoccupied with how the past unfolded and with your responsibility for capturing it. You become instead interested in evoking, in the present moment, a future impression in your reader, viewer, or listener. It is the present that suddenly interests you, and how the present can unfold in the future: what can become of your work, in what unique and novel ways it can reverberate with people, what social change or intellectual fascination it can inspire, what impressions it can animate, what surprises it can generate, what expectations it can violate, what new stories it can generate.

This involves recognising the performative quality of words themselves and the intersubjective means by which knowledge is co-created by writer and reader (Anderson, 2014). In this way, research accounts should empower readers to make their own meanings and interpretations, through showing experience, rather than telling (Anderson, 2014). For Anderson, whilst there are limits as to what can be achieved through re-presenting subjective experiences, there are also generative possibilities achieved by the 'shared understanding of the representative function of language' and 'connections of experience, emotions, and solidarity' (2014, p. 30). It is a shift from aiming to *explain* how something might 'feel' to instead attempting to expressively *evoke* how something might 'feel'.

More-than-representational scholars might seek to develop a lively style of communicating with words that creates itself in the image of the very practices it presents. This involves a change from the formal 'realis mood' of much academic writing – which attempts to be authoritative, logical and definitive in its statements - to an 'irrealis mood' in academic writing that can be purposefully uncertain, can create senses of unreality, surreality, possibility and hope, and can help subject matter come to life. Together these moods represent a movement away from a bland professional academic language that is bare of expression, performance and unpredictability, to a language shared by those who work in the arts – sometimes written, sometimes spoken aloud – that is filled with and actively celebrates these things.

More-Than-Representational Methodological Futures

Of course, these approaches should not be sought out for their sheer novelty (Osborne and Jones, 2017), nor put on a pedestal as providing any form of inherent access to authenticity (Merriman, 2014). As Dewsbury (2010, p. 325) notes 'a well conceived set of interview questions might well be far more effective at capturing the tension of the performing body as witnessed by the body of the interviewee'. There are multiple ways of exploring and evoking practice and performance. The experimentalism involved in more-than-representational methodologies involves experimenting with existing approaches as much as with the innovative.

Indeed, 'finding' the more-than-representational might simply involve making the best of the methods and data already at our disposal. For example, listening to and reading narratives obtained through traditional research methods whilst being attentive to and drawing out immediate, active and sensory content. This content is very often present, no matter what the original theory, research priorities or questions might have been, simply because it is an important part of life. In other words, if one is receptive and prepared to look, this content might be detectable in data, it having been present in the moment it was live, lived and received (Ducey, 2007).

More-than-representational methods are as much a part of a researcher's toolkit in addressing geographical problems, whether relating to social justice or sustainability, as those methods more closely associated with humanistic studies – albeit highlighting how such issues have come to be experienced in different ways. Indeed, the ambition of a more-than-representational approach is to make human experience – and the way that people do things – easier to understand (Andrews et al., 2014). It is about finding different ways in which experiences, bodies, and practices can come to be known (Dennis, 2020). Telling different kinds of stories differently can bring about new relationships; new kinds of bodies can be known and made knowable, and as a result, new modes of responding to such bodies can become possible and imagined within interventions, policies, and politics (Dennis, 2020). Dennis' (2020) mobilisation of more-than-representational methods to evoke the different spacetimes of 'drugged bodies' in order that more hospitable ways of living with drugs might be produced, is a powerful example of how more-than-representational methods can create change in the world toward practical ends.

Certainly, the creativity and playfulness associated with more-than-representational methods brings challenges in engaging policymakers and practitioners. For example, Gorman and Cacciatore (2020) note that whilst a focus on 'energies' may not enamour healthcare providers keen to see quantifiable evidence about community health initiatives, such does offer a route to take seriously aspects of how such places are sensed, shaped, and shared – aspects that fundamentally inform how health and wellbeing are practiced. The more-than-representational researcher must do additional work of translation and rationalisation in order to highlight to stakeholders what can be done with the more-than-representational evidence generated, how this might be applied to inform and change the world. Like life itself, the work of more-than-representational methods is ongoing.

At the same time, some authors, like Horton and Kraftl (2005), interested in the more-than-representational have also challenged research to be 'more-than-useful', that geography should be interesting, enabling, inspiring, vibrant, exciting, affecting, and cool in ways that conventional understandings of 'usefulness' often prohibit. They argue that much is lost when 'usefulness' is centred and prioritised.

More-than-representational methods hold great potential for geographers of all subdisciplines. As Andrews and Duff (2019, p. 130) have argued, a more-than-representational approach 'might help better understand and show the vitality in 'therapeutics'; the movement and sensation in 'walk-ability'; the energy and taste in 'foodscapes'; the affect in 'caring environments'; the life in 'lifecourse''. Indeed, one could speculate that more-than-representational forms of inquiry are increasingly required to be able to engage with the type of societies that are ever emerging – faster moving, forever connected and relationally entangled, and increasingly posthuman (Andrews, 2019; Andrews and Duff, 2019). More-than-representational styles of approaching, thinking, and doing geography offer a mechanism to bring a sharper focus to worlds in which humans are consistently, and ever increasingly, exposed to affective forces and where it is increasingly difficult to separate the human and non-human.

References

Anderson, B. (2011) Non-representational theory. *The Dictionary of Human Geography* Gregory D, Johnston R, Pratt G, et al., (eds). Malden, Mass: Blackwell.

Anderson, B. (2019) Cultural geography II: The force of representations. *Progress in Human Geography* 43(6). SAGE Publications Ltd: 1120–1132. DOI: 10.1177/0309132518761431

Anderson, J. (2014) Exploring the space between words and meaning: Understanding the relational sensibility of surf spaces. *Emotion, Space and Society* 10: 27–34. DOI: 10.1016/j.emospa.2012.11.002

Andrews, G.J. (2011) 'I had to go to the hospital and it was freaking me out': Needle phobic encounter space. *Health & Place* 17(4): 875–884.

Andrews, G.J. (2017) From post-game to play-by-play: Animating sports movement-space. Progress in Human Geography 41(6). SAGE Publications Ltd: 766–794. DOI: 10.1177/0309132516660207

Andrews, G.J. (2018) *Non-Representational Theory & Health: The Health in Life in Space-Time Revealing.* Routledge.

Andrews, G.J. (2019) Health geographies II: The posthuman turn. Progress in Human Geography 43(6). SAGE Publications Ltd: 1109–1119. DOI: 10.1177/0309132518805812

Andrews, G.J. and Drass, E. (2016) From 'The pump'to 'Senescence'. In: Fenton, N.E. and Baxter, J. (eds) *Practicing Qualitative Methods in Health Geographies.* Oxon and New York: Routledge, p. 211.

Andrews, G.J. and Duff, C. (2019) Matter beginning to matter: On posthumanist understandings of the vital emergence of health. *Social Science & Medicine* 226: 123–134. DOI: 10.1016/j.socscimed.2019.02.045

Andrews, G.J., Chen, S. and Myers, S. (2014) The 'taking place' of health and wellbeing: Towards non-representational theory. *Social Science & Medicine* 108: 210–222.

Balayannis, A. (2019) Routine Exposures: Reimaging the Visual Politics of Hazardous Sites. *GeoHumanities* 5(2): 572–590. DOI: 10.1080/2373566X.2019.1624189

Barnes, A. (2019) Geo/Graphic Design. In: Boyd CP and Edwardes C (eds) *Non-Representational Theory and the Creative Arts*. Singapore: Springer Singapore. DOI: 10.1007/978-981-13-5749-7

Barratt, P. (2012) 'My magic cam': A more-than-representational account of the climbing assemblage. *Area* 44(1): 46–53. DOI: 10.1111/j.1475-4762.2011.01069.x

Bear, C., Wilkinson, K. and Holloway, L. (2017) Visualizing human-animal-technology relations: Field notes, still photography, and digital video on the robotic dairy farm. *Society & Animals* 25(3). Brill: 225–256.

Beljaars, D. (2020) Towards compulsive geographies. *Transactions of the Institute of British Geographers* 45(2): 284–298. DOI: 10.1111/tran.12349

Bell, S.L., Phoenix C, Lovell R, et al., (2015) Using GPS and geo-narratives: A methodological approach for understanding and situating everyday green space encounters. *Area* 47(1). John Wiley & Sons, Ltd: 88–96. DOI: 10.1111/area.12152

Bondi, L. (2005) Making connections and thinking through emotions: Between geography and psychotherapy. *Transactions of the Institute of British Geographers* 30(4): 433–448. DOI: https://doi.org/10.1111/j.1475-5661.2005.00183.x

Boyd, C.P. and Edwardes C (eds) (2019) *Non-Representational Theory and the Creative Arts*. Singapore: Springer Singapore. DOI: 10.1007/978-981-13-5749-7.

Britton, E. and Foley, R. (2020) Sensing Water: Uncovering Health and Well-Being in the Sea and Surf. *Journal of Sport and Social Issues*. SAGE Publications Inc: 0193723520928597. DOI: 10.1177/0193723520928597

Buller, H. (2014) Animal geographies II: Methods. *Progress in Human Geography* 39(3): 374–384.

Cadman, L. (2009) Non-Representational Theory/Non-Representational Geographies. In: *International Encyclopedia of Human Geography*. Elsevier, pp. 456–463. DOI: 10.1016/B978-008044910-4.00717-3

Carolan, M.S. (2008) More-than-Representational Knowledge/s of the Countryside: How We Think as Bodies. *Sociologia Ruralis* 48(4): 408–422. DOI: 10.1111/j.1467-9523.2008.00458.x

Carpiano, R.M. (2009) Come take a walk with me: The "Go-Along" interview as a novel method for studying the implications of place for health and well-being. *Health & Place* 15(1): 263–272.

Coen, S.E. (2016) What Can Participant-Generated Drawing Add to Health Geography's Qualitative Palette? In: *Practicing Qualitative Methods in Health Geographies*. Oxon and New York: Routledge.

Coen, S.E., Tillmann, S., Ergler, C.R., et al., (2018) Playing with Poetry: Poetic Representation of Research in Children's Geographies of Nature and Adventurous Play. *GeoHumanities* 4(2): 557–575. DOI: 10.1080/2373566X.2018.1516956

Colls, R. (2012) Feminism, bodily difference and non-representational geographies. *Transactions of the Institute of British Geographers* 37(3). John Wiley & Sons, Ltd: 430–445. DOI: 10.1111/j.1475-5661.2011.00477.x

Crang, M. (2005) Qualitative methods: There is nothing outside the text? Progress in Human Geography 29(2). SAGE Publications Ltd: 225–233. DOI: 10.1191/0309132505ph541pr

Dennis, F. (2020) Mapping the Drugged Body: Telling Different Kinds of Drug-using Stories. *Bo-dy & Society* 26(3). SAGE Publications Ltd: 61–93. DOI: 10.1177/1357034X20925530

Denton, H. and Aranda, K. (2020) The wellbeing benefits of sea swimming. Is it time to revisit the sea cure? *Qualitative Research in Sport, Exercise and Health* 12(5): 647–663. DOI: 10.1080/2159676X.2019.1649714

Dewsbury, J. (2010) Performative, non-representational, and affect-based research: Seven injunctions. In: DeLyser D, Herbert S, Aitken S, et al., (eds) *The SAGE Handbook of Qualitative Geography*. London: SAGE Publications, pp. 321–334.

Dewsbury, J-D., Harrison, P., Rose, M., et al., (2002) Enacting geographies. *Geoforum* 33(4): 437–440.

Doel, M.A. (2010) Representation and Difference. In: *Taking-Place: Non-Representational Theories and Geography*. Farnham: Ashgate.

Douglas, K. and Carless, D. (2013) An Invitation to Performative Research. *Methodological Innovations Online* 8(1): 53–64. DOI: 10.4256/mio.2013.0004

Dowling, R., Lloyd, K. and Suchet-Pearson, S. (2017) Qualitative methods II: 'More-than-human' methodologies and/in praxis. *Progress in Human Geography*, 41(6): 823–831.

Dubow, J. (2011) Representation. *The Dictionary of Human Geography* Gregory D, Johnston R, Pratt G, et al., (eds). Malden, MA: Blackwell.

Ducey, A. (2007) More than a job: Meaning, affect, and training health care workers. In: *The Affective Turn: Theorizing the Social.* Durham: Duke University Press, pp. 187–208.

Gallagher, M. and Prior, J. (2014) Sonic geographies Exploring phonographic methods. *Progress in Human Geography* 38(2): 267–284.

Gilmartin, M. (2017) Representation. In: *International Encyclopedia of Geography*, pp. 1–6. DOI: 10.1002/9781118786352.wbieg0230

Ginn, F. (2017) Posthumanism. *International Encyclopedia of Geography.* pp. 1–9. https://doi.org/10.1002/9781118786352.wbieg0414

Gorman, R. (2017a) Smelling Therapeutic Landscapes: Embodied Encounters within Spaces of Care Farming. *Health & Place* 47: 22–28.

Gorman, R. (2017b) Therapeutic landscapes and non-human animals: The roles and contested positions of animals within care farming assemblages. *Social & Cultural Geography* 18(3): 315–335.

Gorman, R. and Cacciatore, J. (2020) Care-farming as a catalyst for healthy and sustainable lifestyle choices in those affected by traumatic grief. *NJAS - Wageningen Journal of Life Sciences* 92: 100339. DOI: 10.1016/j.njas.2020.100339

Gorman, R., Farsides, B. and Gammidge, T. (2022). Stop-motion storytelling: Exploring methods for animating the worlds of rare genetic disease. *Qualitative Research.* SAGE Publications: 14687941221110168. DOI: 10.1177/14687941221110168.

Greenhough, B. (2010) Vitalist Geographies: Life and the More-Than-Human. In: *Taking-Place: Non-Representational Theories and Geography.* Farnham: Ashgate.

Hawkins, H. (2011) Dialogues and Doings: Sketching the Relationships Between Geography and Art. *Geography Compass* 5(7): 464–478. DOI: 10.1111/j.1749-8198.2011.00429.x

Hocknell, S. (2016) Chewing the Fat: "Unpacking" Distasteful Encounters. *Gastronomica* 16(3): 13–18. DOI: 10.1525/gfc.2016.16.3.13

Horton, J. and Kraftl, P. (2005) For more-than-usefulness: Six overlapping points about Children's Geographies. *Children's Geographies* 3(2): 131–143. DOI: 10.1080/14733280500161503

Kirksey, S. and Helmreich, S. (2010) The emergence of multispecies ethnography. *Cultural Anthropology* 25(4): 545–576.

Latham, A. (2003) Research, Performance, and Doing Human Geography: Some Reflections on the Diary-Photograph, Diary-Interview Method. Environment and Planning A: Economy and Space 35(11). SAGE Publications Ltd: 1993–2017. DOI: 10.1068/a3587

Latham, A. and McCormack, D.P. (2009) Thinking with images in non-representational cities: Vignettes from Berlin. *Area* 41(3): 252–262. DOI: 10.1111/j.1475-4762.2008.00868.x

Longhurst, R., Ho, E. and Johnston, L. (2008) Using 'the body' as an 'instrument of research': Kimch'i and pavlova. *Area* 40(2): 208–217.

Lorimer, H. (2005) Cultural geography: The busyness of being 'more-than-representational'. *Progress in Human Geography* 29(1). SAGE Publications Ltd: 83–94. DOI: 10.1191/0309132505ph531pr

Lorimer, H. (2019) Dear departed: Writing the lifeworlds of place. *Transactions of the Institute of British Geographers* 44(2). John Wiley & Sons, Ltd: 331–345. DOI: 10.1111/tran.12278

Lorimer, J. (2010) Moving image methodologies for more-than-human geographies. *Cultural Geographies* 17(2): 237–258. DOI: 10.1177/1474474010363853

MacLure, M. (2013) Researching without representation? Language and materiality in post-qualitative methodology. *International Journal of Qualitative Studies in Education* 26(6): 658–667. DOI: 10.1080/09518398.2013.788755

McCormack, D.P. (2013) *Refrains for Moving Bodies.* Duke University Press. DOI: 10.2307/j.ctv11smm6b

Merriman, P. (2014) Rethinking Mobile Methods. *Mobilities* 9(2): 167–187. DOI: 10.1080/17450101.2013.784540

Müller, M. (2015) More-Than-Representational Political Geographies. In: Agnew J, Mamadouh V, Secor AJ, et al., (eds) *The Wiley Blackwell Companion to Political Geography.* Chichester, UK: John Wiley & Sons, Ltd, pp. 407–423. DOI: 10.1002/9781118725771.ch30

O'Connell, R. (2013) The use of visual methods with children in a mixed methods study of family food practices. *International Journal of Social Research Methodology* 16(1): 31–46. DOI: 10.1080/13645579.2011.647517

Osborne, T. and Jones, P.I. (2017) Biosensing and geography: A mixed methods approach. *Applied Geography* 87: 160–169. DOI: 10.1016/j.apgeog.2017.08.006

Pacini-Ketchabaw, V., Taylor, A. and Blaise, M. (2016) Decentring the human in multispecies ethnographies. In: Taylor C and Hughes C (eds) *Posthuman Research Practices in Education.* New York: Palgrave Macmillan, pp. 149–167.

Philo, C. (1995) Animals, geography, and the city: Notes on inclusions and exclusions. *Environment and Planning D: Society and Space* 13(6): 655–681.

Pine, A.M. and Kuhlke, O. (2013) *Geographies of Dance: Body, Movement, and Corporeal Negotiations.* New York: Lexington Books.

Pink, S. (2009) *Doing Sensory Ethnography.* London: SAGE Publications.

Pitt, H. (2015) On showing and being shown plants – a guide to methods for more-than-human geography. *Area* 47(1): 48–55. DOI: 10.1111/area.12145

Raynor, R. (2019) Speaking, feeling, mattering: Theatre as method and model for practice-based, collaborative, research. *Progress in Human Geography* 43(4): 691–710. DOI: 10.1177/0309132518783267

Richmond, C.A.M. (2016) Applying decolonizing methodologies in environment-health research: A community-based film project with Anishinabe communities. In: *Practicing Qualitative Methods in Health Geographies.* pp. 153–166. Oxon: Routledge.

Roe, E. and Greenhough, B. (2014) Experimental partnering: Interpreting improvisatory habits in the research field. *International Journal of Social Research Methodology* 17(1): 45–57. DOI: 10.1080/13645579.2014.854014

Sheller, M. (2015) Vital Methodologies: Live Methods, Mobile Art, and Research-Creation. In: Vannini P (ed.) *Non-Representational Methodologies: Re-Envisioning Research.* 1st ed. Oxon: Routledge. DOI: 10.4324/9781315883540

Simpson, P. (2017) Nonrepresentational Theory. In: *International Encyclopedia of Geography*, pp. 1–4. DOI: 10.1002/9781118786352.wbieg0273

Smith, T.A., Laurier, E., Reeves, S., et al., (2020) "Off the beaten map": Navigating with digital maps on moorland. *Transactions of the Institute of British Geographers* 45(1): 223–240. DOI: 10.1111/tran.12336

Spinney, J. (2011) A Chance to Catch a Breath: Using Mobile Video Ethnography in Cycling Research. *Mobilities* 6(2): 161–182. DOI: 10.1080/17450101.2011.552771

Spinney, J. (2015) Close encounters? Mobile methods, (post)phenomenology and affect. *Cultural Geographies* 22(2): 231–246. DOI: 10.1177/1474474014558988

Taylor, A. and Pacini-Ketchabaw, V. (2015) Learning with children, ants, and worms in the Anthropocene: Towards a common world pedagogy of multispecies vulnerability. *Pedagogy, Culture & Society* 23: 1–21. DOI: 10.1080/14681366.2015.1039050

Thorpe, H. and Rinehart, R. (2010) Alternative sport and affect: Non-representational theory examined. *Sport in Society* 13(7–8): 1268–1291. DOI: 10.1080/17430431003780278

Thrift, N. (2000) Afterwords. *Environment and Planning D: Society and Space* 18(2). SAGE Publications Ltd STM: 213–255. DOI: 10.1068/d214t

Thrift, N.J. (1996) Spatial Formations. Sage. https://dx.doi.org/10.4135/9781446222362

Thrift, N.J. (2007) *Non-Representational Theory: Space, Politics, Affect.* International Library of Sociology. Milton Park, Abingdon, Oxon; New York, NY: Routledge.

Tsing, A. (2010) Arts of Inclusion, or How to Love a Mushroom. *Manoa* 22(2): 191–203.

Vannini, P. (2015) *Non-Representational Methodologies: Re-Envisioning Research.* 1st ed. Oxon: Routledge. DOI: 10.4324/9781315883540

Whatmore, S. (2002) *Hybrid Geographies: Natures Cultures Spaces.* Sage Publications.

Wylie, J. (2007) *Landscape.* Oxon: Routledge.

Wylie, J. (2010) Cultural geographies of the future, or looking rosy and feeling blue. *Cultural Geographies* 17(2): 211–217. DOI: 10.1177/1474474010363852

31

DEAR FEMINIST COLLECTIVE

How Does One Take Up Slow Scholarship (in the Midst of Crises)?

Jenna M. Loyd, Stepha Velednitsky, Ileana I. Diaz,
Sameera Ibrahim, Carla Giddings, Kela Caldwell,
Roberta Hawkins, Alison Mountz and Anne Bonds

Slow scholarship involves working with a feminist ethics of care for others, prioritized over demands for maximized productivity with measurable outcomes. How do principles of feminist slow scholarship inform geographical research methodologies? As a group of feminist geographers comprising five graduate students and four tenured faculty across five universities in Canada and the United States, this question emerged as significant and pressing in the midst of the COVID-19 pandemic and mass uprisings for racial justice. Some of us were teaching research methods when our schools went virtual, while others were designing or commencing new research projects when the deadly virus, and disorganized responses to it, consumed our days. We struggled to locate ourselves and our research in times that felt alternatively chaotic and continuous, slow and accelerated. We found inspiration from Arundhati Roy's (2020) notion that the "pandemic is a portal," giving rise to the question of how we might take this moment to collectively reflect on (methodologies for) different futures (Ho & Maddrell, 2021).

The abruptness, immediacy, and scale of the pandemic underscores the university's deep inequalities that reflect broader society: its reliance upon – and devaluation of – social reproduction, its organization around ableist expectations, and its intersecting settler colonial, racial, and gendered groundings (Batacharya & Wong, 2018; Gutiérrez y Muhs et al., 2012; Hamer & Lang, 2015; Peake & Mullings, 2016; Titchkosky, 2010). For people paid to care for others – in health care, residential facilities, transportation, and beyond – work did not come to a halt, and these "essential workers" faced heightened risks of exposure and death (Islam & Netto 2020). Within other sectors, like education and childcare, work often "pivoted" to home, magnifying the imbalances of care work (Power, 2020). The multiple crises of social reproduction have widened and intensified existing gendered and racial labor inequalities, almost immediately erasing some of the small gains made by women, still largely responsible for care work, at home and within academia (Weedon et al., 2016; Malisch et al., 2020). But COVID-19 is not the only crisis. Racism in the academy has garnered heightened attention and collective action amidst sustained protests over policing and systemic anti-Black racism (Bellamy & Mosley, 2020; Roberts, 2020). For then-president of the Social Science Research Council, Alondra Nelson (2020), this moment necessitates a question, "How can we continue to invest in

DOI: 10.4324/9781003038849-35

[research] projects that sustain the description of gross inequality without offering prescriptions for change?" For us, a feminist ethic and praxis of slow scholarship is a care-full response.

This chapter engages with feminist slow scholarship as praxis, one that involves "reflection and action upon the world in order to change it" (Freire, 1970, p. 36). In the following section, we discuss our previous work with feminist slow scholarship, reflecting on critiques and highlighting how we wish to further develop the concept. We then discuss feminist slow scholarship's relationship to feminist care ethics and their methodological implications. After that, we situate the politics of slowness within temporalities of crisis, emphasizing how centering care can shift temporalities of everyday life and research. The remainder of the chapter is organized around three principles fundamental to feminist slow scholarship: feminist care ethics, attention to temporality, and embodied emotional labor. We write from the position of an imagined Feminist Advice Column Collective (FACC) responding to questions about how feminist slow scholarship might inform doing, planning, or teaching about research.

Feminist Slow Scholarship: Some Background

In 2015, some of the authors of this chapter co-authored an article advocating for feminist slow scholarship (Mountz et al., 2015). In that article, we questioned the neoliberal restructuring of the university that manifests as a simultaneous speed-up of labor, growing precarity, and uneven landscape of adjunctification and austerity. These dynamics reinforce each other and accentuate the care and activist work that women and people of color (not mutually exclusive) carry disproportionately within academia: mentoring, service, "diversity work" (Ahmed, 2012), and efforts to challenge oppressive practices. Other scholars writing at the time also used the term "slow university" or "slow professor" to analyze the neoliberal terrain of the university (e.g., Berg & Seeber, 2016). We distinguished our analysis with aspiration for an academy in which everyone could learn, write, and research more sustainably and freely by advocating for *collective* action to transform the classed, racialized, and gendered injustices of labor in the academy. We named feminist care ethics (themselves inherently relational and attentive to transforming power relationships) as fundamental to making feminist slow scholarship a reality.

The response to our call for *a feminist politics of resistance through collective action* gained momentum that we did not anticipate. It was taken up internationally by writers, researchers, librarians, students, and university administrators (Lopez & Gillespie, 2016; Askins & Blazek, 2017; Black, 2018; Thorogood, et al., 2019; Caretta & Faria, 2020; Anguelovski et al., 2020; Kolliinjjiivvadi et al., 2020; Hamilton et al., 2021; Save Our Slides N.D.). These works speak to the subversive potential of slow scholarship as a *collective politic*, as well as the pressing need to advance *intersectional* praxis about slowness, which we did not adequately discuss, and seek to do here.

Some have criticized the idea of slow scholarship, arguing that the desire for a slower pace of work is tethered to a romantic view of the 'pre-neoliberal' university and that the pursuit of slowness is inherently elitist (Carrigan & Vostal, 2016; Meyeroff & Noterman, 2019). For Meyerhoff and Noterman, the problem with the university is deeper than work speed-up and necessitates "a more thoroughly decolonial, anti-racist historicizing of the university's spacetimes" (2019, p. 220). In the feminist spirit of reflexivity, we seek to build from this critique; a deeper understanding of the colonial and racialized foundations of the university is necessary. Still, we view collective efforts to challenge increased work expectations and to create care-full time as ethical and essential. Indeed, we foreground the necessity of collectively transforming the conditions of our labor, and hence conditions of knowledge production precisely because we have had intimate experiences with the unjust power relations in the academy. We seek to better understand how our efforts to create time and space for care and critical feminist work are connected to movements to end the violences of colonialism, racial capitalism, and heteropatriarchy.

Feminist Slow Scholarship Can Inform Research Methodologies

Research methodologies reflect decisions about how to combine epistemology with methods for gathering and analyzing research materials. Principles of feminist slow scholarship are more expansive than a research methodology because we advocate for collective action to transform the injustices of knowledge production and labor in the academy. Yet, feminist slow scholarship can inform research methodologies because the transformations we call for also inform the research process. We consider ourselves to "be" feminist researchers because we understand knowledge as power-laden and not neutral, an element of liberatory social change. We engage in reflexive practices, paying attention to the politics of representation, and recognizing how our embodiment shapes our research (Leavy & Harris, 2019).

The feminist ethics of care we seek to enact through feminist slow scholarship is a praxis that takes seriously the imperatives of care and caring labor, unsettles binary and essentialist thinking, centers the voices of the marginalized, and challenges interlocking forms of oppression, including critical intersectional analyses (Crenshaw, 1989; Lawson, 2007; Lopez, 2019; Raghuram, 2019). Care and caring practices are fundamentally racialized and shaped by (settler) colonial geographies that delineate global divisions of caring labor, differentiate who gives and receives care, and what that care looks like (Raghuram, 2019). Lopez argues that "[w]hile neoliberal ethics begin from the narrow presumption of individuality and equality of opportunity, devoid of concern for historical, social, political, and economic situatedness, care ethics offers a *counter*-ethics that begins from within a relational analysis which demands an attention to situatedness and a recognition of the interrelatedness of lives" (2019, p. 831, our emphasis). Such challenges can take many forms and involve a range of creative practices, including participatory methods (Cahill, 2007), refusing extractive research practices (Tuck & Yang, 2014; Coddington, 2017), continuously working through questions of accountability and reciprocity (TallBear, 2014; Ybarra, 2014), and rejecting theories, models, and forms of knowledge that create harms and/or reinforce the status quo (Daigle & Ramírez, 2019). This work takes time.

Emphasizing feminist slow scholarship in the research process makes space to infuse a feminist ethics of embodied care into the ways we enact research. Preissle (2006) argues that feminist ethics of care offer contextual, relational, and embodied ways of engaging with the principle-based ethics of respect for persons, beneficence, and justice. We build on feminist theorists who have challenged the premise of disembodied theory and research (Sevenhuijsen, 1998) to engage feminist care ethics that articulate the significance of the body to gendered practices of maintenance and nurturance (Fisher & Tronto, 1990). For example, addressing the racialized and gendered embodiments of academic acceleration, feminist slow scholarship stresses the ways in which "the effects of the neoliberal university are written on the body" in the form of insomnia, substance abuse, and illness (Mountz et al., 2015, p. 1245). Conversely, "[s]lowing down involves resisting neoliberal regimes of harried time by working with care while also *caring for ourselves and others*" (19, original emphasis). As such, feminist slow scholarship attends to the corporeal effects of overlapping crises, making time and space for embodied knowledges to guide our decisions and actions as researchers.

Rethinking Temporalities of Crisis and Care

Within Western contexts, disasters and catastrophes are often presented as unique, individual moments, as aberrations of the normal. In this framework, the temporality of crisis is short, neither protracted nor situated within social, economic, and environmental conditions that may have produced the crisis. For Bisoka (2020), COVID-19 has *not* become such an 'event' because it is being treated as a 'logistical challenge' by universities and funders in the Global North, while the precarities of daily life have continued for researchers in the Global South. Similarly, Whyte (Potawatomi Nation) (2021) questions crisis narratives surrounding climate change. He argues that the ubiquity of urgency and unprecedentedness in Western notions of crisis have been a vehicle for *advancing* colonial practices. The rush to respond to an 'urgent' issue invites a presentist framework, leading decision-makers to reach for "solutions that can occur quickly, maintain the current state of affairs, lack any sense of realism, and further entrench power" (11). For Bisoka, Whyte, and Meyerhoff and Noterman, presentist framings of crisis obscure colonialism as the longer process at work, and hence underscore the imperative of decolonizing both knowledge production and crisis. For us, 'slowing things down' amid a push to 'rush,' 'pivot' and 'be nimble' is a necessary aspect of feminist ethics and research.

For Gilmore, "[c]risis is not objectively bad or good; rather, it signals systemic change whose outcome is determined through struggle" (2007, p. 54), including how we characterize the temporality of both crisis and struggle. Disability justice advocates, for example, have questioned the 'novelty' of requests to wear face masks, wash hands, and arrange accommodations for virtual involvement, all of which have comprised accessibility demands from sick and disabled people for decades (Irresistible, 2020). To return to the pre-pandemic 'normal' would represent failures to sustain more inclusionary spaces and to engage with the "chronic nature of human vulnerability" (Nelson, 2020; Spencer, Mullings & Nadasen, 2020). Vulnerability should be the opening through which we can imagine our communities and research practices emerging from the pandemic. Disability justice scholar Piepzna-Samarasinha, for example, argues that care work that centers the shifting intimacies and temporalities of our (always) changing bodies is fundamental to building movements that are truly healing and inclusive. These include practicing and valuing mutual care and sociality – holding community, sharing food and resources, and childcare – and moving "at the rate of the person who needs the longest to get there" (Irresistible, 2020). Disability justice frameworks show how politics of time and politics of care are interconnected, insights that should inform feminist research practice.

With these insights in mind, we explore how a feminist praxis of slow scholarship can inform research design and methodologies. This section, written by an imagined Feminist Advice Column Collective (FACC), addresses specific elements of feminist ethics of care, politics of time, and the embodied aspects of care work. We posit these elements as central to feminist slow scholarship. Thinking through the interconnections of these three elements, our responses draw from both individual experiences and collective efforts.

Dear FACC: How Do We Practice a Feminist Ethics of Care Throughout the Research Process?

Dear Reader: Great question! We offer some thoughts derived from Carla's care-full attempts to attend to the "situatedness" and "interrelatedness of lives" (Lopez, 2019, p. 831) as part of doctoral research on experiences of care and belonging through a Private Sponsorship of Refugees (PSR) program in a mid-sized Canadian city.

CARLA: *I was working with community interpreters to co-create care maps with resettled refugee families that visually depict the everyday people, places, and practices of care during re-settlement. The research was well underway when COVID-19 precautions and closures abroad raised ethical questions about whether and how to proceed.*

One interpreter expressed concerns about unforeseen childcare needs, while another shared her own health vulnerabilities and limited technology access. As a team, we considered exacerbated vulnerabilities for participants: language barriers to health care information, housing shortages, variable technology access, transportation interruptions, school closures, caregiving responsibilities, income loss, and increasing demands on partnering organizations serving newcomers. By mid-March, we collaboratively paused face-to-face interviews – a month before the official University halt to all in-person research.

As the pandemic continued, I considered my "toolbox of options" (Billo & Hiemstra, 2013). Forging ahead was impossible for our team and participants, yet stopping research altogether would also have consequences. Collectively, we discussed the financial realities and community connections for each of us. These conversations revealed our different positions in a settler colonial system and our diverse interconnections with the communities involved in the research. As a white, middle-class, Canadian PhD candidate in my final year of funding, taking a leave of absence made sense financially and professionally. I could put my funding on pause, resume work as a healthcare professional, then return to the project as it was originally conceived. The community interpreters shared their experiences of being racialized newcomers embedded within the communities we were conducting research with, living in a nation-state that does not recognize their foreign credentials, and being hired through a local settlement organization as independent contractors. For them, stopping the research would mean a loss of their income during an already precarious time as well as possible tensions within their social networks. Long-term interruption of research could also jeopardize honoraria for participants, research outcomes in the form of reports for community organizations, and advocacy for policy changes to support family reunification efforts.

Carla's decision-making demonstrates how attention to temporality in feminist slow scholarship informs research methodology. Despite the hectic and seemingly urgent need to either rush through a set of interviews or stop research altogether, she slowed down enough to engage with interpreters, community members and partnering organizations, and listened to emergent needs, concerns and new everyday realities. She was then able to prioritize longstanding relationships and honor the wellbeing of everyone involved. This is one way that a researcher might enact a feminist ethic of care – one that attends to intersectional power dynamics and embodiment.

Dear FACC: How Do We Make Time For Slow Scholarship When We're Working in a System That Privileges and Prioritizes Speed?

Dear Reader: Another valuable question! While writing this chapter, we spent a lot of time discussing how feminist slow scholarship does not track onto time in a singular way. Sameera and Kela's experiences as graduate students in the early stages of their research projects show that practicing a feminist ethics of care offers insights for individual and collective efforts to 'make time,' which is to say create possibilities.

SAMEERA: *Like many graduate students who do qualitative research, COVID-19 upended my carefully laid out summer research plans for my master's thesis. Rather than immediately pivoting to remote qualitative methods, I chose a different path. I spent the summer and subsequent year thinking more carefully about how to draw on other methodological approaches that didn't require 'accessing' the communities I work with, whether in-person or through a Zoom connection. As I further reflected on the meaning of slow scholarship and feminist ethics in my own life this past year, I realized how this "privileged pause" propelled new insights and created new priorities for my training within not only my department and institution, but also the discipline of geography.*

During a time when we were physically apart and under crisis conditions, graduate students in the department came together to demand additional flexibility in our program deadlines beyond the institutional timeframe for when COVID-19 would be 'resolved.' As the racial reckoning of summer 2020 came to pass, my colleagues and I formed an abolitionist collective to advocate for not only police and prison abolition at the institutional level, but also to demand that our department collectively reflect on and address the racist practices operating within it. I also became heavily involved with my graduate student union's mutual aid efforts to collectively organize funds and material goods like diapers and food staples for our community. In 'normal' times, anything that didn't fall under the strict rubric of professional development, including departmental organizing and care work broadly conceived, would be considered mere distractions on my limited time to progress through my degree program at a satisfactory pace. As it became clearer that conditions would not quickly return to 'normal,' I found myself less concerned with my thesis and more engaged with efforts to ensure my fellow colleagues had access to the support they needed to survive not only the pandemic, but what came after.

Prior to the pandemic, I only used feminist research methods and feminist care ethics to think through what I owe to the communities I research in faraway places, but this pause prompted me to encounter these same questions and dilemmas closer to home and within my own department. This time has clarified for me that it is impossible for one to 'just' focus on one's own research. The reckoning that COVID-19 and Black Lives Matter prompted opened up new modes of engagement and care relations with fellow students, faculty, and community members that might otherwise not have unfolded.

KELA: *It should not be a surprise that Sameera and I also became friends at the intersection of these collective efforts and through the navigation of an alienating system. And like Sameera, the pandemic also allowed me a privileged pause. The pause I experienced coincided with the transitional adjustments I had to work through as I looked to see myself as a graduate student and adult. Accompanying the excitement and potentials of graduate life, I was also working to understand my own insecurities and the feelings that left me stuck regarding my thesis work and progress. I had no confidence in my ideas or how to articulate them clearly on paper. I had no idea how to be a geographer, a Black one at that. Privileged as it may be, this was a pause I required to make space for myself. It was a pause I was too scared to advocate for, but that my own advisor advocated for on my behalf. This pause allowed me time to discover a writing process, reframe my thesis, and build confidence in my own abilities and potential as a graduate student new to the field of geography and the spaces of the academy.*

Most importantly, however, was a recognition of the care and support I received and advocated for -- other graduate students, my advisor, family, friends, therapy, teaching, learning, and reading -- encompassing a multitude of dynamic care networks, methods, and practices that have sustained and guided me through this pause, graduate school, and beyond. One lens to view these networks of care may be through privilege. However, reframing this care and support through the lens of a feminist praxis and intentional method of living, acknowledges and takes seriously the time required for learning and refining humanizing practices, indicating that these systems of care are what is required for survival, pandemic or not.

Graduate school can be atomizing and lonely. However, both Sameera's and Kela's experiences speak to how a once-lonely environment emerged as a space for building collective and empowering networks of care, support, and sustainability. Inspired by intersecting crises and organizing emerging from the pandemic and Black Lives Matter protests, Kela and Sameera found new ways to engage in care relations. These practices highlight the need for time to enact feminist research. They also demonstrate how graduate students' advocacy for more time became a necessary collective act of care to push back against the academy's temporal logics.

In this next reflection, Roberta builds on the observation that the temporalities of feminist slow scholarship and the neoliberal university are often at odds. This observation allows exploration of ethical dilemmas that can emerge when we prioritize urgency and immediate action in response to crisis.

ROBERTA: *When my university shut down suddenly, I was about to begin instructing 15 graduate students in an International Development methodology class. This course was meant to prepare students for 'the field,' in many cases to conduct in-depth, ethnographic research abroad. Instead of the intensive, collaborative two-week workshop I had spent months planning, we were thrust into a world of virtual, asynchronous communication. In addition to dealing with the stressors of health crises, uncertainty, and drastically different everyday lives (including care-giving responsibilities), students had to deal with cancelled research and travel plans. Their master's research plans screeched to a halt as travel bans took effect and in-person research was prohibited by the University. Financial stress loomed large as many students needed to complete their research that summer to avoid paying extra tuition. Many students turned to secondary data, such as policy documents or social media. This introduced a whole new set of ethical questions for us around how one can best consider positionality and power dynamics when doing secondary data analysis. In many cases, it also introduced a set of ethical dilemmas about whether research should still be pursued on topics of concern to more vulnerable communities when their voices wouldn't necessarily be incorporated into the research process. The unevenness of technology, time availability, and mobility were stark. The pressure from the University to assign course grades and proceed 'as normal' with teaching illustrated how poorly the institution could adapt to crisis. As an instructor and mother suddenly swamped with my own stress and care responsibilities and struggling to support students with these kinds of decisions, I felt inadequate and drained. With limited time and mental resources, I tried to prioritize supporting students as people first (rather than researchers) and then to offer guidance on research decisions whenever possible.*

Feminist slow scholarship aims to enable deep ethical and relational engagement, but what do we do when these engagements are challenged? One of the difficulties that accompanied the pandemic was balancing the time required to think deeply, to read, reflect, and make more complicated research or teaching-related decisions. Our vignettes illustrate that sometimes the answer is to not quickly change course or not to do research at all if we cannot do it in a way that aligns with our ethics. Sometimes the answer involves ignoring the urgency and sitting with decisions longer, whenever possible, and giving ourselves time to think through their impacts. However, as several of us write, in many cases ignoring the timeframe carries material consequences for researchers and participants. Being able to push against institutional pressures helps, but this is not something we can do alone. We need to work with one another to create slow temporalities and spaces for thinking and connecting.

Dear FACC: How is Slowing Down Relevant to Practices of Embodied Care Work, Both Within and Beyond Formal Scholarship?

Dear Reader: The relationship between slowness and care has manifested differently for each of us, but the common thread is that the embodied demands of care work – as researchers and human beings surviving a global pandemic – have forced us to slow down aspects of academic work. Conversely, the pressure to still produce research "outputs" has taken a significant toll on our bodies and minds. Alison considers the embodied challenges of trying to work from home with children during the pandemic.

> ALISON: *In my impossible morning, by 10:00 am, I've tried to persuade my five-year old to sit in front of a glitchy national anthem and a land acknowledgment he doesn't yet understand by convincing him to roll Play-Doh on his desk. This required my physical presence and constant cajoling, but only lasted ten minutes at which point I set up a letter-based scavenger hunt for him and my mother to do together on Facetime. But now he is distracted from this and wants to participate in an online PE class with his older sister. This will last a little longer and he will lose interest and I'll cajole him into something else.*
>
> *During this time, I watch several email requests and demands roll angrily into my inbox where they join a mountain of others, all of which elicits something beyond frustration, something more akin to rage. Writing on the tiny screen of the only device left for my use – my phone – I tell students, deans, editors, everyone – that it will have to wait until I get out of full-time kindergarten. I occasionally draft exasperated letters to funding agencies and university administrators suggesting they delay deadlines and cancel three-hour webinars. People write back: what alternative can I offer? In reality I have no time or energy to imagine alternatives or solutions. I am exhausted. We're all just surviving. Our kids, out of school and separated from friends, need us for everything: they need our minds to learn, our bodies to cuddle and nourish, our care to attend to their own emotions. I love my kids while I rage at the outside world that has no idea what has happened to our lives.*

For Alison, the demands of work and pandemic parenting on her body have exacerbated existing conditions of speed, forcing a type of slowing down that is in direct conflict with the demands of paid work. For Stepha, slowing down has served as an essential practice for managing embodied trauma, the effects of which were highlighted during the pandemic.

> STEPHA: *Last summer, I worked with migrant live-in caregivers, translating and documenting their stories of abuse at the hands of their employers. Some of the women had been confined in their employers' homes for months at a time; they described sleep deprivation, migraines, and hand tremors as embodied effects of their experiences. I began to wonder how an awareness of somatic trauma might inform my research methods.*
>
> *Seven months later, the pandemic reached my university. I saw myself and those around me demonstrating signs of somatic dysregulation – alternating between anxious "activation" and dissociative "deactivation." I knew that the caregivers I'd worked with were going through particularly trying*

times, taking care of elders in lockdown conditions that increased their vulnerability to domestic vio-
lence. I, on the other hand, had secure housing, an income, and saved money. Even so, my own
embodied experiences of quarantining brought me closer to an understanding of what caregivers had
been experiencing long before the pandemic – extended confinement in the home. Wanting to develop
more skills that would help me in working with both my local and field site communities, I enrolled
in somatic training.

In the course of my training, I learned that traumatized people often tell their stories in ways
that re-traumatize them. This means that researchers may be exacerbating harm simply by asking
interlocutors to tell their stories. This realization has fundamentally transformed my approach to
structuring interviews, focus groups, and participatory research. Somatic practices of slowing down the
conversation and "titrating" through engaging with grounding resources can help to reduce the harmful
potentials of ethnographic research.

In both of our cases, we found that the temporality of the academy was fundamentally at odds
with our embodied commitments to care. Whether engaging in care work in our homes,
communities, or at our research sites, we found that slowness was an essential component for
developing relationships built on feminist ethics of care.

Conclusion

As the world faced myriad political, economic, and ecological crises, we set out to articulate
what it means to design and carry out research framed by feminist slow scholarship. We emphasize feminist care ethics, a politics of time, and embodiment as elemental dimensions of feminist
slow scholarship. Like other feminist methodologies and research ethics, these are interwoven
into research design and process, not linear protocols. Moreover, they are fundamentally tied to
broader projects of liberatory engagement with the university. We can work collectively to push
back on the commodification of knowledge by refusing to produce findings in ways that harm
people with whom we work. We can resist capitalist timelines by collectively carving out more
time for reflection and conversation to produce work that is more aligned with our values. And
we can challenge the university's disembodiment of knowledge by attuning ourselves to the
emotional and the embodied. Attention to the embodiment of care work – for children, loved
ones, students and colleagues – during the pandemic underscores how *all* research rests on social
reproduction. Feminists draw attention to the value of and imperative for creating more equitable and just conditions for this work.

A feminist politics of slowness offers research the imperative of engaging the politics of
time. We draw inspiration from feminist, decolonial, and disability justice frameworks that
foreground temporality and collective efforts to make space-time necessary to engaging in care.
These efforts may appear small, not up to the task of much-needed institutional and societal
change. However, recognition of the imperative for care in conditions of uneven but shared
vulnerability also suggests less "monumental," more prosaic collective responses that can grapple
with "continuities of older forms of distancing and exclusion" (Bhan et al., 2020, p. 1). As
Spencer, Mullings, and Nadasen (2020) write, "COVID-19 makes a return to 'normal' impossible and presents an opportunity for imagining a new world. Indeed, a transformative vision
has the potential to erupt from the ashes." In this sense, feminist slow scholarship offers a pro-
spective horizon, a way of doing care-full research as part of collective endeavors to change the
university and world around us.

References

Ahmed, S. (2012) *On being included: Racism and diversity in institutions*. Durham, NC: Duke University Press.

Anguelovski, I., Corbera, E., Honey-Rosés, J. and Ruiz-Mallén, I. (2020) Academia in the time of Covid-19: Our chance to develop an ethics of care. *Barcelona Laboratory for Urban Environmental Justice and Sustainability*. Available from: http://www.bcnuej.org/2020/03/31/academia-in-the-time-of-covid-19-our-chance-to-develop-an-ethics-of-care/ [Accessed 1 June 2021].

Askins, K., and Blazek, M. (2017) Feeling our way: Academia, emotions and a politics of care. *Social & Cultural Geography*. 18 (8), 1086-1105.

Batacharya, S., and Wong, Y.-L. R. (eds.) (2018) *Sharing breath: Embodied learning and decolonization*. Edmonton, Canada: AU Press.

Bellamy, P. L. and Mosley, D. V. (2020). A tale of two towers: An open letter and call to action. *Academics for Black Survival and Wellness*. Available from: https://www.academics4blacklives.com/call-to-action/a4bl-call-to-action/call-to-action-our-letter-to-academics [Accessed 1 June 2021].

Berg, M. and Seeber, B. K. (2016) *The slow professor: Challenging the culture of speed in the academy*. Toronto, Canada: University of Toronto Press.

Bhan, G., Caldeira, T, Gillespie, K. and Simone, A. (2020) The pandemic, Southern urbanisms and collective life. *Society & Space*. Available from: https://www.societyandspace.org/articles/the-pandemic-southern-urbanisms-and-collective-life [Accessed 1 June 2021].

Billo, E. and Hiemstra, N. (2013) Mediating messiness: Expanding ideas of flexibility, reflexivity, and embodiment in fieldwork. *Gender, Place & Culture*. 20 (3), 313–328.

Bisoka, A. N. (2020) Disturbing the aesthetics of power: Why Covid-19 is not an "event" for fieldwork-based social scientists. *Social Science Research Council*. Available from: https://items.ssrc.org/covid-19-and-the-social-sciences/social-research-and-insecurity/disturbing-the-aesthetics-of-power-why-covid-19-is-not-an-event-for-fieldwork-based-social-scientists/ [Accessed 1 June 2021].

Black, A. (2018) Responding to longings for slow scholarship: Writing ourselves into being. In: Black, A. and Garvis, S. (eds.) *Women activating agency in academia: Metaphors, manifestos and memoir*. New York: Routledge, pp. 23–34.

Cahill, C. (2007) The personal is political: Developing new subjectivities through participatory action research. *Gender, Place & Culture*. 14 (3), 267–292.

Caretta, M. A., and Faria, C. V. (2020) Time and care in the "lab" and the "field": Slow mentoring and feminist research in geography. *Geographical Review*. 110 (1–2), 172–182.

Carrigan, M. and Vostal, F. (2016) Not so fast! A critique of the 'slow professor.' *University Affairs*. Available from: https://www.universityaffairs.ca/opinion/in-my-opinion/not-so-fast-a-critique-of-the-slow-professor/ [Accessed 1 June 2021].

Coddington, K. (2017) Voice under scrutiny: Feminist methods, anticolonial responses, and new methodlogical tools. *The Professional Geographer*. 69, 314–320.

Crenshaw, K. (1989) Demarginalizing the intersection of race and sex: A Black feminist critique of antidiscrimination doctrine, feminist theory and antiracist policies. *The University of Chicago Legal Forum*. 1989, 139–167.

Daigle, M. and Ramírez, M. (2019) Decolonial geographies. In: Antipode Editorial Collective (eds.) *Key words in radical geography*. New York: Wiley, pp. 78–84.

Fisher, B. and Tronto, J. (1990) Toward a feminist theory of caring. In: Abel, E. K. and Nelson, M. K. (eds.) *Circles of care: Work and identity in women's lives*. Albany, NY: SUNY Press, pp. 35–62.

Freire, P. (1970) *Pedagogy of the oppressed*. Harmondsworth, Middlesex: Penguin Books.

Gilmore, R. W. (2007) *Golden gulag: Prisons, surplus, crisis, and opposition in globalizing California*. Berkeley: University of California Press.

Gutiérrez y Muhs, G., Niemman, Y. F., González, C. G., and Harris, A. P. (eds.) (2012) *Presumed incompetent: The intersections of race and class for women in academia*. Boulder: University Press of Colorado.

Hamer, J. F. and Lang, C. (2015) Race, structural violence, and the neoliberal university: The challenges of inhabitation. *Critical Sociology*. 41 (6), 897–912. https://doi.org/10.1177%2F0896920515594765

Hamilton, T., Hawkins, R., and Walton-Roberts. M. (Forthcoming, 2021) Canon, legacy or imprint: A feminist reframing of intellectual contribution. In: Black, A. L. and Dwyer, R. (eds.) *Reimagining the academy: shifting towards kindness, connection, and an ethics of care*. New York: Palgrave MacMillan.

Ho, E. L.-E., and Maddrell, A. (2021) Intolerable intersectional burdens: A COVID-19 research agenda for social and cultural geographies. *Social & Cultural Geography*. 22 (1), 1–10.

Irresistible (2020) Organizing in a pandemic: Disability justice wisdom. *Irresistible* (podcast). Available from: https://irresistible.org/podcast/61 [Accessed 1 June 2021].

Islam, F. and Netto, G. (2020) "The virus does not discriminate": Debunking the myth: The unequal impact of COVID-19 on ethnic minority groups. *Radical Statistics*. 126. Available from: https://www.radstats.org.uk/no126/IslamNetto126.pdf [Accessed 1 June 2021].

Kolinjivadi, V., Van Hecken, G., Casolo, J., Abdulla, S. and Blomqvist, R. E. (2020) Towards a non-extractive and care-driven academy. *Global Working Group Beyond Development*. Available from: https://beyonddevelopment.net/towards-a-non-extractive-and-care-driven-academy/ [Accessed 1 June 2021).

Lawson, V. (2007) Geographies of care and responsibility. *Annals of the Association of American Geographers*. 97 (1), 1–11.

Lopez, P. J. (2019) Toward a care ethical approach to access to health care in neoliberal times. *Gender, Place & Culture*. 26 (6), 830–846.

Lopez, P. J. and Gillespie, K. (2016) A love story: For 'buddy system' research in the academy. *Gender, Place & Culture*. 23 (12), 1689–1700.

Malisch, J. L., Harris, B. N., Sherrer, S. M., Lewis, K. A., Shepherd, S. L., McCarthy, P. C., ... and Ramalingam, L. (2020) Opinion: In the wake of COVID-19, academia needs new solutions to ensure gender equity. *Proceedings of the National Academy of Sciences*. 117 (27), 15378–15381.

Meyerhoff, E. and Noterman, E. (2019) Revolutionary scholarship by any speed necessary: Slow or fast but for the end of this world. *ACME: An International Journal for Critical Geographies*. 18 (1), 217–245.

Mountz, A., Bonds, A., Mansfield, B., Loyd, J., Hyndman, J., Walton -Roberts, M., Basu, R., Whitson, R., Hawkins, R., Hamilton, T., Curran, W. (2015) For slow scholarship: A feminist politics of resistance through collective action in the neoliberal university. *ACME: An International Journal for Critical Geographies*. 14 (4), 1235–1259.

Mountz, A., Bonds, A., Mansfield, B., Loyd, J., Hyndman, J., Walton -Roberts, M., Basu, R., Whitson, R., Hawkins, R., Hamilton, T., Curran, W. (2016) All for slow scholarship and slow scholarship for all. *University Affairs*. Available from: https://www.universityaffairs.ca/opinion/in-my-opinion/slow-scholarship-slow-scholarship/ [Accessed 1 June 2021].

Nelson, A. (2020) Society after pandemic. *Social Science Research Council*. Available from: https://items.ssrc.org/covid-19-and-the-social-sciences/society-after-pandemic/ [Accessed 1 June 2021].

Peake, L. and Mullings, B. (2016) Critical reflections on mental and emotional distress in the academy. *ACME: An International Journal for Critical Geographies*. 15 (2), 253–284.

Power, K. (2020) The COVID-19 pandemic has increased the care burden of women and families. *Sustainability: Science, Practice, and Policy*. 16(1), 67–73.

Preissle, J. (2007) Feminist research ethics. In: Hesse-Biber S (ed) *Handbook of feminist research: Theory and praxis*. Sage Publications, London, pp. 515–532.

Raghuram, P. (2019) Race and feminist care ethics: Intersectionality as method. *Gender, Place & Culture*. 26 (5), 613–637.

Roberts-Crew, J. (2020) White academia: Do better. *The Faculty*. Available from: https://medium.com/the-faculty/white-academia-do-better-fa96cede1fc5 [Accessed 1 June 2021].

Roy, A. (2020) The pandemic is a portal. *Financial Times*. Available from: https://www.ft.com/content/10d8f5e8-74eb-11ea-95fe-fcd274e920ca [Accessed 1 June 2021].

Save Our Slides (N.D.) Facebook group. Available from: https://www.facebook.com/Save-Our-Slides-563251040474916/ [Accessed 1 June 2021].

Sevenhuijsen, S. (1998) *Citizenship and the ethics of care: Feminist considerations on justice, morality and politics*. New York: Routledge Press.

Spencer, R. Y., Mullings, L., and Nadasen, P. (2020) The fire this time: A statement from Scholars for Social Justice. *Boston Review*. Available from: https://bostonreview.net/articles/robyn-c-spencer-leith-mullings-premilla-nadasen-fire-time/ [Accessed 1 June 2021].

TallBear, K. (2014) Standing with and speaking as faith: A feminist-indigenous approach to inquiry. *Journal of Research Practice*. 10 (2), N17-N17.

Thorogood, J., Faulkner, S., and Warner, L. (2019) Slow strategies for student (and staff) engagement. *Student Engagement in Higher Education Journal*. 2 (2), 105–123.

Titchkosky, T. (2010) The not-yet-time of disability in the bureaucratization of university life. *Disability Studies Quarterly*. 30 (3/4), 1–14.

Tuck, E. and Yang, K.W. (2014) R–Words: Refusing research. In: Paris, D. & Winn, M. T. (eds.) *Humanizing research: Decolonizing qualitative inquiry with youth and communities.* London: SAGE Publications, pp. 223–248.

Weeden, K., Cha, Y., and Bucca, M. (2016) Long work hours, part-time work, and trends in the gender gap in pay, the motherhood wage penalty, and the fatherhood wage premium. *RSF: The Russell Sage Foundation Journal of the Social Sciences.* 2 (4), 71–102.

Whyte, K. (2021). Against crisis epistemology. In: Hokowhitu, B., Moreton-Robinson, A., Tuhiwai-Smith, L., Larkin, S. and Anderson, C. (eds.) *Routledge Handbook of Critical Indigenous Studies.* London: Routledge, pp. 52–64.

Ybarra, M. (2014) Don't just pay it back, pay it forward: From accountability to reciprocity in research relationships. *Journal of Research Practice.* 10 (2), N5.

32

REFINING RESEARCH METHODOLOGIES TO MAKE A DIFFERENCE IN POLICY

Carolyn DeLoyde and Warren Mabee

Introduction

Human geographers around the world are conducting research that is relevant to policy challenges faced by society. For example, Henri Lefebvre's idea of 'right to the city' and David Harvey's work with regard to the ability of collective power to reshape urbanization have shaped public policy designed to increase social equity within the urban public realm (Jonas et al., 2015). Critical topics including health, environment, social equity, and education are widely represented in human geography literatures, and include findings that speak to critical policy challenges. Importantly, research in this discipline is usually framed in geographic terms, which provides a readily accessible context with which the average reader is more likely to be familiar, and which tends to line up with jurisdictions for which policymakers are responsible. Despite these factors, the impact of geography on public policy has been limited over the years (Martin, 2001). For researchers who wish to design their investigations to provide information that can be taken up and used in policy, it is worthwhile reflecting on the challenges that policymakers face.

Policy making has been described as 'part science and part art' (Rosewell, 2017). Today's policymakers need to work with complex data while balancing a number of competing demands including stakeholder needs, budgetary constraints, and the broader public interest (Valaitis et al., 2020). As policymakers struggle with problems that are increasingly complex, the data that they require to make decisions becomes more arcane. This is because science has become increasingly specialized, and the tools being applied have become more intricate, making it harder for non-specialists to evaluate or understand the data being presented. Put simply, while human knowledge has advanced at a tremendous rate, the ways in which this knowledge can be applied to daily life is less obvious. This creates a challenge: it can be very difficult for policymakers to take up the findings of research programs.

Over the years, different concepts have been proposed to better understand how research findings, or evidence, are used to inform policy. Gaining recognition since the mid-1990s and termed evidence-based policy making (EBPM) this approach is used to create policies based on knowledge and research. The 1997 introduction and use of the term by the Blair government in the United Kingston, recognized that the creation of policy based on available knowledge and research ('evidence') was generally accepted (De Marchi et al., 2016) but not

DOI: 10.4324/9781003038849-36

considered new. EBPM, also considered a 'rationalist' or 'engineering' approach, suggests that facts uncovered by research will be picked up by policymakers directly and acknowledges that policymakers are constantly revising and updating their plans, and that they have full access to, and understanding of, the findings that scientists are bringing forward. Perhaps recognizing that this is not always the case, the 'enlightened' (or constructivist approach) has been proposed as an alternative. While this model also holds to the idea that facts will be used, it suggests that the ways in which new ideas filter into policy are more indirect and complex (e.g., Ekici, 2002; Weiss, 1979). The 'constructivist' approach recognizes the complexities of human experiences and subjective interpretation (Newman, 2017). By acknowledging the importance of alternative routes by which data inform policy, this model recognizes that researchers can benefit from embracing a variety of dissemination models, which increase the chances of new ideas being taken up.

Some might assume from this discussion that communication is the primary problem, and that more inclusive dissemination strategies – such as publication of laypersons summaries, or media outputs – will ensure that more research can be included in policy through both direct and indirect pathways. Recent work challenges this assumption, particularly with the introduction of the strategic model of evidence-based policymaking. This model suggests that policymakers are becoming increasingly selective ('strategic') in the use of research findings and tend to seek out findings that support their own agenda (e.g., Hawkes et al., 2012). Clearly there are dangers in simply relying on decision makers to take up new ideas; researchers need to adopt strategies that go beyond communication and dissemination of their findings and begin to address fundamental concerns of research design.

It has been argued that the complex nature of research translation and knowledge transfer from researcher to policymaker requires a move toward enabling an approach more akin to knowledge interaction (Smith and Joyce, 2012). Interaction such as engaging researchers with policymakers, and decision makers at the earliest stages of a project helping to define the actual research questions being addressed. Engagement and interaction such as this can facilitate knowledge translation and ultimate research uptake into policy decisions. While this engagement approach appears clear, obstacles exist owing to the complexities of knowledge transfer. Identification of these obstacles (as highlighted below) and the design of methods to overcome them (as shown in Figure 32.1) can begin a transition to a system of 'research informed' policy design. Research design concerns such as that highlighted in recent work by Klein and Juhola (2014), explores barriers that exist when policymakers attempt to use research findings. Their work identified a series of specific bottlenecks, which the authors of this chapter have generalized as follows:

1. The theoretical scenarios used by researchers do not relate to the 'reality' that policymakers face;
2. The scale at which research operates does not match political realities:
    a. There is a mismatch between spatial scale of research and scales at which policymakers act;
    b. There is a mismatch between the timelines at which research reports vs. the timelines in which policymakers tend to plan and act;
3. Research findings are not presented in the right context:
    a. The response to uncertainty is usually to 'wait and see', rather than enact new policy;
    b. Research findings must compete with other priorities in policymakers' decisions.

A careful consideration of these bottlenecks reveals important considerations for experimental design. Indeed, addressing all of these bottlenecks requires considerable effort and affects all aspects of the research enterprise. The first and second points speak to experimental design and

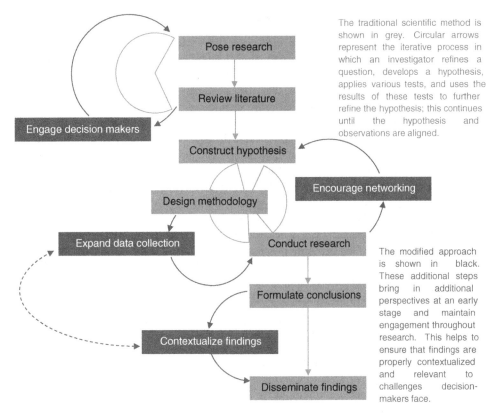

Figure 32.1 Modifying the scientific method to facilitate better policy uptake.

Source: Authors' creation.

initial project scoping. The remaining point deals with the questions being asked, the analytical framework that is applied, and the context in which results are presented from research. It follows that research methodologies need to be refined if the intention is to deliver research findings that resonate with – and are usable by – policymakers.

In Figure 32.1, four modifications to the traditional scientific method are proposed to better facilitate the use of research findings in policy. In the remainder of this chapter, each of these proposed modifications are discussed. This chapter will use examples of geographic research in areas such as land use, energy development, and health geographies in order to illustrate how researchers can modify traditional approaches to research design.

A quick note on nomenclature: this chapter uses the terms 'decisionmaker' and 'policymaker'. While the two are somewhat interchangeable, in this context 'decisionmaker' refers to a broader selection of stakeholders in this area, including representatives from community groups, non-governmental organizations, and corporations in addition to government representatives, who have input into the policymaking process. 'Policymaker' is used to refer to individuals responsible for developing and implementing policies that can be enforced across a jurisdiction; while these individuals are usually government or public sector workers, the term can also apply to a high-ranking official in any organization who has responsibility for policy delivery.

Engage Decisionmakers

In following the traditional scientific method, researchers start by posing questions and using the literature to understand how others have approached these issues. This allows the investigator to see if anyone else has already answered their question, and what methodologies other researchers have employed in doing so. Through an iterative process, a novel question will emerge, allowing the investigator to then move to the stage of constructing and testing their hypothesis. Scientists today heavily favour peer-reviewed literature for this task.

It is important to understand this process because the heavy reliance on academic antecedents can make it harder to ultimately use the outcomes in policy. This approach encourages researchers to use language and concepts that resonate with previously posed questions, reinforcing past practice. The process also encourages simplification of the problem statement in order to create a novel, unique, and manageable question. Taken together, the effect of developing this question can be a shift of focus away from the initial policy challenge towards a simplified, academically focused problem. Even when the original questions are asked in reference to specific policy challenges, the literature that is used to refine and shape this question skews to academic assessments.

Policymakers facing complex social or ecological challenges tend to have big questions that parallel those being asked by human geographers (e.g., 'how can we improve health in different societal groups', or 'how can we reduce emissions?'). While these questions are difficult enough to answer on their own, policymakers face additional difficulties in that their solutions must work within a complex set of constraints. In effect, each additional constraint has the result of modifying policymakers' questions in substantive ways (e.g., 'how can we improve health in different societal groups, while not incurring higher costs?' or 'how can we reduce emissions, without shutting down our existing industry?'). To further complicate things, some of these constraints may be unwritten or unacknowledged; it may not be politically expedient to reveal each consideration that is in effect, but they will still be considered by policymakers when they are determining ultimate solutions.

Human geographers who wish to ensure that their findings are useful in policy should consider engaging policymakers and decisionmakers in the initial, scoping phase of research. Understanding desired policy outcomes at this stage can help researchers better align their research questions with policy goals (Welfle et al., 2020). Engagement at this early phase allows the investigator to understand not only the appropriate question, but the constraints that are in place from the decisionmaker's perspective. By sharing questions early and responding to decisionmakers when they provide input, investigators will be able to develop overarching questions that reflect current policy challenges. Regular engagement at this phase will even address unwritten constraints, as investigators can observe the responses of decisionmakers to different kinds of questions and will be able to better understand their issues.

Researchers should give thought to governments or organizations that policymakers and decisionmakers represent; this is generally informed by the type of policy that researchers hope to influence. Policy tends to work in a nested and hierarchical fashion, with national and/or subnational policy describing overarching frameworks, while regional and/or local governments provide more site-specific policy documents (Seasons, 2021). Researchers need to consider the level at which policy operates and target decisionmakers who are active at that level.

As the research question is developed, input from decisionmakers can also inform a broader literature review. The literature review is essential in the research process; it provides the theoretical underpinning and assists to place research within the larger context of the geographical canon and geographic theories. While traditional, peer-reviewed academic sources remain critical, investigators need to also consider official policy documents. Decisionmakers engaged in the process of defining research questions can provide useful advice in steering investigators towards appropriate policy resources, and – depending on the investigator's familiarity with policy – may also provide some support in interpreting these documents. Researchers need to take care to use the most recent versions of policies, as these documents are often subject to regular reviews and updates. Policy documents include materials beyond official policies, laws, or strategies; background documents including technical studies, discussion papers, staff reports and public consultation documents also need to be reviewed and may provide valuable context.

Early engagement of decisionmakers has been shown to be beneficial in practice. For example, while conducting health policy research in Pakistan, Hawkes et al., (2012) stressed the need for early engagement with decisionmakers in order to better refine the questions being carried out. Doing so created a research program that more closely addressed some of the policy challenges being faced. Similarly, research on climate change adaptation carried out in Canada highlights the risk of gaps between research outcomes and policy development without the input of decisionmakers at an early stage of work (Dany et al., 2016). These examples are particularly powerful because these research areas face similar challenges of large data requirements coupled with high uncertainty; decisionmakers need to be comfortable with the way in which questions are being asked and how these challenges are taken into consideration.

Allowing decisionmakers to provide input into the development of initial research questions need not be interpreted as giving up control over the research program. Indeed, incorporating the perspective of individuals responsible for policy development also runs the risk of introducing political agendas into the research process, which is inappropriate. Hawkes et al., (2012) commented the critical need to engage stakeholders or decisionmakers that exist outside the 'policy elite'. The latter categories might include representatives from ENGOs, communities, or other groups that (a) represent marginalized communities, and (b) inform but do not make critical policy (Hawkes et al., 2012). It is useful, therefore, to consider a broad series of consultations with decisionmakers from a variety of groups at the earliest phase of research.

Engaging decisionmakers at an early stage begins to address some of the important bottlenecks identified in the introduction. Early discussions and incorporation of policy-related literature allows questions to be asked which acknowledge other priorities for which policymakers need to account; it also begins to inform issues of scale and timelines by ensuring that the research question be pitched at appropriate levels. Importantly, this engagement begins to ensure that the theoretical framework in which researchers work shares commonality with the reality in which policymakers operate.

Expand Data Collection

After defining the research question, the research project can shift to developing and testing the hypothesis. This phase of research design and implementation, when informed by the consultation described in the previous section, may result in an expanded data collection scope. Additional data may be required to situate the research at appropriate spatial or temporal scales, or to provide critical information for the decision-making process.

The types of additional data required at this phase are highly contextual, depending on the subject matter being explored. For example, researchers in China specifically considered what data policymakers needed to build better policy when designing a study on healthcare worker training. This work confirmed that additional information was needed, beyond that typically collected in this type of research. What was needed was not just data on the outcomes of the program, but information on the resources required to implement training, the indirect impacts of training, and the sustainability and scalability of these training exercises to other settings (Wu et al., 2018).

Without appropriate data collection, researchers may not be able to provide the context that decisionmakers require to inform policy. Even where discrete case studies are used, researchers need to be deliberate about incorporating the right additional data to provide context to decisionmakers. This need was highlighted by recent work from Valentine (2014), who assessed best practices for case study work as released by the International Renewable Energy Agency (IRENA); Valentine argues that it is not enough to simply delineate a specific place or case and recommends additional work to better inform policy outcomes.

Data on program cost is one of the primary requirements that policymakers identify when they seek to implement research findings (Moseley et al., 2013). It can be very difficult; however, for researchers to gather or present useful economic data. Very few articles published in the last decade directly address these issues. Authors including Iheukwumere et al., (2021), Kazemian et al., (2020), and Tschakert et al., (2017) provide examples of comprehensive works that have included data collection around costs of program delivery. However, the relative paucity of studies that incorporate this type of data speaks to the difficulty in including data on the costs of policy options. It is important in research design that data collection is targeted and focused to deliver maximum returns; this emphasizes the importance of including policymakers in the research design phase, as they can highlight the data that is essential to support their decision-making process. If geographers seek to impact policy development, assembling data on costs can increase the likelihood of their findings being taken up by decision makers.

Encourage Networking

During the research process, it is essential that the research team remain in close contact with stakeholders, including policymakers that may use the findings of their research. This can be a difficult requirement for researchers, as it requires the investigator(s) and their team to create a structure that allows stakeholders access to the work as it occurs. Ideally, this type of interaction allows stakeholders to inform iterations of the research as it takes place; this allows end users to fine-tune the research, and particularly the ways in which data are being collected and assessed, to better suit stakeholder needs.

Networking at this stage is critical because, if done correctly, it allows a variety of voices to participate, essentially creating a reality that is closer to what policymakers and decisionmakers' experience (Meehan et al., 2018). The stakeholders engaged at this stage of the research may include some of the policymakers that are consulted in the earliest phases of research, but at this point in the research it is important that a wide variety of stakeholders be consulted. Researchers need to treat this stage as an opportunity to create linkages not only with the organization or government that is developing a policy of interest, but also with the bodies that regularly interact with them. Doing so brings a variety of perspectives into the research in an iterative fashion (Pallett and Chilvers, 2015). These networks can stretch from local to global, depending upon the policy area of interest (Prince, 2012).

In British Columbia, researchers investigated the use of integrated knowledge translation (IKT) to improve the uptake of their findings into public health policy. The study incorporated a knowledge broker embedded within the research team to help improve knowledge about the findings of the project among the different stakeholders; the outcomes of the study emphasized benefits including improved contextual knowledge and better understanding of the potential community level impacts of policy instruments (Mendell and Richardson, 2021). Researchers working in this space, particularly with complex networks of actors that influence or guide policy, might consider adding a dedicated knowledge broker to the team.

There can be a risk with including non-researchers directly in the research program. Some researchers have commented on the ways in which knowledge is used by different stakeholders and noted that while some practitioners and policymakers report better uptake of research findings when they are included in this fashion, others balk at using findings that contradict their own lived and professional experiences (Wathen et al., 2011). Researchers have, however, found that incorporating policymakers and decisionmakers in these aspects of research is critical in building trust and suggest that better policy emerges from these processes (Scott, 2015).

Expanding the network that is engaged in actual research has multiple benefits with respect to addressing critical bottlenecks in the science-policy interface. As discussed, the plurality of voices engaged in the discussion will ensure that the 'reality' that researchers face is a closer approximation of the environment that policymakers and decisionmakers inhabit. It also can help researchers address critical issues of project scope and timelines and ensure that the researcher is more closely aligned to the scales at which policymakers act. Finally, the introduction of different voices ensures that researchers pay more attention to the context in which policymakers and decisionmakers are required to act.

Contextualize Results

Much of the data that is generated by researchers today can be difficult to understand. Researchers have highlighted the need to carefully design dissemination in order to get their messages across. This can be difficult and there may not be much guidance available in the literature itself on how to design these dissemination tactics. There has been substantive focus on 'open science' – the strategy of publishing in open-source publications that present no barriers, such as subscriptions, to potential readers – as a means of gaining more readers and making work more accessible. Critical assessment of open science as a dissemination technique, however, points to the need for researchers to carefully tailor their writing in order to increase both transparency and reader engagement (Leonelli et al., 2015). There is limited guidance in the literature on just how researchers should go about doing so (e.g., Sullivan et al., 2014).

One area that may assist researchers in presenting their work are the principles of 'outcomes research', which requires work to be relevant, practical, and applied. Krumholz suggests that the adoptions of an outcomes research approach as part of research design early in the process can result in contributions that are more directly related to policy development (Krumholz, 2008).

Others, such as Pabst (2021), stress the importance of presenting research findings in the context in which policymakers work, and suggest that researchers seeking to inform policy fail when they do not recognize the context in which policy operates. This directly addresses one of the bottlenecks that Klein and Juhola (2014) identify. Work by researchers such as Moseley et al., (2013) suggest that vested interests, including special interest groups, industry associations, and community groups, hold a powerful amount of sway in decision-making processes, and that this context needs to be addressed when presenting and analyzing research findings that might be used in new policy development. These factors need to be reported on actively in written and oral dissemination strategies. An active dissemination strategy needs to include presentations beyond academic audiences; professional meetings and community meetings are also useful outputs and can convey these findings more effectively.

A dissemination strategy designed for policymakers needs to address uncertainty in research findings. Typically, researchers present their findings with a nod to areas of uncertainty or research limitations. As reported by Klein and Juhola (2014), when decisionmakers are faced with uncertainty there can be a paralyzing effect; in essence, no action is seen to be better than the wrong action. This has been seen in almost any area in which human geographers may work, but this phenomenon has been particularly noted in research related to health (e.g., Hinchliffe, 2001) and environment (e.g., Dovan and Oppenheimer, 2015).

The best ways to overcome uncertainty in reporting on research is to pay particular attention to educating the reader as to how evidence has been assembled, and what the evidence means. Researchers should strive to explain the ways in which data have been assembled and how meaningful they are, and how to interpret uncertainty in their data. Good examples can be seen in discussions of water management (Scott et al., 2013) and diet and health management (Lee et al., 2019). Guidance on the ways in which methodologies and results are explained can be found in the networks and expert consultation discussed in previous sections.

Tailoring dissemination strategies to (a) present research findings that describe situations of interest to policymakers, and (b) are proactive in engaging different stakeholder audiences, would help to address critical bottlenecks in getting science into policy. Importantly, the dissemination strategy should contextualize results to better allow policymakers to assess new opportunities against competing priorities.

Conclusions

Human geographers who are working to design research projects that can inform policy can take advantage of a number of modifications to the scientific method, as described in the preceding sections. Implementing some of these options may overcome specific bottlenecks, which have been found to limit the uptake of research findings in policy applications.

Research projects should deliver outcomes that reflect the political realities that policymakers face. This can best be achieved by engaging policymakers, and decisionmakers in general, at the earliest stages of a project to help define the actual research questions being addressed. As the project moves forward, it is helpful to maintain or (if possible) expand these networks to ensure that a variety of different voices related to the policymaking process have input into the evolving research program. This ensures that different considerations are incorporated into the methodology. Data collection may be modified based on the input from these groups, which in turn will help address other critical bottlenecks in the research-policy connection. Bailey (2008) emphasizes the importance of stakeholder engagement in developing climate policy, and stresses that the role of engagement when addressing issues as disputed as the potential impacts of climate change.

Part of addressing political realities is scoping projects so that they reflect spatial and temporal scales, which make sense to policymakers. Small case studies may be appealing to researchers but have little application to policymakers who are working with large and heterogeneous jurisdictions. Researchers who are seeking to influence policy need to design their projects to address regions or timelines that can directly inform policy; input from policymakers and decisionmakers can help provide guidance on scope.

Research findings should be disseminated in a fashion that is accessible by policymakers. This includes presentation of important data that provide context to the project; often the most desired data is related to the economics of potential policy changes, but it can also include data related to the logistics of implementation of these policies. Collecting and presenting this data should be guided by and informed through networking with decisionmakers and other key stakeholders. Presentation of research findings should also be mindful of the political context in which potential policy changes need to co-exist and acknowledge competing priorities. Finally, all research findings should include a discussion about uncertainty and help guide the potential user of these findings in understanding the research methodology that was undertaken.

The guidance provided in this chapter, taken together, represents fairly major changes to experimental design. In practice, of course, researchers may have to pick and choose based on the feasibility of making these changes. Early-stage researchers may be less likely to focus on the needs of policymakers, although it should be noted that increasingly academic performance measures used to assess research performance include multiple forms of dissemination and engagement, which means that adopting the findings of this work may not necessarily decrease measures of productivity. Each of these suggestions should improve the potential use of research findings in policy and help geographers (and other researchers) better connect with the decision-making community. The unique position of human geographers enables them to utilize the concept of landscape to study critical issues facing our society today – issues around culture, society, resources, and environment – which have proven almost impossible to address using past practices (Brace and Geoghegan, 2011). For those developing solutions to these issues, applying the approaches described in this chapter increase the likelihood of seeing research turn into action.

References

Bailey, I. (2008) Geographical work at the boundaries of climate policy: A commentary and complement to Mike Hulme. *Transactions – Institute of British Geographers*, 33(3), 420–423, https://doi.org/10.1111/j.1475-5661.2008.00310.x

Brace, C. and Geoghegan, H. (2011) Human geographies of climate change: Landscape, temporality, and lay knowledges. *Progress in Human Geography*, 35(3), 284–302, https://doi.org/10.1177/0309132510376259

Dany, V., Bajracharya, B., Lebel, L., Regan, M., & Taplin, R. (2016) Narrowing gaps between research and policy development in climate change adaptation work in the water resources and agriculture sectors of Cambodia. *Climate Policy*, 16(2), 237–252. https://doi.org/10.1080/14693062.2014.1003523

De Marchi, G., Lucertini, G., and Tsoukiàs, A. (2016) From evidence-based policy making to policy analytics. *Annals of Operations Research*, 236(1), 15–38. https://doi.org/10.1007/s10479-014-1578-6

Donovan, A. and Oppenheimer, C. (2015) Modelling risk and risking models: The diffusive boundary between science and policy in volcanic risk management. *Geoforum*, 58, 153–165, https://doi.org/10.1016/j.geoforum.2014.11.005

Ekici, A. (2002) Putting consumer voice back in public policy: An enlightenment model approach. In S. Broniarczyk and K. Nakamoto, eds., *NA - Advances in Consumer Research Volume 29*, Association for Consumer Research: Vandosta, GA, pp. 377–385.

Hawkes, S., Zaheer, H., Tawil, O., O'Dwyer, M. and Buse, K. (2012) Managing research evidence to inform action: Influencing HIV policy to protect marginalized populations in Pakistan. *Global Public Health*, 7(5), 482–494, https://doi.org/10.10180/17441692.2012.663778

Hinchliffe, S. (2001) Indeterminacy in-decisions - science, policy and politics in the BSE (Bovine Spongiform Encephalopathy) crisis. *Transactions of the Institute of British Geographers*, 26(2), 182–204, https://doi.org/10.1111/1475-5661.00014

Iheukwumere, O., Moore, D. and Omotayo, T. (2021) A meta-analysis of multi-factors leading to performance challenges across Nigeria's state-owned refineries. *Applied Petrochemical Research*, https://doi.org/10.1007/s13203-021-00272-0

Jonas, A., McCann, E., and Thomas, M. (2015) *Urban geography: A critical introduction* . London: Wiley Blackwell, 384 pp.

Kazemian, P., Costantini, S., Neilan, A., Resch, S., Walensky, R., Weinstein, M. and Freedberg, K. (2020) A novel method to estimate the indirect community benefit of HIV interventions using a microsiulation model of HIV disease. *Journal of Biomedical Informatics*, 107, 103475, https://doi.org/10.1016/j.jbi.2020.103475

KIein, R. and Juhola, S. (2014) A framework for Nordic actor-oriented climate adaptation research. *Environmental Science & Policy*, 40, https://doi.org/10.1016/j.envsci.2014.01.011

Krumholz, H. (2008) Outcomes research: Generating evidence for best practice and policies. *Circulation*, 118(3), 309–318, https://doi.org/10.1161/CIRCULATIONAHA.107.690917

Lee, Y., Mozaffarian, D., Sy, S., Huang, Y., Liu, J., Wilde, P., Abrahams-Gessel, S., Jardim, T., Gaziano, T. and Micha, R. (2019) Cost-effectiveness of financial incentives for improving diet and health through Medicare and Medicaid: A microsimulation study. *PLOS Medicine*, 16(3), https://doi.org/10.1371/journal.pmed.1002761

Leonelli, S., Spichtinger, D. and Prainsack, B. (2015) Sticks and carrots: Encouraging open science at its source. *Geo-Geography and Environment*, 2(1), 12–16, https://doi.org/10.1002/geo2.2

Martin, R. (2001) Geography and public policy: The case of the missing agenda. *Progress in Human Geography*, 25(2), 189–210, https://doi.org/10.1191/030913201678580476

Meehan, K., Klenk, N. and Mendez, F. (2018) The geopolitics of climate knowledge mobilization: Transdisciplinary research at the science-policy interface(s) in the Americas. *Science Technology & Human Values*, 43(5), 759–784, https://doi.org/10.1177/0162243917745601

Mendell, J. and Richardson, L. (2021) Integrated knowledge translation to strengthen public policy research: A case study from experimental research on income assistance receipt among people who use drugs. *BMC Public Health*, 21(1), https://doi.org/10.1186/s12889-020-10121-9

Moseley, C., Kleinert, H., Sheppard-Jones, K. and Hall, S. (2013) Using research evidence to inform public policy decisions. *Intellectual and Developmental Disabilities*, 51(5), 412–422, https://doi.org/10.1352/1934-9556-51.5.412

Newman, J. (2017) Deconstructing the debate over evidence-based policy. *Critical Policy Studies*, 11(2), 211–226, https://doi.org/10.1080/19460171.2016.1224724

Pabst, A. (2021) Rethinking evidence-based policy. *National Institute Economic Review*, 255, 85–91, https://doi.org/10.1017/nie.2021.2

Pallett, H. and Chilvers, J. (2015) Organizations in the making: Learning and intervening at the science-policy interface. *Progress in Human Geography*, 39(2), 146–166, https://doi.org/10.1177/0309132513518831

Prince, R. (2012) Policy transfer, consultants and the geographies of governance. *Progress in Human Geography*, 36(2), 188–203, https://doi.org/10.1177/0309132511417659

Rosewell, B. (2017) Complexity science and the art of policy making. In: J. Johnson, A. Nowak, P. Ormerod, B. Rosewell, YC. Zhang, eds., *Non-Equilibrium Social Science and Policy. Understanding Complex Systems*, New York: Springer, https://doi.org/10.1007/978-3-319-42424-8_11

Scott, C., Meza, F., Varady, R., Tiessen, H., McEvoy, J., Garfin, G., Wilder, M., Farfan, L., Pablos, N. and Montana, E. (2013) Water security and adaptive management in the arid Americas. *Annals of the Association of American Geographers*, 103(2), 280–289, https://doi.org/10.1080/00045608.2013.754660

Scott, M. (2015) Re-theorizing social network analysis and environmental governance: Insights from human geography. *Progress in Human Geography*, 39(4), 449–463, https://doi.org/10.1177/0309132514554322

Seasons, M. (2021) *Evaluating urban and regional plans from theory to practice*. Vancouver: UBC Press, 248 pp.

Smith, K. and Joyce, K. (2012) Capturing complex realities: Understanding efforts to achieve evidence-based policy and practice in public health. *Evidence and Policy*, 8(1), 57–78, https://doi.org/10.1332/174426412X6201371

Sullivan, S., Coyle, D. and Wells, G. (2014) What guidance are researchers given on how to present network meta-analyses to end-users such as policymakers and clinicians? A systematic review. *PLOS One*, 9(12), https://doi.org/10.1371/journal.pone.0113277

Tschakert, P., Barnett, J., Ellis, N., Lawrence, C., Tuana, N., New, M., Elrick-Barr, C., Pandit, R. and Pannell, D. (2017) Climate change and loss, as if people mattered: Values, places, and experiences. *Wiley Interdisciplinary Reviews - Climate Change*, 8(5), https://doi.org/10.1002/wcc.476

Valaitis, R., Wong, S., MacDonald, M., Martin-Misener, R., O'Mara, L., Meagher-Stewart, D., Isaacs, S., Murray, N., Baumann, A., Burge, F., Green, M., Kaczorowski, J. and Savage, R. (2020) Addressing quadruple aims through primary care and public health collaboration: Ten Canadian case studies. *BMC Public Health*, 20(1), https://doi.org/10.1186/s12889-020-08610-y

Valentine, S. (2014) Gradualist best practice in wind power policy. *Energy for Sustainable Development*, 22, 74–84, https://doi.org/10.1016/j.esd.2013.11.003

Wathen, C., Sibbald, S., Jack, S. and MacMillan, H. (2011) Talk, trust and time: A longitudinal study evaluating knowledge translation and exchange processes for research on violence against women. *Implementation Science*, 6, https://doi.org/10.1186/1748-5908-6-102

Welfle, A., Thornley, P. and Röder, M. (2020) A review of the role of bioenergy modelling in renewable energy research & policy development. *Biomass & Bioenergy*, 136, https://doi.org/10.1016/j.biombioe.2020.105542

Weiss, C. (1979) The many meaning of research utilization. *Public Administration Review*, 39, 426–431.

Wu, S., Legido-Quigley, H., Spencer, J., Coker, R. and Khan, M. (2018) Designing evaluation studies to optimally inform policy: What factors do policy-makers in China consider when making resource allocation decisions on healthcare worker training programmes? *Health Research Policy and Systems*, 16, https://doi.org/10.1186/s12961-018-0292-2

INDEX

activist/action research: capturing emotion in 111; in economic geographies 284–290; increasing focus on 107; scholarly publication, difficulty of 324

affect: affect vs. emotion 111, 114; COVID-19 and 111, 119; inequality and affective practices 375; learning to be affected 387; power to change economic subjectivities 289; *see also* affective landscapes; emotional geographies; non-representational theory

affective landscapes: capturing emotion in place 117; Indigenous peoples and 109, 111; landscapes of fear 111; non-affective landscapes 119; place capture 112, 117, 118; relational understanding of 117; shaped by power structures and cultural norms 109; *see also* emotional geographies

Álvares, Domingos 188–189

American Community Survey (ACS) 208–209

anarchist geographies: academic trend-mongering 326; Black anarchism 332; contemporary resurgence of anarchism 323, 326; core and adjacent concepts of anarchism 325; creative imagination in research methods 328–329; decolonizing anarchism 332; dissemination of research 330–331; ethics of care 326–327; everyday acts, focus on 325, 328, 329; feminism, anarchist 325–326; intersectional struggles 331–332; mutual aid, research as practice of 329; nineteenth-century foundations 323; positivist scientific method, limits of 328; posthuman futures 332; researcher/researched hierarchy, need to resist 326, 329; research in and beyond the academy 329–330

animal geographies: "absent presence" of nonhuman animals 260, 264; animal-based industries 258, 260; anthropocentrism 260,

263–264; beneficence 263; blending of methods 261–262; companion nonhuman animals 258; critical anthropomorphism 262, 264; cultural texts, analysis of 259–260, 264; emergence of 257; ethics of researching nonhuman animals 262–264, 265; ethnographic methods, attempts to expand 261, 264; ethology 261, 262, 264; historical animal geographies 259, 260; impacts of human political-economic systems 260; interviews, focus groups and surveys 258–259, 264; mechanomorphism 262; multispecies ethnography 260–261, 264; non-maleficence 263; partiality of all methods 264; researcher positionality 259, 260; shared more-than-human worlds 259, 262; sociocultural orderings of nonhuman animals 259, 262; subjectivity and agency of nonhuman animals 262, 264, 265; unmediated knowledge, impossibility of 264; visual methods 261; voluntary participation 263–264

Anthropocene: 135, 140–141, 290, 331

anthropocentrism 87, 134, 260, 263–264

archives: Anthropocene as archive 140–141; architecture of 175–176; assemblage theory and 138–141; Black geographies and 190–191; digitization priorities, hierarchies reinforced by 177; geographical imaginaries produced by 174, 180; historic space obscured by digitization 178–179; India Office 175; as instituting imaginaries 173, 176; locational power of 176–177, 180; movement of materials 176; nation states, supported by archives 175, 180; official vs. informal collections 174; proprietary digital platforms 178; provenance, attention to 174; queer geographies and 129–131; reading along or against the grain 173, 177, 178; Royal Geographical Society 178; Sauer and 26–27;

search algorithms, opaque basis of 179, 180; theorizing absence 174; written vs. oral and material 174

arts-based research 67

assemblage theory: activist research 137–138, 140, 141–142; agonism 136; Anthropocene as archive 140–141; anthropocentrism, refusal of 134, 138, 141; archives 138–141; assemblage ethnography 135–138; as autoethnography 136, 141; capacities, analysis of 136; coherent rational subject, challenge to idea of 135; definition of 134; heritage as material-discursive process 140; netnography 137; progressive potential of 136, 141–142; spatio-temporal scales, multiplicity of 134, 141; state as assemblage 138; synchronic emergence 135–136, 139; *see also* political geography

Australian Aboriginals: country model 111; kinship concept 140; yarning 155–156

autoethnography 111

Barcelona, technological sovereignty in 202–203

behavioural geography 64, 65

Bergmann, Gustav 14–15

Black geographies: as anti-racist political project 185–186, 189, 192; Black agency in examination of traditionally invisible spatialities 188–189; cartographic activism 190–191; disadvantageous experience, study of 184–185; embodied practices and knowledge 191; emergence of 184, 185; epistemological critique 190; Eurocentrism, breaking with 190; interdisciplinarity 191; metonymic reason, critique of 191–192; mobilization of relational sense of place 187–188, 192; multiplicity of 185, 192; naming of places 187; non-traditional archives 190–191; *quilombismo* 187; racialization of spaces 186–187; segregation, analytical approaches to 186–187; theoretical creativity of 186

Brazil: Álvares, Domingos 188–189; cartographic activism 190–191; quilombos 187; right to landed tenure 187–188; "scientific geography" 184

Budapest ghetto: Historical GIS of 51–53; International and Pest ghettos 51; "invisible walls" 54; spatial stability of 53

Bunge, Bill 10, 16–18, 20, 43, 308

Canada: climate change adaptation research 411; integrated knowledge translation 412–413; Private Sponsorship of Refugees program 398–399; summer camp for disabled youth 313–315; youth-performed play space assessments 314–318; *see also* Canadian First Nations

Canadian First Nations: Anishinaabe Mino-Bimaadiziwin 146; colonialist use of research

on 86; participatory geographical projects 365; sharing circles 156; underrepresented in academia 85

carceral geographies 364–365

Caribbean storytelling: Anancy stories 164; calypso as talking back to colonial power 165; Caribbean radical imagination 163; Creole 164, 168; cultural maroonage, storytelling as 163, 165; historical memory, storytelling as 164; interpretation, challenges and ethics of 168–169, 170; "library in crisis," African diaspora as 164; liming 166, 169–170; silences as strategic technologies 166; testimonios, oral histories as 167; translation and editing of 168; *see also* storytelling

cartography: of Budapest ghetto 51–55; critical cartography 57; deep maps 56; geographic information systems and 49–50; as window into cultures 47, 57; *see also* uncertainty

central place theory 41

children's geographies: action research frameworks 312; benefits of youth research participation 312; capacity building for child participants 318; deep participation 311; degrees of research participation 310–311; growing role of children in research process 308; intersecting identity characteristics 310; interviews 313–314; minimizing child–adult power imbalances in research participation 315; mixing a variety of participation methods 312, 318; neurodiverse youth, frequent exclusion of 314; new social studies of childhood (NSCC) movement 308–309; Photovoice method at Camp Neuro 313–314; play essential for children's well-being 306–307; social factors hindering children's research participation 309–310; social geography, rise of 308; tokenistic participation of children 310; traditional sidelining of children in research and decision-making 306, 308, 314; unique social agency of children 309; youth-performed play space assessments 314–318

climate change 348, 398

Combahee River Collective 70, 71, 72, 74, 75

community geography 365, 378

corpus linguistics 56, 57

COVID-19 pandemic: affective encounter restricted by 119; as affective event 111; anarchist geography and 323; Bluetooth technology in contact tracing 240; data collection in New Zealand 235; digital online living intensified by 111, 117; disruption to academic work 399–404; #geendorhout 253; impact on urban life 271, 275; as "logistical challenge" in Global North 398; as portal 395; social and economic inequalities worsened by 342; transformative vision, potential to

stimulate 403; university inequalities laid bare by 395; virtual platforms for education ushered in by 298
Crenshaw, Kimberlé 73, 75–76
cultural geography 381, 387

decolonial and anticolonial methodologies: in Indigenous geographies 144, 146, 148, 150–151, 153; in New Zealand environmental research *149*, 154; *see also* storytelling
descriptive methodology 9, 10, 13; *see also* Schaefer–Hartshorne debate
development geography 252
digital geographies: agential abilities of digital systems 198–199; broad range of inquiry 196; counter-mapping 200; cyberspace, theorizing of 197; "death of geography" 197; deployment of algorithms in practice, analysis of 200–201, 204; digital divide 199, 239; digital turn 196; in disability geography 253; everyday life, focus on 197–198; genealogy of digital assemblages 201, 204; material resource demands of digital capitalism 198; narrative methods 200; participatory GIS (PGIS) 201, 204; posthumanism and 198, 199, 202; proxying 200; space-time relations, digital reordering of 197; spatial metaphors of computer networking 197; technical apparatuses, analysis of 200, 202, 203, 204; technological sovereignty in Barcelona 202–203; traditional ethnographic methods 202–203, 204
disability geography: ableism 245, 248, 252; behavioural approaches 247; critical social theory approaches 247; development geography and 252; digital methods 253; eugenicist ideology 253; feminist approaches 247, 249; fluidity and contingency of disability 244, 252; humanist–positivist divide 247–248; identifying as disabled 245; Indigenous epistemes 252; intersectionality 252; interviews 249; medi cal vs. social models of disability 245–246; mixing quantitative and qualitative methods 248–250, 252, 253; queer approaches 252; positivist early approaches 247; posthumanist approaches 247; quantitative to qualitative shift 244, 246; relational models of disability 246; researcher positionality 250–252

economic geographies: action research 284–290, 291; affect, power to change economic subjectivities 289; capitalocentrism, critique of 280, 281, 282, 283, 289, 291; collective ethical action 288; community economies research 279, 280, 284, 288, 289–290, 291; critical realism 279; diverse economies research 279, 280–281, 282–292; diverse economies table 283; "iceberg" model 280, 283, 284;

Indigenous approaches, making space for 291, 292; loose and rich description approach 282–283; Marxist approaches 279; more-than-human action research 291–292; performativity of research 288, 291; quantitative mid-century methods 279; self-reparative research 289
education, geographies of: adult education, need for more study of 302; as broad and inclusive sub-discipline 295; cartoon analysis 297; child-centered approach 297, 302; co-construction of knowledges 297; data as architecture of governance 298; digital information, complexities of analysing 298; emergent and existing spatial connections 297; emotionality of learning and teaching 301–302; geography education, study of 300–302; institutional vs. transnational focuses 299–300; interview methods 296–297, 298, 301; inward vs. outward-looking approaches 299; methodological nationalism 300; multi-scalar research 298–300, 302; multi-sited ethnography 300; New Social Studies of Childhood paradigm 297; participatory methods 297; qualitative methodologies, prioritizing of 296, 302; quantitative methods 297–298; reflexivity required of researchers 297; social media analysis 296; transition pedagogy 301; virtual platforms, COVID-19 as stimulus to 298; visual techniques in teaching 301
Eliot, T.S. 27, 36n5
embodied engagement 65–66
emotional geographies: affect vs. emotion 111, 114; of aspiration 296; critical turn and 110; cultural turn and 110, 112; digital tools 113, 115, 116, 117; disabled subjects 116; "emotional turn" 110, 111; emotion crucial to understanding of place and space 109; empathic co-listening 112; encounter methods 115–117; ethics of 118; feminist geography, influence of 110, 113; field notes 114; happiness mapping 118; humanistic forerunners of 110; in Indigenous geographies 109, 117; kinaesthetic emotions 116; mental health and wellbeing 111; "more-than" affective approach 110, 111, 116, 117, 118; non-representational geographies as enabler of 110; observational methods 114–115; poststructuralism, alignment with 111; queer geographies and 113; refugee experience 113, 115; relationality 111, 116, 117–118; story methods 112–113; temporality 118; value of 117–119; visual methods 114–115; *see also* affective landscapes; psychotherapeutic geographies
environmental determinism 10–11, 25, 80
epistemicide 186, 190, 281
epistemologies: anarchistic 328; in Black geographies 188, 190, 191; of GIS 50;

Euro–Western 146, 162, 166, 170; feminist 249; humanist 260; Indigenous 144, 146, 153; intersectionality and 70; Māori 147; multiple 292; ontology and 281; participatory research and 360; of settler-colonial societies 73; storytelling and 162, 166, 170; in urban studies 275
Ethiopia 348
ethnography: in animal geographies 261, 264; assemblage ethnography 135–138; autoethnography 111, 136, 141; in digital geographies 202–203, 204; lived experience and 373, 375–376; multi-sited ethnography 300; multispecies ethnography 260–261, 264, 386–387

feminist geographies: acclimatization to white supremacist educational institutions 74–75, 76; Black/intersectional feminism crucial to 71, 79; disability geography and 247; economic geographies and 280; emotional geography influenced by 110, 113; geographical sub-disciplines influenced by 105; marginalization of women in geographical profession 73; omnipresence of intersectionality 77; political geography critiqued by 135; rejection of quantitative methodologies 44; travel writing 174; *see also* intersectionality; slow scholarship; womanism
focus groups: in animal geographies 258, 264; in children's geographies 315, 317; in emotional geographies 114; more-than-representational approaches and 388; in participatory research 359; as qualitative method 7, 62, 112; sharing circles vs. 156; traumatized participants 403; women-only 355
Freire, Paulo 360, 362

Garrison, William 19–20
geographic information science (GIScience): abstract definition of space 48; anthrospace 56; deep maps 56; inability to explain data 47, 55; interdisciplinarity of 50, 53–54, 55; kernel density analysis 53; participatory GIS (PGIS) 201; platial GIS 55–57; qualitative spatial reasoning (QSR) 56–57; as "science behind the systems" 50; *see also* geographic information systems; uncertainty
geographic information systems (GIS): cartography and 49–50; as computer technology 49; critical GIS 50, 55, 79; in disability studies 248; inaccessibility to the visually impaired 253; rise of 44, 45, 49; *see also* Budapest ghetto; geographic information science; uncertainty
geohumanities: as creative experimentation 97, 98, 100; critical reflexivity, need for 100;

difficulty of defining 96–97; Euro-Anglo domination of 100–101; *GeoHumanities* journal 3, 100; methodologies of 97–98; perennial intertwining of geographical and humanistic endeavors 95–96, 100; performative methods 99; relational practices 98, 99–100; rise of as concept 96–97; textual methods 98–99; transdisciplinarity 96, 97; visual methods 99; *see also* humanism
German foundations of geography 11–12
Ghana 165, 310, 348, 349, 352, 354
Giuliani, Rudi 341
Glasgow 344

Hālau 'Ōhi'a 87–88
Hartshorne, Richard: career of 13; German influences on 11–12; *Nature of Geography* 11–12, 15, 17, 18; *Perspective on the Nature of Geography* 18; regional geography, commitment to 10, 12, 18; *see also* Schaefer–Hartshorne debate
Harvey, David 43, 279, 407
Hawai'i 86–88
health and medical geography: activity spaces 232; agent-based models 240; big data, problems with 232–234, 238; blending qualitative and quantitative approaches 230, 237; combining individual and aggregate datasets 230; digital divide 239; exposome 230, 240; interdisciplinary reach 227; modifiable areal unit problem (MAUP) 232; neighbourhood interventions 237; physical and natural environmental exposure 228, 230, 240; pilotism 237; protection of privacy 232, 235; reproducibility 237; shift from aggregate to individual data 228, 240; social exposure 230, 239–240; socio-ecological models, emergence of 246; Spatial Microsimulation 240; time geography and 232; uncertain geographic context problem (UGCoP) 232
historical geographies: Sauer and 27; *see also* archives
Holocaust: ghettoization 51; Holocaust Geographies Collaborative 49; scale as key concept 49, 53; *see also* Budapest ghetto
human geography, definition of 48
humanism 64–65, 109, 110, 246; *see also* geohumanities

Indigenous geographies: anonymizing research participants, critique of 151; decolonizing methods and methodologies 144, 146, 148, 150–151, 153, 363; elevating Indigenous voices and perspectives 84–86, 144, 291; four Rs (relational accountability, reciprocal appropriation, respectful representation, rights and regulations) 145, 146, 148, 152, 153, 155; holism 147; Indigenous-non-Indigenous research framework interfaces 148–153; Indigenous research leadership 89–90, 91, 361, 366; marginalized by Euro-Western ontologies 145–146; more-than-human beings, interconnection with 145, 146–147; participatory approaches 365; place-specific and culture-specific knowledge 146; quantitative vs. qualitative methodologies 152, 153; relational ontologies 146–147; research ethics 145–146; research paradigms 87–88, 145, 146–148; research sovereignty 88, 89; as sites of anticolonial resistance 146; usefulness of research for Indigenous communities 84–86, 145, 363; *see also* Canadian First Nations; Indigenous peoples; Māori

Indigenous peoples: affective landscapes and 109, 111; collectivist relations 147; colonial appropriation of Indigenous knowledge 87, 89, 174; colonialist roots of geography 84, 361; damage-centered narratives 2, 85–86; deficit framings of 84, 89; disability geography and 252; Indigenous intelligence 87, 91; kincentric ontologies 87, 90; marginalized in geography profession 83–84, 86; oral traditions 153; underrepresented in academia 85; usefulness of research for Indigenous communities 84–86; *see also* Indigenous geographies; race

inequality: affective practices 375; COVID-19, compounded by 372; food poverty 375–376; identity dilemmas 375; ill effects on society 372, 377; increase in recent decades 372, 377; reinforced by representations 381; rooted in socio-historical structures 372; shame and stigma 372, 375; social housing and 372–373; underestimated in lay perceptions 372; *see also* lived experience, integrating in research on inequalities

intersectionality: activist movements of mid-twentieth century 74–75; anarchism and 323, 331; as analysis of cumulative relations of power 70, 71, 77; Black women as key protagonists of 70, 71, 72–76, 78, 80; deconstructing scales of power 77; in disability geography 252; disruption of single-axis thinking 78; ending violence and injustice 78–79; human rights as goal of 76; innovative methods required by 78; misuse and misunderstanding of 77, 78; queer geographies and 124; slow scholarship and 396; in third-wave academic feminism 75–76; unlearning of prevailing geographical imaginaries 79–80

interviews: children 313; in disability geography 249; in education geographies 296–297, 298; "go along" 65–66, 114, 116, 297, 373, 374; hīkoi method 154; Indigenous critique of Western methods 153–154; in more-than-representational approaches 388;

semi-structured 64, 374; sharing circles 156; Talaona method 155; testimonio 78, 167; yarning 155–156

Isard, Walter 42, 43

Keali'ikanaka'oleohaililani, Kekuhi 87–88
Kenya 349, 352, 353
Kroeber, Alfred L. 27
Kropotkin, Peter 323, 330
Kuhn, Thomas 40, 62

Ladakhi, educational experiences of 296
Leighly, John 11, 36
lived experience: concept of 371–372; dynamic nature of 374; as embodied space of perception 371, 377; as intersubjective space 371–372, 375; non-representational aspects of 374; *see also* lived experience, integrating in research on inequalities

lived experience, integrating in research on inequalities: advanced marginality 375; attachments to place, importance of 371; ethnography 373, 375–376; food poverty 375–376; "go along" interviews 373, 374; identity dilemmas, negotiation of 375; long-term ties to partners and communities 378; policy informed by 378; qualitative longitudinal research 373; resource intensiveness of 377; retaining critical engagement 377–378; visual methods 376–377

logical positivism 14, 15, 19, 40, 43, 105
Louis, Renee Pualani 363
Lowie, Robert H. 27

Malawi 349, 353, 354, 355
Manchester 270–271
Māori: decolonizing environmental research *149*, 154; hīkoi 154; Kaupapa Māori 146, 147–148, 282; mana motuhake 88–89; Māori realities overwritten by colonial "realities" 281; mātauranga Māori 152–153; mauri 281–282; rangatiratanga 88; Te Ao Māori 147; Treaty of Waitangi 147; Waka-Taurua framework 148, *150*

Marsh, George Perkins 29–30
Marxism 64, 94
"mess" in social science 281, 282, 291, 292
methodological turns: creative turn 3, 95; critical turn 110, 308; cultural turn 110, 112, 135, 349, 358; digital turn 196; embodied engagement turn 65–66; emotional turn 110; linear trajectory vs. 61–62; poststructuralist turn 3; qualitative turn 62, 64; queer turn 126; relational turn 274; turn, concept of 61; visual turn 66–67; *see also* Quantitative Revolution

more-than-representational theory: arts-based approaches 387; creative and playful methods 385, 391; description of 382–383;

experimentalism 390; interviews and conversation 388; learning to be affected 387; methodological challenges of 384–385; more-than-human, engagement with 384, 386; more-than-useful research 391; multispecies ethnography 386–387; new technologies 388–389; new ways of writing 389–390; nomadic consciousness 383; performative practices 384, 385, 388; presenting research to stakeholders 391; sensory ethnographies 385–386; traditional methods 390; *see also* non-representational theory

Morrison, Toni 80

natural language processing 56, 57

neoliberalism: anarchist critique 323; care ethics as counter to 397; in Caribbean 163; disrupting the narratives of 289; in higher education spaces 330, 331, 338, 359, 396, 397; restriction and medicalization of difference 253; rollback of mid-century equality gains 75; slowing down as resistance to 397; temptation to over-focus on 283; withdrawal of state social services 343, 345

New Zealand: community co-production of economic tools 290; COVID-19 data collection 235; decolonizing environmental research *149*, 154; Integrated Data Infrastructure 230–232; Kids in the City project 310; smart city technology in Christchurch 234; Te Arawa energy generation activities 283, **286–287**; #WellconnectedNZ 238; *see also* Māori

netnography 137

Nin, Anaïs 3

non-representational theory (NRT): anarchism and 326; coherent rational subject, challenge to 135; disabled subjectivities 247; emotion and affect, interaction of 111, 381–382; "go along" interviews 374; as not anti-representation 382; as umbrella term 382; *see also* more-than-representational theory

ontologies: anthropocentric 260; assemblage theory and 134, 141; epistemology and 281; Euro-Western 146; in feminist geographies 73; of GIS 50; Indigenous 145, 153; intersectionality and 70; kincentric 90; multiple 282, 291, 292; posthuman 88, 91; relational 87, 90, 105, 141, 146

participatory geographies: balancing research power 359–360; building community research capacity 359, 360; community geography, growth of 365; contact zones 364; decolonial stimulus to 361; definition of 359; emergence of 1, 3, 358, 362–363; emotional geographies and 114; as equitable engagement to produce positive change 358, 359, 363; Eurocentric

knowledge hegemony of universities 358; failure to challenge colonial power formations 89–90, 366; financial resources, distribution of 365; homogenizing notions of community 362; immersive field schools 364; Indigenous geographies and 363, 365; International Participatory Research Network 360–361; invited vs. popular spaces 363–364; origins in Global South communities 360–361; participatory GIS 364; participatory action mapping 364; participatory development 361–362, 366; Participatory Geographies Working Group 363; relational geohumanities as 100; researcher involvement in community activities 360; spatial dimensions of participation 363; technological innovations 366; as techno-managerialist tool 362; white cis-gender academics engaging with "other" 360

place: as dynamic entity 48, 373; platial GIS 55–56; social relations and 55

pluriverse 281–282

policy, refining research methodology to influence: bottlenecks in science–policy interface 408–409; constructivist approach 408; contextualizing results 413–414; data collection, expansion of 411–412, 414; engaging decisionmakers in early phases of research 410–411, 414; evidence-based policy making (EBPM) 407–408; increasing complexity of policy making 407; integrated knowledge translation 412–413; knowledge interaction 408; limited past policy influence of geography 407; networking with stakeholders 412–413, 414; open science 413; outcomes research 413; policy documents, inclusion in literature review 410–411; research informed policy design 408; scoping of projects 413, 414; strategic uptake of EBPM 408; targeting policymakers at appropriate level of hierarchy 410; uncertainty, contextualizing for lay people 413–414, 415; vested interest groups, power of 413

political geography: cultural geography, impact of 135; ethnographic broadening of 135; feminist critique of 135; objectivity, expectations of 141; *see also* assemblage theory

posthumanism 87, 88, 106, 196, 198, 199, 202, 247, 248, 381, 383, 384

post-qualitative research 1, 383–384

poststructuralism 3, 110, 111, 271, 125, 126, 161, 358, 383

Processo das Culturas Negras 188

psychotherapeutic geographies 111, 115, 117, 118

Quantitative Revolution: applied mathematics, advances in 40, 42; as basis for qualitative counterrevolutions 45, 94; computing, advances in 40, 45; critiqued as capitalist 43; critiqued for eclipsing individual experience 43–44; geographic information systems, rise of 44, 45; geometry and spatial order 41–42; holistic view of regions, difficulty of 43; logical positivism, limited influence of 40, 43; philosophy of science, influence of 40; model-based approach, rise of 40; rational and predictable behaviour, belief in 42; spatial behaviour 42–43; structure–agency debate 43; theoretical justification of 43; urban and economic geographers as leading proponents 41, 42

queer geographies: archival work 129–131; complex story methods 113; disability geography and 252; ephemera as methodological resource 128–129; *Gay and Lesbian Atlas* 128–129; geographies of sexualities as predecessor of 124, 131; intersectionality 124; in/visibility, strategic 127–128; messy methodologies 130–131; as ongoing project 127; "queer" as evolving contested concept 123; queering geography 123; refusal as politics and method 127–128

race: anti-racist geographies, rise of 184; Black women as intersectional activists 70, 71, 72–75; colorblindness as reactionary tool 75; Racial Geography 183–185; racialized standpoints marginalized in geography 77; racism in the academy 395; *see also* Black geographies; Indigenous geographies; Indigenous peoples

radical pedagogy 360

Reclus, Élisée 323, 327

regional geography *see* Schaefer–Hartshorne debate

Rent (musical) 128

research methodology: conventional understandings of 1; creativity–conformity balance 68; design in course of execution 68; disruptive potential 3; experimentation, embrace of 2; interdisciplinarity 227; methods vs. methodology 7; ontological reflection on 1–2; positionality, importance of 62; research decisions as political acts 2; Western Science construct, criticized as 1; *see also* methodological turns

rural research challenges: adapting to changing socio-cultural practices 351; assumptions of rural sameness, as trap 350–351; building trust with gatekeepers 352–353; building trust with respondents 351–352; climate change impacts on food security 348; divergent perspectives, need to include 352; environmental development projects 348; gender norms 355; increase in rural research 349; insider/outsider space, fluidity of 353; migration, as difficulty for longitudinal research 351, 354, 356; more-than-human dimensions 350; navigating

socio-cultural norms 354–356; performance of rurality 350, 351, 356; population decline 348; positionality and subjectivity, need to reflect on 351, 353–354; respondent apathy from overexposure to researchers 351; rural, contested concept of 350–351; rural–urban connections, need to encompass 350; virtual geographies 356

Samoa 155

Sauer, Carl O.: aesthetic qualities of landscape 26; anthropology, interest in 27; archival work 27; art, geography as 35; colonialism, critiques of 35–36; cultural landscape concept 24, 25, 27, 28; cultural turn, anticipation of 26; *Early Spanish Main* 29, 35–36; education and career 25; "Education of a Geographer" 29; environmental possibilism 25; field work 28–29; "Foreword to Historical Geography" 27, 37n18; historical geography 27, 28–29; history of geographic thought 29; *Ibero-Americana* 29; inspirational legacy 24; *Man in Nature* 29–35, 36; *Morphology of Landscape* 24, 25–26; "Outline for Field Work" 24; *Seventeenth-Century North America* 24; "Theme of Plant and Animal Destruction in Economic History" 35

scale: contested dynamic of 299; definition of 48–49; as key concern of geography 77, 298

Schaefer, Fred K.: career of 13–14; death of 15; "Exceptionalism in Geography" 15, 16–17; logical positivism 14, 15, 19; *see also* Schaefer–Hartshorne debate

Schaefer–Hartshorne debate: Hartshorne's enraged responses to Schaefer 17; idiographic vs. nomothetic approach 10, 15, 21n1, 35; regional uniqueness, exceptionalist implications of 13, 15, 16, 18; Schaefer's influence on younger generation 18–20; systematic vs. regional geography 10–11, 12, 15–20

Semple, Ellen Churchill 25, 36n3, 36n4, 80

sexualities, geographies of: "coming out" of 125; foregrounding of gay and lesbian subjects and spaces 124–125, 126; intersectionality, neglect of 125, 126; methodologically essentialist approaches 125, 126, 131; as predecessor of queer geographies 124, 126, 131; qualitative methods 125; "straight" space revealed as construction 126

Singapore, education in 296

slow scholarship: collective action, advocacy for 396, 397; care for participants in research process 399, 403; criticized as nostalgic 396; disability justice frameworks 398, 403; embodied care work 402–403; embodied knowledges, making time for 397, 403; feminist ethics of care 395, 396, 397; intersectional praxis 396; presentist framings of crisis, colonialism obscured by 398; as response to unjust academic power relations 396–397; speed, privileged and prioritized in academic system 399–402, 403

smart cities 234–235

Smith, Linda Tuhiwai 77, 161, 281, 361

Sotomayor, Antonio 31, 34

space: anthrospace 56; contested and dynamic 374; definition of 48; qualitative spatial reasoning (QSR) 56–57

"space cadets" 17–18, 19, 20, 21n13, 39

storytelling: autopoeisis 163; building emancipatory futures 170; capacity to reinforce oppressive ideologies 161; as challenge to Western epistemologies 162; listener, role of 167; recovering knowledge in settler colonial contexts 162; re-emergence as decolonial tool 161; struggling for freedom in Black radical contexts 162; as traditional medium for colonized peoples 162; *see also* Caribbean storytelling

sub-disciplinary structure of geography 105, 106

Sunderland 340, 342

systematic geography *see* Schaefer–Hartshorne debate

Thünen, J.H. von 41

time geography 232

Truth, Sojourner 73

uncertainty: aggregation 209, 210; American Community Survey (ACS) 208–209; attribute 207; bivariate mapping 210–212; choropleth maps 212; coefficient of variation 208–209; composite symbol maps 212; heterogeneity 208; interaction of sources of 207; locational 207; map classification 212–214; margins of error 208; measurement error models 222; modifiable areal unit problem (MAUP) 209–210, 232; proportional symbol maps 212; regionalization 209; scale effect 210; side-by-side maps 210; small sample size 208; social media data 223; spatial autocorrelation 214–220, 221–222; specification error 207, 220–222; uncertainty geographic context problem (UGCoP) 223,

232; visualization of 210–220; volunteered geographic information 223; zoning effect 210

United Kingdom: Brexit referendum 340; higher education (im)mobilities in 298; Scottish independence referendum 340; social inequality 378

United Nations Convention on the Rights of the Child (UNCRC) 307, 308, 310

United States, social inequality in 378

urban geographies: boundaries as expressions of power 275; comparative urbanism 272–274; comparison from below 273; comparison from through 273; hierarchical categorization of cities 272; inadequacy of conventional comparative methods 273; less-known cities stigmatized as backward 273; multiplex, cities as 275; on-going urbanization of planet 275; open, porous and inter-connected nature of cities 274; political use of comparison 274; proliferation of paradigms 275; relational urbanism 274–275; restrictive pre-conditions for city comparisons 272–273; standardized case comparison 272; successive paradigms of 271; trans-territorial webs of connectivity 274–275; urban anthropological studies 272; as well-established sub-discipline 271

urban research challenges: COVID-19, social and economic inequalities worsened by 342; divisive politicians 341; institutional gatekeepers, negotiating with 343–344; institutions, key roles in urban life 342–343, 345; politically sensitive research topics 339–340; positionalities in the field 343–344; refugees and asylum seekers, social tensions around 341–342; researchers perceived as outsiders by researched 341; research outputs of use to gatekeeper institutions 345; sectarianism, participant denial of 341

Varenius, Bernhardus 10

visual methodologies 66–67, 99, 114–115, 301; *see also* uncertainty, visualization of

womanism 70, 72, 74, 75, 76

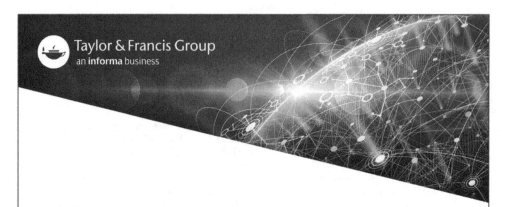